MULTIVARIABLE CALCULUS

MULTIVARIABLE CALCULUS

Produced by the Consortium based at Harvard and funded by a National Science Foundation Grant. All proceeds from the sale of this work are used to support the work of the Consortium.

William G. McCallum
University of Arizona

Daniel Flath
University of South Alabama

Andrew M. Gleason
Harvard University

Sheldon P. Gordon
Suffolk County Community College

Brad G. Osgood
Stanford University

Deborah Hughes-Hallett
Harvard University

Douglas Quinney
University of Keele

Jeff Tecosky-Feldman
Haverford College

Joe B. Thrash
University of Southern Mississippi

Thomas W. Tucker
Colgate University

John Wiley & Sons, Inc.
New York Chichester Brisbane Toronto Singapore

Dedicated to Bob, Janny, and Lisel for their warmth and wizardry.

This material is based upon work supported by the National Science Foundation under Grant No. DUE-9352905.

ISBN: 0-471-30450-6

Printed in the United States of America

10 9 8 7 6 5 4 3 2 1

0.1 PREFACE

Calculus is one of the greatest achievements of the human intellect. Inspired by problems in astronomy, Newton and Leibniz developed the ideas of calculus 300 years ago. Since then, each century has demonstrated the power of calculus to illuminate questions in mathematics, the physical sciences, engineering, and the social and biological sciences.

Calculus has been so successful because of its extraordinary power to reduce complicated problems to simple rules and procedures. Therein lies the danger in teaching calculus: it is possible to teach the subject as nothing but the rules and procedures – thereby losing sight of both the mathematics and of its practical value. With the generous support of the National Science Foundation, our group set out to create a new calculus curriculum that would restore that insight. This book is the second stage in that endeavor. The first stage is our single variable text.

Basic Principles

The two principles that guided our efforts in developing the single variable book remain valid. The first is our prescription for restoring the mathematical content to calculus:

> **The Rule of Three:** *Every topic should be presented geometrically, numerically and algebraically.*

We continually encourage students to think and write about the geometrical and numerical meaning of what they are doing. It is not our intention to undermine the purely algebraic aspect of calculus, but rather to reinforce it by giving meaning to the symbols. In the homework problems dealing with applications, we continually ask students what their answers mean in practical terms.

The second principle, inspired by Archimedes, is our prescription for restoring practical understanding:

> **The Way of Archimedes:** *Formal definitions and procedures evolve from the investigation of practical problems.*

Archimedes believed that insight into mathematical problems is gained by investigating mechanical or physical problems first.[1] For the same reason, our text is problem driven. Whenever possible, we start with a practical problem and derive the general results from it. By practical problems we usually, but not always, mean real world applications. These two principles have led to a dramatically new curriculum – more so than a cursory glance at the table of contents might indicate.

---

[1] . . . I thought fit to write out for you and explain in detail . . . the peculiarity of a certain method, by which it will be possible for you to get a start to enable you to investigate some of the problems in mathematics by means of mechanics. This procedure is, I am persuaded, no less useful even for the proof of the theorems themselves; for certain things first became clear to me by a mechanical method, although they had to be demonstrated by geometry afterwards because their investigation by the said method did not furnish an actual demonstration. But it is of course easier, when we have previously acquired, by the method, some knowledge of the questions, to supply the proof than it is to find it without any previous knowledge. >From *The Method*, in *The Works of Archimedes* edited and translated by Sir Thomas L. Heath (Dover, NY)

Technology

In multivariable calculus, even more so than in single variable calculus, computer technology can be put to great advantage to help students learn to think mathematically. For example, looking at surface graphs and contour diagrams is enormously helpful in understanding functions of many variables. Furthermore, the ability to use technology effectively as a tool in itself is of the greatest importance. Students are expected to use their judgment to determine where technology is useful.

However, the book does not require any specific software or technology, and we have accomodated those without access to sufficiently powerful technology by providing supplementary master copies for overhead slides, showing surface graphs, contour diagrams, parametrized curves, and vector fields. Ideally, students should have accesss to technology with the ability to draw surface graphs, contour diagrams, and vector fields, and to calculate multiple integrals and line integrals numerically. Failing that, however, the combination of handheld graphing calculators and the overhead transparencies is quite satisfactory, and has been used successfully by test sites.

What Student Background is Expected?

Students using this book should have successfully completed a course in single variable calculus. It is not necessary for them to have used the single variable book from the same consortium in order for them to learn from this book.

The book is thought-provoking for well-prepared students while still accessible to students with weaker backgrounds. Providing numerical and graphical approaches as well as the algebraic gives students another way of mastering the material. This approach encourages students to persist, thereby lowering failure rates.

Content

Our approach to designing this curriculum was the same as the one we took in our single variable book: we started with a clean slate, and compiled a list of topics that we thought were fundamental to the subject, after discussions with mathematicians, engineers, physicists, chemists, biologists, and economists. In order to meet individual needs or course requiremens, topics can easily be added or deleted, or the order changed.

Chapter 11: Functions of Many Variables

Chapter 11 introduces functions of many variables from several points of view, using surface graphs, contour diagrams, and tables. This chapter is as crucial for this course as Chapter 1 is for the single variable course; it gives students the skills to read graphs and contour diagrams and think graphically, to read tables and think numerically, and to apply these skills, along with their algebraic skills, to modeling the real world. Attention is paid to the notion of varying one variable in the domain independently of the others; it is important that the student thoroughly digest this notion from both graphical and numerical point of view, before being exposed to the ideas of partial derivatives and gradients. In preparation for the notion of local linearity, particular attention is paid to linear functions, from all points of view.

Chapter 12: A Fundamental Tool: Vectors

Chapter 12 introduces the key concept of a vector from a thoroughly geometric point of view. The fundamental operations with vectors are defined in a geometric, coordinate-free way. After finishing this chapter, the student should understand vectors as geometrical objects, not merely as lists of numbers. The representation of vectors in terms of coordinates is presented as a useful way of manipulating vectors algebraically, not as a definition. This is important for later chapters, where vector quantities such as the gradient are introduced in a geometric way (i.e., as quantities having magnitude and direction) rather than in terms of coordinates.

Chapter 13: Differentiating Functions of Many Variables

Chapter 13 introduces the basic notions of partial derivative, directional derivative, and gradient. In keeping with the spirit of the single variable book, the key notion of local linearity is introduced as a framework into which all the different notion of derivatives can be logically placed. This also makes possible a development of the notion of the differential, which is important for some applications such as thermodynamics. The multivariable chain rule and higher order partial derivatives are also covered, and the chapter concludes with a brief introduction to partial differential equations.

Chapter 14: Optimization

Chapter 14 presents applications to optmization. Both constrained and unconstrained optimization are covered, the former from a geometric point of view using contour diagrams.

Chapter 15: Integrating Functions of Many Variables

Chapter 15 presents the multivariable definite integral. The two-dimensional integral is motivated geometrically by considering the problem of estimating total population from a contour diagram for population density, using finer and finer grids. Tables are used to reinforce the notion numerically. Three-variable integrals and integrals in other systems of coordinates are also considered.

Chapter 16: Motion in Space

Chapter 16 considers parametric equations in two and three dimensions from a physical point of view, as representing motion in space. This allows for a geometric development of the velocity and acceleration vectors; the algebraic way of obtaining these vectors in terms of coordinates is then presented. The chapter leads to an appendix with one of the original, and still the most inspiring, applications of calculus; the derivation of Kepler's laws of motion from Newton's laws.

Chapter 17: Vector Fields and Line Integrals

Chapter 17 introduces the notion of a vector field, and of line integrals along vector fields. To build intuition, emphasis is placed on physical interpretations, such as flow or force fields. Line integrals are introduced as Riemann sums, without coordinates or parametrizations of the paths. Some time is spent building intuition using diagrams of vector fields with paths superimposed, before the method of calculating line integrals using parametrizations is introduced. There is a special section on the important class of conservative vector fields.

Chapter 18: Calculus of Vector Fields

Chapter 18 introduces the notions of divergence and curl in a coordinate-free way; the divergence as the outflow per unit volume, and curl as the directional rotation per unit area. The coordinate representations of these quantities are then derived. This is particularly important when considering how to represent these quantities correctly in other coordinate systems; it is the geometric, not the algebraic, representation that must be the guide. The chapter includes a discussion of the divergence theorem and Stokes theorem, with emphasis on physical applications.

Supplementary Materials

- **Instructor's Manual** with teaching tips, calculator programs, some overhead transparency masters and sample exams.
- **Instructor's Solution Manual** with complete solutions to all problems.
- **Student's Solution Manual** with complete solutions to every other odd-numbered problem.
- **Answer Manual** with brief answers to all odd-numbered problems.
- **MultiGraph, Preliminary Version** for windows based surface plotting software.

Our Experiences

In the process of developing the ideas incorporated in this book, we have been conscious of the need to test the materials thoroughly in a wide variety of institutions serving many different types of students. Consortium members have used previous versions of the book at a broad range of institutions. During the 1993-94 academic year, we were assisted by colleagues at over fifty schools around the country who class-tested the book and reported their experiences and those of their students. This diverse group of schools used the book in semester and quarter systems, in computer labs, small groups, and traditional settings, and with a number of different technologies. We appreciate the valuable suggestions they made, which we have tried to incorporate into this draft version of the text.

Acknowledgements

Thanks to Carole Anderson, Kevin Anderson, Ralph Baierlein, Roxann Batiste, Jerrie Beiberstein, Paul Blanchard, Otto Bretscher, John Brillhart, Ruvim Breydo, Chris Bowman, Will Brockman, Edward Chandler, Phil Cheifetz, C. K. Cheung, Robert Condon, Eric Connally, Josh Cowley, John Drabicki, Bill Dunn, Bill Faris, George Fennemore, Katy Flint, Leonid Friedlander, Deborah Gaines, Liwei Gao, Nikki Grant, Marty Greenlee, John Hagood, Robert Hanson, Angus Hendrick, Tricia Hersh, Randy Ho, Greg Holmberg, Sharon Hurst, Robert Indik, Utith Inprasit, Adrian Iovita, Jack Jackson, Jerry Johnson, Millie Johnson, Georgia Kamvosoulis, Joe Kanapka, Alex Kasman, Matthias Kawski, Mike Klucznik, Dmitri Kountourgiannis, Robert Kuhn, Kam Kwong, Ted Laetsch, Sylvain Laroche, Janny Leung, Dave Levermore, Weiye Li, Li Liu, Carlos Lizzaraga, Patti Frazer Lock, John Lucas, Alex Mallozzi, Ricardo Martinez, Mark McConnell, Dan McGee, Karen Millstone, Kathy Mosher, David Mumford, Marshall Mundt, Don Myers, Jeff Nelson, Alan Newell, Huy Nguyen, John Olson, Myriam Oviedo, Ed Park, Howard Penn, Tony Phillips, Jessica Polito, Steve Prothero, Amy Rabb-Liu, Wayne Raskind, Renee Robles, David Royster, W. R. Salzman, Bill Schultz, Barbara Shipman, Michael Stringer, Noah Syroid, Mike Tabor, Sulian Tay, Tepache, Denise Todd, Jose Torres, Elias Toubassi, Jerry Uhl, Doug Ulmer, Steve Uurtamo, Faye Villalobos, Alice Wang, Joseph Watkins, Eric Wepsic, Steve Wheaton, Maciej Wojtkowski, Xianbao Xu, and Bruce Yoshiwara.

William G. McCallum	Sheldon P. Gordon	Jeff Tecosky-Feldman
Deborah Hughes-Hallett	Brad G. Osgood	Joe B. Thrash
Daniel E. Flath	Douglas Quinney	Thomas W. Tucker
Andrew M. Gleason		

To Students: How to Learn from this Book

- This book may be different from other math textbooks that you have used, so it may be helpful to know about some of the differences in advance. This book emphasizes at every stage the *meaning* (in practical, graphical or numerical terms) of the symbols you are using. There is much less emphasis on "plug-and-chug" and using formulas, and much more emphasis on the interpretation of these formulas than you may expect. You will often be asked to explain your ideas in words or to explain an answer using graphs.

- The book contains the main ideas of multivariable calculus in plain English. Your success in using this book will depend on your reading, questioning, and thinking hard about the ideas presented. Although you may not have done this with other books, you should plan on reading the text in detail, not just the worked examples.

- There are very few examples in the text that are exactly like the homework problems. This means that you can't just look at a homework problem and search for a similar-looking "worked out" example. Success with the homework will come by grappling with the ideas of calculus.

- Many of the problems that we have included in the book are open-ended. This means that there may be more than one approach and more than one solution, depending on your analysis. Many times, solving a problem relies on common sense ideas that are not stated in the problem but which you will know from everyday life.

- This book assumes that you have access to a graphing calculator or computer; preferably one that can draw surface graphs, contour diagrams, and vector fields, and can compute multivariable integrals and line integrals numerically. There are many situations where you may not be able to find an exact solution to a problem, but you can use a calculator or computer to get a reasonable approximation. An answer obtained this way is usually just as useful as an exact one. However, the problem does not always state that a calculator is required, so use your judgement.

- This book attempts to give equal weight to three methods for describing functions: graphical (a picture), numerical (a table of values) and algebraic (a formula). Sometimes you may find it easier to translate a problem given in one form into another. For example, if you have to find the maximum of a function, you might use a contour diagram to estimate its approximate position, use its formula to find equations that give the exact position, then use a numerical method to solve the equations. The best idea is to be flexible about your approach: if one way of looking at a problem doesn't work, try another.

- Students using this book have found discussing these problems in small groups very helpful. There are a great many problems which are not cut-and-dried; it can help to attack them with the other perspectives your colleagues can provide. If group work is not feasible, see if your instructor can organize a discussion session in which additional problems can be worked on.

- You are probably wondering what you'll get from the book. The answer is, if you put in a solid effort, you will get a real understanding of one of the most important accomplishments of the millennium – calculus – as well as a real sense of how mathematics is used in the age of technology.

x

Table of Contents:

CHAPTER ELEVEN

FUNCTIONS OF MANY VARIABLES

Many quantities depend on more than one variable: the amount of food grown depends on the amount of rain and the amount of fertilizer used; the rate of a chemical reaction depends on the temperature and the pressure of the environment in which it proceeds; and the rate of fallout from a volcanic explosion depends on the distance from the volcano and the time since the explosion. In this chapter we will see the many different ways of looking at functions of many variables.

11.1   FUNCTIONS OF TWO VARIABLES

If you are planning to take out a 5-year loan to buy a car, you need to calculate what your monthly payment will be; this depends on both the amount of money you borrow and the interest rate on the loan. Both of these quantities can vary separately; the loan amount could change while the interest rate remains the same, and the interest rate could change while the loan amount remains the same. To calculate your monthly payment you need to know both. If the monthly payment is m, the loan amount is L, and the interest rate is $r\%$, then we express the fact that m is a function of L and r by writing:

$$m = f(L, r).$$

This is just like the function notation of one-variable calculus. The variable m is called the dependent variable, and the variables L and r are called the independent variables. The letter f stands for the *function* or rule that gives the value of m corresponding to given values of L and r. A function of two variables can be represented pictorially by graphs and contour diagrams, numerically by a table of values, or algebraically by a formula. In this section we will give examples of each of these three ways of viewing a function.

Graphical Example: A Weather Map

Figure 11.1 shows a weather map such as might be seen in the daily newspaper. What information does it convey? It is displaying the predicted high temperature, T, in degrees Fahrenheit ($°F$), at any point in the US on that day. The curving lines on the map (called isotherms) separate the country into zones, according to whether T is in the 60s, 70s, 80s, 90s, or 100s. Notice that neighboring zones are close in temperature; for example, the 80s zone has a common boundary with zones in the 70s and 90s, but not with any zone in the 60s or 100s. The reason for this is that the temperature varies continuously from place to place, without sudden jumps. As we cross from the 80s zone into the 90s zones, the temperature goes from the high 80s, through 90, to the low 90s. The point where we cross the isotherm separating the zones is the point where the temperature is exactly 90°F. The temperature is 90°F all along this isotherm; in fact, to plot the isotherm, a meteorologist puts a curve through all the points where the temperature is 90°F.

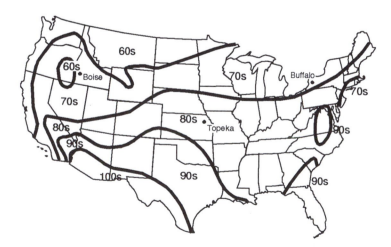

Figure 11.1: Weather map showing predicted high temperatures, T, for June 30, 1992.

Example 1   Estimate the value of T in Boise, Idaho; Topeka, Kansas; and Buffalo, New York.

Solution   Boise and Buffalo are in the 70s region, and Topeka is in the 80s region. Thus, the temperature in Boise and Buffalo is between 70 and 80; the temperature in Topeka is between 80 and 90.

In fact, we can say more. Although both Boise and Buffalo are in the 70s, Boise is quite close to the $T = 70$ isotherm, whereas Buffalo is quite close to the $T = 80$ isotherm. So we guess that the temperature is in the low 70s in Boise, and the high 70s in Buffalo. Topeka is more or less halfway in between the $T = 80$ isotherm and the $T = 90$ isotherm. Thus, we guess that the temperature in Topeka is in the mid 80s. In fact, the high temperatures for that day were 71°F for Boise, 79°F for Buffalo, and 86°F for Topeka.

The high temperature T illustrated by the weather map is a function of (that is, depends on) location. The location of a point in the US is given by two variables, often longitude and latitude, but possibly, miles east-west and miles north-south of a fixed point, say, Topeka. Thus, T is a function of two variables and the weather map is a way of visualizing that function.

Terminology

The weather map for high temperatures T in the US is *not* called the "graph" of the function T. Recall that the graph of a function of one variable, $y = f(x)$, uses a vertical axis for the dependent variable y and a horizontal axis for the independent variable x. The graph of a function of two variables, $z = f(x, y)$, needs a vertical axis for the dependent variable z and two axes for the independent variables x and y. Thus, the graph of a function of two variables is drawn in three dimensional xyz-space. We will study the graph of function of two variables in Section 11.3. The weather map in Figure 11.1 is actually called a *contour map* or *contour diagram*. We will study contour maps in Section 11.4.

Numerical Example: Beef Consumption

Suppose you are a beef producer and you want to know how much beef people will buy. This will depend on how much money people have and on the price of beef. Thus, the consumption of beef, C (in pounds per week per household) is a function of household income, I (in thousands of dollars per year), and the price of beef, p (in dollars per pound). In functional notation, we write:

$$C = f(I, p).$$

Table 11.1 shows some values of this function. Values of p are shown across the top, values of I are shown down the left side, and corresponding values of $f(I, p)$ are given in the table.[1]

TABLE 11.1   *Beef consumption in the US (in pounds per household per week)*

		Price of beef, p ($/lb)			
		3.00	3.50	4.00	4.50
Household Income Per Year, I ($1000)	20	2.65	2.59	2.51	2.43
	40	4.14	4.05	3.94	3.88
	60	5.11	5.00	4.97	4.84
	80	5.35	5.29	5.19	5.07
	100	5.79	5.77	5.60	5.53

[1]Richard G. Lipsey, *An Introduction to Positive Economics 3rd Ed.*, Weidenfeld and Nicolson, London, 1971

Notice how this differs from the table of values of a one-variable function, where a single row or column is enough to list the values of the function. Here many rows and columns are needed because the function has a value for every pair of values of the independent variables. We list one of the independent variables horizontally (in this case p), the other vertically (in this case I), and fill in the values corresponding to each combination in the table. For example, to find the value of $f(40, 3.50)$, we look in the row corresponding to $I = 40$ under $p = 3.50$, where we find the number 4.05. Thus,

$$f(40, 3.50) = 4.05.$$

This means that if a household's income is \$40,000 a year and the price of beef is \$3.50/lb, the family will buy an average of 4.05 lbs of beef per week.

Example 2   Find the amount M that each household spends on beef (in dollars per household per week), as a function of the price of beef and household income.

Solution   The amount of money spent on beef is the amount of beef bought times the price. Thus

$$M = pC = pf(I, p).$$

To calculate the values of M we need to multiply each value of f by the corresponding value of p, that is, we need to multiply each entry in Table 11.1 by the price at the top of the column. This yields Table 11.2.

TABLE 11.2   *Amount of money spent on beef (\$/household/week)*

| | | Price of beef, p (\$/lb) | | |
	3.00	3.50	4.00	4.50
20	7.95	9.07	10.04	10.94
40	12.42	14.18	15.76	17.46
60	15.33	17.50	19.88	21.78
80	16.05	18.52	20.76	22.82
100	17.37	20.20	22.40	24.89

Household income, I (\$1000)

Algebraic Examples: Formulas

In both the weather map and beef consumption examples, there was no formula for the underlying function. That is usually the case for real-life data. On the other hand, for many idealized models or situations in physics or geometry, there are exact formulas.

Example 3   Give a formula for the function

$$M = f(B, t)$$

where M is the amount of money in a bank account t years after an initial investment of B dollars, if interest is accrued at a rate of 5% per year
(a) Compounded annually   (b) Compounded continuously

Solution   (a) Annual compounding means that

$$M = f(B, t) = B(1.05)^t.$$

(b) Continuous compounding means that

$$M = f(B, t) = Be^{0.05t}.$$

Example 4   A cylinder with closed ends has a radius r and a height h. If its volume is V and surface area is A, find formulas for the functions $V = f(r, h)$ and $A = g(r, h)$.

Solution   The volume is the area of the circular base, πr^2, times the height, h:

$$V = f(r, h) = \pi r^2 h.$$

The surface area is the sum of the areas of the circular top and bottom, both πr^2, and the area of the side, which is the circumference of the bottom, $2\pi r$, times the height h. Thus,

$$A = g(r, h) = \pi r^2 + \pi r^2 + 2\pi rh = 2\pi r^2 + 2\pi rh.$$

A Useful Strategy: Allow Only One Variable at a Time to Vary

To use what we know about functions of one variable to study functions of two or more variables, let one variable vary at a time and hold the others fixed, giving a function of one variable.

A Vibrating Guitar String

Suppose you pluck a guitar string and watch it vibrate. If you took snapshots of the guitar string at one second intervals, you might get something like Figure 11.2.

We can analyze the motion of the guitar string using a function of two variables. Think of the guitar string stretched tight along the x-axis from $x = 0$ to $x = \pi$. Each point on the string has an x value, $0 \leq x \leq \pi$. As the string vibrates, each point on the string moves back and forth on either side of the x-axis. The ends of the string at $x = 0$ and $x = \pi$ remain stationary, while the point at the middle of the string moves the most. Let $f(x, t)$ be the displacement at time t of the point on the string located x units from the left end. Then a possible equation for $f(x, t)$ is given by

$$f(x, t) = \cos t \sin x, \quad 0 \leq x \leq \pi.$$

To see why this function models the vibrating guitar string, we will see what happens when we allow only one of the quantities x and t to vary at a time.

Figure 11.2: A vibrating guitar string:
$f(x, t) = \cos t \sin x$

Example 5   Relate the functions $f(x, 0)$ and $f(x, 1)$ to the vibrating string. Relate the functions $f(0, t)$ and $f(1, t)$ to the vibrating string.

Solution   The function $f(x, 0) = \cos 0 \sin x = \sin x$ gives the displacement of each point of the string when time is held fixed at $t = 0$. The function $f(x, 1) = \cos 1 \sin x = 0.54 \sin x$ gives the displacement of each point of the string at time $t = 1$. Graphing $f(x, 0)$ and $f(x, 1)$ gives in each case an arch of the sine curve, the first with amplitude 1 and the second with amplitude 0.54. For each different fixed value of t, we get a different snapshot of the string, each one a sine curve with amplitude given by the value of $\cos t$. The result looks like the sequence of snapshots shown in Figure 11.2.

The function $f(0,t) = \cos t \sin 0 = 0$ gives the displacement of the left end of the string as time varies. Since that point remains stationary, the displacement is zero as we expected. The function $f(1,t) = \cos t \sin 1 = 0.84 \cos t$ gives the displacement of the point at $x = 1$ unit along the string as time varies. Since $\cos t$ oscillates back and forth between 1 and -1, this point moves back and forth with maximum displacement of 0.84 in either direction. Notice the displacements will be greatest at $x = \pi/2$ where $\sin x = 1$.

Example 6   Describe the motion of the guitar strings whose displacements are given by
(a) $g(x,t) = \cos 2t \sin x$.   (b) $h(x,t) = \cos t \sin 2x$.

Solution   (a)   For g, our snapshots for fixed values of t are still one arch of the sine curve. The amplitudes, which are governed by the $\cos 2t$ factor, now change twice as fast as before. That is, the string is vibrating twice as fast.

(b)   For $h(x,t) = \cos t \sin 2x$, the vibration of the string is more complicated. If we hold t fixed at $t = 0$, the snapshot now shows one full period, a crest and a trough, of the sine curve. Now the center of the string, $x = \pi/2$, remains stationary just like the end points. The vibration of this string is shown in Figure 11.3.

Figure 11.3: Another vibrating string:
$h(x,t) = \cos t \sin 2x$

The vibrating string of Examples 5 and 6 on page 5 is called a *standing* wave, since the crest of the sine wave and the stationary points always stay in the same location on the string. A *traveling* wave (more like a wave in the ocean) is described in Problems 13-16.

The Beef Data

If a function is given by a table of values, such as the beef consumption data, we can allow one variable to vary at a time. For example, if we hold the income, I, fixed at 40, then to see what happens as the price, p, varies, we look at the row $I = 40$. This row gives the function $f(40, p)$, some of whose values are shown in Table 11.3.

TABLE 11.3   *Beef consumption by households making $40,000*

p	3.00	3.50	4.00	4.50
$f(40,p)$	4.14	4.05	3.94	3.88

Note that since we have fixed a specific value of I, we now have a function of one variable that shows how much beef will be bought by people who earn \$40,000 a year as a function of the price of beef. From Table 11.3 we see that $f(40, p)$ decreases as p increases. The other rows tell the same story; in each fixed income group, the consumption of beef goes down as the price increases. This makes sense; the more beef costs, the less beef a household can afford to buy. Thus, f is a decreasing function of price, p, for each fixed value of income, I.

Example 7   If the price of beef is held at a particular level, how does beef consumption vary as a function of household income?

Solution   Each price corresponds to a column in Table 11.1. For example, the column corresponding to $p = 3.00$ gives the function $h(I) = f(I, 3.00)$; it tells you how much beef a household will buy at \$3.00/lb, as a function of its annual income. Looking at the column from the top down, you can see that it is an increasing function of I. Again, this is true in each column. This says that at any fixed price level for beef, the consumption goes up as household income goes up, which makes sense. Thus, f is an increasing function of I for each value of p.

Weather Map

What happens on the weather map in Figure 11.1 when we allow only one variable at a time to vary? For example, suppose we moved along the east-west line through Topeka. Suppose x represents miles east or west of Topeka and y represents miles north or south. We keep y fixed at 0 and let x vary. Along this line, the high temperature T goes from the 60s along the west coast, to the 70s in Nevada and Utah, to the 80s in Topeka to the 90s just before the east coast, then returns to the 80s. The graph looks like that shown in Figure 11.4.

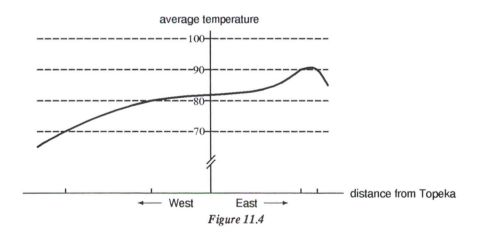

Figure 11.4

Terminology

Given a function f of two variables x and y, the function we get by holding y fixed at some value and letting x vary is called a *section* of f with y fixed.   If we hold x constant instead, we get a *section* of f with x fixed. If the fixed value for x is $x = c$, we speak of *sections* of f with $x = c$.

Thus in the vibrating string examples, the functions $f(x, 0)$ and $f(x, 1)$ are sections of f with $t = 0$ and $t = 1$. In the beef example, as we look across the $I = 40$ row of the table, we are looking at values for the f section with $I = 40$. And in the weather map example, the graph shown in Figure 11.4 is the graph of the T section with y fixed at the latitude of Topeka, Kansas.

Problems for Section 11.1

1. Use the weather map in Figure 11.1 on page 2 to give the range of daily high temperatures on June 30, 1992 for (a) Pennsylvania   (b) North Dakota   (c) California.

2. Sketch the graph of the high temperature T on a line north-south through Topeka in the weather map of Figure 11.1.

3. Referring to Figure 11.1, graph the high temperature in a north-south and an east-west line through Boise, Idaho.

4. Sketch the graph of the bank account function f of Example 3 on page 4, holding B fixed at three different values and letting only t vary. Then sketch the graph of f holding t fixed at three different values and letting only B vary.

5. Consider the acceleration due to gravity, g, at a height h above the surface of a planet of mass m.
   (a) Is g an increasing or decreasing function of h? How do you know?
   (b) Is g an increasing or decreasing function of m? How do you know?

6. Consider a function giving the number, n, of new cars sold in a year as a function of the price of new cars, c, and of the average price of gas, g.
   (a) If c is held constant, is n an increasing or decreasing function of g? Why?
   (b) If g is held constant, is n an increasing or decreasing function of c? Why?

7. You are planning a long driving trip, and your principal cost will be gasoline.
   (a) Make a table showing how the daily fuel cost varies as a function of the price of gasoline (in dollars per gallon) and the number of gallons you buy each day.
   (b) If your car can go 30 miles on each gallon of gasoline, make a table showing how your daily fuel cost varies as a function of your travel distance and the price of gas.

8. Consider the standing wave given by

$$h(x, t) = \sin x \cos t, \quad 0 \le x \le \pi.$$

   (a) Sketch graphs of h versus x for fixed t values, $t = 0, \pi/4, \pi/2, 3\pi/4, \pi$. In each case only consider $0 \le x \le \pi$.
   (b) This is an example of a standing wave that you might get when you pluck a guitar string and let it vibrate. Can you see why?
   (c) What is the effect of replacing x by $2x$ in the equation for $h(x, t)$?
   (d) What is the effect of replacing t by $2t$ in the equation for $h(x, t)$?

9. Table 11.4 shows the wind-chill factor as a function of wind speed and temperature. The wind-chill factor is a temperature which tells you how cold it feels, as a result of the combination of wind and temperature.

TABLE 11.4 *Wind-chill factor*

		Temperature (degrees Fahrenheit)							
		35	30	25	20	15	10	5	0
Wind (mph)	5	33	27	21	16	12	7	0	-5
	10	22	16	10	3	-3	-9	-15	-22
	15	16	9	2	-5	-11	-18	-25	-31
	20	12	4	-3	-10	-17	-24	-31	-39
	25	8	1	-7	-15	-22	-29	-36	-44

(a) If the temperature is 0° Fahrenheit, and the wind speed is 15 mph, how cold does it feel?

(b) If the temperature is 35° Fahrenheit, how fast would the wind need to blow to make it feel like 22° Fahrenheit?

(c) If the temperature is 25° Fahrenheit, how fast would the wind need to blow to make it feel like 20° Fahrenheit?

(d) If the wind is blowing at 15 mph, what temperature feels like 0° Fahrenheit?

10. Table 11.5 shows the heat index as a function of temperature and humidity. The heat index is a temperature which tells you how hot it feels as a result of the combination of the two. Heat exhaustion is likely to occur when the heat index reaches 105.

TABLE 11.5 *Heat index*

		Temperature (degrees Fahrenheit)								
		70	75	80	85	90	95	100	105	110	115
Relative humidity (%)	0	64	69	73	78	83	87	91	95	99	103
	10	65	70	75	80	85	90	95	100	105	111
	20	66	72	77	82	87	93	99	105	112	120
	30	67	73	78	84	90	96	104	113	123	135
	40	68	74	79	86	93	101	110	123	137	151
	50	69	75	81	88	96	107	120	135	150	
	60	70	76	82	90	100	114	132	149		

(a) If the temperature is 80° F and the humidity is 50%, how hot does it feel?

(b) At what humidity does 90° F feel like 90° F?

(c) Make a table showing the approximate temperature when heat exhaustion is a danger, as a function of humidity.

(d) Can you explain why the heat index is sometimes above the actual temperature and sometimes below it?

For Exercises 11 and 12, refer to Table 11.1.

11. Make a table of the proportion P of household income spent on beef per week as a function of price and income.

12. Express P in terms of the original function $f(I, p)$ which gave consumption as a function of p and I.

Suppose the audience is doing the wave in a large crowded stadium. The wave is a ritual in which members of the audience stand up and down in such a way as to create a wave that moves around the stadium. Normally, a single wave travels all the way around the stadium, but we will assume there is a continuous sequence of waves. What sort of function describes the motion of the audience? We consider just one row of spectators and look for a function which describes the motion of each individual in the row. This will be a function of two variables: x (the seat number), and t (the time, in seconds). For each value of x and t, let $h(x, t)$ be the height in feet above the ground of the head of the spectator x seats in from the aisle, at time t seconds. Suppose that

$$h(x, t) = 5 + \cos(0.5x - t).$$

In Problems 13-16, we investigate this function by looking at sections of h with x held fixed as t varies and then with t held fixed as x varies.

13. Explain the significance of the sections of h with $x = 2$ and with $t = 5$.

14. Show that the graph of the section of h with $x = 7$ is the same shape as the graph of the section of h with $x = 2$.

15. Use the result of Problem 14 to find the speed of the wave.

16. By considering different sections of h with t fixed, confirm the speed of the wave is 2 seats per second.

17. Suppose the function for the stadium wave in Problems 13–16 were $h(x, t) = 5 + \cos(x - 2t)$. How does this wave compare with the original wave? What is its speed?

18. Suppose the stadium wave in Problems 13–16 were moving in the opposite direction, right-to-left instead of left-to-right. Give a possible formula for h.

11.2   A TOUR OF THREE-DIMENSIONAL SPACE

Cartesian Coordinates in Three-Space

The way we describe points in the plane by giving x- and y-coordinates can be extended to three-dimensional space. Imagine three coordinate axes meeting at the *origin*: a vertical axis, and two horizontal axes at right angles to each other. Figure 11.5 is a perspective drawing. Think of the xy-plane as being horizontal, while the z-axis extends vertically above and below the plane. The labels x, y, and z show which part of each axis is positive. You can specify a point in 3-space by giving its coordinates (x, y, z) with respect to these axes. Think of the coordinates as instructions telling you how to get to the point; starting at the origin, going x units in the direction parallel to the x-axis, y units in the direction parallel to the y-axis, and z units in the direction parallel to the z-axis.

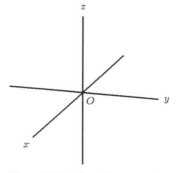

Figure 11.5: Coordinate axes in 3-space.

The coordinates can be zero or negative; a zero coordinate means "don't move in this direction," and a negative coordinate means "go in the negative direction along this axis." For example, the origin has coordinates $(0, 0, 0)$, since you get there from the origin by doing nothing at all.

Example 1   Describe the position of the points with coordinates $(1, 2, 3)$ and $(0, 0, -1)$.

Solution   We get to the point $(1, 2, 3)$ by starting at the origin, going 1 unit along the x-axis, 2 units in the direction parallel to y-axis, and 3 units up in the direction parallel to the z-axis (see Figure 11.6).
    To get to $(0, 0, -1)$, we don't move at all in the x and y directions, and we move 1 unit in the negative z direction. So the point is on the negative z-axis (see Figure 11.6).

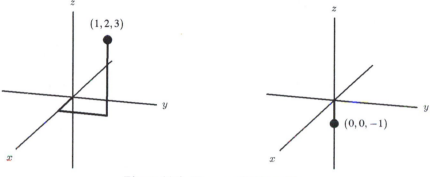

Figure 11.6: How coordinates work

Example 2   You start at the origin, go along the y-axis a distance of 2 units in the positive direction, and then move vertically upward a distance of 1 unit. What are your coordinates?

Solution   You started at the origin with coordinates $(0, 0, 0)$. When you went along the y-axis your y-coordinate increased to 2, and when you went vertically your z-coordinate increased to 1; your x-coordinate hasn't changed because you have not moved at all in the x direction. So your coordinates are $(0, 2, 1)$.

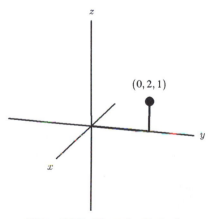

Figure 11.7: The point $(0, 2, 1)$

It is often helpful to picture a three dimensional coordinate system in terms of a room you are in. The origin is a corner at floor level where two walls meet the floor. The vertical or z-axis is the intersection of the two walls; the x- and y-axes are the intersections of each wall with the floor. Points with negative coordinates lie behind a wall in the next room or below the floor.

Graphing Equations in 3-Space

Just as with two variables, we can graph equations involving the variables x, y, and z in three-dimensional space.

Example 3   What do the graphs of the equations $z = 0$, $z = 3$, and $z = -1$ look like?

Solution   Graphing an equation means drawing the set of points in space whose coordinates satisfy the equation. So to graph $z = 0$ we need to visualize the set of points whose z-coordinate is zero. If your z-coordinate is 0, then you must be at the same vertical level as the origin, or, in other words, you must be in the horizontal plane containing the origin. On the other hand, since your x- and y-coordinates can be anything, you can get to any point in this plane by moving a certain distance in the x direction and a certain distance in the y direction. So the graph of $z = 0$ is the middle plane in Figure 11.8.

If your z-coordinate is 3, then again you must always stay at the same level, but this time the level is 3 units higher than before. Thus the graph of $z = 3$ is parallel to the graph of $z = 0$, but three units above it. Similarly, the graph of $z = -1$ is a plane parallel to $z = 0$, but one unit below it.

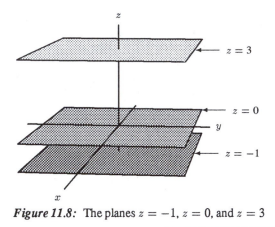

Figure 11.8:   The planes $z = -1$, $z = 0$, and $z = 3$

The plane $z = 0$ contains the x- and y-coordinate axes, and hence is called the xy-coordinate plane, or xy-plane for short. In Example 3, we saw that all the points with $z = 3$ are three units above the xy-plane, and all the points with $z = -1$ are one unit below it. Thus, the z-coordinate of a point may be thought of as the distance from the xy-plane (with the distance being negative if the point lies below the plane).

There are two other coordinate planes. The yz-plane contains both the y- and the z-axes, and the xz-plane contains both the x- and the z-axes. See Figure 11.9. The x-coordinate gives distance from the yz-plane; the y-coordinate gives distance from the xz-plane.

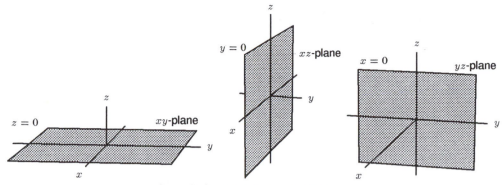

Figure 11.9: The three coordinate planes

The z-coordinate of a point is the distance from the point to the xy-plane and the y-coordinate is the distance from the xz-plane, and the x-coordinate is the distance from the yz-plane. To see this, imagine a box a units deep, b units long, and c units high, with one corner at the origin, as in Figure 11.10. By following along the edges in the figure, you can see that the corner of the box diagonally opposite the origin is the point (a, b, c), since you get there by going a units along the x-axis, b units parallel to the y-axis, and c units parallel to the z-axis. Picture this in terms of the room you are in. On the other hand, since the box has depth a you can see that the point (a, b, c) is a units from the yz-plane; since it has length b you can see that the point is b units from the xz-plane; and since it has height c you can see that the point is c units from the xy-plane.

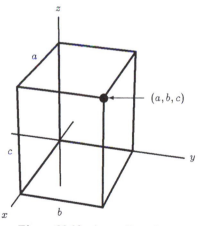

Figure 11.10: A coordinate box

Example 4   Which of the points $A = (1, -1, 0)$, $B = (0, 3, 4)$, $C = (2, 2, 1)$, and $D = (0, -4, 0)$ lies closest to the xz-plane? Which point lies on the y-axis?

Solution   Since the y-coordinate gives the distance to the xz-plane, the point A lies closest to that plane, because it has the smallest y-coordinate in absolute value. (It actually lies to the left of the plane, since its y-coordinate is negative.)

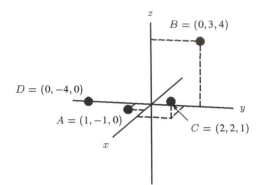

Figure 11.11: Which point lies closest to the
xz-plane? On the y-axis?

To get to a point on the y-axis, you move along the y-axis, but you don't move at all in the x or z directions; thus a point on the y-axis has both its x- and z-coordinates equal to zero. The only point of the four that satisfies this is D, so it is the only one on the y-axis. Figure 11.11 shows the points.

As a general rule, if a point has one of its coordinates equal to zero, it lies in one of the coordinate planes, and if it has two of its coordinates equal to zero, then it lies on one of the coordinate axes. The origin has all three coordinates equal to zero, that is, it is the point $(0, 0, 0)$.

Example 5   You are 2 units below the xy-plane and in the yz-plane. What are your coordinates?

Solution   Since you are 2 units below the xy-plane, your z-coordinate is -2. And since you are in the yz-plane, your x-coordinate is 0. Your y-coordinate can be anything. Thus you are at the point $(0, y, -2)$, where y can be anything. The set of all such points forms a line parallel to the y-axis, 2 units below the xy-plane, and in the yz-plane. See Figure 11.12.

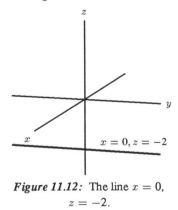

Figure 11.12: The line $x = 0$,
$z = -2$.

Example 6   You are standing at the point $(4, 5, 2)$, looking at the point $(0.5, 0, 3)$. Are you looking up or down?

Solution   The point you are standing at has z-coordinate 2, while the point you are looking at has z-coordinate 3; hence, it is higher than you and you are looking up.

Example 7   Imagine that the yz-plane in Figure 11.5 on page 10 is a page of this book. Describe the region behind the page.

Solution   The positive part of the x-axis pokes out of the page; moving in the positive x direction brings you out in front of the page. The region behind the page corresponds to negative values of x, and so it is the set of all points in three-dimensional space satisfying the inequality $x < 0$. Think of it as all the pages behind the one you are looking at.

Distance

In 2-space, the formula for the distance between two points (x, y) and (a, b) is given by

$$\text{Distance} = \sqrt{(x - a)^2 + (y - b)^2}.$$

In 3-space, the formula is similar. The distance, d, between two points (x, y, z) and (a, b, c) is

$$\text{Distance} = \sqrt{(x - a)^2 + (y - b)^2 + (z - c)^2}.$$

Both formulas come from Pythagoras' theorem.

Example 8   Find the distance between $(1, 2, 1)$ and $(-3, 1, 2)$.

Solution   The formula gives a distance of

$$D = \sqrt{(-3 - 1)^2 + (1 - 2)^2 + (2 - 1)^2} = \sqrt{18} = 4.24.$$

Example 9   Find an expression for the distance from the origin to the point (x, y, z).

Solution   The origin has coordinates $(0, 0, 0)$, so the distance from the origin to a point (x, y, z)

$$D = \sqrt{(x - 0)^2 + (y - 0)^2 + (z - 0)^2} = \sqrt{x^2 + y^2 + z^2}.$$

Example 10   Which of the points $P = (1, 2, 1)$ and $Q = (2, 0, 0)$ is closest to the origin?

Solution   The point P is $\sqrt{1^2 + 2^2 + 1^2} = \sqrt{6} = 2.45$ units from the origin, and Q is $\sqrt{2^2 + 0^2 + 0^2} = 2$ units from the origin. Since $2 < \sqrt{6}$, the point Q is closer.

Example 11   Find the equation for a sphere of radius 1 with center at the origin.

Solution   The sphere consists of all points (x, y, z) whose distance from the origin is 1, that is, which satisfy the equation

$$\sqrt{x^2 + y^2 + z^2} = 1.$$

This is an equation for the sphere. If we square both sides we get the equation in the more pleasant looking form

$$x^2 + y^2 + z^2 = 1.$$

Note that this represents the surface of the sphere. The solid region enclosed by the sphere is represented by the inequality $x^2 + y^2 + z^2 < 1$.

Problems for Section 11.2

1. Which of the points $A = (1.3, -2.7, 0)$, $B = (0.9, 0, 3.2)$, $C = (2.5, 0.1, -0.3)$ is closest to the yz-plane? Which one lies on the xz-plane? Which one is farthest from the xy-plane?

2. Which of the points $A = (23, 92, 48)$, $B = (-60, 0, 0)$, $C = (60, 1, -92)$ is closest to the yz-plane? Which one lies on the xz-plane? Which one is farthest from the xy-plane?

3. You are at the point $(-1, -3, -3)$ facing the yz-plane. You walk 2 units forward, turn left, and walk for another 2 units. Are you in front of or behind the yz-plane? Are you to the left or to the right of the xz-plane? Are you above or below the xy-plane? Assume the axes are oriented as in Figure 11.5 on page 10.

4. You are at the point $(3, 1, 1)$ facing the yz-plane. You walk 2 units forward, turn left, and walk for another 2 units. Are you in front of or behind the yz-plane? Are you to the left of or to the right of the xz-plane? Are you above or below the xy-plane? Assume the axes are oriented as in Figure 11.5 on page 10.

Sketch the graphs of the equations in Problems 5–7.

5. $x = -3$
6. $y = 1$
7. $z = 2$ and $y = 4$

8. Find a formula for the shortest distance between a point (a, b, c) and the y-axis.

9. Describe the set of points whose distance from the x-axis is 2.

10. Describe the set of points whose distance from the x-axis equals the distance from the yz-plane.

11. Which two of the three points $(1, 2, 3)$, $(3, 2, 1)$ and $(1, 1, 0)$ are closest to each other?

12. A cube is located such that its top four corners have the coordinates of $(-1, -2, 2)$, $(-1, 3, 2)$, $(4, -2, 2)$ and $(4, 3, 2)$. Give the coordinates of the center of the cube.

13. A rectangular solid lies with its length parallel to the y-axis, and its top and bottom faces parallel to the plane $z = 0$. If the center of the object is at $(1, 1, -2)$ and it has a length of 13, a height of 5 and a width of 6, give the coordinates of all eight corners and draw the figure labeling the eight corners.

14. Which of the points $(-3, 2, 15)$, $(0, -10, 0)$, $(-6, 5, 3)$ and $(-4, 2, 7)$ is closest to $(6, 0, 4)$?

15. You are standing on the point $(3, 4, -2)$. North is in the negative x-direction, East is in the positive y-direction, and up in the positive z-direction.

    (a) What coordinates will you be at after you move North 3 units, up 5 units, West 7 units, down 2 units, East 7 units, and South 2 units?

    (b) After completing part (a), will you be looking up or down to the point $(6, 7, 7)$?

16. On a set of x, y, and z axes oriented as in Figure 11.5 on page 10, draw a straight line through the origin and lying in the xz-plane and such that if you move along the line with your x-coordinate increasing, your z-coordinate is decreasing.

17. On a set of x, y and z axes oriented as in Figure 11.5 on page 10, draw a straight line through the origin and lying in the yz-plane and such that if you move along the line with your y-coordinate increasing, your z-coordinate is increasing.

18. Find the equation of the sphere of radius 5 centered at the origin.

19. Find the equation of the sphere of radius 5 centered at $(1, 2, 3)$.

20. Given the sphere

$$(x - 1)^2 + (y + 3)^2 + (z - 2)^2 = 1,$$

   (a) Find the equation of the circles (if any) where the sphere intersects each coordinate plane.

   (b) Find the points (if any) where the sphere intersects each coordinate axis.

11.3 GRAPHS OF FUNCTIONS OF TWO VARIABLES

Visualizing the Beef Data

Suppose you wanted to plot the beef consumption data in Table 11.1 on page 3. You can't plot it on a pair of axes the way you plot functions of one variable, because in this case there are two variables to plot on the horizontal axis, namely, household income, I, and price of beef, p. So you need two horizontal axes, an I-axis and a p-axis. To have two horizontal axes you need to work in three dimensions. Imagine putting Table 11.1 down on the horizontal plane formed by these two axes, so that the entries line up with the corresponding numbers on the axes; for example, the entry $f(20, 3.00) = 2.65$ should line up with $I = 20$ on the I-axis and $p = 3$ on the p-axis. To plot the figures for beef consumption C given in the table, we put above each table entry a vertical bar with height equal to the value of the entry. This gives the three-dimensional bar graph pictured in Figure 11.13.

Notice that for any fixed value of I, the heights of the bars decrease as p increases. This is the same behavior we noted before when we said that, for fixed I, f is a decreasing function of p. Similarly, for any fixed price p, the heights increase as I increases, since f is an increasing function of I.

The tops of the bars in the bar graph form a rather bumpy surface, just as the tops of the bar graphs of a one-variable bar graph form a rather bumpy curve. If you plotted data at more and more

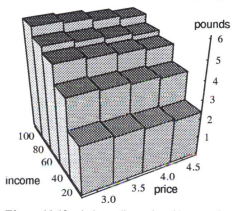

Figure 11.13: A three-dimensional bar graph

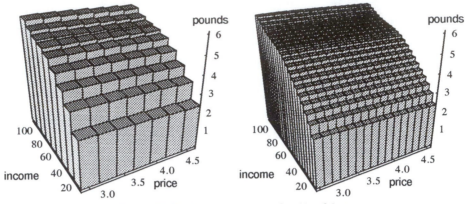

Figure 11.14: More and more refined beef data

intermediate values, you would get more and more bars crowding together, until their tops appear to form a smooth surface. See Figure 11.14.

What Is the Graph of a Function of Two Variables, $f(x, y)$?

Recall that the graph of a function of one variable, $f(x)$, is the collection of all points in 2-space of the form (x, y) such that $y = f(x)$. Consider, the graph of a function of two variables, say $f(x, y)$. The graph of f is all points in 3-space of the form (x, y, z) such that $z = f(x, y)$. For example, $(-2, 3, 13)$ is on the graph of $z = x^2 + y^2$ since $(-2)^2 + 3^2 = 13$.

The Graph of the Function $f(x, y) = x^2 + y^2$ is a Surface

To draw the graph of f we want to "connect points" just as we did for a function of one variable. We first need a table of values of f, such as in Table 11.6.

TABLE 11.6   *Table of values of $f(x, y) = x^2 + y^2$*

		\multicolumn{7}{c}{y}					
		-3	-2	-1	0	1	2	3
	-3	18	13	10	9	10	13	18
	-2	13	8	5	4	5	8	13
	-1	10	5	2	1	2	5	10
x	0	9	4	1	0	1	4	9
	1	10	5	2	1	2	5	10
	2	13	8	5	4	5	8	13
	3	18	13	10	9	10	13	18

Look at this table for a minute. Suppose we allow only one variable at a time to change. If we keep x fixed at 0 and let y vary, we get the middle row of the table: a sequence of squared numbers decreasing to 0 and then increasing. This is because $f(0, y) = y^2$. The row corresponding to $x = -2$ is the same as the row for $x = 0$ except 4 has been added to each value. This is because $f(-2, y) = 4 + y^2$. The columns of the table behave the same way as the rows because x and y appear in the equation $f(x, y) = x^2 + y^2$ in exactly the same way.

Now we have to plot the points and connect them. We connect the point $x = 1, y = 2$ to the points $x = 0, y = 2$ and $x = 2, y = 2$ and to the points $x = 1, y = 1$ and $x = 1, y = 3$. In other

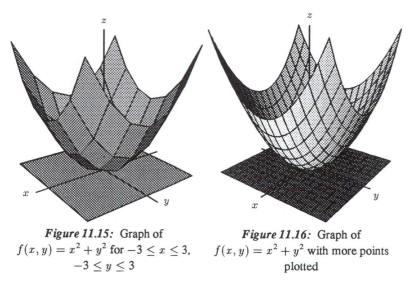

Figure 11.15: Graph of
$f(x,y) = x^2 + y^2$ for $-3 \le x \le 3$,
$-3 \le y \le 3$

Figure 11.16: Graph of
$f(x,y) = x^2 + y^2$ with more points
plotted

words, we connect up points moving across the rows or up and down the columns in the table of values. The result is called a *wire-frame* picture of the graph. Filling in between the wires gives a surface. That is the way a computer drew the wire-frame graph of $f(x,y) = x^2 + y^2$ shown in Figure 11.15. We get a better picture by choosing a smaller step-size. Figure 11.16 shows the result of plotting more points in each direction.

You should check to see if the graph makes sense. Notice that the graph dips down to the origin since $x = 0, y = 0, z = 0$ satisfies $z = x^2 + y^2$. Observe that if x is held fixed and y is allowed to vary, the graph dips down and then goes back up, just like the rows of Table 11.6. Similarly, if y is held fixed and x is allowed to vary, the graph dips down and then goes back up, just like the columns of Table 11.6.

Graphs of Functions Related to $f(x,y) = x^2 + y^2$

Once we know what the graph of $f(x,y) = x^2 + y^2$ looks like, we can visualize the graphs of some closely related functions.

Example 1   Describe the graphs of

(a) $f(x,y) = 3 + x^2 + y^2$ (b) $g(x,y) = 5 - x^2 - y^2$ (c) $h(x,y) = x^2 + (y-1)^2$

Solution   We already know from Figure 11.16 that the graph of $x^2 + y^2$ looks like a bowl with its base at the origin. From this we can work out what the graphs of f, g, and h will look like.

(a) $f(x,y) = 3 + x^2 + y^2$ is $x^2 + y^2$ with 3 added, so its graph looks like the graph of $x^2 + y^2$, but raised by 3 units. See Figure 11.17.

(b) $-x^2 - y^2$ is the negative of $x^2 + y^2$, so the graph of $-x^2 - y^2$ is an upside down bowl. Adding 5 raises the graph by 5 units, so the graph of $g(x,y) = 5 - x^2 - y^2$ looks like an upside down bowl whose vertex is at $(0,0,5)$, as in Figure 11.18.

(c)   The graph of $h(x,y) = x^2 + (y-1)^2$ looks like a bowl whose vertex is at $x = 0, y = 1$, since that is where $h(x,y) = 0$, as in Figure 11.19.

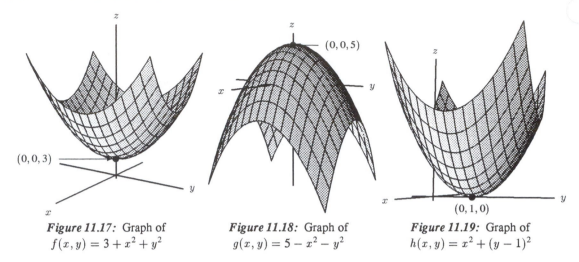

Figure 11.17: Graph of
$f(x, y) = 3 + x^2 + y^2$

Figure 11.18: Graph of
$g(x, y) = 5 - x^2 - y^2$

Figure 11.19: Graph of
$h(x, y) = x^2 + (y - 1)^2$

Example 2   Describe the graph of $z = e^{-(x^2+y^2)}$.

Solution   Since the exponential function is always positive, the graph lies entirely above the xy-plane. From the graph of $x^2 + y^2$ we see that $x^2 + y^2$ is zero at the origin and gets larger as you move farther from the origin in any direction. Thus $e^{-(x^2+y^2)}$ is 1 at the origin, and gets smaller as you move from the origin in any direction. It can't go below the xy-plane; instead it flattens out, getting closer and closer to the plane. We say the surface is *asymptotic* to the xy-plane. See Figure 11.20.

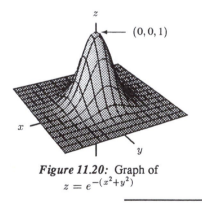

Figure 11.20: Graph of
$z = e^{-(x^2+y^2)}$

Sections and the Graph of a Function

The graph of a two-variable function can be understood in terms of the sections of the function, that is, the functions we get by keeping one of the two variables fixed. For example, for the function $f(x, y) = x^2 + y^2$, if we connect the points corresponding to the $x = 2$ row in Table 11.6, we get a curve on the graph of f that represents the section of f with $x = 2$. This is the same as the curve we get by intersecting the graph of f with the plane perpendicular to the x-axis at $x = 2$.

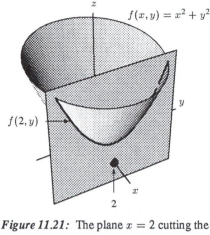

Figure 11.21: The plane $x = 2$ cutting the
surface $z = f(x, y)$

(See Figure 11.21.) That is the reason we call the function $f(2, y)$ the section of f with $x = 2$, in the sense of a cross-section or slice. Figure 11.22 shows graphs of other sections of f with x fixed; Figure 11.23 shows graphs of sections with y fixed. Together these sections generate the wire-frame surface in Figure 11.24.

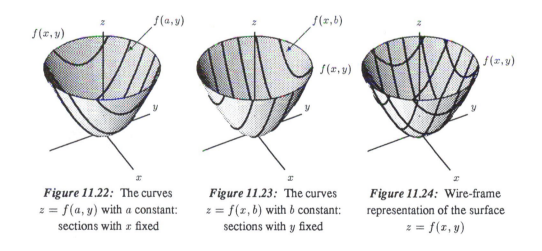

Figure 11.22: The curves $z = f(a, y)$ with a constant: sections with x fixed

Figure 11.23: The curves $z = f(x, b)$ with b constant: sections with y fixed

Figure 11.24: Wire-frame representation of the surface $z = f(x, y)$

Example 3   Describe the sections of the function $g(x, y) = x^2 - y^2$ with x fixed and then with y fixed. Use these sections to describe the shape of the graph of g.

Solution   We can do this numerically by looking at the table of values of g shown in Table 11.7. The sections with y fixed correspond to the columns of the table. For the $y = 0$ column, the values of y are all squared numbers that decrease to 0 and then increase. The column for $y = 2$ is the same except that 4 is subtracted from each entry. Thus, each section with y fixed gives a parabola opening upwards. The minimum value for the section corresponding to $y = a$ is $z = -a^2$. We could also deduce this algebraically by noticing that $g(x, 0) = x^2$, and $g(x, 2) = x^2 - 4$, and $g(x, a) = x^2 - a^2$.

TABLE 11.7   *Table of values of* $g(x, y) = x^2 - y^2$

		-3	-2	-1	0	1	2	3
	-3	0	5	8	9	8	5	0
	-2	-5	0	3	4	3	0	-5
	-1	-8	-3	0	1	0	-3	-8
x	0	-9	-4	-1	0	-1	-4	-9
	1	-8	-3	0	1	0	-3	-8
	2	-5	0	3	4	3	0	-5
	3	0	5	8	9	8	5	0

What about sections with x fixed? These correspond to the rows of the table. The sections with x fixed are parabolas opening downwards with a maximum of $z = a^2$, if the section is at $x = a$. Again, algebra tells us the same thing: $g(0, y) = -y^2$, and $g(a, y) = a^2 - y^2$. Figure 11.25 shows the graphs of sections with y fixed all drawn on one xz-plane; Figure 11.26 shows the graphs of sections with x fixed.

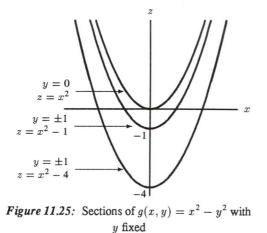

Figure 11.25: Sections of $g(x, y) = x^2 - y^2$ with y fixed

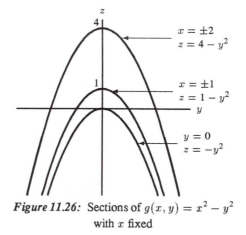

Figure 11.26: Sections of $g(x, y) = x^2 - y^2$ with x fixed

The graph of g is shown in Figure 11.27. Notice the upward opening parabolas in the x-direction and the downward opening parabolas in the y-direction. We say that the surface is saddle-shaped.

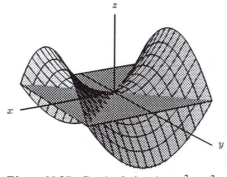

Figure 11.27: Graph of $g(x, y) = x^2 - y^2$.

Example 4 Describe the sections with t fixed and with x fixed of the vibrating guitar string function

$$f(x, t) = \cos t \sin x, \quad 0 \le x \le \pi,$$

and use them to understand the graph of f.

Solution The sections with t fixed describe snapshots of the string at different instants as discussed on page 5. Graphs of these sections can be seen in the graph of f shown in Figure 11.28. Every plane perpendicular to the t-axis intersects the surface in one arch of a sine curve. The arch is highest at the $t = 0$ section, flat at $t = \pi/2$, lowest at $t = \pi$, flat at $t = 3\pi/2$, and high again at $t = 2\pi$. The sections with x fixed show how a single point on the string moves as time goes by. Notice in Figure 11.28 that the sections with $x = 0$ and $x = \pi$ are flat lines since the endpoints of the string don't move. The section with $x = \pi/2$ is a cosine curve, because the midpoint of the string moves continually back and forth.

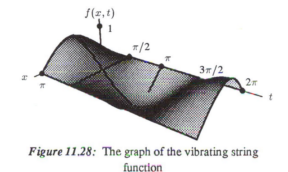

Figure 11.28: The graph of the vibrating string function

Example 5 Suppose the graph in Figure 11.29 shows your happiness as a function of love and money. Describe in words your happiness:
  (a) As a function of money, with love fixed.
  (b) As a function of love, with money fixed.
  (c) Draw the graphs of two different sections with love fixed and two different sections with money fixed.

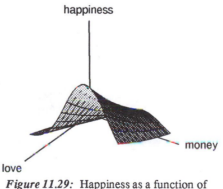

Figure 11.29: Happiness as a function of love and money.

Solution (a) As money increases, with love fixed, your happiness goes up, reaches a maximum and then goes back down. Evidently, there is such a thing as too much money.

(b) On the other hand, as love increases, with money fixed, your happiness keeps going up.

(c) A section with love fixed will show your happiness as money increases; the curve goes up to a maximum then back down, as in Figure 11.30. The higher section, showing more overall happiness, corresponds to a larger amount of love, because as love increases so does happiness. Figure 11.31 shows two sections with money fixed. Happiness increases as love increases. We cannot say, however, which section corresponds to a larger fixed amount of money, because as money increases happiness can either increase or decrease.

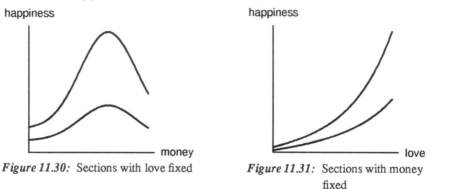

Figure 11.30: Sections with love fixed   *Figure 11.31:* Sections with money fixed

Linear Functions

You may be able to guess the shape of the graph of a linear function of two variables. (It's a plane.) Let's look at an example.

Example 6 Describe the graph of $f(x, y) = 1 + x - y$.

Solution The sections with x fixed form the family of straight lines $z = 1 + a - y$. These lines all lie in planes parallel to the yz-plane, and slope downward in the y-direction. Similarly, the sections with y fixed form the family of straight lines $z = 1 + x - b$, which lie in planes parallel to the xz-plane and slope upward in the x-direction. Since all the sections are straight lines, you might expect the graph to be a flat plane, sloping down in the y-direction and up in the x-direction. This is indeed the case. See Figure 11.32.

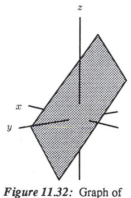

Figure 11.32: Graph of the plane $z = 1 + x - y$

When One Variable is Missing: Cylinders

Suppose we graph an equation like $z = x^2$ which has one variable missing. What does the graph look like?

The sections with x fixed graph as horizontal lines of the form $z =$ constant. The sections with y fixed all graph as the same parabola, namely $z = x^2$. Thus, if you let y vary up and down the y-axis, this parabola sweeps out the trough-shaped surface shown in Figure 11.33.

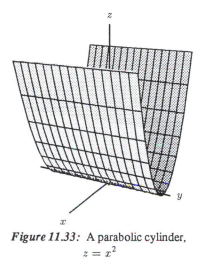

Figure 11.33: A parabolic cylinder,
$$z = x^2$$

This surface is called a *parabolic cylinder*, because it is formed from a parabola in the same way that an ordinary cylinder is formed from a circle; it has a parabolic cross-section instead of a circular one.

Problems for Section 11.3

1. The surface in Figure 11.34 shows a graph of the function $z = f(x, y)$ for positive x and y.

    (a) Suppose y is fixed and positive. Does z increase or decrease as x increases? Sketch a graph of z against x.

    (b) Suppose x is fixed and positive. Does z increase or decrease as y increases? Sketch a graph of z against y.

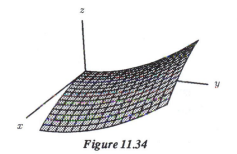

Figure 11.34

For Problems 2–5, use a computer to sketch a function whose graphs have the given shapes. In each case make a print-out of the graph showing the equation used to generate it. Include the axes in your sketch.

2. A cone of circular cross-section opening downward and with its vertex at the origin.

3. A bowl which opens upward and has its vertex at 5 on the z-axis.

4. A plane which has its x, y, and z intercepts all positive.

5. A parabolic cylinder opening upward from along the line $y = x$ in the xy-plane.

6. For each of the following functions, decide whether it could be a bowl, a plate, or neither. Consider a plate to be any fairly flat surface and a bowl to be anything that could hold water, assuming the positive z-axis is up.

    (a) $z = x^2 + y^2$ (b) $z = 1 - x^2 - y^2$ (c) $x + y + z = 1$
    (d) $z = -\sqrt{5 - x^2 - y^2}$ (e) $z = 3$

7. For each function in Problem 6 sketch:

    (a) Sections with x fixed at $x = 0$ and $x = 1$.
    (b) Sections with y fixed at $y = 0$ and $y = 1$.

8. Match the following functions with their graphs.

    (a) $z = \dfrac{1}{x^2 + y^2}$ (b) $z = -e^{-x^2 - y^2}$
    (c) $z = x + 2y + 3$ (d) $z = -y^2$
    (e) $z = x^3 - \sin y$

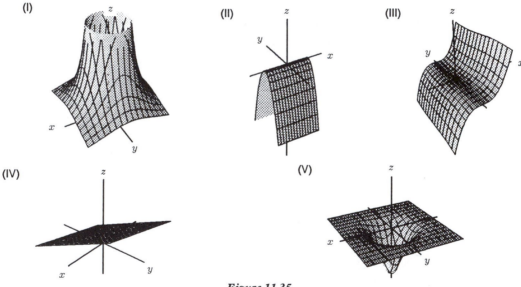

Figure 11.35

9. You like pizza and you like cola. Which of the following graphs represents your happiness as a function of how many pizzas and how much cola you have if

    (a) There is no such thing as too many pizzas and too much cola?
    (b) There is such a thing as too many pizzas and too much cola?
    (c) There is such a thing as too much cola but no such thing as too many pizzas?

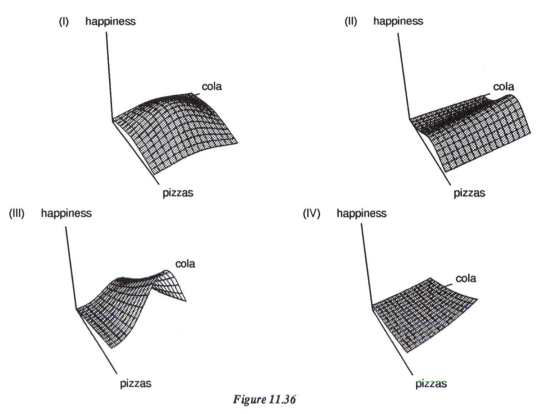

Figure 11.36

10. For each graph in Problem 9 draw:
    (a) two sections with pizza fixed.
    (b) two sections with cola fixed.

11. You are planning a social event. The success of the event is a function of the number of males and the number of females that attend. Which of the following graphs best fits this function if your event is:
    (a) A bridal shower?   (b) A party at a singles dating club?

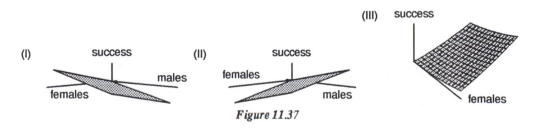

Figure 11.37

12. For each graph in Problem 11, draw two sections with the number of males fixed and two sections with the number of females fixed.

13. Match the following descriptions of a company's success with the corresponding graphs.
    (a) Although we aren't always totally successful, it seems that the amount of money invested doesn't matter. As long as we put hard work into the company our success will increase.

(b)   No matter how much money or hard work we put into the company, we just couldn't make a go of it.

(c)   Our success is measured in dollars, plain and simple. More hard work won't hurt, but it also won't help.

(d)   The company seemed to take off by itself. Its success seems unstoppable.

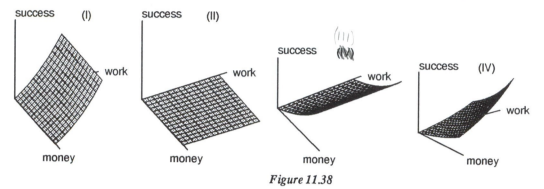

Figure 11.38

14.   Below are graphs of $z = f(x, b)$ for $b = -2, -1, 0, 1, 2$.

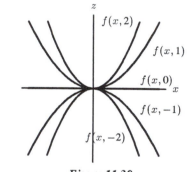

Figure 11.39

Which of the following graphs of $z = f(x, y)$ best fits this information?

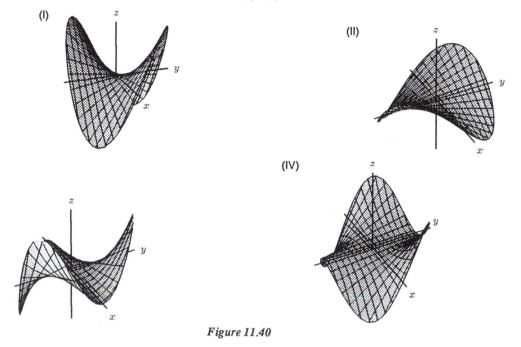

Figure 11.40

15. Imagine a single wave traveling along a canal. Suppose x is the distance from the beginning of the canal, t is the time, and z is the height of the water above the equilibrium level. The graph of z as a function of x and t is shown in Figure 11.41.
    (a) Draw the profile of the wave for $t = -1, 0, 1, 2$. (Show the x-axis to the right and the z-axis vertically.)
    (b) Is the wave traveling in the direction of increasing or decreasing x?
    (c) Sketch a surface representing a wave traveling in the opposite direction.

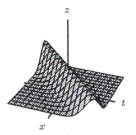

Figure 11.41

16. Use a computer to draw the graph of the vibrating string function:

$$g(x, t) = \cos t \sin 2x, \quad 0 \le x \le \pi, \quad 0 \le t \le 2\pi$$

Explain the shape of the graph using sections with t fixed and with x fixed.

17. Use a computer to draw the graph of the standing wave function:

$$h(x, t) = 3 + \cos(x - 0.5t), \quad 0 \le x \le 2\pi, \quad 0 \le t \le 2\pi.$$

Explain the shape of the graph using sections with t fixed and with x fixed.

18. Consider the function f given by $f(x, y) = y^3 + xy$. Draw graphs of sections with:
    (a) x fixed at $x = -1$, $x = 0$, and $x = 1$.   (b) y fixed at $y = -1$, $y = 0$, and $y = 1$.

19. A swinging pendulum consists of a mass at the end of a string. At one moment the string makes an angle x with the vertical and the mass has speed y. At that time, the energy, E, of the pendulum, is given by the expression[2]

$$E = 1 - \cos x + \frac{y^2}{2}.$$

    (a) Consider the surface representing the energy. Sketch a cross-section of the surface:
        (i)Perpendicular to the x-axis at $x = c$. (ii)Perpendicular to the y-axis at $y = c$.
    (b) For each of the graphs in Figures 11.42 and 11.43 use your answer to part (a) to decide which is the x-axis and which is the y-axis and to put reasonable units on each one.

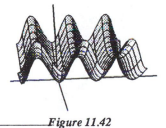

Figure 11.42   *Figure 11.43*

[2]Adapted from *Calculus in Context* by James Callahan, et al, W.H.Freeman NY 1993

11.4 CONTOUR DIAGRAMS

Graphs are useful for seeing the general behavior of a function, but because they are perspective drawings of three-dimensional objects, it is difficult to estimate numerical values from them, or even to interpret the function's behavior from them. In practice, functions of two variables are often represented in either tabular form or in contour diagrams, such as the weather map on page 2.

One of the most common examples of a contour diagram is a topographical map like that shown in Figure 11.44. Such a map tells the elevation at every point in the region and is the best way of getting an overall picture of the terrain: where the mountains are, where the flat areas are. You have probably seen topographical maps on the wall of a classroom; frequently it is colored with green for the lower elevations and brown, red, or even white for the higher elevations.

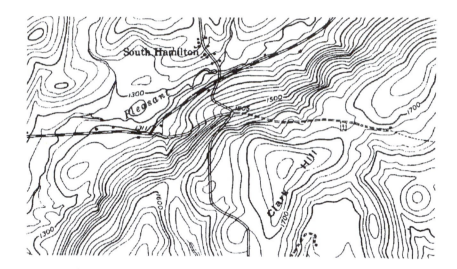

Figure 11.44: A topographical map

The curves in a topographical map separating lower elevations from higher elevations are called *contour lines* or simply *contours* because they outline the contour or shape of the land. Because every point along a contour curve has the same elevation, contour curves are also called *level curves*. The more closely spaced the contours, the steeper the terrain; the more widely spaced the contours, the flatter the terrain. (Provided, of course, that the elevation between contours varies by a constant amount.) Certain features have distinctive characteristics. A mountain peak is surrounded by

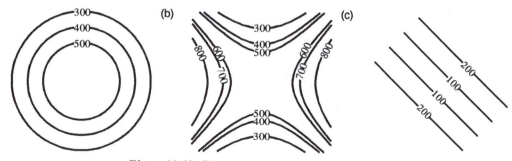

Figure 11.45: Features of a topographical map

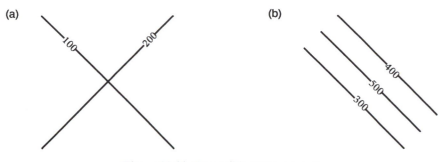

Figure 11.46: Impossible contour curves

concentric circular contour curves as on the left of Figure 11.45a. A pass in a range of mountains has contours that look like Figure 11.45b. A long valley has contour curves that look like parallel lines indicating the rising elevations on both sides of the valley (see Figure 11.45c); a long ridge of mountains has the same type of contour curves, only the elevations decrease on both sides of the ridge. Notice that the elevation numbers on the contour curves are just as important as the curves themselves.

There are some things contour curves cannot do. Two contours corresponding to different elevations cannot cross each other as shown in Figure 11.46a. If they did, the point of intersection of the two curves would have two different elevations, which is impossible (assuming the terrain has no overhangs). For the same reason, a contour line cannot be skipped. See Figure 11.46. If the terrain is continuous, we must cross a 400 contour in going from 300 to 500.

More Examples

A topographical map is a contour diagram where elevation is the function of interest, but we can use contour diagrams to give us a picture of any function of two variables. The weather map of Section 11.1 had temperature as the underlying function for a contour diagram. We could also use barometric pressure, or dew point. The radar pictures of rain or snow that you see on TV weather reports are contour diagrams where the underlying function is intensity of precipitation. The population density of foxes in England, radioactivity levels in the region surrounding Chernobyl, the proportion of Democratic voters in the state of California, all of these could be pictured with contour diagrams. In each of these examples the two independent variables just represent the position of a point in the same region (the US, England, California), but one can even use contour diagrams when the independent variables measure different things, as in the following examples.

Example 1 Figure 11.47 is a contour diagram giving the corn production $f(R, T)$ in the US as a function of the total rainfall, R, in inches, and average temperature, T, in degrees Fahrenheit, during the growing season. At the present, $R = 15$ inches and $T = 76°$ and the diagram shows what would happen if the climate were to change. Production is measured as a percentage of the present production; thus the contour through $R = 15, T = 76$ has value 100. Explain in words what the contour diagram shows.

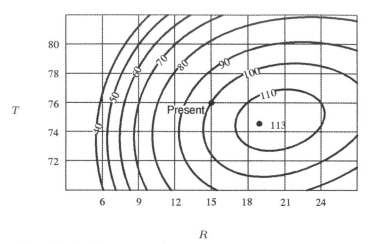

Figure 11.47: Corn production as a function of rainfall and temperature

Solution   First, notice what happens to corn production if the temperature stays fixed at the present value of 76, but the rainfall changes. As one would expect, if there is a drought, that is if rainfall is less, corn production goes down: as we stay on the $T = 76$ line and move to the left, the values on the contour lines go down so that, for example, at $R = 12, T = 76$ corn production is about 85% of the present production. Conversely, if rainfall increases, that is, as we move to the right from $R = 15, T = 76$, corn production increases, reaching a maximum of more than 110% when $R = 21$, and then decreases (too much rainfall floods the fields). If instead rainfall remains at the present value and temperature increases, corn production decreases; a 2° increase causes a 10% drop in production. This makes sense since hotter temperatures lead to greater evaporation and hence drier conditions, even with rainfall constant at 15 inches. Similarly, a decrease in temperature leads to a very slight increase in production, reaching a maximum of around 102% when $T = 74$, followed by a decrease (the corn won't grow if it is too cold). Notice that the diagram shows the optimal climate for corn production has rainfall around 20 inches and temperature around 75°.

Example 2   Figure 11.48 shows the contour diagram for the vibrating string function studied in Section 11.1:

$$f(x, t) = \cos t \sin x, 0 \le x \le \pi$$

Explain the diagram in terms of t-sections and x-sections of f.

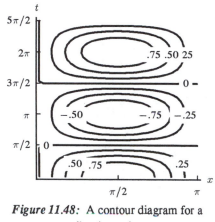

Figure 11.48: A contour diagram for a vibrating string

Solution   To read off the t-sections of f, we fix a t value and move horizontally across the diagram looking at the values on the contours. For $t = 0$, as we move from the left at $x = 0$ to the right at $x = \pi$, we cross contours of $0.25, 0.50, 0.75$ and reach a maximum at $x = \pi/2$, and then decrease back to 0. That is because if time is fixed at $t = 0$, $f(x, 0)$ is the displacement of the string: no displacement at $x = 0$ and $x = \pi$ and greatest displacement at $x = \pi/2$. For larger values of t, the t-section we get as we move along a horizontal line crosses fewer contours and reaches a smaller maximum value: the string is becoming less curved. At time $t = \pi/2$, the string is straight so we see a value of 0 all the way across the diagram, namely a contour with value 0. For $t = \pi$, the string has vibrated to the other side and the displacements are negative as we read across the diagram reaching a minimum at $x = \pi/2$.

The x-sections of f are read vertically. At $x = 0$ and $x = \pi$, we see vertical contours of value 0 because the end points of the string have 0 displacement no matter what time it is. The section for $x = \pi/2$ is found by moving vertically up the diagram at $x = \pi/2$. As we expect, the contour values are largest at $t = 0$, zero at $t = \pi/2$, and a minimum at $t = \pi$.

Notice that the spacing of the contours is also important. For example, for the $t = 0$ section, contours are most closely spaced at the end points at $x = 0$ and $x = \pi$ and most spread out at $x = \pi/2$. That is because the shape of the string at time $t = 0$ is a sine curve, which is steepest at the end points and relatively flat in the middle. Thus the contour diagram should show the steepest terrain at the end points and flattest terrain at the midpoint.

Finding Contours Algebraically

Algebraic equations for the contours of a function f are easy to give if we have an algebraic equation for $f(x, y)$. In general, the equation for the contour with value c is given by:

$$f(x, y) = c$$

Example 3   Find equations for the contours of $f(x, y) = x^2 + y^2$ and use them to draw a contour diagram for f. Relate the contour diagram to the graph of f.

Solution   The contour with value c is given by

$$f(x, y) = x^2 + y^2 = c$$

This is the equation of a circle of radius $\sqrt{c}$. Thus the contours at an elevation of $c = 1, 2, 3, 4, \ldots$ are all circles centered at the origin of radius $1, \sqrt{2}, \sqrt{3}, 2, \ldots$. The contour diagram is shown in Figure 11.49. The graph of f is shown in Figure 11.50. The contour diagram shows that if we take the graph of f and intersect it with a horizontal plane c units above the xy-plane, we get the contour at level c, namely a circle of radius $\sqrt{c}$. Notice that the graph of f is bowl-shaped and gets steeper and steeper as we move further away from the origin. The contour diagram shows this since the contours become more closely packed as we move further from the origin; for example, four contours $(c = 1, 2, 3, 4)$ intersect the x-axis between $x = 1$ and $x = 2$ but six contours $(c = 4, 5, 6, 7, 8, 9)$ intersect the x-axis between $x = 2$ and $x = 3$.

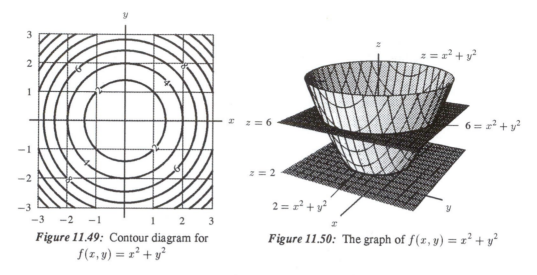

Figure 11.49: Contour diagram for
$f(x, y) = x^2 + y^2$

Figure 11.50: The graph of $f(x, y) = x^2 + y^2$

Example 4   Draw a contour diagram for $f(x, y) = \sqrt{x^2 + y^2}$ and relate it to the graph of f.

Solution   The contour at level c is given by

$$f(x, y) = \sqrt{x^2 + y^2} = c$$

This is a circle, just as in the previous example, but here the radius is c instead of $\sqrt{c}$. Now if the level c increases by 1, the radius of the contour increases by 1. This means the contours are equally spaced concentric circles (see Figure 11.51) and do not become closely packed further from the origin as they did in the previous example. Thus the graph of f should have the same constant slope as we move away from the origin (see Figure 11.52), making it a cone rather than a bowl as in the previous example.

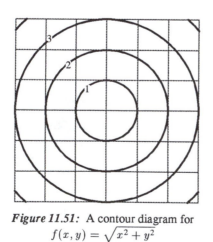

Figure 11.51: A contour diagram for
$f(x, y) = \sqrt{x^2 + y^2}$

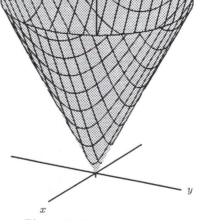

Figure 11.52: The graph of
$f(x, y) = \sqrt{x^2 + y^2}$

Example 5 Draw a contour diagram for $f(x, y) = 2x + 3y + 1$.

Solution The contour at level c is given by the equation $2x + 3y + 1 = c$. We can rewrite this in slope intercept form as $y = -(2/3)x + (c - 1)/3$. Thus the contours are parallel lines all with slope $-2/3$. The y-intercept for the contour at level c is $(c - 1)/3$; each time c increases by 3, the y-intercept moves up by 1. The contour diagram is shown in Figure 11.53.

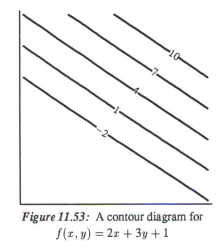

Figure 11.53: A contour diagram for
$f(x, y) = 2x + 3y + 1$

Contour Diagrams and Tables

Sometimes you can get an idea of what the contour map of a function looks like from its table.

Example 6 Use a table of values of $f(x, y) = x^2 - y^2$ to sketch its contour diagram.

Solution One striking feature of the values of the function in Table 11.8 are the zeros along the diagonals. This occurs because $x^2 - y^2 = 0$ along the lines $y = x$ and $y = -x$. So the zero contour consists of these two lines. In the triangular regions of the table that lie to the right and left of both the

TABLE 11.8 *Table of values of $f(x, y) = x^2 - y^2$*

	3	0	-5	-8	-9	-8	-5	0
	2	5	0	-3	-4	-3	0	5
	1	8	3	0	-1	0	3	8
y	0	9	4	1	0	1	4	9
	-1	8	3	0	-1	0	3	8
	-2	5	0	-3	-4	-3	0	5
	-3	0	-5	-8	-9	-8	-5	0
		-3	-2	-1	0	1	2	3

x

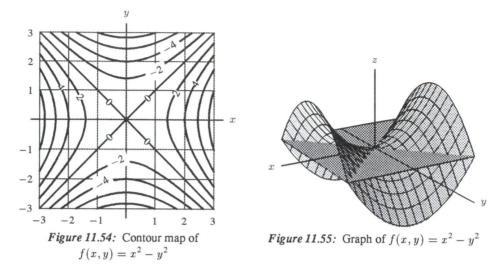

Figure 11.54: Contour map of $f(x, y) = x^2 - y^2$

Figure 11.55: Graph of $f(x, y) = x^2 - y^2$

diagonals, the entries are positive. Thus, in the contour diagram, the positive contours will lie in the triangular regions to the right and left of the lines $y = x$ and $y = -x$. Further, the table shows that the numbers on the left are the same as the numbers on the right; thus each contour will have two pieces, one on the left and one on the right. See Figure 11.54. As we move away from the origin, along the x-axis, we cross contours corresponding to successively larger values. On the saddle-shaped graph of $f(x, y) = x^2 - y^2$ shown in Figure 11.55, this corresponds to climbing out of the saddle along one of the ridges. Similarly, the negative contours occur in pairs in the top and bottom triangular regions and get more and more negative as we go out along the y-axis. This corresponds to descending from the saddle along the valleys that are submerged below the xy-plane in Figure 11.55. Notice that we could get the contour diagram instead by graphing the family of hyperbolas $x^2 - y^2 = 0, \pm 2, \pm 4, \ldots$.

Using Contour Diagrams: The Cobb-Douglas Production Function

Suppose you are running a small printing business, and decide to expand because you have more orders than you can handle. How should you expand? Should you start a night shift and hire more workers? Should you buy more expensive but faster computers which will enable the current staff to keep up with the work? Or should you do some combination of the two?

Obviously, the way such a decision is made in practice involves many other considerations — such as whether you could get a suitably trained night shift, or whether there are any faster computers available that your current staff could use. Nevertheless, you might model the quantity of work produced by your business as a function of two variables: your total number N of workers, and the total value V of your equipment.

How would you expect such a production function to behave? In general, having more equipment and more workers enables you to produce more. However, increasing equipment without increasing the number of workers will increase production a bit, but not beyond a point. (If equipment is already lying idle, having more of it won't help.) Similarly, increasing the number of workers without increasing equipment will increase production, but not past the point where the equipment is fully utilized, as any new workers would have no equipment available to them.

Example 7   Which of the contour diagrams in Figures 11.56 and 11.57 best models the behavior described above?

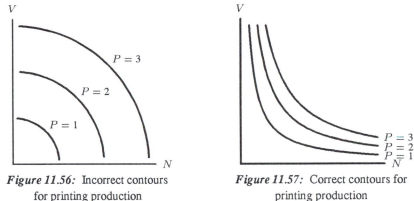

Figure 11.56: Incorrect contours for printing production

Figure 11.57: Correct contours for printing production

Solution   First look at the contour diagram on the left. Fixing V at a particular value and letting N increase means moving to the right on the contour diagram. As you do so, you cross ever larger contours, so that it seems that production will increase forever. On the other hand, on the contour diagram on the right, as you move in the same direction you find yourself moving more and more parallel to the contours, crossing them less and less frequently. Therefore, production increases more and more slowly as N increases while V is held fixed. Similarly, if you hold N fixed and let V increase, the contour diagram on the left shows production increasing at a steady rate, whereas the one on the right shows production increasing, but at a decreasing rate. In addition, in the diagram on the right, if N is zero or very small (namely for points on or near the V-axis), the value of P is zero or very small, no matter how big V is. This is because if you have no workers at all, you can't print anything, no matter what the value of your equipment is. The diagram on the left shows print production increasing even when $N = 0$. Thus, the one on the right fits best.

Formula for a Production Function

Production functions with the qualitative behavior we want are often approximated by formulas of the form

$$P = f(N, V) = cN^{\alpha}V^{\beta}$$

where P is the total quantity produced and c, α, and β are positive constants with $0 < \alpha < 1$ and $0 < \beta < 1$.

Example 8   Show that the contours of the function $P = cN^{\alpha}V^{\beta}$ have approximately the right shape to fit the contour diagram in Figure 11.57.

Solution   The contours are the curves where P is equal to a constant value, say P_0, i.e., where

$$cN^{\alpha}V^{\beta} = P_0.$$

Solving for V we get

$$V = \left(\frac{P_0}{c}\right)^{1/\beta} \cdot N^{-\alpha/\beta}.$$

Thus, as a function of N, V is a power function with a negative exponent, and hence its graph has the required shape.

The Cobb-Douglas Production Model

In 1928, Cobb and Douglas used a similar function to model the production of the entire US economy in the first quarter of this century. Using government estimates of P, the total yearly production between 1899 and 1922; of K, the total capital investment over the same period and of L, the total labor force, they found that P is well approximated by

---

The Cobb-Douglas Production Formula

$$P = 1.01 L^{0.75} K^{0.25}.$$

This function turns out to model the US economy surprisingly well, both for the period on which it was based, and indeed for some time afterwards.

Returns to Scale

A general Cobb-Douglas production function is one of the form

$$P = cL^{\alpha}K^{\beta}.$$

An important economic question is what happens to production if labor and capital are both "scaled up"? For example, does production double if both labor and capital are doubled? To find out, suppose P_0 is the production given by L_0 and K_0, so that

$$P_0 = f(L_0, K_0) = cL_0^{\alpha}K_0^{\beta}.$$

We want to know what happens to production if L_0 is increased to $2L_0$ and K_0 is increased to $2K_0$:

$$\begin{aligned}
P = f(2L_0, 2K_0) \\
= c(2L_0)^{\alpha}(2K_0)^{\beta} \\
= c2^{\alpha}L_0^{\alpha}2^{\beta}K_0^{\beta} \\
= 2^{\alpha+\beta}cL_0^{\alpha}K_0^{\beta} \\
= 2^{\alpha+\beta}P_0.
\end{aligned}$$

Thus, doubling L and K has the effect of multiplying P by $2^{\alpha+\beta}$. Notice that if $\alpha + \beta > 1$, then $2^{\alpha+\beta} > 2$, if $\alpha + \beta = 1$, then $2^{\alpha+\beta} = 2$, and if $\alpha + \beta < 1$, then $2^{\alpha+\beta} < 2$. Thus,

- if $\alpha + \beta > 1$, then doubling L and K more than doubles P,
- if $\alpha + \beta = 1$, then doubling L and K exactly doubles P,
- if $\alpha + \beta < 1$, then doubling L and K less than doubles P.

Economists describe this situation by saying that

- $\alpha + \beta > 1$ gives *increasing returns to scale*,
- $\alpha + \beta = 1$ gives *constant returns to scale*,
- $\alpha + \beta < 1$ gives *decreasing returns to scale*.

Problems for Section 11.4

For the functions in Problems 1–15, sketch a contour diagram with at least four labeled contours. Describe the contours in words and how they are spaced.

1. $f(x, y) = x + y$
2. $f(x, y) = 2x - y$
3. $f(x, y) = xy$

4. $f(x, y) = x^2 + y^2$
5. $f(x, y) = x + y + 1$
6. $f(x, y) = 3x + 3y$

7. $f(x, y) = -x - y$
8. $f(x, y) = x^2 + y^2 - 1$
9. $f(x, y) = -x^2 - y^2 + 1$

10. $f(x, y) = x^2 + 2y^2$
11. $f(x, y) = \sqrt{x^2 + 2y^2}$
12. $f(x, y) = y - x^2$

13. $f(x, y) = 2y - 2x^2$
14. $f(x, y) = e^{-x^2 - y^2}$
15. $f(x, y) = \cos(\sqrt{x^2 + y^2})$

16. Figure 11.58 graphs the monthly payment you must make on a 5-year car loan as a function of the interest rate and the amount you borrow. Suppose the interest rate is 13% and that you decide to borrow $6,000.

    (a) What is your monthly payment?
    (b) If interest rates drop to 11%, how much more can you borrow without increasing your monthly payment?
    (c) Make a table of how much you can borrow without increasing your monthly payment, as a function of the interest rate.

17. Figure 11.59 shows a contour map of a hill with two paths, A and B.

    (a) On which path, A or B, will you have to climb more steeply?
    (b) On which path, A or B, will you probably have a better view of the surrounding countryside? (Assuming trees do not block your view.)
    (c) Near which path is there more likely to be a river?

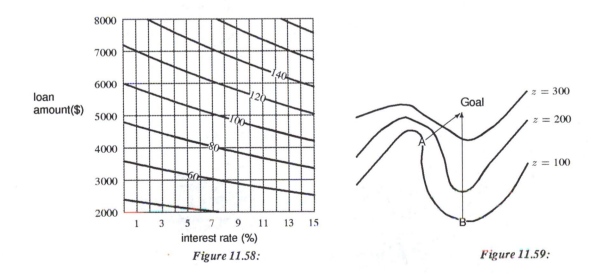

Figure 11.58:   Figure 11.59:

18. Each of the contour diagrams in Figure 11.60 shows population density in a certain region of a city.
    Match contour diagrams and locations if the middle contour is:
    (a) A highway.   (b) An open sewage canal.   (c) A railroad line.
    Many different matchings are possible. Pick any reasonable one and justify your choice.

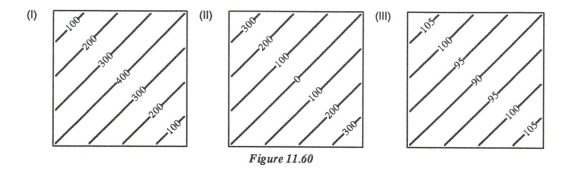

Figure 11.60

19. Each of the contour diagrams in Figure 11.61 shows population density in a certain region.
    Match contour diagrams and locations if the center of the diagram is:
    (a) A city.   (b) A lake.   (c) A power plant.
    Many different matchings are possible. Pick any reasonable one and justify your choice.

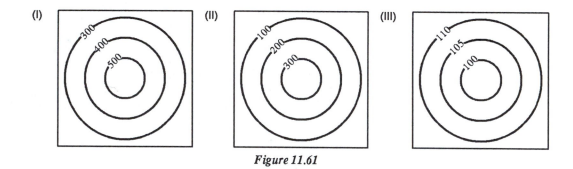

Figure 11.61

For each of the surfaces in Problems 20–22, draw a possible set of level curves or a contour map, marked with reasonable z values. (Note: There are many possible answers to these problems.)

20.   21.   22.

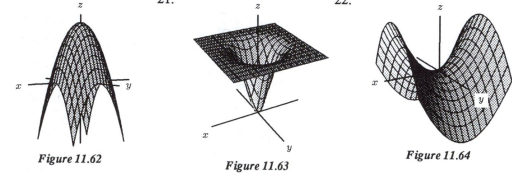

Figure 11.62   *Figure 11.63*   *Figure 11.64*

23. Figure 11.65 shows the density of the fox population P (in foxes per square mile) for southern England. Draw two different vertical sections and two different horizontal sections of the population density P.

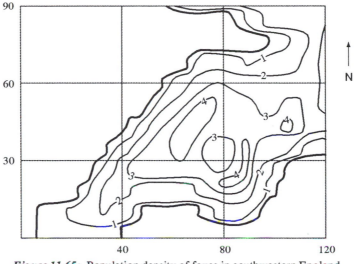

Figure 11.65: Population density of foxes in southwestern England

24. Use a computer to sketch a contour diagram for the vibrating string function

$$f(x, y) = \cos t \sin 2x, \qquad 0 \le x \le \pi, \ 0 \le t \le \pi.$$

Use $c = -2/3, -1/3, 0, 1/3, 2/3$. (You will not be able to do this algebraically.)

25. On page 10, Problems 13–16, we introduced the traveling wave function

$$h(x, t) = 5 + \cos(0.5x - t).$$

Draw a contour diagram using $c = -2/3, -1/3, 0, 1/3, 2/3$ for this function. Explain how your contour diagram relates to the sections of h discussed in Problem 15 on page 29. Where are the contours most closely spaced? Most widely spaced?

26. Draw contour diagrams for each of the pizza-cola-happiness graphs given Problem 9 on page 26.

27. The cornea is the front surface of the eye. Corneal specialists use a TMS, or Topographical Modeling System, to produce a "map" of the curvature of the eye's surface. A computer analyzes light reflected off the eye and draws level curves joining points of constant curvature. The regions between these curves are colored different colors. The first two pictures in Figure 11.66 are cross-sections of eyes with constant curvature, the smaller being about 38 units and the larger about 50 units. For contrast, the third eye has varying curvature.

 (a) Describe in words how the TMS map of an eye of constant curvature will look.

 (b) Draw the TMS map of an eye with the cross-section in Figure 11.67. Assume the eye is circular when viewed from the front, and the cross-section is the same in every direction. Put reasonable numeric labels on your level curves.

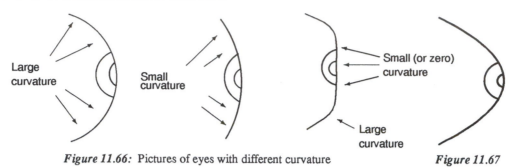

Figure 11.66: Pictures of eyes with different curvature   **Figure 11.67**

28. A manufacturer sells two goods, one at a price of $3000 a unit and the other at a price of $12000 a unit. Suppose a quantity q_1 of the first good and q_2 of the second good are sold at a total fixed cost of $4000 to the manufacturer.

    (a) Express the manufacturer's profit, π, as a function of q_1 and q_2.
    (b) Sketch curves of constant profit in the q_1q_2-plane for $\pi = 10000$, $\pi = 20000$, and $\pi = 30000$ and the break-even curve $\pi = 0$.

29. Match the tables in (a) - (d) with the contour diagrams in (I) - (IV).

    (a)

		x		
		-1	0	1
	-1	2	1	2
y	0	1	0	1
	1	2	1	2

    (b)

		x		
		-1	0	1
	-1	0	1	0
y	0	1	2	1
	1	0	1	0

    (c)

		x		
		-1	0	1
	-1	2	0	2
y	0	2	0	2
	1	2	0	2

    (d)

		x		
		-1	0	1
	-1	2	2	2
y	0	0	0	0
	1	2	2	2

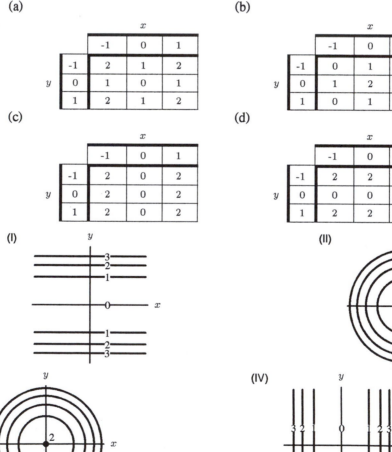

Figure 11.68

30. Match the surfaces (a) to (e) below with the corresponding contour diagrams (I) to (V).

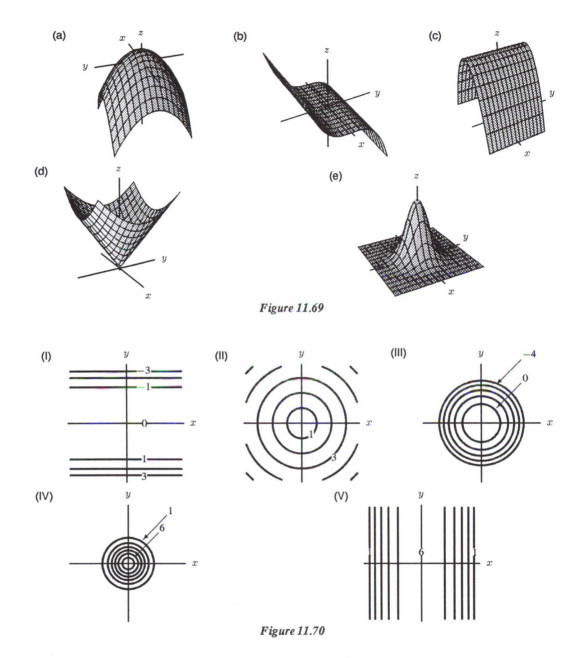

Figure 11.69

Figure 11.70

31. A city on an island has a large rectangular central park. Draw a possible contour diagram showing light intensity at night as a function of position. Label your contours with values between 0 and 1, where 0 represents total darkness and 1 represents daylight illumination.

32. The map in Figure 11.71 is the undergraduate senior thesis of Professor Robert Cook, Director of Harvard's Arnold Arboretum. It shows level curves of the function giving the species density of breeding birds at each point in the US, Canada and Mexico.

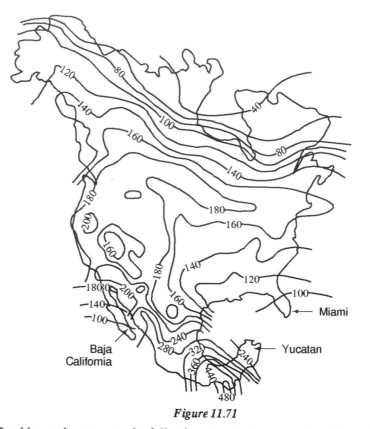

Figure 11.71

Looking at the map, are the following statements true or false? Explain your answers.

(a)   Moving from south to north across Canada, the species density increases.

(b)   The species density in the area around Miami is over 100.

(c)   In general, peninsulas (for example, Florida, Baja California, the Yucatan) have lower species densities than the areas around them.

(d)   The greatest rate of change in species density with distance is in Mexico. If you think this is true, mark the point and direction which give the maximum and explain why you picked the point and direction you did.

33.   The temperature T (in °C) at any point in the region $-10 \leq x \leq 10, -10 \leq y \leq 10$ is given by the function
$$T(x, y) = 100 - x^2 - y^2.$$

(a)   Sketch isothermal curves for $T = 100°C, T = 75°C, T = 50°C, T = 25°C$, and $T = 0°C$.

(b)   Suppose a heat-seeking bug is put down at any point on the xy-plane. In which direction should it move to increase its temperature fastest? How is that direction related to the level curve through that point?

34.   Let f be the linear function $f(x, y) = c + mx + ny$, where c, m, n are constants and $m, n \neq 0$.

(a)   Show that all the contours of f are lines of slope $-m/n$, if $n \neq 0$.

(b)   Show that
$$f(x + n, y - m) = f(x, y),$$
for all x and y.

(c)   Explain the relation between parts (a) and (b).

35. Identify the contour diagrams (I)–(V) and the surfaces (F)–(K) corresponding to the equations
(a)–(e). Assume that each contour diagram is shown in a square window.

    (a) $z = \sin x$ (b) $z = xy$ (c) $z = e^{-(x^2 + y^2)}$ (d) $z = 1 - 2x - y$

    (e) $z = x^2 + 4y^2$

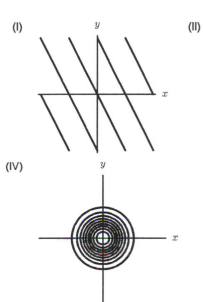

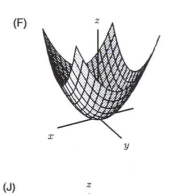

Figure 11.72

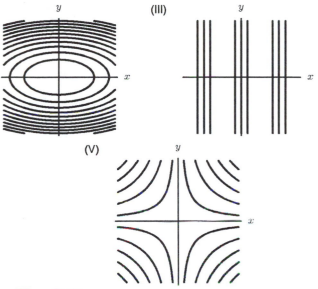

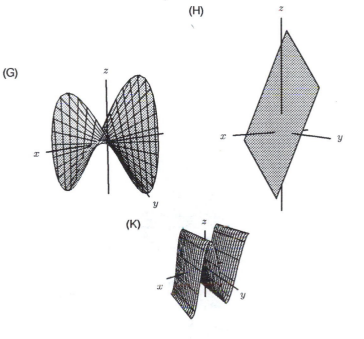

Figure 11.73

36.

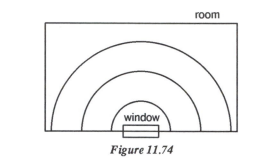

Figure 11.74

Figure 11.74 shows the level curves of the temperature H in a room near a recently opened window. Label the three level curves with reasonable values of H if the house is in the following locations.

(a)   Minnesota in winter (where winters are harsh).

(b)   San Francisco in winter (where winters are mild).

(c)   Houston in summer (where summers are hot).

(d)   Oregon in summer (where summers are mild).

37.   Match each Cobb-Douglas production function with the correct graph and the correct statements. (Note: A graph or statement can be used more than once.)

(a) $F(L, K) = L^{0.25} K^{0.25}$ (D)   Tripling each input triples output.

(b) $F(L, K) = L^{0.5} K^{0.5}$ (E)   Quadrupling each input doubles output.

(c) $F(L, K) = L^{0.75} K^{0.75}$ (G)   Doubling each input almost triples output.

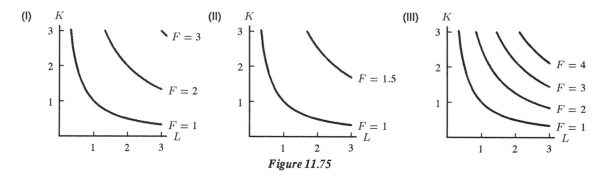

Figure 11.75

38.   Figure 11.76 shows the contours of the temperature along one wall of a room through a day, with time indicated as on a 24-hour clock.

   The room has a heater located at the left-most corner of the wall, and one window in the wall. The heater is controlled by a thermostat a couple of feet from the window.

(a)   Where is the window?

(b)   When is the window open?

(c)   When is the heat on?

(d)   Draw graphs of the temperature in the room at 6 am, at 11 am, at 3 pm (15 hours) and at 5 pm (17 hours).

(e)   Draw a graph of the temperature as a function of time at the heater, at the window and midway between them.

(f)   Can you explain why the temperature at the window at 5 pm (17 hours) is less than at 11 am (11 hours)?

(g)   What temperature do you think the thermostat is set to? Why?

(h)   Where is the thermostat?

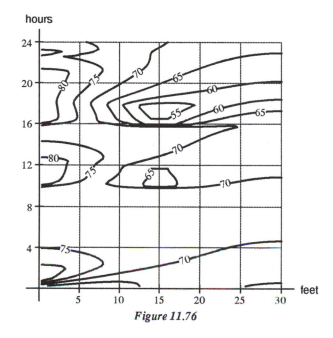

Figure 11.76

39.   Figure 11.77 shows the contours of light intensity as a function of location and time in a microscopic wave-guide.

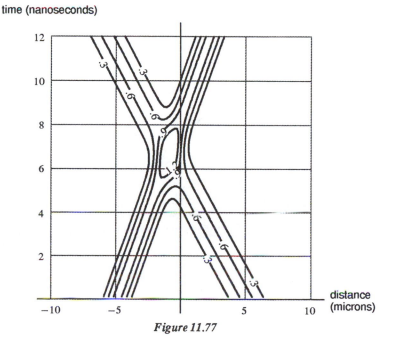

Figure 11.77

(a) Draw graphs showing intensity as a function of location at times 0, 2, 4, 6, 8, and 10 nanoseconds.

(b) If you could create an animation showing how the graph of intensity as a function of location varies as time passes, what would it look like?

(c) Draw a graph of intensity as a function of time at locations $-5, 0$ and 5 microns.

(d) Describe what the light beams are doing in this wave-guide.

11.5   LINEAR FUNCTIONS

What is a Linear Function of Two Variables?

In one-variable calculus we saw that linear functions played a basic role: most functions are locally linear in the sense that if you zoom in on the graph of the function, it looks more and more linear. In the next chapter we will see that for functions of two variables, the same role is played by functions whose graphs are planes. These functions are also called *linear functions*.

What Makes a Plane Flat?

What is it about a function $f(x, y)$ that makes its graph $z = f(x, y)$ a plane? Linear functions of *one* variable have straight line graphs because they have constant slope. No matter where you are on the graph, you will always find that the slope is the same; a given increase in the x-coordinate will always give the same increase in the y-coordinate; equivalently, the ratio $\Delta y/\Delta x$ is constant.

When you are on a plane, the situation is a bit more complicated. If you imagine walking around on a tilted plane, you can see that the slope is not always the same: it depends what direction you are walking in. You could be walking in the uphill direction, or you could choose a direction that keeps you at the same level. However, it is still true that no matter where you are on the plane, the slope is the same as long as you choose the same direction. If you walk in the direction parallel to the x-axis, you will always find yourself walking up or down with the same slope; the same thing is true if you walk in the direction parallel to the y-axis. In other words, the slope ratios $\Delta z/\Delta x$ and $\Delta z/\Delta y$ are both constant.

Example 1   You are on a plane that cuts the z-axis at $z = 5$, has slope 2 in the x direction and slope -1 in the y direction. What is the equation of the plane?

Solution   Knowing the equation of the plane is the same thing as knowing a formula for the z-coordinate in terms of the x- and y-coordinates. So suppose you want to know the z-coordinate when you are directly above the point (x, y) in the xy-plane. Think how you get to that point starting from the point above the origin. First you walk x units in the x direction; since the slope in the x direction is 2, this means that your height increases by $2x$. Then you walk y units in the y direction; since the slope in the y direction is -1, this means that your height decreases by y units. In total, your height has changed by $2x - y$ units. Since you started at $z = 5$ on the z-axis, your height is now $5 + 2x - y$. So your z-coordinate is $5 + 2x - y$. Thus, the equation for the plane is

$$z = 5 + 2x - y.$$

Another way of saying this is that the plane is the graph of the function

$$f(x, y) = 5 + 2x - y.$$

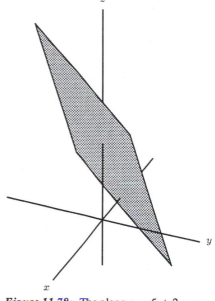

Figure 11.78: The plane $z = 5 + 2x - y$

For any linear function, if you know the z intercept, the slope in the x direction, and the slope in the y direction, then you know the equation of the function. This is just like the equations of lines in the one-variable case, except that there are two slopes instead of one.

If a plane has z intercept c, its slope in the x direction is m and its slope in the y direction is n, then its equation is

$$z = c + mx + ny.$$

It is the graph of the linear function

$$f(x, y) = c + mx + ny.$$

Just as in 2-dimensional space a line is determined by 2 points, so in 3-dimensional space a plane is determined by 3 points.

Example 2 Find the equation of the plane that passes through the three points $(1, 0, 1), (1, -1, 3)$, and $(3, 0, -1)$.

Solution The first two points have the same x-coordinate, so we can use them to find the slope of the plane in the y-direction. As the y-coordinate changes from 0 to -1, the z-coordinate changes from 1 to 3, so the y slope is $\Delta z / \Delta y = (3 - 1)/(-1 - 0) = -2$. The first and third points have the same y-coordinate, so we can use them to find the x slope; it is $\Delta z / \Delta x = (-1 - 1)/(3 - 1) = -1$. So the plane has equation

$$z = -x - 2y + c.$$

Because the plane passes through $(1, 0, 1)$, we get

$$1 = -1 - 0 + c,$$

so $c = 2$. Thus, the equation of the plane is

$$z = -x - 2y + 2.$$

You should check that this equation is also satisfied by the points $(1, -1, 3)$ and $(3, 0, -1)$.

The previous example was made easier by the fact that two of the points had the same x-coordinate and two of them had the same y-coordinate. An alternative method to solve this problem is to substitute the x, y, and z values for each of the three given points into the equation $z = c + mx + ny$. The resulting three equations in c, m, n can then be solved simultaneously. See Problem 21.

Linear Functions from a Numerical Point of View

To avoid flying planes with too many empty seats, airlines sell some tickets at full price, and some at a discount. Table 11.9 shows the airline's revenue in dollars from tickets sold on a particular route, as a function of the number of full price tickets sold and the number of discount tickets.

TABLE 11.9   *Revenue from ticket sales (dollars)*

		Full price tickets			
		100	200	300	400
	200	39,700	63,600	87,500	111,400
	400	55,500	79,400	103,300	127,200
Discount tickets	600	71,300	95,200	119,100	143,000
	800	87,100	111,000	134,900	158,800
	1000	102,900	126,800	150,700	174,600

Looking down each column, you see that the revenue jumps by \$15,800 for each extra 200 discount tickets. Thus, each column is a linear function of the number of discount tickets sold. Furthermore, each column has the same slope, which is $15,800/200 = 79$ dollars/ticket. In practical terms, this is the price of a discount ticket. It is the same no matter which column you are in. Looking now at the rows we see that each row also has the same slope, 239, which is the price of a full fare. Putting together the two prices we can write down a formula for the total revenue R from f full fares and d discount tickets:

$$R = 239f + 79d.$$

For example, if the airline sells 100 full fares and 400 discount fares, its revenue should be

$$R = 100 \cdot 239 + 400 \cdot 79 = 55,500.$$

This agrees with the value given in the table.

Thus, R is a linear function of f and d. The way you can tell this by looking at the table of values is that every row is linear with the same slope, and every column is linear with the same slope. The row slope is the f slope and the column slope is the d slope.

In general, you can recognize a **linear function** from its table by the following features
- each row and each column is linear
- all the rows have the same slope
- all the columns have the same slope (although this may be different from the slope of the rows).

Example 3   Table 11.10 gives some values of a linear function. Fill in the blank entry and give a formula for the function.

TABLE 11.10   *A linear function*

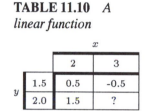

		x	
		2	3
y	1.5	0.5	-0.5
	2.0	1.5	?

Solution   In the first row the function decreases by 1 when x goes from 2 to 3. Since it is linear, it must decrease by the same amount in the second row. So the missing entry is 0.5, and the x slope is -1. The y slope is 2, since in each column the function increases by 1 when y increases by 0.5. Thus, the equation is

$$f(x, y) = c - x + 2y.$$

From the table we get $f(2, 1.5) = 0.5$. Putting this into the equation, we get

$$0.5 = c - 2 + 2 \cdot 1.5 = c + 1,$$

and so $c = -0.5$. Therefore, the equation is

$$f(x, y) = -0.5 - x + 2y.$$

What Does the Contour Diagram of a Linear Function Look Like?

Consider the function for airline revenue given in Table 11.9. Its formula is

$$R = 239f + 79d,$$

where f is the number of full fares and d is the number of discount fares. Figure 11.79 gives the contour diagram for this function.

Notice that each contour is a straight line, and that all the contours have the same slope. What is the practical significance of the slope? Suppose you are on the contour $R = 100{,}000$; that means you are looking at combinations of ticket sales that will yield $100,000 in revenue. If you move down and to the right on the contour, your f-coordinate increases and your d-coordinate decreases,

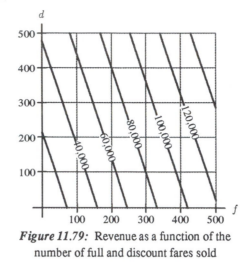

Figure 11.79: Revenue as a function of the
number of full and discount fares sold

so you are selling more full fares and fewer discount fares. If you move up and to the left, it's the other way around. This makes sense, because to receive a fixed revenue of $100,000, then you must sell more full fares if you sell fewer discount fares. The exact trade-off depends on the slope of the contour; looking at the diagram we can see that each contour has a slope of about -3. This means that for a fixed amount of revenue, you must sell 3 discount fares for each full fare that you lose. This can also be seen by comparing prices. Each full fare brings in $239; to earn the same amount in discount fares you need to sell $239/79 = 3.02 \approx 3$ fares. Since the price ratio is independent of how many of each type of fare you sell, this slope remains constant over the whole contour map; that's why the contours are all parallel straight lines.

The other thing to notice about the contour map is that the contours are evenly spaced. This means that no matter which contour you are on, a fixed increase in one of the variables will always take you the same distance to the next contour. This also makes sense in terms of revenue; it says that no matter how many fares you have sold, an extra fare, whether full or discount, will bring the same revenue as it did before.

Example 4   Find equations for the linear functions whose contour diagrams are graphed below.

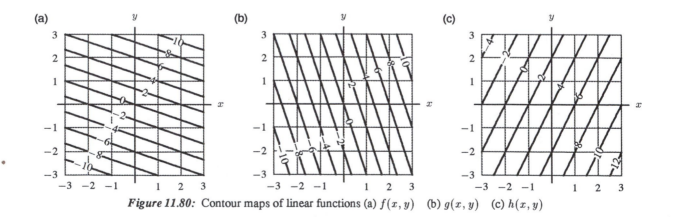

Figure 11.80: Contour maps of linear functions (a) $f(x, y)$ (b) $g(x, y)$ (c) $h(x, y)$

Solution In each diagram the contours correspond to values of the function that are 2 units apart, i.e., there are contours for $-2, 0, 2$, etc. (a) Moving two units in the y direction we cross three contours; i.e., a change of 2 in y changes the function by 6, so the y-slope is $\Delta f/\Delta y = 6/2 = 3$. Similarly, a move of 2 in the x direction crosses one contour line and changes the function by 2; so the x-slope is $\Delta f/\Delta x = 2/2 = 1$. Hence, $f(x, y) = c + x + 3y$. We see from the diagram that $f(0, 0) = 0$, so $c = 0$. Therefore, $f(x, y) = x + 3y$. (b) The function is the other way around. It has an x slope of 3 and a y slope of 1, so it is $c + 3x + y$. Again $g(0, 0) = 0$, so the formula is $g(x, y) = 3x + y$. (c) The function decreases as y increases: each increase of y by 2 takes you down one contour and hence changes the function by 2, so the y slope is -1. The x slope is 2, so the formula is $c + 2x - y$. From the diagram we see that $h(0, 0) = 4$, so $c = 4$. Therefore, the formula for this linear function is $h(x, y) = 4 + 2x - y$.

Problems for Section 11.5

1. Suppose that z is a linear function of x and y with x-slope of 2 and y-slope of 3. A change of 0.5 in x and -0.2 in y produces what change in z? If $z = 2$ when $x = 5$ and $y = 7$, what is the value of z when $x = 4.9$ and $y = 7.2$?

2. Find the equation of the linear function $z = c + mx + ny$ whose graph contains the points $(0, 0, 0)$, $(0, 2, -1)$, and $(-3, 0, -4)$.

3. Find the equation of the plane through the points $(4, 0, 0)$, $(0, 3, 0)$ and $(0, 0, 2)$.

4. Find an equation for the plane containing the line in the xy-plane where $y = 1$, and the line in the xz-plane where $z = 2$.

5. Find the equation of the linear function $z = c + mx + ny$ whose graph intersects the xz-plane in the line $z = 3x + 4$ and intersects the yz-plane in the line $z = y + 4$.

6. Find the equation of the linear function $z = mx + ny + c$ whose graph intersects the xy-plane in the line $y = 3x + 4$ and contains the point $(0, 0, 5)$.

7. Is the function represented in the following table linear?

		v			
		1.1	1.2	1.3	1.4
	3.2	11.06	12.06	13.07	14.07
	3.4	11.75	12.82	13.89	14.95
u	3.6	12.44	13.57	14.70	15.83
	3.8	13.13	14.33	15.52	16.71
	4.0	13.82	15.08	16.34	17.59

For Problems 8–10, find equations for the linear functions with the given tables.

8.

	y: -1	0	1	2
x: 0	1.5	1	0.5	0
1	3.5	3	2.5	2
2	5.5	5	4.5	4
3	7.5	7	6.5	6

9.

	y: 1	1.1	1.2	1.3
x: 2.2	7.9	7.5	7.1	6.7
2.4	8.2	7.8	7.4	7.0
2.6	8.5	8.1	7.7	7.3
2.8	8.8	8.4	8.0	7.6

10.

	y: 10	20	30	40
x: 100	3	6	9	12
200	2	5	8	11
300	1	4	7	10
400	0	3	6	9

Problems 11–13 each contain a partial table of values for a linear function. Fill in the blanks.

11.

	y: 0.0	1.0
x: 0.0		1.0
2.0	3.0	5.0

12.

	y: 3.0	4.0	5.0
x: 1.0	5.0		7.0
2.0		9.0	
3.0			

13.

	y: -1.0	0.0	1.0
x: 2.0	4.0		
3.0		3.0	5.0

For Problems 14–16, find equations for the linear functions with the given contour diagrams.

14.   **15.**   **16.**

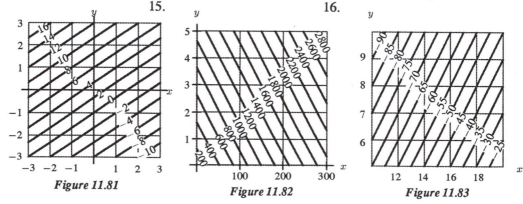

Figure 11.81   Figure 11.82   Figure 11.83

It is difficult to draw the graph of a linear function by hand. One trick that works if the x, y, and z-intercepts of the graph are positive is to plot the intercepts and join them by a triangle as shown in Figure 11.84; this at least gives the part of the graph in the octant where $x \geq 0$, $y \geq 0$, $z \geq 0$. If the intercepts are not all positive, the same method works if the x, y, and z-axes are drawn from a different perspective. Try this method to sketch the graphs of the linear functions in Problems 17–19:

17. $z = 6 - 2x - 3y$ **18.** $z = 4 + x - 2y$ **19.** $z = 2 - 2x + y$

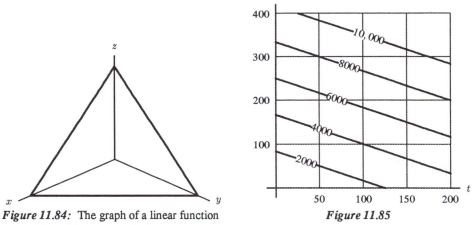

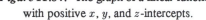

Figure 11.84: The graph of a linear function with positive x, y, and z-intercepts.

Figure 11.85

20. Figure 11.85 shows the revenue of a music store as a function of the number t of tapes and the number c of compact discs that it sells. What is the price of tapes? What is the price of compact discs?

21. Find the equation of the plane that passes through the points $(1, 1, 3)$, $(-1, 2, 2)$, and $(0, 3, 3)$. [Hint: Write the general form for a linear function, substitute the given points into it, and solve for the coefficients.]

22. A manufacturer makes two products out of two raw materials. Let q_1, q_2 be the quantities sold of the two products, p_1, p_2 their prices, and m_1, m_2 the quantities purchased of the two raw materials. Which of the following functions do you expect to be linear, and why? In each case, assume that all variables except the ones mentioned are held fixed.

    (a) Expenditure as a function of m_1 and m_2.
    (b) Revenue as a function of q_1 and q_2.
    (c) Revenue as a function of p_1 and q_1.

23. Give an example of a nonlinear function $f(x, y)$ such that all the x sections and all the y sections are straight lines.

24. A college admissions office uses the following linear equation to predict the grade point average of an incoming student:

$$z = 0.003x + 0.8y - 4$$

where z is the predicted college GPA on a scale of 0 to 4.3, and x is the sum of the student's SAT Math and SAT Verbal on a scale of 400 to 1600, and y is the student's high school GPA on a scale of 0 to 4.3. The college admits students whose predicted GPA is at least 2.3.

    (a) Will a student with SAT's of 1050 and high school GPA of 3.0 be admitted?
    (b) Will every student with SAT's of 1600 be admitted?
    (c) Will every student with a high school GPA of 4.3 be admitted?
    (d) Draw a contour diagram for predicted GPA z with $400 \leq x \leq 1600$ and $0 \leq y \leq 4.3$. Shade the points corresponding to students who will be admitted.
    (e) Which is more important, an extra 100 points on the SAT or an extra 0.5 of high school GPA?

11.6   FUNCTIONS OF MORE THAN TWO VARIABLES

In applications of calculus, functions of any number of variables can arise. The density of matter in the universe can be described by a function of three or four variables, since it takes three variables to specify a point in space, and if you want to study the evolution of this density over time, you need to add the fourth variable, time. The calculus we shall learn applies to functions of arbitrarily many variables. The main difference between these and functions of two variables is that it is hard to visualize the graph of a function of more than two variables. The graph of a function of one variable is a curve in 2-space, the graph of a function of two variables is a surface in 3-space, and so the graph of a function of three variables would be a solid in 4-space, and so on. We can't visualize 4-space, or any higher dimensional space, and so we won't talk about the graphs of functions of three or more variables.

On the other hand, the idea of sectioning a function by keeping one variable fixed allows us to get visual and numerical pictures of a function of three variables. It is also possible to give contour diagrams for these functions, only now the contours are surfaces in three-dimensional space.

Graphical Examples

One way to analyze a function of two variables $f(x, y)$ is to keep one of the variables x or y fixed; that is, look at the constant x or constant y sections of the function. We can do the same thing for a function of three variables $f(x, y, z)$. Keep one of the variables, say z, fixed and view f as a function of the remaining two variables x and y. For each fixed value, $z = c$, we then would have a contour diagram or a graph or even a table of values for the two-variable function $f(x, y, c)$. As c varies, we get a whole collection of contour diagrams or graphs or tables.

Example 1   A cone-shaped pond 30 feet at its deepest point and 200 feet across is infested with algae. Suppose the density of algae at a point z feet deep and x and y feet east-west and north-south of the center of the pond is given by

$$f(x, y, z) = \frac{1}{10}(50 + \sqrt{x^2 + y^2})(30 - z),$$

where density is measured in ounces per cubic feet. Draw contour diagrams for f at the surface of the pond and at a depth of 10 feet. Explain how f varies with depth and distance from the center of the pond.

Solution   At the surface of the pond, $z = 0$, so we want a contour diagram for the $z = 0$ section:

$$f(x, y, 0) = \frac{1}{10}(50 + \sqrt{x^2 + y^2})(30 - 0)$$

The contours are circles, since $f(x, y, 0)$ is constant when $\sqrt{x^2 + y^2}$ is constant. The density is $\frac{1}{10}(50)(30) = 150$ at the center of the pond and decreases by 30 for every 10 feet increase in the distance $\sqrt{x^2 + y^2}$ from the center of the pond. The contour diagram is shown in Figure 11.86a. For $z = 10$, we have

$$f(x, y, 10) = \frac{1}{10}(50 + \sqrt{x^2 + y^2})(30 - 10)$$

Now the center of the pond at depth 10 feet has density 100 and the density increases by 20 per every 10 feet increase in $\sqrt{x^2 + y^2}$. The contour diagram for the $z = 10$ section is shown in Figure 11.86b. Evidently, there is more algae near the surface than deeper down in the lake and more algae near the shoreline of the lake than near the center.

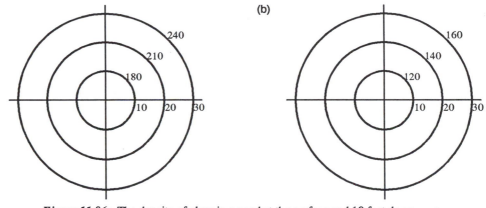

Figure 11.86: The density of algae in a pond at the surface and 10 feet down

The temperature at different points in a lake could be analyzed in a similar way by giving contour diagrams for the temperature at different fixed depths. For weather forecasting, it is important to know the barometric pressure and temperature not only at the surface of the earth but also aloft in the atmosphere above the earth. On TV, you may have heard about a "surface" low pressure system or an "upper level" low pressure system (upper level lows influence the jet stream). The surface and upper level maps studied by meteorologists are just contour diagrams for different altitude sections of the pressure or temperature function.

A Numerical Example

In the previous examples, we viewed the sections of a three-variable function $f(x, y, z)$ as contour diagrams or graphs of a two-variable function. We can also view the sections as tables of values.

Example 2 In Section 11.1, the consumption of beef as a function of the price p of beef and household income I was given in a table of values. The consumption C of beef is also a function of the price q of a competing meat, say chicken, as well as the price of beef and household income, that is, $C = f(I, p, q)$. This function can be given by a collection of tables, one for each different value of the price q of chicken. Tables 11.11 and 11.12 show two different q sections for the consumption function f. Explain how the two tables are related.

Table 11.11: $q = 1.50$

		p	
	3.00	3.50	4.00
20	2.65	2.59	2.51
I 60	5.11	5.00	4.97
100	5.79	5.77	5.68

Table 11.12: $q = 2.00$

		p	
	3.00	3.50	4.00
20	2.75	2.51	2.13
I 60	5.21	4.98	4.89
100	5.78	5.77	5.65

Solution When the price q of chicken falls, there are two possible effects. People might buy more chicken and less beef because chicken is cheaper, or people might buy the same amount of chicken and more beef with the money they save on chicken. One might expect that the second effect is more important when the price of beef is low. Thus, for households of income \$20,000 (that is, $I = 20$), more beef is consumed when $q = 1.50$ than when $q = 2.00$ if the beef price is low, say $p = 3.00$. However, if the beef price is high, say $p = 4.00$, less beef is consumed when $q = 1.50$ than when $q = 2.00$. The other effect we would expect to see is that households with high income are less sensitive to price. Thus, the $I = 100$ rows in the two tables do not differ nearly so much as the $I = 20$ rows.

Level Surfaces

For a function of two variables you draw the contour diagram by laying out a plane with an axis for each of the variables, and joining together the points where the function has a particular value. The function is constant along the contours.

For a function of three variables, we have to lay out three axes, so the contour diagram of a function of three variables will live in 3-space. But the idea is basically the same; you join together all the points where the function has a particular value. Only when you do this you won't get a curve, you will get a surface, called a *level surface* of the function.

Example 3   What do the level surfaces of the function $f(x, y, z) = x^2 + y^2 + z^2$ look like?

Solution   Each level surface corresponds to a particular value of the function, just as each contour in a contour diagram corresponds to a particular value of the function. The level surface corresponding to value 3 is the set of all points where $f(x, y, z) = 3$, i.e.,

$$x^2 + y^2 + z^2 = 3.$$

This is the equation of a sphere of radius $\sqrt{3}$, with center at the origin. Similarly, the level surface corresponding to the value 2 is the sphere with radius $\sqrt{2}$. The other level surfaces will be concentric spheres of either larger or smaller radii; the larger the radius, the larger the corresponding value of the function. See Figure 11.87. The level surfaces become more closely spaced as we move further from the origin. The situation is analogous to the contour diagram for $g(x, y) = x^2 + y^2$ studied in Section 11.4 with spheres in three-space taking the place of circles in the plane.

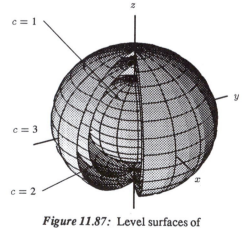

Figure 11.87: Level surfaces of
$f(x, y, z) = x^2 + y^2 + z^2.$

The level surfaces of a function will be nested together in some way, so it is more difficult to draw them. You have to have cut-outs to show the inner surfaces. You should imagine the level surfaces of $x^2 + y^2 + z^2$ to be the concentric layers of an onion, each one labeled with a different number, starting from small numbers in the middle and getting larger as you go out.

Example 4   What do the level surfaces of $f(x, y, z) = x^2 + y^2 - z^2$ look like?

Solution   You might recall that the two-variable quadratic function $g(x, y) = x^2 - y^2$, studied in Section 11.4, has a saddle-shaped graph and three types of contours depending on the value of c in the contour equation $x^2 - y^2 = c$: a hyperbola opening right-left when $c > 0$, a hyperbola opening up-down when $c < 0$, and a pair of intersecting lines when $c = 0$. Similarly, the three-variable quadratic function $f(x, y, z) = x^2 + y^2 - z^2$ also has three types of level surfaces depending on the value of c in the equation $x^2 + y^2 - z^2 = c$.

Suppose that $c > 0$, say $c = 1$. Rewrite the equation as $x^2 + y^2 = z^2 + 1$ and think of what happens as we section the surface perpendicular to the z axis by holding z fixed. The result is a circle, $x^2 + y^2 =$ constant, of radius at least 1 (since the constant $z^2 + 1 \geq 1$). The circles get larger as z gets larger. If we took the $x = 0$ section instead we would get the hyperbola $y^2 - z^2 = 1$. The result is shown in Figure 11.88a.

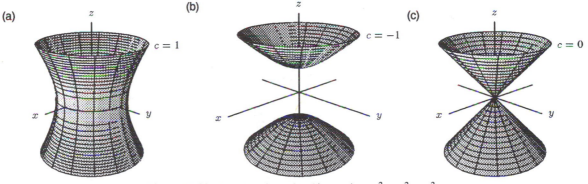

(a)   (b)   (c) $c = 1$ $c = -1$ $c = 0$

Figure 11.88: Level surfaces for $f(x, y, z) = x^2 + y^2 - z^2$

Suppose instead $c < 0$, say $c = -1$, then the z-sections, $x^2 + y^2 = z^2 - 1$, are again circles except that the radii shrink all the way to 0 at $z = \pm 1$ and between $z = -1$ and $z = 1$ there are no sections at all. The result is shown in Figure 11.88b.

When $c = 0$, we get the equation $x^2 + y^2 = z^2$. Again the z-sections are circles, this time with the radius shrinking down to exactly 0 when $z = 0$. The resulting surface, shown in Figure 11.88c, should look familiar: it is the cone $z = \sqrt{x^2 + y^2}$ studied in Section 11.4, together with a second lower cone $z = -\sqrt{x^2 + y^2}$. Notice that the surfaces in Figure 11.88a will be outside the cone in Figure 11.88c whereas the surfaces in Figure 11.88b will be inside the cone.

Example 5   What do the level surfaces of $f(x, y, z) = x^2 + y^2$ and $g(x, y, z) = y - z$ look like?

Solution   The level surface of f corresponding to the constant c is the surface consisting of all points satisfying the equation

$$x^2 + y^2 = c.$$

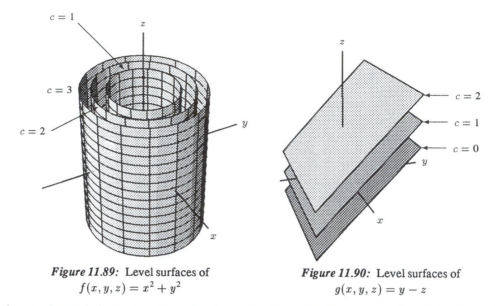

Figure 11.89: Level surfaces of
$f(x, y, z) = x^2 + y^2$

Figure 11.90: Level surfaces of
$g(x, y, z) = y - z$

In the xy-plane this is the equation of a circle of radius $\sqrt{c}$ with center at the origin. Since there is no z-coordinate in the equation, z can take any value whatsoever, and so in 3-space this is the equation of the surface you get by starting with a circle in the xy-plane and letting the z-coordinate of each point roam up and down. Thus, you get a circular cylinder around the z-axis, of radius $\sqrt{c}$. The level surfaces are concentric cylinders; the narrow ones near the z-axis are the ones where f has small values, and the wider ones are where it has larger values. See Figure 11.89.

The level surface of g corresponding to the constant c is the surface

$$y - z = c.$$

This time there is no x variable, so we can visualize this surface as the one you get by taking each point on the straight line $y - z = c$ in the yz-plane and letting x roam back and forth. This is a plane parallel to the x-axis which cuts the yz-plane diagonally. See Figure 11.90.

Level Surfaces, Graphs, and Implicit Functions

We have used 3-space in two different ways. First we used it to visualize graphs of functions of two variables. The xy-plane was the domain of the function, and the z-axis was used to record the values of the function at various points in the xy-plane. The resulting graph is a surface. The second way we used 3-space was to represent the domain of a function of three variables; in that case all three coordinates of a point are arguments of a function. There is no fourth axis on which to record the value of the function. Instead, we must visualize the function by drawing its level surfaces.

So surfaces come up in two different ways; as graphs of functions of two variables, and as level surfaces of functions of three variables. Is there a connection between the two ways?

Consider the function $f(x, y, z) = x^2 + y^2 + z^2$, whose level surfaces we drew in Figure 11.87. Take the level surface where the function has the value 1, defined by the equation

$$x^2 + y^2 + z^2 = 1.$$

The level surface is the sphere of radius 1. The equation may be thought of as giving z implicitly in terms of x and y. If we solve it for z, we get

$$z = \pm\sqrt{1 - x^2 - y^2}.$$

This is really two functions, the one you get by choosing the plus sign and the one you get by choosing the minus sign. The graph of the first one is the upper hemisphere of the level surface, and the graph of the second one is the lower hemisphere.

What this example shows is that if you have a level surface of a function $f(x, y, z)$, it can often be thought of as the graph of a function of two variables, or at least a piece of it can. If your original surface is the level surface where f has the value c, then the implicit function in question is the function $z = g(x, y)$ that you get by solving

$$f(x, y, z) = c$$

for z. Of course, as with implicit functions in the one-variable case, there may be more than one solution and you may not be able to find a formula for the solution.

Linear Functions of More than Two Variables

What does a linear function of three variables look like? Recall that the contour diagram of a linear function of two variables looked like equally spaced parallel lines. The level surfaces of a linear function of three variables look like a bunch of equally spaced parallel planes. Just as in the case of a function of two variables, the rate of change of a linear function of three variables is the same no matter where you are, as long as you point in the same direction. Suppose that a linear function has rate of change a in the x direction, b in the y direction, and c in the z direction. Suppose its value at $(0, 0, 0)$ is d. Then if you start at the origin and go x_0 units in the x direction, y_0 units in the y direction, and z_0 units in the z direction, the value of the function will change by $ax_0 + by_0 + cz_0$. So your value at (x_0, y_0, z_0) is $d + ax_0 + by_0 + cz_0$. Thus, the general equation for the linear function is

$$f(x, y, z) = ax + by + cz + d.$$

Example 6   Give sections of the linear function $f(x, y, z) = 2x - y + 3z + 4$ with z fixed, first in terms of tables, then in terms of contour diagrams. $z = 2x - y + 7$

Solution   Tables 11.13, 11.14, and 11.15 give three different sections of f with $z = 1, 2, 3$. Notice that each individual table is linear with the same x and y slopes in each table. Moreover, each table is obtained by adding the same constant, 3, to the previous table.

TABLE 11.13 $z = 1$

	y		
	0	1	2
x 0	7	6	5
x 1	9	8	7
x 2	11	10	9

TABLE 11.14 $z = 2$

	y		
	0	1	2
x 0	10	9	8
x 1	12	11	10
x 2	14	13	12

TABLE 11.15 $z = 3$

	y		
	0	1	2
x 0	13	12	11
x 1	15	14	13
x 2	17	16	15

Contour diagrams for sections of f, with $z = 1, 2, 3$, are given in Figure 11.91. Notice that each contour diagram has lines with the same slope and same spacing. Only the labeling on the lines changes from diagram to diagram. The line through the origin in the $z = c$ section has the label $f(0, 0, c) = 3c + 4$. Another way to look at these diagrams is to concentrate on one label, say 5, and watch how the line with that label moves from one diagram to the next.

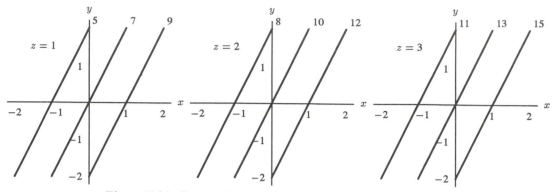

Figure 11.91: Contour diagrams for three sections of a linear function

A Catalog of Surfaces

For later reference, here is a small catalog of the surfaces we have encountered.

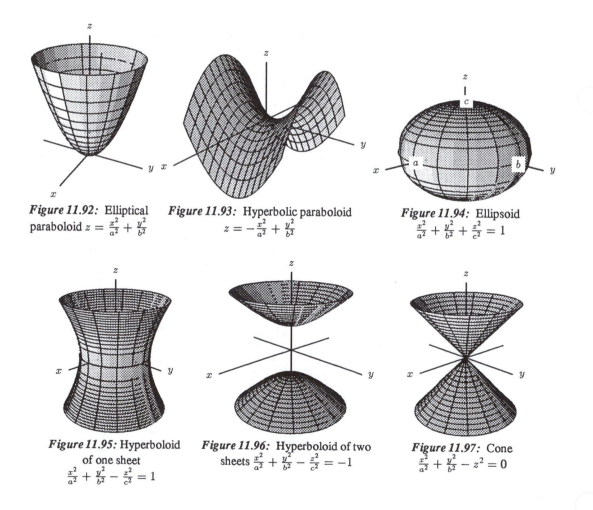

Figure 11.92: Elliptical paraboloid $z = \frac{x^2}{a^2} + \frac{y^2}{b^2}$

Figure 11.93: Hyperbolic paraboloid $z = -\frac{x^2}{a^2} + \frac{y^2}{b^2}$

Figure 11.94: Ellipsoid $\frac{x^2}{a^2} + \frac{y^2}{b^2} + \frac{z^2}{c^2} = 1$

Figure 11.95: Hyperboloid of one sheet $\frac{x^2}{a^2} + \frac{y^2}{b^2} - \frac{z^2}{c^2} = 1$

Figure 11.96: Hyperboloid of two sheets $\frac{x^2}{a^2} + \frac{y^2}{b^2} - \frac{z^2}{c^2} = -1$

Figure 11.97: Cone $\frac{x^2}{a^2} + \frac{y^2}{b^2} - z^2 = 0$

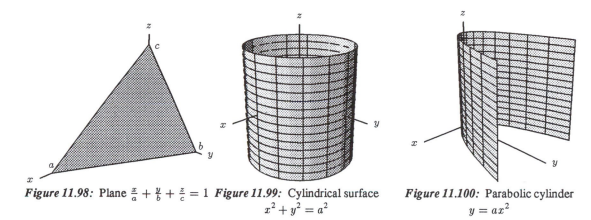

Figure 11.98: Plane $\frac{x}{a} + \frac{y}{b} + \frac{z}{c} = 1$ *Figure 11.99:* Cylindrical surface   *Figure 11.100:* Parabolic cylinder

$$x^2 + y^2 = a^2 \qquad\qquad y = ax^2$$

(These are viewed as equations in three variables x, y and z)

Problems for Section 11.6

1. Hot water is entering a rectangular swimming pool at the surface of the pool in one corner. Give a contour diagram for the temperature of the pool at the surface and three feet below the surface.

2. Give tables for $I = 20$ and $I = 100$ sections of the beef consumption function of Example 2. Comment on what you see in the tables. Do the same for $p = 3.00$ and $p = 4.00$ sections.

3. Draw contour diagrams for three different sections with t fixed of the function

$$f(x, y, t) = \cos t \cos \sqrt{x^2 + y^2}, \ 0 \leq \sqrt{x^2 + y^2} \leq \pi/2.$$

4. Match the functions whose equations are given below with the sets of level surfaces in Figure 11.101.

   (a) $f(x, y, z) = y^2 + z^2$
   (b) $h(x, y, z) = x^2 + z^2$.

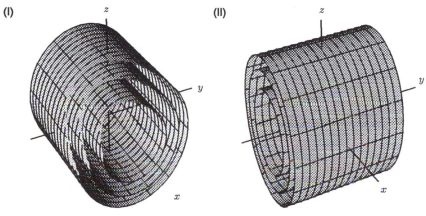

Figure 11.101

5. Describe the level surfaces of $f(x, y, z) = \sin(x + y + z)$.

6. Describe the level surfaces of $e^{-(x^2 + y^2 + z^2)}$.

7. The height (in feet) of the water above the bottom of a pond at time t is given by the function $h(x, y, t) = 20 + \sin(x + y - t)$, where x and y are Cartesian coordinates arranged so that the positive y-axis is north and the positive x-axis is east, and where t is in seconds. By considering contour diagrams for different t sections (snapshots) of f, describe the motion of the water surface in the pond.

8. Figure 11.102 shows contour diagrams of temperature in degrees Fahrenheit in a room at three different times. Describe the heat flow in the room. What could be causing this?

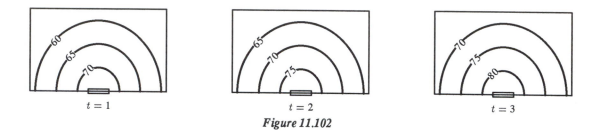

$t = 1$ $t = 2$ $t = 3$

Figure 11.102

9. Give the linear function $f(x, y, z) = ax + by + cz + d$ that has these partial table of values:

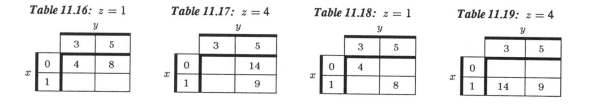

Table 11.16: $z = 1$

x \ y	3	5
0	4	8
1		

Table 11.17: $z = 4$

x \ y	3	5
0		14
1		9

Table 11.18: $z = 1$

x \ y	3	5
0	4	
1		8

Table 11.19: $z = 4$

x \ y	3	5
0		
1	14	9

10. Give the linear function $f(x, y, z) = ax + by + cz + d$ that has the contour diagrams for two z-sections shown in Figures 11.103 and 11.104:

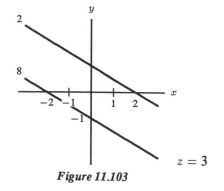

$z = 3$

Figure 11.103

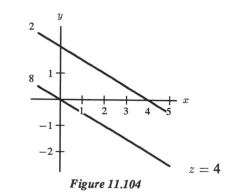

$z = 4$

Figure 11.104

11. In Problem 9, suppose the two tables had only three values filled in. Could you determine f? Suppose the two tables have four values filled in. Can you always determine f? How about five values?

12. What do the level surfaces of $f(x, y, z) = x^2 - y^2 + z^2$ look like? [Hint: Use y-sections instead of z-sections.]

Use the catalog of surfaces to identify the surfaces in Problems 13–22:

13. $-x^2 - y^2 + z^2 = 1$
14. $x^2 - y^2 - z^2 = 1$
15. $-x^2 + y^2 - z^2 = 0$

16. $x^2 + y^2 - z = 0$
17. $x^2 + z^2 = 1$
18. $x^2 - y^2 + z = 0$

19. $x^2 + y^2/4 + z^2 = 1$
20. $x + y = 1$
21. $(x - 1)^2 + y^2 + z^2 = 1$

22. $x + y + z = 1$

23. Describe the surface $x^2 + y^2 = (2 + \sin z)^2$. In general, if $f(z) \geq 0$ for all z, describe the surface $x^2 + y^2 = (f(z))^2$.

REVIEW PROBLEMS FOR CHAPTER ELEVEN

1. A point has coordinates (a, b, c) where a, b, c are all positive. Interpret a, b, and c as distances from the coordinate planes. Illustrate your answer with a picture.

2. Sketch the graph of the cylinder volume function $f(r, h) = \pi r^2 h$ first by keeping h fixed, then by keeping r fixed.

3. Describe the set of points whose x coordinate is 2 and whose y coordinate is 1.

4. Find the center and radius of the sphere with equation

$$x^2 + 4x + y^2 - 6y + z^2 + 12z = 0.$$

5. Find the equation of the plane through the points $(0, 0, 2), (0, 3, 0), (5, 0, 0)$.

6. Find an equation for the plane containing the line in the xy-plane defined by $y = 3x + 1$, and the point $(1, 0, 3)$.

7. Find the equation of the plane that intersects the xy-plane in the line $y = 2x + 2$ and contains the point $(1, 2, 2)$.

Decide if the statements in Problems 8–12 must be true, might be true, or could not be true.

8. The level curves corresponding to $z = 1$ and $z = -1$ cross at the origin.

9. The level curve $z = 1$ consists of the circle $x^2 + y^2 = 2$, and the circle $x^2 + y^2 = 3$, but no other points.

10. The level curve $z = 1$ consists of two lines which intersect at the origin.

11. If $z = e^{-(x^2+y^2)}$, there is a level curve for every value of z.

12. If $z = e^{-(x^2+y^2)}$, there is a level curve through every point (x, y).

For each of the functions in Problems 13–16, make a contour plot in the region $-2 < x < 2$ and $-2 < y < 2$. Describe the shape of the contour lines.

13. $z = \sin y$ **14.** $z = 3x - 5y + 1$ **15.** $z = 2x^2 + y^2$ **16.** $z = e^{-2x^2 - y^2}$

17.   Suppose you are in a room 30 feet long with a heater at one end. In the morning the room is $65° F$. You turn on the heater, which quickly warms up to $85° F$. Let $H(x, t)$ be the temperature x feet from the heater, t minutes after the heater is turned on. Figure 11.105 shows the contour diagram for H.

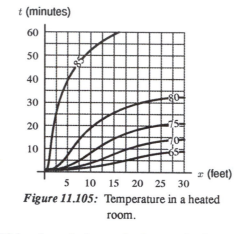

t (minutes)

Figure 11.105: Temperature in a heated room.

How warm is it 10 feet from the heater 5 minutes after it was turned on? 10 minutes after it was turned on?

18.   Using the contour diagram in Figure 11.105, sketch the graphs of the one-variable functions $H(x, 5)$ and $H(x, 20)$. Interpret the two graphs in practical terms, and explain the difference between them.

19.   Consider the function $z = \cos \sqrt{x^2 + y^2}$.

(a)   Sketch the level curves of this function.

(b)   Sketch a cross-section through the surface $z = \cos \sqrt{x^2 + y^2}$ in the plane containing the x- and z-axes. Put units on your axes.

(c)   Sketch the cross-section through the surface $z = \cos \sqrt{x^2 + y^2}$ in the plane containing the z-axis and the line $y = x$ in the xy-plane.

20.   Draw the contour diagrams for the functions $f(x, y) = (x - y)$ and $g(x, y) = (x - y)^2$ for $-3 \leq x \leq 3$, $-3 \leq y \leq 3$. Where does each function attain its maximum value on this region?

21.   You are an anthropologist observing a native ritual. Sixteen people arrange themselves along a bench, with all but the three on the far left side seated. The first person is standing with her hands at her side, the second is standing with his hands raised and the third is standing with her hands at her side. At some unseen signal, the first one sits down, and everyone else copies what his neighbor to the left was doing one second earlier. Every second that passes, this behavior is repeated until all are once again seated.

(a)   Draw graphs at several different times showing how the height depends upon the distance along the bench.

(b)   Graph the location of the raised hands as a function of time.

(c)   What US ritual is most closely related to what you have observed?

22. Table 11.20 shows the predictions of one simple model on how average yearly ultra-violet (UV) exposure might vary with the year and the latitude.

TABLE 11.20 *Ultra-violet exposure.*

	Year								
	1970	1975	1980	1985	1990	1995	2000	2005	2010
90	0.00	0.00	0.00	0.00	0.00	0.00	0.00	0.00	0.00
80	0.00	0.00	0.00	0.00	0.00	3.03	5.79	6.37	6.52
70	0.00	0.00	0.00	0.00	0.00	0.02	3.14	6.80	8.21
60	0.01	0.01	0.01	0.01	0.01	0.01	0.12	1.78	3.41
50	0.08	0.08	0.08	0.08	0.08	0.08	0.08	0.08	0.33
40	0.25	0.25	0.25	0.25	0.25	0.25	0.25	0.25	0.25
30	0.49	0.49	0.49	0.49	0.49	0.49	0.49	0.49	0.49
20	0.74	0.74	0.74	0.74	0.74	0.74	0.74	0.74	0.74
10	0.93	0.93	0.93	0.93	0.93	0.93	0.93	0.93	0.93
Latitude 0	1.00	1.00	1.00	1.00	1.00	1.00	1.00	1.00	1.00
-10	0.93	0.93	0.93	0.93	0.93	0.93	0.93	0.93	0.93
-20	0.74	0.74	0.74	0.74	0.74	0.74	0.74	0.74	0.74
-30	0.49	0.49	0.49	0.49	0.49	0.49	0.49	0.49	0.49
-40	0.25	0.25	0.25	0.25	0.25	0.25	0.25	1.39	2.31
-50	0.08	0.08	0.08	0.08	0.08	0.08	2.41	5.68	7.27
-60	0.01	0.01	0.01	0.01	0.01	1.41	7.64	10.82	11.97
-70	0.00	0.00	0.00	0.00	0.08	6.36	10.36	11.46	11.80
-80	0.00	0.00	0.00	0.00	3.69	6.32	6.71	6.80	6.82
-90	0.00	0.00	0.00	0.00	0.00	0.00	0.00	0.00	0.00

(a) Graph the sections of UV exposure for the years 1970, 1990 and 2000.

(b) Produce a table showing what latitude has the most severe exposure to UV, as a function of the year.

(c) What is a possible explanation for this phenomenon?

CHAPTER TWELVE

A FUNDAMENTAL TOOL: VECTORS

In one-variable calculus we were able to represent quantities such as velocity by numbers. Once we enter the world of more than one variable, we can no longer do this. To specify the velocity of a moving object, we need to say how fast it is moving, and also what direction it is moving in. In this chapter we will learn how to use *vectors* to represent quantities that have direction as well as magnitude.

12.1 THE GEOMETRIC DEFINITION OF A VECTOR

Suppose you are the pilot of a private plane and are planning a flight from Dallas to Pittsburgh. There are two things you must know: How long the flight is (so you have enough fuel to make it) and in what direction to go (so you don't miss Pittsburgh). Thus, you must specify both the *direction* and the amount or *magnitude* of the displacement.

> A quantity specified by a magnitude and a direction is called a **vector**; it is represented by an arrow whose length is the magnitude of the vector and which points in the direction of the vector.

In comparison, a quantity specified only by a number, but no direction, is called a *scalar*.[1] For instance, the weight of the plane at take-off and the time taken by the flight from Dallas to Pittsburgh are both scalar quantities.

In this book, vectors are written with an arrow over them, $\vec{v}$, to distinguish them from scalars. The magnitude, or length, of a vector $\vec{v}$ is written $\|\vec{v}\|$. Other books use a bold v to denote a vector.

The displacement between the cities is represented by an arrow with its tail at Dallas and its tip at Pittsburgh, as in Figure 12.1. If we had two separate pairs of cities, each the same distance apart measured along parallel lines in the same direction, then the displacement vectors are the same length, parallel, and in the same direction. Albuquerque, NM, and Oshkosh, Wisconsin are one such pair, while Los Angeles, and Buffalo, South Dakota are another. Then we say that the vectors between corresponding cities are the same, even though they are not in the same position. See Figure 12.1. In other words:

> Vectors which point in the same direction and have the same length are considered to be the same vector, even if they do not start at the same point.

Figure 12.1: Vector displacements between cities

---

[1] So named by W. R. Hamilton because they are merely numbers on the *scale* from $-\infty$ to ∞.

Displacement versus Distance

The distance from one object to another is represented by a number; it has no direction, only magnitude, and thus is a scalar. On the other hand, when we talk about the displacement from one object to another, as between cities, for example, we mean both the distance and the direction, so displacement is a vector. The magnitude of the displacement vector is the distance, so a displacement vector between two objects is represented graphically by drawing an arrow from the first object to the second. See Figure 12.1.

Velocity versus Speed

Although the words "speed" and "velocity" are often used interchangeably in everyday language, we distinguish between them in mathematics. The speed of a moving body tells us how fast it is moving, say 50 mph. The speed is just a number; it is therefore a scalar. The velocity, on the other hand, tells us both how fast the body is moving and the direction of motion. For instance, if a car is heading northeast at 50 mph, then its velocity is 50 mph represented by an arrow in a northeast direction. The speed, 50 mph, is equal to the magnitude of the vector. The length of the arrow indicates the magnitude of the vector; the direction of motion is given by the direction of the arrow.

Example 1   A car is traveling north at a speed of 60 mph, while a plane above is flying horizontally south-west at a speed of 300 mph. Draw the velocity vectors of the car and the plane.

Solution   Figure 12.2 shows the velocity vectors. The plane's velocity vector is five times as long as the car's, because its speed is five times greater ($300 = 5 \cdot 60$), and speed is the magnitude of velocity.

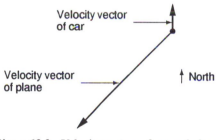

Figure 12.2: Velocity vectors of car and plane

The Geometric Definition of Addition and Subtraction of Vectors

Suppose NASA is controlling a robot explorer on Mars. NASA sends a command to have the robot move 75 meters in a certain direction. Then NASA sends a second message to have the robot move 50 meters in a second direction as shown in Figure 12.3.

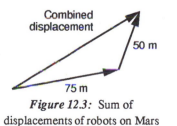

Figure 12.3: Sum of displacements of robots on Mars

Where does the robot end up as a result of these two commands? Each movement can be thought of as a vector, first $\vec{v}$ and then $\vec{w}$. The combined displacement has a length and a direction and is said to be the sum of these two displacement vectors, $\vec{v} + \vec{w}$.

> The **sum**, $\vec{v} + \vec{w}$, is the combined displacement represented by the third side of the triangle in Figure 12.4.

Suppose now that NASA is controlling two different robots, starting out from the same location. One moves along a displacement vector $\vec{v}$ and the second along a displacement vector $\vec{w}$. What is the position of the second robot relative to the first? We want the displacement vector between the two robots. We draw the vector and call it $\vec{x}$, as shown in Figure 12.5. Notice that $\vec{v} + \vec{x} = \vec{w}$ so that $\vec{x}$ is the difference: $\vec{x} = \vec{w} - \vec{v}$. Note that $\vec{w} - \vec{v}$ is the vector pointing from the tip of $\vec{v}$ to the tip of $\vec{w}$.

> The **difference**, $\vec{w} - \vec{v}$, is the vector which when added to $\vec{v}$ gives $\vec{w}$, that is, such that $\vec{w} = \vec{v} + (\vec{w} - \vec{v})$. See Figure 12.5.

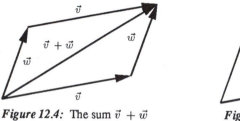

Figure 12.4: The sum $\vec{v} + \vec{w}$

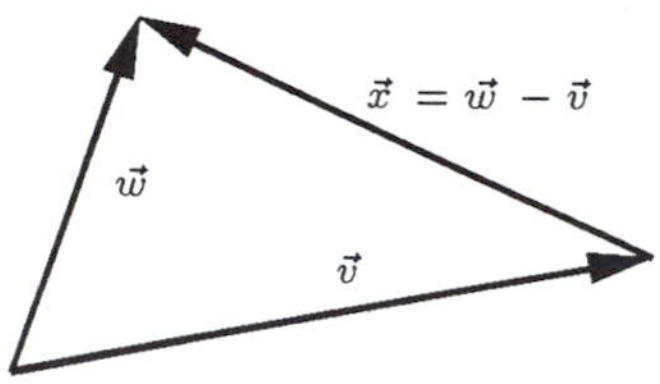

Figure 12.5: The difference $\vec{w} - \vec{v}$

Example 2   A riverboat is moving with a velocity, $\vec{v}$, of 8 km/hr relative to the water. In addition, there is a current, $\vec{c}$, of 1 km/hr at an angle to $\vec{v}$. See Figure 12.6. What is the physical significance of $\vec{v} + \vec{c}$?

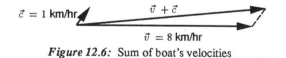

Figure 12.6: Sum of boat's velocities

Solution   The vector $\vec{v}$ shows how the boat is moving relative to the water, while $\vec{c}$ shows how the water is moving relative to the riverbed. Thus, $\vec{v} + \vec{c}$ shows how the boat is moving relative to the riverbed.

The Geometric Definition of Multiplication by a Scalar

If $\vec{v}$ represents a displacement, the vector $2\vec{v}$ represents a displacement of twice the magnitude in the same direction. If $\vec{v}$ is a velocity, then $-2\vec{v}$ represents a velocity of twice the magnitude in the opposite direction. Similarly, $0.5\vec{v}$ represents a vector in the same direction as $\vec{v}$ but half the length. See Figure 12.7.

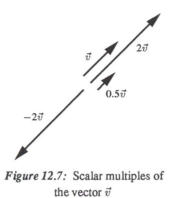

Figure 12.7: Scalar multiples of
the vector $\vec{v}$

Definition of Multiplication by a Scalar

If λ is a scalar and $\vec{v}$ is a vector, the **scalar multiple of $\vec{v}$ by** λ, written $\lambda\vec{v}$, is the vector with the following properties:

- The vector $\lambda\vec{v}$ is parallel to the vector $\vec{v}$, pointing in the same direction if $\lambda > 0$, and in the opposite direction if $\lambda < 0$.

- The magnitude of $\lambda\vec{v}$ is $|\lambda|$ times the magnitude of $\vec{v}$, i.e.,

$$||\lambda\vec{v}|| = |\lambda|\,||\vec{v}||.$$

Note that $|\lambda|$ represents the absolute value of the scalar λ while $||\lambda\vec{v}||$ represents the magnitude of the vector $\lambda\vec{v}$.

Properties of Addition and Scalar Multiplication

For any vectors $\vec{u}$, $\vec{v}$, and $\vec{w}$ and any scalars α and β, we have the following results.

Commutativity
1. $\vec{u} + \vec{v} = \vec{v} + \vec{u}$

Associativity
2. $(\vec{u} + \vec{v}) + \vec{w} = \vec{u} + (\vec{v} + \vec{w})$
3. $\alpha(\beta\vec{v}) = (\alpha\beta)\vec{v}$

Distributivity
4. $(\alpha + \beta)\vec{v} = \alpha\vec{v} + \beta\vec{v}$
5. $\alpha(\vec{v} + \vec{w}) = \alpha\vec{v} + \alpha\vec{w}$

Problems 13–17 at the end of this section ask for a justification of these results.

Further Examples of vectors

Acceleration

Another important vector quantity is acceleration. Acceleration, like velocity, is specified by both a magnitude and a direction — for example, the acceleration due to gravity at the surface of the Earth is 9.81 m/sec^2 *downward*.

Force

Force is another vector quantity. Suppose you push on an open door. The result depends not just on how hard you push but also in what direction. Thus, to specify a force we must give its magnitude (or strength) and the direction in which it is acting. For example, the gravitational force exerted on a planet by a star is a vector pointing from the planet in the direction of the star; its magnitude is the strength of the gravitational attraction, which depends on the masses of the two bodies and the distance between them. (Newton's Law of Gravitation gives the exact relation.)

Example 3   Kepler's first two laws of planetary motion state the following:
   I.   Planets follow elliptical orbits with the sun at one focus.
   II. The line joining a planet to the sun sweeps out equal areas in equal times.
   (a) Sketch force vectors representing the gravitational force of the sun on the earth at several different positions in its orbit.
   (b) Sketch the velocity vector of the earth at various points of its orbit.

Solution   (a) Figure 12.8 shows the earth orbiting the sun, with the gravitational force vectors at two different positions. Note that the force vector is larger when the earth is closer to the sun because the gravitational force there is greater.

   (b) The velocity vector always points in the direction of motion of the earth. Thus, the velocity vector must be tangent to the ellipse. See Figure 12.9. Second, the velocity vector is longer at points of the orbit where the planet is moving quickly, and shorter at points where it is moving slowly, because the magnitude, or length, of the velocity vector represents the speed.

   Kepler's Second Law enables us to determine when the earth is moving quickly and when it is moving slowly. The meaning of the second law is illustrated in Figure 12.9. Over a fixed period of time, say one month, the line joining the earth to the sun sweeps out a sector having a certain area. Figure 12.9 shows two sectors swept out in two different one-month time-intervals. The second law says that the areas of the two sectors are the same. For the areas to be the same, the earth must move farther in a month when it is close to the sun than when it is far from the sun. Thus, the earth moves faster when it is close to the sun and slower when it is far away.

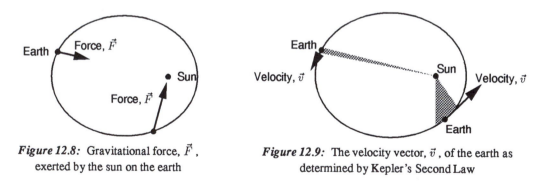

Figure 12.8: Gravitational force, $\vec{F}$, exerted by the sun on the earth

Figure 12.9: The velocity vector, $\vec{v}$, of the earth as determined by Kepler's Second Law

Problems for Section 12.1

In Problems 1–4 which of the following quantities is a vector and which is a scalar?

1. The distance from Seattle to St. Louis.

2. The population of the US.

3. The magnetic field at a point on the earth's surface.

4. The temperature at a point on the earth's surface.

5. Given the vectors $\vec{v}$ and $\vec{w}$ in Figure 12.10, draw the following vectors:
   (a) $\vec{v} + \vec{w}$.   (b) $\vec{v} - \vec{w}$.   (c) $2\vec{v}$.   (d) $2\vec{v} + \vec{w}$.   (e) $\vec{v} - 2\vec{w}$.

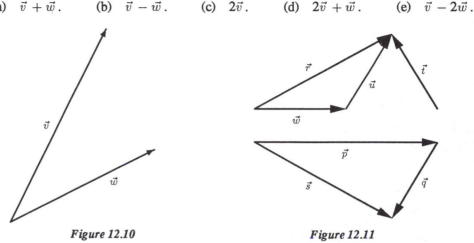

Figure 12.10   Figure 12.11

6. The vectors $\vec{w}$ and $\vec{u}$ are shown in Figure 12.11. Match the vectors $\vec{p}, \vec{q}, \vec{r}, \vec{s}, \vec{t}$ with five of the following vectors: $\vec{u} + \vec{w}, \vec{u} - \vec{w}, \vec{w} - \vec{u}, 2\vec{w} - \vec{u}, \vec{u} - 2\vec{w}, 2\vec{w}, -2\vec{w}, 2\vec{u}, -2\vec{u}, -\vec{w}, -\vec{u}$.

7. Two adjacent sides of a regular hexagon are given as the vectors $\vec{u}$ and $\vec{v}$ in Figure 12.12. Label the remaining sides in terms of $\vec{u}$ and $\vec{v}$.

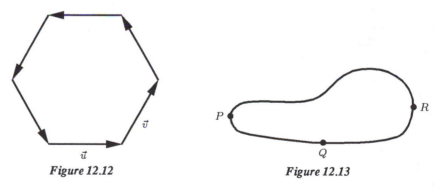

Figure 12.12   Figure 12.13

8. A car drives clockwise around the course in Figure 12.13, slowing down at the curves and speeding up along the straight portions of the track. Sketch velocity vectors at the points P, Q, and R.

9. A racing car drives clockwise around the track shown in Figure 12.13 at a constant speed. At what point on the track does the car have the longest acceleration vector, and in roughly what direction is it pointing? Remember that acceleration is the rate of change of velocity.

10. If the two diagonal lines of a quadrilateral bisect each other, use vectors to show that this quadrilateral must be a parallelogram.

11. Show that the medians of a triangle intersect at a point.

12. Show that the lines joining the centroid of a tetrahedron and the opposite vertex meet at a point.

Use the geometric definition of addition, subtraction, and of scalar multiplication to explain each of the properties in Problems 13–17.

13. $\vec{u} + \vec{v} = \vec{v} + \vec{u}$

14. $(\alpha + \beta)\vec{u} = \alpha\vec{u} + \beta\vec{u}$

15. $\alpha(\vec{u} + \vec{v}) = \alpha\vec{u} + \alpha\vec{v}$

16. $(\vec{u} + \vec{v}) + \vec{w} = \vec{u} + (\vec{v} + \vec{w})$

17. $\alpha(\beta\vec{u}) = (\alpha\beta)\vec{u}$

12.2   THE NUMERICAL REPRESENTATION OF A VECTOR

The geometric definition of a vector in the previous section gives a clear picture of what a vector *is*, but it is not always easy to use in calculations. There is, however, another method of representing a vector.

Suppose we want to describe the positions of atoms in a molecule. Typically, the atoms are arranged in a three-dimensional configuration. We pick one atom as our point of reference, or center, and draw a set of axes with this atom at the origin. See Figure 12.14. Then the displacement of an atom from the center is given by the coordinates of the atom. If the molecule is a regularly shaped pyramid of side length 2 units, then the coordinates of the atoms at A, B, C are as shown in Figure 12.14.

Since we move from O to A by moving 2 units along the x-axis, we say that the displacement vector $\vec{w}$ is represented by

$$\vec{w} = (2, 0, 0),$$

where the 2 gives the displacement in the x direction and the 0s give the displacement in the y and z directions. Notice that the same notation, $(2, 0, 0)$, is used both for the coordinates of the point A and for the displacement vector for O to A. Similarly, to get to C from O, we must move 1 unit in the x direction, $1/\sqrt{3}$ in the y direction, and $2\sqrt{2/3}$ in the z direction, so we say the vector $\vec{v}$ is represented by

$$\vec{v} = \left(1, \frac{1}{\sqrt{3}}, 2\sqrt{\frac{2}{3}}\right).$$

Again, the same notation is used for the coordinates of the point C and for the displacement vector from O to C. The entries in the vector $\vec{v}$ and $\vec{w}$ are called *components* and represent the part of the displacement in each coordinate direction.

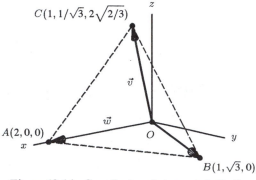

Figure 12.14: Coordinates of atoms at A, B, C

Example 1   Represent the velocity vectors of the car and the plane in Example 1 on page 71 using components. Use a coordinate system in which north is the positive y-axis, east is the positive x-axis, and the positive z-axis is up.

Solution   The car is traveling north at 60 mph, so the component of its velocity in the y-direction is 60 and the component in the x-direction is 0. Since it is traveling horizontally, the z-component is also 0. So its velocity vector is given by

$$\vec{v} = (0, 60, 0).$$

The plane's velocity vector also has z-component equal to zero. Since it is traveling southwest, its y and x components are negative (north and east are positive). To find the components, note that the plane travels 300 miles in an hour. This represents a distance of $300/\sqrt{2} \approx 212$ miles in the westerly direction and the same distance in the southerly direction. See Figure 12.15. Thus, the velocity vector of the plane is given by

$$\vec{v} = (-212, -212, 0).$$

Of course if the car were climbing a steep hill or if the plane was descending for a landing, then the z-component would be nonzero.

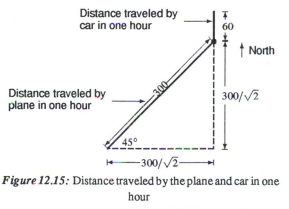

Figure 12.15: Distance traveled by the plane and car in one hour

Vectors in Two, Three, and n Dimensions

The vectors representing the displacement between atoms in a molecule are vectors in three dimensions: they are represented by arrows in 3-space and they have three components. A two-dimensional vector is a vector in the plane and can be represented by two components. For example, $\vec{a} = (1, 2)$ and $\vec{b} = (-2, 1)$ are shown in Figure 12.16. In addition, using components, we can define a vector in n dimensions as a string of n numbers, or n components. Thus, a vector in n dimensions can be represented by an expression of the form

$$\vec{c} = (c_1, c_2, \ldots, c_n).$$

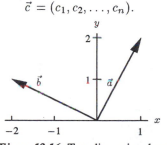

Figure 12.16: Two-dimensional vectors

Why Would We Want Vectors in n Dimensions?

Vectors in two and three dimensions can be used to model displacement, velocities, or forces. But what about vectors in n dimensions? There is another interpretation of n-dimensional vectors (or n-vectors) which is useful: they can be thought of as listing n different quantities. Even a 3-vector is also a way of keeping three quantities organized — for example, the displacements parallel to the x, y, and z axes. Similarly

$$\vec{c} = (c_1, c_2, \ldots, c_n)$$

is a way of keeping n different quantities organized. For example, a *consumption* vector ,

$$\vec{q} = (q_1, q_2, \ldots, q_n)$$

shows the quantities $q_1, q_2, \ldots, q_n$ consumed of each of n different goods. A *price* vector

$$\vec{p} = (p_1, p_2, \ldots, p_n)$$

contains the prices of n different commodities. A *population* vector $\vec{N}$ might show the number of children and adults in a population:

$$\vec{N} = (\text{number of children, number of adults}),$$

or if we are interested in a more detailed breakdown of ages, we might give the number in each ten-year age bracket in the population (up to age 110) in the form

$$\vec{N} = (N_1, N_2, N_3, N_4, \ldots, N_{10}, N_{11})$$

where N_1 is the population aged 0–10, and N_2 is the population aged 10–20, and so on.

In 1907-8, Hermann Minkowski used vectors with four components when he introduced *space-time coordinates*, whereby each event is assigned a vector position $\vec{v}$ with four coordinates, three space and one time:

$$\vec{v} = (x, y, z, t).$$

Unit Vectors

A *unit* vector is a vector whose magnitude is 1. In two dimensions, there are two special unit vectors, one along the positive x-axis, called $\vec{i}$, and one along the positive y-axis, called $\vec{j}$. In components, $\vec{i} = (1, 0)$ and $\vec{j} = (0, 1)$. See Figure 12.17. In three dimensions, there are three special unit vectors shown in Figure 12.18: $\vec{i} = (1, 0, 0)$, $\vec{j} = (0, 1, 0)$, and $\vec{k} = (0, 0, 1)$.

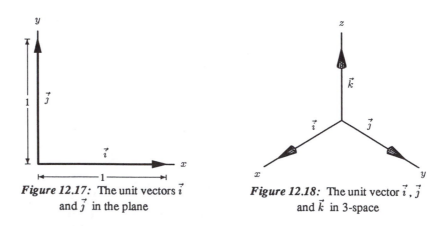

Figure 12.17: The unit vectors $\vec{i}$ and $\vec{j}$ in the plane

Figure 12.18: The unit vector $\vec{i}$, $\vec{j}$ and $\vec{k}$ in 3-space

The Zero Vector

The vector whose magnitude is zero is called the zero vector and written $\vec{0}$. In two dimensions $\vec{0} = (0, 0)$; in three dimensions $\vec{0} = (0, 0, 0)$.

Displacement Vectors

The **displacement vector**, $\overrightarrow{P_1 P_2}$, from the point $P_1 = (x_1, y_1, z_1)$ to the point $P_2 = (x_2, y_2, z_2)$ shown in Figure 12.19 is given by

$$\overrightarrow{P_1 P_2} = (x_2 - x_1, y_2 - y_1, z_2 - z_1)$$

For example, the vector from the point $P_1 = (2, 4, 6)$ to the point $P_2 = (3, 7, 10)$ is

$$\overrightarrow{P_1 P_2} = (3 - 2, 7 - 4, 10 - 6) = (1, 3, 4).$$

If the initial point is the origin, then the displacement vector is called the position vector. In general, the position vector gives the position of the point with regard to the origin. A displacement vector, on the other hand, gives the change in coordinates from one point to another.

Position Vectors: the Displacement of a Point from the Origin

In general, we say that the point (x_0, y_0, z_0) in space has *position vector* $\vec{r}_0 = (x_0, y_0, z_0)$. See Figure 12.20. Thus, the vector $\vec{r} = (x, y, z)$ is a position vector with its tail at the origin and its tip at the point (x, y, z).

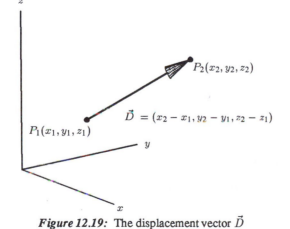

Figure 12.19: The displacement vector $\vec{D}$

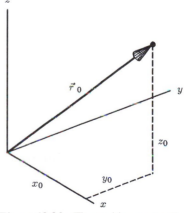

Figure 12.20: The position vector $\vec{r}_0$

How Do We Find the Magnitude of a Vector Given in Components?

For a vector $\vec{v}$ represented by an arrow,

Magnitude of $\vec{v}$ = $\|\vec{v}\|$ = Length of the arrow.

How do we compute the magnitude $\|\vec{v}\|$ if $\vec{v}$ is given in components? For a two-dimensional vector $\vec{v} = (v_1, v_2)$, as in Figure 12.21, the magnitude is given by

$$\|\vec{v}\| = \sqrt{v_1^2 + v_2^2}.$$

Similarly, for a three-dimensional vector, $\vec{v} = (v_1, v_2, v_3)$,

$$\|\vec{v}\| = \sqrt{v_1^2 + v_2^2 + v_3^2},$$

and for $\vec{v} = (v_1, v_2, \ldots, v_n)$, we define

$$\|\vec{v}\| = \sqrt{v_1^2 + v_2^2 + \cdots + v_n^2}.$$

For instance, if $\vec{v} = (3, -4, 5)$, then $\|\vec{v}\| = \sqrt{3^2 + (-4)^2 + 5^2} = \sqrt{50}$.

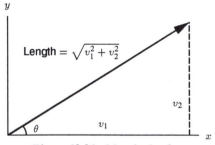

Figure 12.21: Magnitude of a two-dimensional vector

Addition and Subtraction of Vectors Using Components

Suppose we want to know exactly how fast and in what direction the boat in Example 2 on page 72 is moving. If we are given the two vectors in components, we need a way to compute their sum in components.

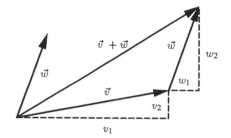

Figure 12.22: Sum $\vec{v} + \vec{w}$ in components

Suppose two vectors $\vec{v}$ and $\vec{w}$ are:

$$\vec{v} = (v_1, v_2),$$
$$\vec{w} = (w_1, w_2).$$

Figure 12.22 shows that the first component of the sum $\vec{v} + \vec{w}$ is the combined displacement $v_1 + w_1$, and the second component is $v_2 + w_2$. Thus,

$$\vec{v} + \vec{w} = (v_1, v_2) + (w_1, w_2) = (v_1 + w_1, v_2 + w_2).$$

For instance, if $\vec{v} = (5, 3)$ and $\vec{w} = (2, 6)$ then $\vec{v} + \vec{w} = (7, 9)$ Similarly, for vectors in 3-space,

$$\vec{v} + \vec{w} = (v_1, v_2, v_3) + (w_1, w_2, w_3) = (v_1 + w_1, v_2 + w_2, v_3 + w_3),$$

and for vectors in n-space,

$$\vec{v} + \vec{w} = (v_1, v_2, \ldots, v_n) + (w_1, w_2, \ldots, w_n) = (v_1 + w_1, v_2 + w_2, \ldots, v_n + w_n).$$

Example 2   Suppose the boat in Example 2 on page 72 is moving with a velocity $\vec{v}$, relative to the water, given by

$$\vec{v} = (8, 0)\text{km}/h,$$

and that the velocity of the current is

$$\vec{c} = (0.6, 0.8)\text{km}/h.$$

(Notice that $\|\vec{c}\| = \sqrt{0.6^2 + 0.8^2} = 1$, the speed given in Example 2.)

(a)   What is the speed of the boat relative to the riverbed?

(b)   What angle does the velocity of the boat relative to the riverbed make with the vector $\vec{v}$? What does this angle tell you in practical terms?

Solution   (a)   We start by finding the velocity, $\vec{w}$, of the boat relative to the riverbed. It is given by

$$\vec{w} = \vec{v} + \vec{c} = (8, 0) + (0.6, 0.8) = (8.6, 0.8).$$

The speed of the boat relative to the riverbed is the magnitude of $\vec{w}$:

$$\text{Speed} = \|\vec{w}\| = \sqrt{8.6^2 + 0.8^2} \approx 8.64 \text{km}/hr.$$

(b) Since the velocity, $\vec{v} = (8, 0)$, is parallel to the x-axis, we want to find the angle between $\vec{w}$ and the x-axis. From Figure 12.23, this angle is

$$\theta = \arctan\left(\frac{0.8}{8.6}\right) \approx 0.093 \text{ radians} \approx 5.3°.$$

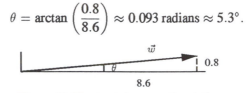

Figure 12.23: Angle between $\vec{v}$ and $\vec{w}$

In practical terms, this angle tells you that if you set your boat parallel to the x-axis at 8 km/hr, the current will take you about 5° off course.

Example 3   The vector $\vec{I}$ represents the number of copies, in thousands, made by each of four copy centers in the month of December, and $\vec{J}$ represents the number of copies made at the same four copy centers during the previous eleven months (the "year-to-date"). If $\vec{I} = (25, 211, 818, 642)$, and $\vec{J} = (331, 3227, 1377, 2570)$, compute $\vec{I} + \vec{J}$. What does this sum represent?

Solution   The sum is

$$\vec{I} + \vec{J} = (25 + 331, 211 + 3227, 818 + 1377, 642 + 2570) = (356, 3438, 2195, 3212).$$

Each term in $\vec{I} + \vec{J}$ represents the sum of the number of copies made in December plus those in the previous eleven months, that is, the total number of copies made during the entire year at each copy center.

Multiplication of a Vector by a Scalar Using Components

Suppose we are given the vector $\vec{v}$ in components:

$$\vec{v} = (v_1, v_2).$$

Then Figure 12.24 shows that the vector $2\vec{v}$ has components given by

$$2\vec{v} = (2v_1, 2v_2).$$

Similarly, $-5\vec{v} = (-5v_1, -5v_2)$. In general, if λ is a scalar, then

$$\lambda\vec{v} = \lambda(v_1, v_2, \ldots, v_n) = (\lambda v_1, \lambda v_2, \ldots, \lambda v_n).$$

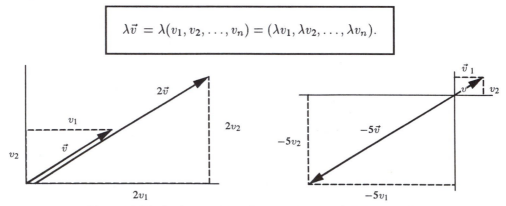

Figure 12.24: Scalar multiples of vectors showing $\vec{v}$, $2\vec{v}$ and $-5\vec{v}$

Example 4

(a) A particle is moving with velocity $\vec{v}$, when it hits a barrier at a right angle and bounces straight back, with its speed reduced by 20%. Express its new velocity in terms of the old one.

(b) The price vector $\vec{p} = (p_1, p_2, p_3)$ represents the prices in dollars of three goods. Write a vector which gives the prices of the same goods in cents.

Solution

(a) The new velocity $= -0.8\vec{v}$, where the negative sign expresses the fact that the new velocity is in the direction opposite to the old.

(b) The prices in cents are $100p_1$, $100p_2$, and $100p_3$ respectively, so the new price vector is $(100p_1, 100p_2, 100p_3) = 100\vec{p}$.

Constructing a Unit Vector

It is often helpful to find a unit vector in the same direction as a given vector $\vec{v}$. Suppose that $||\vec{v}|| = 10$; a unit vector in the same direction as $\vec{v}$ will be $\dfrac{1}{10}\vec{v}$. In general, a unit vector in the direction of any nonzero vector $\vec{v}$ is

$$\vec{u} = \frac{\vec{v}}{||\vec{v}||}.$$

Example 5   Find a unit vector, $\vec{u}$, in the direction of the vector $\vec{v} = (1, 3)$.

Solution   If $\vec{v} = (1, 3)$, then $||\vec{v}|| = \sqrt{1^2 + 3^2} = \sqrt{10}$.

Thus, $\vec{u} = \dfrac{\vec{v}}{\sqrt{10}} = \dfrac{1}{\sqrt{10}}(1, 3) = \left(\dfrac{1}{\sqrt{10}}, \dfrac{3}{\sqrt{10}}\right) = (0.32, 0.95)$.

Writing Vectors Using $\vec{i}, \vec{j}, \vec{k}$

Consider the vector $\vec{v}$ in the plane from the origin to the point $P = (3, 2)$. Figure 12.25 shows that $\vec{v}$ can be written as

$$\vec{v} = 3\vec{i} + 2\vec{j}$$

where 3 is the $\vec{i}$-component and 2 is the $\vec{j}$-component of $\vec{v}$. Similarly, if $\vec{w}$ is a vector from the origin to the point $(2, 5, 7)$ in 3-space, then we can write

$$\vec{w} = 2\vec{i} + 5\vec{j} + 7\vec{k}.$$

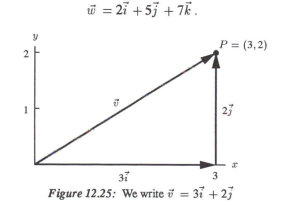

Figure 12.25: We write $\vec{v} = 3\vec{i} + 2\vec{j}$

Any vector in two or three-space can be written in this way using $\vec{i}$, $\vec{j}$, and $\vec{k}$.

If $\vec{v} = (v_1, v_2)$, then $\vec{v}$ has a component of v_1 along the x-axis and v_2 along the y-axis, so we write

$$\vec{v} = v_1\vec{i} + v_2\vec{j}.$$

We say that we have *resolved* $\vec{v}$ into $\vec{i}$ and $\vec{j}$ components. Similarly, if $\vec{w} = (w_1, w_2, w_3)$, then we resolve $\vec{w}$ into three components by writing

$$\vec{w} = w_1\vec{i} + w_2\vec{j} + w_3\vec{k}.$$

How to Resolve a Vector into Components

You may wonder how to find the components of a two-dimensional vector, given its length and direction. Suppose the vector $\vec{v}$ has length v and makes an angle of θ with the x-axis, as in Figure 12.26. If $\vec{v} = v_1\vec{i} + v_2\vec{j}$, the triangle in Figure 12.26 shows

$$v_1 = v \cos\theta$$
$$v_2 = v \sin\theta$$

Thus,

$$\vec{v} = (v \cos\theta)\vec{i} + (v \sin\theta)\vec{j}.$$

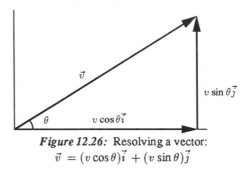

Figure 12.26: Resolving a vector:
$\vec{v} = (v \cos\theta)\vec{i} + (v \sin\theta)\vec{j}$

Example 6   The velocity, $\vec{c}$, of the current in Example 2 is of magnitude 1 and at an angle of about $53°$ to the x-axis. Show that $\vec{c} = 0.6\vec{i} + 0.8\vec{j}$.

Solution   Since $\|\vec{c}\| = 1$,

$$\vec{c} = (1 \cdot \cos 53°)\vec{i} + (1 \cdot \sin 53°)\vec{j} \approx 0.6\vec{i} + 0.8\vec{j}.$$

Choice of Notation for Vectors

We now have two possible notations for a vector

$$\vec{v} = (v_1, v_2) \quad \text{and} \quad \vec{v} = v_1\vec{i} + v_2\vec{j}.$$

Since the first notation can be confused with a point and the second cannot, we will mostly use the second form.

Problems for Section 12.2

By plotting points in the plane, find the components of the vectors in Problems 1–4.

1.   The displacement vector from the point $(-1, 1)$ to the point $(-3, 0)$.

2. The displacement vector from the point $(1.5, 3.2)$ to the point $(0.5, 3.3)$.

3. A vector starting at the point $P = (1, 2)$ and ending at the point $Q = (4, 6)$.

4. A vector starting at the point $Q = (4, 6)$ and ending at the point $P = (1, 2)$.

5. Consider the map in Figure 11.1 on Page 2. If you leave Topeka along the following vectors, does the temperature increase or decrease?
   (a) $\vec{u} = 3\vec{i} + 2\vec{j}$ (b) $\vec{v} = -\vec{i} - \vec{j}$ (c) $\vec{w} = -5\vec{i} - 5\vec{j}$

A cat is sitting on the ground at the point $(1, 4, 0)$ watching a squirrel at the top of a tree. The tree is one unit high and its base is at the point $(2, 4, 0)$. Find the displacement vectors in Problems 6–9.

6. From the origin to the cat.

7. From the bottom of the tree to the squirrel.

8. From the bottom of the tree to the cat.

9. From the cat to the squirrel.

10. Give the components of the two dimensional vectors $\vec{a}$, $\vec{b}$, $\vec{v}$, $\vec{w}$ in Figure 12.27.

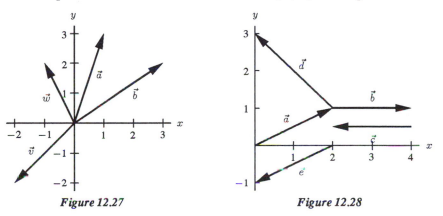

Figure 12.27   Figure 12.28

For Problems 11–15, find the components of the vectors shown in Figure 12.28.

11. $\vec{a}$ 12. $\vec{b}$ 13. $\vec{c}$ 14. $\vec{d}$ 15. $\vec{e}$

16. Find the components and length of each of the vectors shown in Figure 12.29.

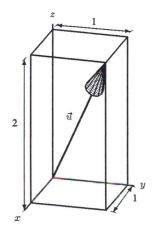
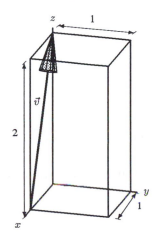

Figure 12.29

17.   Write the components of the vectors in Figure 12.30.

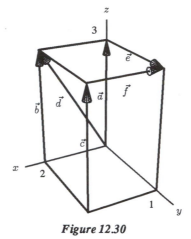

Figure 12.30

For Problems 18–25 perform the indicated computation.

18. $(0.6\vec{i} + 0.2\vec{j} - \vec{k}) + (0.3\vec{i} + 0.3\vec{k})$

19. $(1\vec{i} + 2\vec{j}) + (-3)(2\vec{i} + \vec{j})$

20. $(3\vec{i} - 4\vec{j} + 2\vec{k}) - (6\vec{i} + 8\vec{j} - \vec{k})$

21. $2(0.45\vec{i} - 0.9\vec{j} - 0.01\vec{k}) - 0.5(1.2\vec{i} - 0.1\vec{k})$

22. $(4\vec{i} + 2\vec{j}) - (3\vec{i} - \vec{j})$

23. $(4\vec{i} - 3\vec{j} + 7\vec{k}) - 2(5\vec{i} + \vec{j} - 2\vec{k})$

24. $-4(\vec{i} - 2\vec{j}) - 0.5(\vec{i} - \vec{k})$

25. $\frac{1}{2}(2\vec{i} - \vec{j} + 3\vec{k}) + 3(\vec{i} - \frac{1}{6}\vec{j} + \frac{1}{2}\vec{k})$

For Problems 26–35, perform the following operations on the given vectors:

$$\vec{a} = 2\vec{j} + \vec{k}, \qquad \vec{b} = -3\vec{i} + 5\vec{j} + 4\vec{k} \qquad \vec{c} = \vec{i} + 6\vec{j}$$

$$\vec{x} = -2\vec{i} + 9\vec{j} \qquad \vec{y} = 4\vec{i} - 7\vec{j} \qquad \vec{z} = \vec{i} - 3\vec{j} - \vec{k}$$

26. $\|\vec{z}\|$ 27. $\vec{a} + \vec{z}$ 28. $5\vec{b}$ 29. $2\vec{c} + \vec{x}$ 30. $4\vec{z}$

31. $\|\vec{y}\|$ 32. $5\vec{a} + 2\vec{b}$ 33. $\|\vec{y} - \vec{x}\|$ 34. $3\vec{a}$ 35. $2\vec{a} + 7\vec{b} - 5\vec{z}$

36.   Resolve the following vectors into components:

(a)   The vector in 2-space of length 2 pointing up and to the right at an angle of $\pi/4$ with the x-axis.

(b)   The vector in 3-space of length 1 pointing upward in the xz plane at an angle of $\pi/6$ with the xy plane.

37.   Find a vector that points in the same direction as $\vec{i} - \vec{j} + 2\vec{k}$ but has length 2.

38.   (a)   Find a unit vector starting at the point $P = (1, 2)$ and pointing toward the point $Q = (4, 6)$.

(b)   Find a vector of length 10 pointing in the same direction.

39.   A car is traveling at a speed of 50 km/hr. Assume the positive y-axis is north and the positive x-axis is east. Find the components of its velocity vector (in two-space) if it is traveling in each of the following directions:

(a)   East   (b)   South   (c)   Southeast   (d)   Northwest.

40. Which is traveling faster, a car whose velocity vector is $21\vec{i} + 35\vec{j}$, or a car whose velocity vector is $40\vec{i}$, assuming that the units are the same for both?

41. Shortly after taking off, a plane is climbing in a northwesterly direction through still air at an airspeed of 200 mph, and rising at a rate of 1000 ft/min. Find the components of its velocity vector in a coordinate system in which up is the z-axis, north is the y-axis, and east is the x-axis.

42. Verify that the molecule shown in Figure 12.14 has the property that every atom in the molecule is 2 units away from every other atom.

Find the length of the vectors in Problems 43–46.

43. $\vec{v} = \vec{i} - \vec{j} + 3\vec{k}$

44. $\vec{v} = \vec{i} - \vec{j} + 2\vec{k}$

45. $\vec{v} = 1.2\vec{i} - 3.6\vec{j} + 4.1\vec{k}$

46. $\vec{v} = 7.2\vec{i} - 1.5\vec{j} + 2.1\vec{k}$

47. An object is moving counterclockwise at a constant speed around the circle $x^2 + y^2 = 1$, where x and y are measured in meters. It completes one revolution every minute.
    (a) What is its speed?
    (b) What is its velocity vector 30 seconds after it passes the point $(1, 0)$? Does it change your answer if the object is moving clockwise? Explain.

48. A moving object has velocity vector $50\vec{i} + 20\vec{j}$ in feet per second. Express the velocity in miles per hour.

49. A car is traveling due north at 30 mph approaching a crossroad. On a perpendicular road a police car is traveling west towards the intersection at 40 mph. Suppose that both cars will reach the crossroad in exactly one hour. Find the vector representing the line of sight of the car with respect to the police car.

50. An airplane is heading northeast at an airspeed of 700 km/hr, but there is a wind blowing from the west at 60 km/hr. In what direction does the plane end up flying? What is its speed relative to the ground?

51. An airplane is flying at an airspeed of 600 km/hr in a cross-wind that is blowing from the northeast at a speed of 50 km/hr. In what direction should the plane head to end up going due east?

52. A moving particle hits a barrier at an angle of 60° and bounces off at an angle of 60° in the opposite direction, as shown in Figure 12.31.

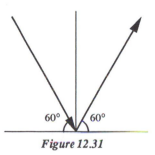

60°   60°

Figure 12.31

If the speed of the particle, v, is reduced by 20 percent on impact, find the velocity of the object after impact.

53. In the game of laser tag, you shoot a harmless laser gun and try to hit a target worn at the waist by other players. Suppose you are standing at the origin of a three dimensional coordinate system and that the xy-plane is the floor. Suppose that waist-high is 3 feet above floor level and that eye level is 5 feet above the floor. Three of your friends are your opponents. One is standing so that his target is 30 feet along the x-axis, the other lying down so that his target is at the point $x = 20$, $y = 15$, and the third lying in ambush so that his target is at a point 8 feet above the point $x = 12$, $y = 30$.

    (a) If you aim carefully with your gun at eye level, find the vector from your gun to each of the three targets.

    (b) If you try a quick draw and shoot approach (shoot from waist height, with your gun one foot to the right of the center of your body as you face along the x-axis), find the vector from your gun to each of the three targets.

12.3   THE DOT PRODUCT

This section introduces the dot product of two vectors, in which we consider the component of one vector in the direction of the other. Before giving the definition, we will look at the example of work.

Definition of Work

In physics, the word "work" has a slightly different meaning from its everyday meaning. In physics, when a force F acts on an object through a distance d, we say the work done, W, by the force is given by

$$W = Fd$$

provided the force and the displacement are in the same direction. For example, if a 1 kg body falls 10 meters under the force of gravity, which is 9.8 newtons, then the work done is

$$W = (9.8 \text{ newtons}) \cdot (10 \text{ meters}) = 98 \text{ joules}.$$

What if the force and the displacement are not in the same direction? If the force and the displacement are perpendicular, no work is done in the technical definition of the word. For example, if you carry a box across the room at the same horizontal height, we say no work is done by gravity because the force of gravity is vertical but the motion is horizontal.

What happens when the force and the displacement are neither parallel nor perpendicular? Then it is the component of the force parallel to the displacement which counts. Suppose the force $\vec{F}$ in Figure 12.32 has magnitude $\|\vec{F}\|$, and the displacement $\vec{d}$ has magnitude $\|\vec{d}\|$. Then we can

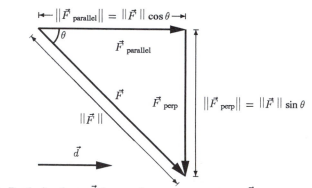

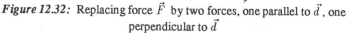

Figure 12.32: Replacing force $\vec{F}$ by two forces, one parallel to $\vec{d}$, one perpendicular to $\vec{d}$

write the force $\vec{F}$ as the sum of two forces, one parallel to $\vec{d}$ of magnitude $\|\vec{F}\| \cos\theta$ and one perpendicular to $\vec{d}$ of magnitude $\|\vec{F}\| \sin\theta$. See Figure 12.32. We define the work done, W, by the component of the force in the direction of the displacement:

$$W = (\|\vec{F}\| \cos\theta)\|\vec{d}\| = \|\vec{F}\| \|\vec{d}\| \cos\theta.$$

Notice that when the vectors are parallel and in the same direction, $\cos\theta = \cos 0 = 1$, and this definition reduces to $W = Fd$. When the vectors are perpendicular, $\cos\theta = \cos\frac{\pi}{2} = 0$, and so $W = 0$.

The Geometric Definition of the Dot Product

The quantity $\|\vec{F}\| \|\vec{d}\| \cos\theta$ is defined to be the *dot product* of $\vec{F}$ and $\vec{d}$, written $\vec{F} \cdot \vec{d}$. In general, for any two vectors $\vec{v}$ and $\vec{w}$ with an angle θ between them:

The **dot product**, $\vec{v} \cdot \vec{w}$, is the scalar defined as follows:

$$\vec{v} \cdot \vec{w} = \|\vec{v}\| \|\vec{w}\| \cos\theta.$$

where θ is the angle between $\vec{v}$ and $\vec{w}$ and $0 \le \theta \le \pi$.

In words:
The dot product is the product of the component of $\vec{v}$ in the direction of $\vec{w}$ with the magnitude of $\vec{w}$. The dot product is also known as the *scalar product*.

Example 1   Suppose the vector $\vec{b}$ is fixed and has length 2; the vector $\vec{a}$ is free to rotate and has length 3. What are the maximum and minimum values of the dot product $\vec{a} \cdot \vec{b}$ as the vector $\vec{a}$ rotates through all possible positions? What positions of $\vec{a}$ and $\vec{b}$ lead to these values?

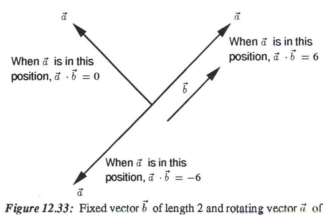

Figure 12.33: Fixed vector $\vec{b}$ of length 2 and rotating vector $\vec{a}$ of length 3

Solution   Since $\vec{a} \cdot \vec{b} = \|\vec{a}\| \|\vec{b}\| \cos\theta = 3 \cdot 2 \cos\theta = 6 \cos\theta$, the maximum value of $\vec{a} \cdot \vec{b}$ occurs when $\vec{a}$ and $\vec{b}$ point in the same direction, so that $\theta = 0$ and $\cos\theta = \cos 0 = 1$. The maximum value of $\vec{a} \cdot \vec{b}$ is therefore 6. The minimum value of $\vec{a} \cdot \vec{b}$ occurs when $\vec{a}$ and $\vec{b}$ point in opposite directions, so that $\theta = \pi$ and $\cos\theta = \cos\pi = -1$, so that the minimum value of $\vec{a} \cdot \vec{b}$ is therefore -6. See Figure 12.33.

Properties of the Dot Product

In order to work with the dot product, we need to know its properties. These we get from the definition:

$$\vec{v} \cdot \vec{w} = \|\vec{v}\|\|\vec{w}\| \cos \theta.$$

> **Properties of the Dot Product: For vectors $\vec{u}$, $\vec{v}$, and $\vec{w}$ and scalar λ,**
>
> 1. $\vec{v} \cdot \vec{w} = \vec{w} \cdot \vec{v}$
> 2. $\vec{v} \cdot (\lambda \vec{w}) = \lambda(\vec{v} \cdot \vec{w}) = (\lambda \vec{v}) \cdot \vec{w}$
> 3. $(\vec{v} + \vec{w}) \cdot \vec{u} = \vec{v} \cdot \vec{u} + \vec{w} \cdot \vec{u}.$

Property 1 says that when calculating the dot product of two vectors, it doesn't matter in which order we take them.

Property 2 says that multiplying one of the vectors by a scalar just multiplies the dot product by the same scalar. If $\lambda > 0$, when one vector is multiplied by λ, the angle between the vectors does not change, but the length of one vector, and hence the dot product, is multiplied by λ. For the case when $\lambda < 0$, see Problem 31 on page 96.

Property 3 is distributivity. To visualize this, think of $\vec{v}$ and $\vec{w}$ as force vectors and $\vec{u}$ as a displacement vector. The work done by $\vec{v} + \vec{w}$ in the direction of $\vec{u}$ is the product of the component of $\vec{v} + \vec{w}$ in the direction of $\vec{u}$ with the magnitude of $\vec{u}$. Figure 12.34 shows that the component of $\vec{v} + \vec{w}$ in the direction of $\vec{u}$ is the sum of the components of $\vec{v}$ and $\vec{w}$. Thus, the work done by $\vec{v} + \vec{w}$ is the sum of the work done by $\vec{v}$ and the work done by $\vec{w}$. Therefore, the dot product of $\vec{v} + \vec{w}$ with $\vec{u}$ is the sum of the dot products of $\vec{v}$ with $\vec{u}$ and of $\vec{w}$ with $\vec{u}$.

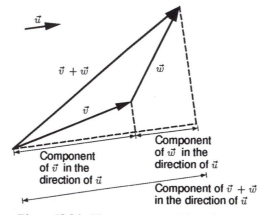

Figure 12.34: The component of $\vec{v} + \vec{w}$ in the direction of $\vec{u}$ is the sum of the components of $\vec{v}$ and $\vec{w}$ in that direction

Perpendicularity, Magnitude, and Dot Products

Two vectors are perpendicular if the angle between them is $\frac{\pi}{2}$ or 90°. Since $\cos \frac{\pi}{2} = 0$, if $\vec{v}$ and $\vec{w}$ are perpendicular, then $\vec{v} \cdot \vec{w} = 0$. Conversely, provided $\vec{v}$ and $\vec{w}$ do not have magnitude 0, the dot product $\vec{v} \cdot \vec{w}$ is only zero when $\theta = \frac{\pi}{2}$ and the vectors are perpendicular. Thus, we have the following result:

Two nonzero vectors $\vec{v}$ and $\vec{w}$ are perpendicular if and only if

$$\vec{v} \cdot \vec{w} = 0$$

For example: $\vec{i} \cdot \vec{j} = 0, \vec{j} \cdot \vec{k} = 0, \vec{i} \cdot \vec{k} = 0$.

If we take the dot product of a vector with itself, then $\theta = 0$ and $\cos \theta = 1$. Thus,

For any vector $\vec{v}$:

$$\vec{v} \cdot \vec{v} = \|\vec{v}\|^2$$

For example: $\vec{i} \cdot \vec{i} = 1, \vec{j} \cdot \vec{j} = 1, \vec{k} \cdot \vec{k} = 1$.

We have seen that vectors can represent quantities that are not geometric. In such cases, it is still possible to have $\vec{v} \cdot \vec{v} = 0$. However, if $\vec{u}$ and $\vec{v}$ are not geometric, we do not describe them as perpendicular. We say that such vectors are *orthogonal*.

Calculation of Dot Product Using Components

How do we compute the dot product of two vectors $\vec{v}$ and $\vec{w}$ given in components? If $\vec{v} = v_1\vec{i} + v_2\vec{j}$ and $\vec{w} = w_1\vec{i} + w_2\vec{j}$ are both vectors in the plane, then using the properties of the dot product:

$$
\begin{aligned}
\vec{v} \cdot \vec{w} &= (v_1\vec{i} + v_2\vec{j}) \cdot (w_1\vec{i} + w_2\vec{j}) \\
&= v_1 w_1 \vec{i} \cdot \vec{i} + v_1 w_2 \vec{i} \cdot \vec{j} + v_2 w_1 \vec{j} \cdot \vec{i} + v_2 w_2 \vec{j} \cdot \vec{j} \\
&= v_1 w_1(1) + v_1 w_2(0) + v_2 w_1(0) + v_2 w_2(1) \\
&= v_1 w_1 + v_2 w_2.
\end{aligned}
$$

For example, if $\vec{v} = 3\vec{i} + 4\vec{j}$ and $\vec{w} = 2\vec{i} - \vec{j}$, then $\vec{v} \cdot \vec{w} = 3 \cdot 2 + 4(-1) = 2$.

A similar calculation gives the following result in 3-dimensions:

If $\vec{v} = v_1\vec{i} + v_2\vec{j} + v_3\vec{k}$ and $\vec{w} = w_1\vec{i} + w_2\vec{j} + w_3\vec{k}$, then

$$\vec{v} \cdot \vec{w} = v_1 w_1 + v_2 w_2 + v_3 w_3$$

By analogy in n-dimensions,

If $\vec{v} = (v_1, v_2, \ldots, v_n)$ and $\vec{w} = (w_1, w_2, \ldots, w_n)$, we define

$$\vec{v} \cdot \vec{w} = v_1 w_1 + v_2 w_2 + \cdots + v_n w_n.$$

Example 2   A video store sells videos, tapes, CD's, and computer games. We define the quantity vector $\vec{q} = (q_1, q_2, q_3, q_4)$, where q_1, q_2, q_3, q_4 denote the quantities sold of each of the goods, and the price vector $\vec{p} = (p_1, p_2, p_3, p_4)$, where p_1, p_2, p_3, p_4 denote the price per unit of each good. What does the dot product $\vec{p} \cdot \vec{q}$ represent?

Solution   The dot product is $\vec{p} \cdot \vec{q} = p_1 q_1 + p_2 q_2 + p_3 q_3 + p_4 q_4$. The quantity $p_1 q_1$ represents the revenue received by the store for the videos, $p_2 q_2$ represents the revenue for the tapes, and so on. The dot product represents the total revenue received by the store for the sale of these four goods.

Example 3   Which pairs from the following list of 3-dimensional vectors are perpendicular to one another?

$$\vec{u} = \vec{i} + \sqrt{3}\vec{k}, \quad \vec{v} = \vec{i} + \sqrt{3}\vec{j}, \quad \vec{w} = \sqrt{3}\vec{i} + \vec{j} - \vec{k}.$$

Solution   Two vectors are perpendicular if and only if their dot product is zero. So we calculate dot products:

$$\vec{v} \cdot \vec{u} = (\vec{i} + \sqrt{3}\vec{j} + 0\vec{k}) \cdot (\vec{i} + 0\vec{j} + \sqrt{3}\vec{k}) = 1 \cdot 1 + \sqrt{3} \cdot 0 + 0 \cdot \sqrt{3} = 1,$$
$$\vec{v} \cdot \vec{w} = (\vec{i} + \sqrt{3}\vec{j} + 0\vec{k}) \cdot (\sqrt{3}\vec{i} + \vec{j} - \vec{k}) = 1 \cdot \sqrt{3} + \sqrt{3} \cdot 1 + 0(-1) = 2\sqrt{3},$$
$$\vec{w} \cdot \vec{u} = (\sqrt{3}\vec{i} + \vec{j} - \vec{k}) \cdot (\vec{i} + 0\vec{j} + \sqrt{3}\vec{k}) = \sqrt{3} \cdot 1 + 1 \cdot 0 + (-1) \cdot \sqrt{3} = 0.$$

So the only two vectors which are perpendicular are $\vec{w}$ and $\vec{u}$.

Projections of Vectors: Components in Arbitrary Directions

When we calculate the work, W, done by a force $\vec{F}$ over a displacement, $\vec{d}$, we use the component of the force in the direction of $\vec{d}$. Using Figure 12.35, we see that

$$\vec{F} = \vec{F}_{\text{perp}} + \vec{F}_{\text{parallel}}.$$

where $\vec{F}_{\text{perp}}$ and $\vec{F}_{\text{parallel}}$ are called the perpendicular and the parallel components of $\vec{F}$ respectively. Now we express $\vec{F}_{\text{perp}}$ and $\vec{F}_{\text{parallel}}$ in terms of dot products. We already know that

$$\left\| \vec{F}_{\text{parallel}} \right\| = \|\vec{F}\| \cos \theta.$$

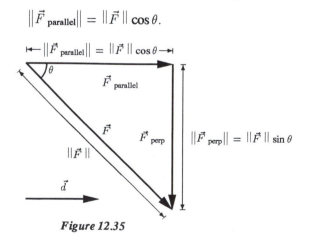

Figure 12.35

If $\vec{d} \neq \vec{0}$, we know that $\vec{d} / \|\vec{d}\|$ is a unit vector parallel to $\vec{d}$, so we have

$$\vec{F}_{\text{parallel}} = \left(\|\vec{F}\| \cos\theta \right) \frac{\vec{d}}{\|\vec{d}\|}$$

Since $\vec{F}_{\text{parallel}}$ is a vector of length $\|\vec{F}\| \cos\theta$ in the direction of $\vec{d}$. Multiplying top and bottom by $\|\vec{d}\|$, we get

$$\vec{F}_{\text{parallel}} = \left(\|\vec{F}\| \cos\theta \right) \frac{\vec{d}\|\vec{d}\|}{\|\vec{d}\|^2} = \left(\frac{\vec{F} \cdot \vec{d}}{\|\vec{d}\|^2} \right) \vec{d}.$$

Thus, we have the following result:

Given two vectors $\vec{F}$ and $\vec{d}$, where $\vec{d} \neq \vec{0}$, we can write, in a unique way,

$$\vec{F} = \vec{F}_{\text{parallel}} + \vec{F}_{\text{perp}}$$

where $\vec{F}_{\text{parallel}}$ is parallel to $\vec{d}$ and $\vec{F}_{\text{perp}}$ is perpendicular to $\vec{d}$. We have

$$\vec{F}_{\text{parallel}} = \left(\frac{\vec{F} \cdot \vec{d}}{\|\vec{d}\|^2} \right) \vec{d}$$

and

$$F_{\text{perp}} = \vec{F} - \vec{F}_{\text{parallel}}$$

Example 4   Figure 12.36 shows the force of the wind on the sail of a sailboat. Find the component of the force in the direction in which the sailboat is traveling.

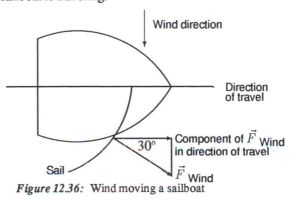

Figure 12.36: Wind moving a sailboat

Solution   The force of the wind on the sail makes an angle of $30°$ with the heading of the boat. Thus, the component of this boat along the heading is $(\cos 30°) \|\vec{F}\| = 0.87\|\vec{F}\|$. Thus, the boat is being pushed forward with about 87% of the total force due to the wind.

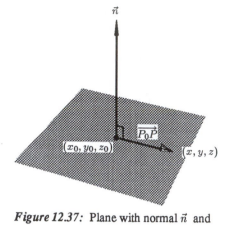

Figure 12.37: Plane with normal $\vec{n}$ and
containing a fixed point (x_0, y_0, z_0)

Using the Dot Product to Write the Equation of a Plane

A *normal vector* to a plane is a vector that is perpendicular to the plane. Using the normal vector
and the dot product, we can write the equation for a plane. Suppose $\vec{n} = a\vec{i} + b\vec{j} + c\vec{k}$ is a vector
normal to a plane. Let (x_0, y_0, z_0) be a fixed point P_0 in the plane, and let (x, y, z) be any other point
P in the plane. Then $\overrightarrow{P_0P} = (x - x_0)\vec{i} + (y - y_0)\vec{j} + (z - z_0)\vec{k}$ is a vector whose head and tail both
lie in the plane.(See Figure 12.37.) Thus, we know that the vectors $\vec{n}$ and $\overrightarrow{P_0P}$ are perpendicular, so
$\vec{n} \cdot \overrightarrow{P_0P} = 0$. Therefore,

$$\vec{n} \cdot \overrightarrow{P_0P} = (a\vec{i} + b\vec{j} + c\vec{k}) \cdot ((x - x_0)\vec{i} + (y - y_0)\vec{j} + (z - z_0)\vec{k}) = 0.$$

Thus, we obtain the following result:

The equation of the plane with normal $\vec{n} = a\vec{i} + b\vec{j} + c\vec{k}$ and containing the point (x_0, y_0, z_0)
is

$$\vec{n} \cdot \overrightarrow{P_0P} = a(x - x_0) + b(y - y_0) + c(z - z_0) = 0.$$

Letting $d = ax_0 + by_0 + cz_0$ (a constant), the equation of the plane can be written in the form

$$ax + by + cz = d.$$

Example 5   Find the equation of the plane perpendicular to $-\vec{i} + 3\vec{j} + 2\vec{k}$ and through the point $(1, 0, 4)$.

Solution   The equation is

$$-(x - 1) + 3(y - 0) + 2(z - 4) = 0,$$

which simplifies to

$$-x + 3y + 2z = 7.$$

Example 6   Find a normal vector to the planes with equation (a) $x - y + 2z = 5$ (b) $z = 0.5x + 1.2y$.

Solution   (a)   Since the components of a normal are the coefficients of x, y, z in the equation of the plane, a normal vector is $\vec{n} = \vec{i} - \vec{j} + 2\vec{k}$.

(b)   Before we can find a normal, we rewrite the equation of the plane in the form

$$0.5x + 1.2y - z = 0,$$

so a normal vector is $\vec{n} = 0.5\vec{i} + 1.2\vec{j} - \vec{k}$.

Problems for Section 12.3

For Problems 1–3, perform the following operations on the given 3-dimensional vectors.

$$\vec{a} = 2\vec{j} + \vec{k} \qquad \vec{b} = -3\vec{i} + 5\vec{j} + 4\vec{k} \qquad \vec{c} = \vec{i} + 6\vec{j} \qquad \vec{y} = 4\vec{i} - 7\vec{j} \qquad \vec{z} = \vec{i} - 3\vec{j} - \vec{k}$$

1. $\vec{c} \cdot \vec{y}$ 2. $\vec{a} \cdot \vec{z}$ 3. $\vec{a} \cdot \vec{b}$

4. Compute the angle between the vectors $\vec{i} + \vec{j} + \vec{k}$ and $\vec{i} - \vec{j} - \vec{k}$.

5. Which of the vectors $(\sqrt{3}\vec{i} + \vec{j})$, $(3\vec{i} + \sqrt{3}\vec{j})$, $(\vec{i} - \sqrt{3}\vec{j})$ are parallel and which are perpendicular?

6. For what values of t are $\vec{u} = t\vec{i} - \vec{j} + \vec{k}$ and $\vec{v} = t\vec{i} + t\vec{j} - 2\vec{k}$ perpendicular? Are there values of t for which $\vec{u}$ and $\vec{v}$ are parallel?

In Problems 7–12, find a normal vector to the given plane.

7. $\pi(x-1) = (1-\pi)(y-z)+\pi$ 8. $2x + y - z = 23$ 9. $1.5x + 3.2y + z = 0$

10. $z = 3x + 4y - 7$ 11. $2(x - z) = 3(x + y)$ 12. $2x + y - z = 5$

In Problems 13–17, find the equation of a plane that satisfies the given conditions.

13. Perpendicular to $\vec{n} = -\vec{i} + 2\vec{j} + \vec{k}$ and through the point $(1, 0, 2)$.

14. Perpendicular to $5\vec{i} + \vec{j} - 2\vec{k}$ and through the point $(0, 1, -1)$.   *What does being ∥ or ⊥ have to do with the final outcome?*

15. Perpendicular to $2\vec{i} - 3\vec{j} + 7\vec{k}$ and through the point $(1, -1, 2)$.

16. Parallel to $2x + 4y - 3z = 1$ and through the point $(1, 0, -1)$.

17. Through the point $(-2, 3, 2)$ and parallel to $3x + y + z = 4$.

18. (a)   Find a vector perpendicular to the plane $z = 2 + 3x - y$.
     (b)   Find a vector parallel to the plane $z = 2 + 3x - y$.

19. Let S be the triangle with vertices $A = (2, 2, 2)$, $B = (4, 2, 1)$, and $C = (2, 3, 1)$.
     (a)   Find the length of the shortest side of S.
     (b)   Find the cosine of the angle BAC at vertex A.

20. The points $(5, 0, 0)$, $(0, -3, 0)$, and $(0, 0, 2)$ form a triangle. Find the lengths of the sides of the triangle and each of its angles.

21. Write $\vec{a} = 3\vec{i} + 2\vec{j} - 6\vec{k}$ as the sum of two vectors, one parallel to $\vec{d} = 2\vec{i} - 4\vec{j} + \vec{k}$ and the other perpendicular to $\vec{d}$.

22. Find the points where the plane

$$z = 5x - 4y + 3$$

crosses each of the coordinate axes. Then find the lengths of the sides and the angles of the triangle formed by these points.

23. Find the angle between the planes

$$5(x - 1) + 3(y + 2) + 2z = 0, \text{ and } x + 3(y - 1) + 2(z + 4) = 0.$$

24. Show that the vectors $(\vec{b} \cdot \vec{c})\vec{a} - (\vec{a} \cdot \vec{c})\vec{b}$ and $\vec{c}$ are perpendicular.

25. (a) If a vector $\vec{v}$ of magnitude v makes an angle α with the positive x-axis, angle β with the positive y-axis, and angle γ with the positive z-axis, then show that $\vec{v} = v \cos \alpha \vec{i} + v \cos \beta \vec{j} + v \cos \gamma \vec{k}$.
    (b) Show that $\cos^2 \alpha + \cos^2 \beta + \cos^2 \gamma = 1$.

26. Find the vector $\vec{v}$ with all the following properties:
    (1) magnitude 10
    (2) angle $45°$ with positive x-axis
    (3) $75°$ with positive y-axis
    (4) positive $\vec{k}$-component

27. Show that if $\vec{u}$ and $\vec{v}$ are two vectors such that

$$\vec{u} \cdot \vec{w} = \vec{v} \cdot \vec{w}$$

for every vector $\vec{w}$, then

$$\vec{u} = \vec{v}.$$

28. A consumption vector of three goods is defined by $\vec{x} = x_1\vec{i} + x_2\vec{j} + x_3\vec{k}$ where x_1 is the quantity of good number one consumed, and so on. Consider a budget constraint represented by the equation $\vec{p} \cdot \vec{x} = k$, where $\vec{p}$ is the price vector of the three goods and k is a constant. Show that the difference between two consumption vectors corresponding to points satisfying the same budget constraint are perpendicular to the price vector $\vec{p}$.

29. A 100-meter dash is run on a track in the direction of the vector $\vec{v} = 2\vec{i} + 5\vec{j}$. The wind velocity $\vec{w}$ is $4\vec{i} + 1\vec{j}$ mph. The rules say that a legal wind speed measured in the direction of the dash must not exceed 3 mph. Will the race results be disqualified due to an illegal wind? Justify your answer.

30. A room is 10 feet high, 12 feet wide, and 20 feet long. Two strings from corners of the ceiling at one end are stretched to the diagonally opposite corners on the floor. What is the cosine of the angle made by the strings as they cross?

31. (a) Using

$$\vec{u} \cdot \vec{v} = \|\vec{u}\| \|\vec{v}\| \cos \theta,$$

show that

$$\vec{u} \cdot (-\vec{v}) = -(\vec{u} \cdot \vec{v}).$$

(Hint: What happens to the angle when you multiply $\vec{v}$ by -1?)

(b) Show for any negative scalar λ that

$$\vec{u} \cdot (\lambda \vec{v}) = \lambda(\vec{u} \cdot \vec{v})$$
$$(\lambda \vec{u}) \cdot \vec{v} = \lambda(\vec{u} \cdot \vec{v}).$$

32. Recall that the dot product can be written as

$$\vec{v} \cdot \vec{w} = \|\vec{v}\| \, \|\vec{w}\| \cos \theta.$$

The result of Problem 33 enables us to use this relationship to define the angle between two vectors in n-dimensions. If $\vec{v}$, $\vec{w}$ are n-vectors, we say

$$\vec{v} \cdot \vec{w} = v_1 w_1 + v_2 w_2 + \cdots + v_n w_n$$

and we define the angle θ by

$$\cos \theta = \frac{\vec{v} \cdot \vec{w}}{\|\vec{v}\| \, \|\vec{w}\|} \quad \text{provided} \ \|\vec{v}\|, \|\vec{w}\| \neq 0.$$

We now use this idea of angle to measure how close two populations are to one another genetically. A table follows showing the relative frequencies of four alleles (variants of a gene) in four populations.

Allele	Eskimo	Bantu	English	Korean
A_1	0.29	0.10	0.20	0.22
A_2	0.00	0.08	0.06	0.00
B	0.03	0.12	0.06	0.20
O	0.67	0.69	0.66	0.57

Let $\vec{a}_1$ be the 4-vector showing relative frequencies in the Eskimo population;

$\vec{a}_2$ be the 4-vector showing relative frequencies in the Bantu population;

$\vec{a}_3$ be the 4-vector showing relative frequencies in the English population;

$\vec{a}_4$ be the 4-vector showing relative frequencies in the Korean population.

The genetic distance between two populations is defined as the angle between the corresponding vectors. Using this definition, is the English population closer genetically to the Bantus or to the Koreans? Explain.[2]

33. Suppose that $\vec{v}$ and $\vec{w}$ are n-dimensional vectors. Consider the following function of t:

$$q(t) = (\vec{v} + t\vec{w}) \cdot (\vec{v} + t\vec{w})$$

(a) Explain why $q(t) \geq 0$ for all real t.

(b) Expand $q(t)$ as a quadratic polynomial in t.

(c) Using the discriminant of the quadratic, show that,

$$|\vec{v} \cdot \vec{w}| \leq \|\vec{v}\| \, \|\vec{w}\|.$$

---

[2]Adapted from Cavalli-Sforza and Edwards, "Models and Estimation Procedures," Am J. Hum. Genet., Vol. 19 (1967), pp. 223-57.

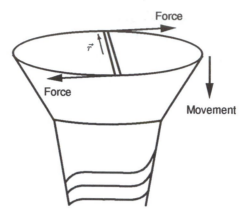

Figure 12.38: Motion of screw is perpendicular to both $\vec{F}$ and $\vec{r}$

12.4 THE CROSS PRODUCT

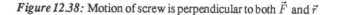

Consider the difference between hammering a nail and twisting a screw into the top of a table. Both involve applying a force, but in two different directions. In the case of the hammer, the force and the movement of the nail are both vertical. In the case of the screw, the force exerted is horizontal but the movement of the screw is perpendicular to both the force, $\vec{F}$, applied to the screw (see Figure 12.38) and to the radius vector $\vec{r}$. In this section we show how to calculate a vector which is perpendicular to two given vectors. The cross product is such a vector.

The Geometric Definition of the Cross Product

We will define the cross product, $\vec{v} \times \vec{w}$, of two vectors $\vec{v}$ and $\vec{w}$ to be perpendicular to both of them. First we will specify the direction. If, for example, $\vec{v}$ and $\vec{w}$ were in the xy-plane, there are two possible normal directions—vertically upward and vertically downward. To choose the correct direction we use the following rule.

The right-hand rule: If you place $\vec{v}$ and $\vec{w}$ so that their tails coincide and curl the fingers of your right hand in an arc from $\vec{v}$ to $\vec{w}$, your thumb points in the direction of $\vec{v} \times \vec{w}$.

Now we define the cross product as follows.

The **cross product**, $\vec{v} \times \vec{w}$, is the vector defined by

$$\vec{v} \times \vec{w} = \|\vec{v}\| \, \|\vec{w}\| \sin\theta \, \vec{n}$$

where $0 \leq \theta \leq \pi$ is the angle between $\vec{v}$ and $\vec{w}$ and $\vec{n}$ is the unit vector perpendicular to $\vec{v}$ and $\vec{w}$ with direction determined by the right-hand rule. Notice that the magnitude of $\vec{v} \times \vec{w}$ is given by

$$\|\vec{v} \times \vec{w}\| = \|\vec{v}\| \, \|\vec{w}\| \sin\theta$$

Note that the cross product of $\vec{v}$ and $\vec{w}$, also called the *vector product*, is a vector, while the dot product $\vec{v} \cdot \vec{w}$ is a scalar. Unlike the dot product, the cross product is only defined for three-dimensional vectors.

Example 1   Find $\vec{i} \times \vec{j}$.

Solution   The vectors $\vec{i}$ and $\vec{j}$ both have magnitude 1 and the angle between them is $\pi/2$. By the right-hand rule, the direction of $\vec{i} \times \vec{j}$ is $\vec{k}$, so we have

$$\vec{i} \times \vec{j} = \left(\|\vec{i}\| \|\vec{j}\| \sin \frac{\pi}{2} \right) \vec{k} = \vec{k}$$

Example 2   For any vector $\vec{v}$, find $\vec{v} \times \vec{v}$.

Solution   Since the angle between $\vec{v}$ and $\vec{v}$ is 0 and $\sin 0 = 0$, we have

$$\vec{v} \times \vec{v} = \vec{0}.$$

Example 3   Suppose $\vec{a}$ is a fixed vector of length 3 in the direction of the positive x-axis and the vector $\vec{b}$ of length 2 is free to rotate in the xy-plane. What are the maximum and minimum values of the magnitude of $\vec{a} \times \vec{b}$? In what direction is $\vec{a} \times \vec{b}$ as $\vec{b}$ rotates?

Solution   The magnitude of $\vec{a} \times \vec{b}$ is given by

$$\|\vec{a} \times \vec{b}\| = \|\vec{a}\| \|\vec{b}\| \sin \theta = 3 \cdot 2 \sin \theta = 6 \sin \theta.$$

Therefore, the maximum possible value of $\|\vec{a} \times \vec{b}\|$ occurs when $\sin \theta = 1$ so $\theta = \pi/2$; that is, when $\vec{a}$ and $\vec{b}$ are perpendicular. The maximum value of $\|\vec{a} \times \vec{b}\|$ is 6. The minimum value of $\|\vec{a} \times \vec{b}\|$ occurs when $\sin \theta = 0$ so $\theta = 0$ or π; that is, when $\vec{a}$ and $\vec{b}$ are parallel. Then $\|\vec{a} \times \vec{b}\|$ is 0.

The direction of $\vec{a} \times \vec{b}$ will be along the positive z-axis when $\vec{b}$ is in the first or second quadrant and along the negative z-axis when $\vec{b}$ is in the third or fourth quadrant.

Properties of the Cross Product

The right-hand rule shows that $\vec{v} \times \vec{w}$ and $\vec{w} \times \vec{v}$ point in opposite directions. The magnitudes of $\vec{v} \times \vec{w}$ and $\vec{w} \times \vec{v}$ are the same, so $\vec{w} \times \vec{v} = -\vec{v} \times \vec{w}$. (See Figure 12.39).

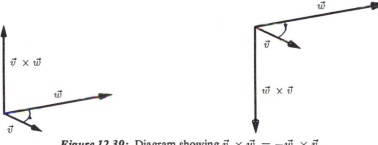

Figure 12.39: Diagram showing $\vec{v} \times \vec{w} = -\vec{w} \times \vec{v}$

This explains the first of the properties listed below. The other two are derived in Problems 19 and 21 at the end of this section.

Properties of the Cross Product: For vectors $\vec{u}$, $\vec{v}$, $\vec{w}$ and scalar λ

1. $\vec{w} \times \vec{v} = -\vec{v} \times \vec{w}$
2. $(\lambda \vec{v}) \times \vec{w} = \lambda(\vec{v} \times \vec{w}) = \vec{v} \times (\lambda \vec{w})$
3. $\vec{u} \times (\vec{v} + \vec{w}) = \vec{u} \times \vec{v} + \vec{u} \times \vec{w}$

Calculation of the Cross Product Using Components

In order to use the cross product in practice, we need to be able to calculate $\vec{v} \times \vec{w}$ when $\vec{v}$ and $\vec{w}$ are given in components. We have already shown that $\vec{i} \times \vec{j} = \vec{k}$ and $\vec{i} \times \vec{i} = \vec{0}$. By similar arguments, we can show $\vec{j} \times \vec{i} = -\vec{k}$, $\vec{k} \times \vec{i} = \vec{j}$, $\vec{j} \times \vec{k} = \vec{i}$, $\vec{j} \times \vec{j} = \vec{0}$, and so on. Thus, we have

$$
\begin{aligned}
\vec{v} \times \vec{w} &= (v_1\vec{i} + v_2\vec{j} + v_3\vec{k}) \times (w_1\vec{i} + w_2\vec{j} + w_3\vec{k}) \\
&= v_1 w_1 \vec{i} \times \vec{i} + v_1 w_2 \vec{i} \times \vec{j} + v_1 w_3 \vec{i} \times \vec{k} \\
&\quad + v_2 w_1 \vec{j} \times \vec{i} + v_2 w_2 \vec{j} \times \vec{j} + v_2 w_3 \vec{j} \times \vec{k} \\
&\quad + v_3 w_1 \vec{k} \times \vec{i} + v_3 w_2 \vec{k} \times \vec{j} + v_3 w_3 \vec{k} \times \vec{k} \\
&= \vec{0} + v_1 w_2 \vec{k} + v_1 w_3(-\vec{j}) + v_2 w_1(-\vec{k}) + \vec{0} + v_2 w_3 \vec{i} + v_3 w_1 \vec{j} + v_3 w_2(-\vec{i}) + \vec{0} \\
&= (v_2 w_3 - v_3 w_2)\vec{i} + (v_3 w_1 - v_1 w_3)\vec{j} + (v_1 w_2 - v_2 w_1)\vec{k}
\end{aligned}
$$

This expression can be more easily remembered by writing it as a determinant:

$$
\vec{v} \times \vec{w} = \begin{vmatrix} \vec{i} & \vec{j} & \vec{k} \\ v_1 & v_2 & v_3 \\ w_1 & w_2 & w_3 \end{vmatrix} = (v_2 w_3 - v_3 w_2)\vec{i} - (v_1 w_3 - v_3 w_1)\vec{j} + (v_1 w_2 - v_2 w_1)\vec{k} .
$$

The properties of determinants are outlined in Appendix B on page 362.

Example 4 Find the cross product of $\vec{v} = 2\vec{i} + \vec{j} - 2\vec{k}$ and $\vec{w} = 3\vec{i} + \vec{k}$ and check that it is perpendicular to both $\vec{v}$ and $\vec{w}$.

Solution

$$
\vec{v} \times \vec{w} = \begin{vmatrix} \vec{i} & \vec{j} & \vec{k} \\ 2 & 1 & -2 \\ 3 & 0 & 1 \end{vmatrix} .
$$

Writing $\vec{v} \times \vec{w}$ as a determinant and expanding it into three two by two determinants, we have

$$
\begin{aligned}
\vec{v} \times \vec{w} &= \begin{vmatrix} \vec{i} & \vec{j} & \vec{k} \\ 2 & 1 & -2 \\ 3 & 0 & 1 \end{vmatrix} = \vec{i} \begin{vmatrix} 1 & -2 \\ 0 & 1 \end{vmatrix} - \vec{j} \begin{vmatrix} 2 & -2 \\ 3 & 1 \end{vmatrix} + \vec{k} \begin{vmatrix} 2 & 1 \\ 3 & 0 \end{vmatrix} \\
&= \vec{i}\,((1)(1) - 0(-2)) - \vec{j}\,(2(1) - 3(-2)) + \vec{k}\,(2(0) - 3(1)) \\
&= \vec{i} - 8\vec{j} - 3\vec{k} .
\end{aligned}
$$

To check that $\vec{v} \times \vec{w}$ is perpendicular to $\vec{v}$, we compute the dot product:

$$\vec{v} \cdot (\vec{v} \times \vec{w}) = (2\vec{i} + \vec{j} - 2\vec{k}) \cdot (\vec{i} - 8\vec{j} - 3\vec{k}) = 2 - 8 + 6 = 0.$$

Similarly

$$\vec{w} \cdot (\vec{v} \times \vec{w}) = (3\vec{i} + 0\vec{j} + \vec{k}) \cdot (\vec{i} - 8\vec{j} - 3\vec{k}) = 3 + 0 - 3 = 0.$$

The Equation of a Plane Through Three Points

The equation of a plane is determined by a point P_0 on it and a normal vector $\vec{n}$. If $P_0 = (x_0, y_0, z_0)$ and $\vec{n} = a\vec{i} + b\vec{j} + c\vec{k}$, we have seen that the equation is

$$a(x - x_0) + b(y - y_0) + c(z - z_0) = 0$$

However a plane can also be determined by three points on it (provided they do not lie on a line). In that case we find the equation of the plane by first determining two vectors in the plane and then finding the normal using the cross product, as in the following example.

Example 5 Find the equation of the plane containing the points $P = (1, 3, 0)$ and $Q = (3, 4, -3)$ and $R = (3, 6, 2)$.

Solution Since the points P and Q are in the plane, the following vector, $\vec{v}$, is in the plane:

$$\vec{v} = \overrightarrow{PQ} = (3 - 1)\vec{i} + (4 - 3)\vec{j} + (-3 - 0)\vec{k} = 2\vec{i} + \vec{j} - 3\vec{k}.$$

The vector $\vec{w} = \overrightarrow{PR}$ is also in the plane:

$$\vec{w} = \overrightarrow{PR} = (3 - 1)\vec{i} + (6 - 3)\vec{j} + (2 - 0)\vec{k} = 2\vec{i} + 3\vec{j} + 2\vec{k}.$$

Thus, a normal vector, $\vec{n}$, to the plane is given by

$$\vec{n} = \vec{v} \times \vec{w} = \begin{vmatrix} \vec{i} & \vec{j} & \vec{k} \\ 2 & 1 & -3 \\ 2 & 3 & 2 \end{vmatrix} = 11\vec{i} - 10\vec{j} + 4\vec{k}.$$

Since the point $(1, 3, 0)$ is on the plane, the equation of the plane is

$$11(x - 1) - 10(y - 3) + 4(z - 0) = 0,$$

which simplifies to

$$11x - 10y + 4z = -19.$$

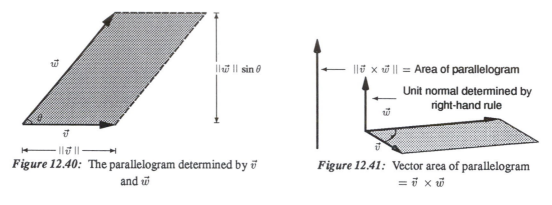

Figure 12.40: The parallelogram determined by $\vec{v}$ and $\vec{w}$

Figure 12.41: Vector area of parallelogram $= \vec{v} \times \vec{w}$

Area and the Cross Product

We now give an interpretation of the cross product, $\vec{v} \times \vec{w}$. Consider the parallelogram formed by the vectors $\vec{v}$ and $\vec{w}$, shown in Figure 12.40. The area of the parallelogram is given by

$$\text{Area} = \text{Base} \cdot \text{Height} = \|\vec{v}\| \cdot \|\vec{w}\| \sin\theta.$$

Thus, as shown in Figure 12.41, we have

$$\text{Area of the parallelogram} = \|\vec{v} \times \vec{w}\|$$

We conclude that the cross product can be visualized as follows:

The cross product $\vec{v} \times \vec{w}$ is a vector with
- Magnitude equal to the area of the parallelogram determined by $\vec{v}$ and $\vec{w}$.
- Direction perpendicular to $\vec{v}$ and $\vec{w}$ and determined by the right-hand rule.

Example 6   Find the area of the parallelogram determined by $\vec{v} = 2\vec{i} + \vec{j} - 3\vec{k}$ and $\vec{w} = \vec{i} + 3\vec{j} + 2\vec{k}$.

Solution   We calculate the cross product:

$$\vec{v} \times \vec{w} = \begin{vmatrix} \vec{i} & \vec{j} & \vec{k} \\ 2 & 1 & -3 \\ 1 & 3 & 2 \end{vmatrix} = (2+9)\vec{i} + (-3-4)\vec{j} + (6-1)\vec{k}$$

$$= 11\vec{i} - 7\vec{j} + 5\vec{k}.$$

The area of the parallelogram determined by $\vec{v}$ and $\vec{w}$ is the magnitude of the vector $\vec{v} \times \vec{w}$:

$$\text{Area} = \|\vec{v} \times \vec{w}\| = \sqrt{11^2 + (-7)^2 + 5^2} = \sqrt{195}.$$

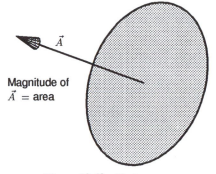

Figure 12.42: Vector area

Definition of Vector Area

Suppose that we want to represent both the magnitude of an area and its orientation in space. For example, consider a solar collector panel. The amount of energy that this panel absorbs depends both on its area and on the angle at which the sun's rays strike it.

If the area we want to represent is flat, then a vector perpendicular to it fixes the orientation of the area in space. (See Figure 12.42.) Thus, we make the following definition:

A flat region in space is represented by a **vector area** $\vec{A}$ such that
- Magnitude of $\vec{A}$ is the magnitude of the area
- Direction of $\vec{A}$ is normal to the area

Since there are two possible directions normal to the surface, or two possible *orientations*, there are two possible vector areas.

Example 7   Find a vector area of the following regions:
  (a)   A rectangle with vertices $(0, 0, 0)$, $(4, 0, 0)$, $(0, 3, 0)$ and $(4, 3, 0)$, oriented upward.
  (b)   The region inside the circle $x^2 + z^2 = 1$ in the plane $y = 4$.

Solution   (a)   The rectangle has sides of 3 and 4, so its area is 12. Since the orientation is upward, the vector area is given by
$$\vec{A} = 12\vec{k}.$$

(b)   The circle has radius 1, so the disc has area $\pi(1)^2 = \pi$. Since the vector area is perpendicular to the disc, it is parallel to the y-axis. Since the orientation has not been specified, there are two possible vector areas
$$\vec{A} = \pi\vec{j} \quad \text{and} \quad \vec{A} = -\pi\vec{j}$$

Example 8 Show that $\left| \vec{a} \cdot (\vec{b} \times \vec{c}) \right|$ is the volume of the parallelopiped with sides formed by $\vec{a}$, $\vec{b}$, and $\vec{c}$.

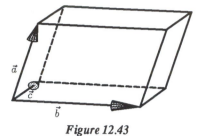

Figure 12.43

Solution If we think of the bases as being formed by the vectors $\vec{b}$ and $\vec{c}$, we have

$$\text{Area of base} = \|\vec{b} \times \vec{c}\|.$$

The vectors $\vec{a}$, $\vec{b}$, and $\vec{c}$ can be arranged either as shown in Figure 12.44 (in which $\vec{b} \times \vec{c}$ and $\vec{a}$ have an angle of less than $\pi/2$ between them) or as in Figure 12.45 (where the angle is more than $\pi/2$). In either case,

$$\text{Height} = \|\vec{a}\| \cos\theta.$$

In the case shown in Figure 12.44 (in which $\vec{a}$, $\vec{b}$, and $\vec{c}$ are called *right-handed*), the triple product, $\vec{a} \cdot (\vec{b} \times \vec{c})$ is positive because the angle, θ, between $\vec{a}$ and $\vec{b} \times \vec{c}$ is less than $\pi/2$. Thus, in this case

$$\text{Volume} = \text{Base} \cdot \text{Height} = \|\vec{b} \times \vec{c}\| \cdot \|\vec{a}\| \cos\theta = (\vec{b} \times \vec{c}) \cdot \vec{a}.$$

For the case shown in Figure 12.45 (in which $\vec{a}$, $\vec{b}$, and $\vec{c}$ are called a *left-handed set*), the triple product $\vec{a} \cdot (\vec{b} \times \vec{c})$ is negative because the angle θ, between $\vec{a}$ and $\vec{b} \times \vec{c}$ is more than $\pi/2$. Thus, in this case we have

$$\begin{aligned}
\text{Volume} = \text{Base} \cdot \text{Height} &= \|\vec{b} \times \vec{c}\| \cdot \|\vec{a}\| \cos\theta \\
&= -\|\vec{b} \times \vec{c}\| \cdot \|\vec{a}\| \cos(\pi - \theta) \\
&= -(\vec{b} \times \vec{c}) \cdot \vec{a} \\
&= \left| (\vec{b} \times \vec{c}) \cdot \vec{a} \right|.
\end{aligned}$$

Therefore, in both cases we have

$$\text{Volume} = \left| \vec{a} \cdot (\vec{b} \times \vec{c}) \right|.$$

Notice that the volume is given by the absolute value of the determinant: (see Problem 21 on page 21)

$$\begin{vmatrix} a_1 & a_2 & a_3 \\ b_1 & b_2 & b_3 \\ c_1 & c_2 & c_3 \end{vmatrix}.$$

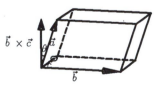

Figure 12.44: The vectors $\vec{a}$, $\vec{b}$, $\vec{c}$ are right-handed

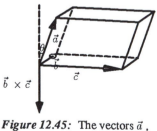

Figure 12.45: The vectors $\vec{a}$, $\vec{b}$, $\vec{c}$ are left-handed

Problems for Section 12.4

In Problems 1–4, find $\vec{a} \times \vec{b}$.

1. $\vec{a} = \vec{i} + \vec{k}$ and $\vec{b} = \vec{i} + \vec{j}$.

2. $\vec{a} = -\vec{i}$ and $\vec{b} = \vec{j} + \vec{k}$.

3. $\vec{a} = \vec{i} + \vec{j} + \vec{k}$ and $\vec{b} = \vec{i} + \vec{j} + -\vec{k}$.

4. $\vec{a} = 2\vec{i} - 3\vec{j} + \vec{k}$ and $\vec{b} = \vec{i} + 2\vec{j} - \vec{k}$.

5. Given $\vec{a} = 3\vec{i} + \vec{j} - \vec{k}$ and $\vec{b} = \vec{i} - 4\vec{j} + 2\vec{k}$, find $\vec{a} \times \vec{b}$ and check that $\vec{a} \times \vec{b}$ is perpendicular to both $\vec{a}$ and $\vec{b}$.

6. If $\vec{v} \times \vec{w} = 2\vec{i} - 3\vec{j} + 5\vec{k}$, and $\vec{v} \cdot \vec{w} = 3$, find $\tan \theta$ where θ is the angle between $\vec{v}$ and $\vec{w}$.

Find an equation of the plane through the points in Problems 7–8.

7. $(1, 0, 0), (0, 1, 0), (0, 0, 1)$.

8. $(3, 4, 2), (-2, 1, 0), (0, 2, 1)$.

In Problems 9–11, give the vector area of the following regions.

9. The rectangle with vertices $(0, 0, 0), (0, 1, 0), (2, 1, 0)$, and $(2, 0, 0)$, oriented so that it faces downward.

10. The circle of radius 2 in the yz-plane, facing in the direction of the positive x-axis.

11. The triangle ABC, where $A = (1, 2, 3)$, $B = (3, 1, 2)$, and $C = (2, 1, 3)$.

12. Find a vector parallel to the intersection of the planes $2x - 3y + 5z = 2$ and $4x + y - 3z = 7$.

13. Find the equation of the plane through the origin which is perpendicular to the line of intersection of the planes in Problem 12.

14. Find the equation of the plane through $P(4, 5, 6)$ which is perpendicular to the line of intersection of the planes in Problem 12.

15. Use the formula for the cross product in components to check that $\vec{a} \times (\vec{b} + \vec{c}) = (\vec{a} \times \vec{b}) + (\vec{a} \times \vec{c})$.

16. In this problem, we approach the idea of finding the cross product algebraically. Let $\vec{a} = a_1\vec{i} + a_2\vec{j} + a_3\vec{k}$ and $\vec{b} = b_1\vec{i} + b_2\vec{j} + b_3\vec{k}$. We seek a vector $\vec{v} = x\vec{i} + y\vec{j} + z\vec{k}$ which is perpendicular to both $\vec{a}$ and $\vec{b}$. Use this requirement to construct two equations for x, y, and z. Eliminate x and solve for y in terms of z. Then eliminate y and solve for x in terms of z. Since z can be any value whatsoever (the direction of $\vec{v}$ is unaffected), select the value for z which eliminates the denominator in the equation you obtained. How does the resulting expression for $\vec{v}$ compare to the formula we derived on page 100?

17. Suppose $\vec{a}$ and $\vec{b}$ are vectors in the xy-plane, such that $\vec{a} = a_1\vec{i} + a_2\vec{j}$ and $\vec{b} = b_1\vec{i} + b_2\vec{j}$ with $0 < a_2 < a_1$ and $0 < b_1 < b_2$.

   (a) Sketch $\vec{a}$ and $\vec{b}$ and the vector $\vec{c} = -a_2\vec{i} + a_1\vec{j}$. Shade the parallelogram formed by $\vec{a}$ and $\vec{b}$.

   (b) What is the relation between $\vec{a}$ and $\vec{c}$? [Hint: Find $\vec{c} \cdot \vec{a}$ and $\vec{c} \cdot \vec{c}$.]

   (c) Find $\vec{c} \cdot \vec{b}$.

   (d) Explain why $\vec{c} \cdot \vec{b}$ gives the area of the parallelogram formed by $\vec{a}$ and $\vec{b}$.

   (e) Verify that in this case $\vec{a} \times \vec{b} = (a_1 b_2 - a_2 b_1)\vec{k}$.

18. If $\vec{a} + \vec{b} + \vec{c} = \vec{0}$, show that

$$\vec{a} \times \vec{b} = \vec{b} \times \vec{c} = \vec{c} \times \vec{a}.$$

Geometrically what does this imply about $\vec{a}$, $\vec{b}$, and $\vec{c}$?

19. If $\vec{v}$ and $\vec{w}$ are nonzero vectors, use the geometric definition of the cross product to explain why

$$(\lambda \vec{v}) \times \vec{w} = \lambda(\vec{v} \times \vec{w}) = \vec{v} \times (\lambda \vec{w})$$

Consider the cases $\lambda > 0$, $\lambda = 0$, and $\lambda < 0$ separately.

20. Use Example 8 on page 104 to show that $\vec{a} \cdot (\vec{b} \times \vec{c}) = (\vec{a} \times \vec{b}) \cdot \vec{c}$ for any vectors $\vec{a}$, $\vec{b}$, and $\vec{c}$.

21. Use the result of Problem 20 to show that the cross product is a linear operator, that is, show that

$$(\vec{a} + \vec{b}) \times \vec{c} = (\vec{a} \times \vec{c}) + (\vec{b} \times \vec{c}).$$

First, use the linearity of the dot product to show that for any vector $\vec{d}$,

$$[(\vec{a} + \vec{b}) \times \vec{c}] \cdot \vec{d} = [(\vec{a} \times \vec{c}) + (\vec{b} \times \vec{c})] \cdot \vec{d}.$$

Next, show that for any vector $\vec{d}$,

$$[((\vec{a} + \vec{b}) \times \vec{c}) - (\vec{a} \times \vec{c}) - (\vec{b} \times \vec{c})] \cdot \vec{d} = 0.$$

Finally, since the above equation is true for all vectors $\vec{d}$, explain why you can conclude that

$$(\vec{a} + \vec{b}) \times \vec{c} = (\vec{a} \times \vec{c}) + (\vec{b} \times \vec{c}).$$

REVIEW PROBLEMS FOR CHAPTER TWELVE

Use the definition to compute the cross products in Problems 1–2.

1. $2\vec{i} \times (\vec{i} + \vec{j})$

2. $(\vec{i} + \vec{j}) \times (\vec{i} - \vec{j})$

Compute the cross products for Problem 3–4.

3. $[(\vec{i} + \vec{j}) \times \vec{i}] \times \vec{j}$

4. $(\vec{i} + \vec{j}) \times [\vec{i} \times \vec{j}]$

5. True or false? $\vec{a} \times \vec{b} = -(\vec{b} \times \vec{a})$ for all $\vec{a}$ and $\vec{b}$. Explain your answer.

6. Find the area of the triangle with vertices $(-2, 2, 0)$, $(1, 3, -1)$, and $(-4, 2, 1)$ using the cross product.

7. An object is attached by an inelastic string to a fixed point and rotates 30 times per minute in a horizontal plane. Show that the speed of the object is constant but the velocity is not. What does this imply about the acceleration?

8. A man wishes to row across a river which is flowing at 4 mph from the east.
   (a) If he wishes to travel the shortest possible distance from the north and can row at 5 mph, which direction should he steer?
   (b) If there is a wind of 10 mph from the southwest, how will this affect his steering?

9. A large ship is being towed by two tugs. The larger tug exerts a force which is 25% greater than the smaller tug and at an angle of 30 degrees. Which direction must the smaller tug pull to ensure that the ship travels due east?

10. An airport is at the point $(200, 10, 0)$ and an approaching plane is at the point $(550, 60, 4)$. Assume that the xy-plane is horizontal, with the x-axis pointing eastward and the y-axis pointing northward. Also assume that the z-axis is upward and that all distances are measured in miles. The plane flies due west at a constant altitude at a speed of 500 mph for half an hour. It then descends at 200 mph, heading straight for the airport.
    (a) Find a vector representation for the velocity of the plane while it is flying at constant altitude.
    (b) Find the coordinates of the point at which the plane starts to descend.
    (c) Find a vector representing the velocity of the plane when it is descending.

11. Is the collection of populations of each of the 50 states a vector or scalar quantity?

12. Find the equation of the plane through the origin which is parallel to $z = 4x - 3y + 8$.

13. Find $||\vec{z}||$ where $\vec{z} = \vec{i} - 3\vec{j} - \vec{k}$.

14. Find a vector normal to the plane $4(x - 1) + 6(z + 3) = 12$.

15. Consider the plane $5x - y + 7z = 21$.
    (a) Find a point on the x-axis on this plane.
    (b) Find two other points on the plane.
    (c) Find a vector perpendicular to the plane.
    (d) Find a vector parallel to the plane.

16. Given the points $P = (1, 2, 3), Q = (3, 5, 7)$, and $R = (2, 5, 3)$, find:
    (a) A unit vector perpendicular to a plane containing P, Q, R.
    (b) The angle between PQ and PR.
    (c) The area of triangle PQR.
    (d) The distance from R to the line through P and Q.

17. Find the equation of the plane through the origin which is parallel to $z = 4x - 3y + 8$.

18. Find a vector parallel to the line of intersection of the planes given by the equations $2x - 3y + 5z = 2$ and $4x + y - 3z = 7$.

19. Find the equation of the plane through the origin which is perpendicular to the line of intersection of the planes in Problem 18.

20. Find the equation of the plane through the point $(4, 5, 6)$ and perpendicular to the line of intersection of the planes in Problem 18.

21. If $\vec{a} = a_1\vec{i} + a_2\vec{j} + a_3\vec{k}$, $\vec{b} = b_1\vec{i} + b_2\vec{j} + b_3\vec{k}$ and $\vec{c} = c_1\vec{i} + c_2\vec{j} + c_3\vec{k}$ are any three vectors in space, show that

$$\vec{a} \cdot (\vec{b} \times \vec{c}) = \begin{vmatrix} a_1 & a_2 & a_3 \\ b_1 & b_2 & b_3 \\ c_1 & c_2 & c_3 \end{vmatrix}.$$

22. Find all vectors $\vec{v}$ in the plane such that $\|\vec{v}\| = 1$ and $\|\vec{v} + \vec{i}\| = 1$, where $\vec{i}$ is the vector $(1, 0, 0)$.

23. Find all vectors $\vec{w}$ in 3 space such that $\|\vec{w}\| = 1$ and $\|\vec{w} + \vec{i}\| = 1$. Describe this set geometrically.

24. The price vector of beans, rice, and tofu is $0.30\vec{i} + 0.20\vec{j} + 0.50\vec{k}$ in dollars per pound. Express it in dollars per ounce.

25. Three people are trying to hold a ferocious lion still for the veterinarian. The lion, in the center, is wearing a collar with three ropes attached to it and each man has hold of a rope. Charlie is pulling in the direction 62° West of North with a force of 350 pounds and Sam is pulling in the direction 43° East of North with a force of 400 pounds. What is the direction and magnitude of the force which must be exerted by Alice on the third rope to counterbalance Sam and Charlie? (Draw a diagram showing all variables that you use.)

26. Using vectors, show that the perpendicular bisectors of a triangle intersect at a point.

27. Find the distance from the point $P = (2, -1, 3)$ to the plane $2x + 4y - z = -1$.

28. Two lines in space are skew if they are not parallel and do not intersect. Determine the minimum distance between two such lines.

29. Find an equation of the plane passing through the three points $(1, 1, 1), (1, 4, 5), (-3, -2, 0)$. Find the distance from the origin to the plane.

CHAPTER THIRTEEN

DIFFERENTIATING FUNCTIONS OF MANY VARIABLES

For a function of one variable, $y = f(x)$, the derivative $dy/dx = f'(x)$ may be thought of as the rate of change of y with respect to x. For a function of two variables $z = f(x, y)$, there is no such thing as *the* rate of change, since each of the independent variables x and y can be held fixed while the other varies. However, we can still consider the rate of change with respect to one or another of the independent variables. In this chapter we will see how to compute and interpret these partial derivatives, and the various ways that they can be combined to give a total picture of the way the function varies.

13.1   THE PARTIAL DERIVATIVE

In one-variable calculus you learned how the derivative measures the rate of change of a function. First we will quickly review this idea.

Temperature in a Metal Rod: a One Variable Problem

Let $m(x)$ be the temperature (in °F) of an unevenly heated metal rod, x feet from its left end. Table 13.1 gives some values of $m(x)$, and Figure 13.1 shows a graph of m. From the table and the graph you can see that the temperature increases as you move from the left of the rod, reaching its maximum at around 4 feet from the left end, after which it starts to decrease.

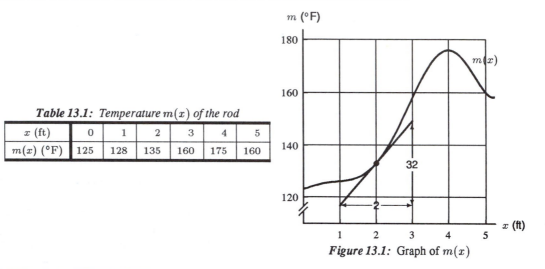

Table 13.1: *Temperature $m(x)$ of the rod*

x (ft)	0	1	2	3	4	5
$m(x)$ (°F)	125	128	135	160	175	160

Figure 13.1: Graph of $m(x)$

Example 1   Estimate the rate of change of m with respect to x at $x = 2$ using Table 13.1 and the graph in Figure 13.1.

Solution   The rate of change of m with respect to x at the point $x = 2$ is the derivative $m'(2)$, which is defined as the limit of a difference quotient:

$$m'(2) = \lim_{h \to 0} \frac{m(2+h) - m(2)}{h}.$$

Choosing $h = 1$ and reading values from Table 13.1, we get

$$m'(2) \approx \frac{m(2+1) - m(2)}{1} = \frac{160 - 135}{1} = 25.$$

Thus, the average rate of change of temperature between $x = 2$ and $x = 3$ is 25 degrees per foot. The instantaneous rate of change of m at $x = 2$ is the slope of the tangent line to the graph, and we can get a better approximation by estimating the slope graphically. From Figure 13.1 we see that the slope is about 16 degrees per foot, so

$$m'(2) \approx 16.$$

The fact that $m'(2)$ is positive means that the temperature of the rod is increasing as you move along the rod past the point $x = 2$ in the direction of increasing x (that is, from left to right on the rod).

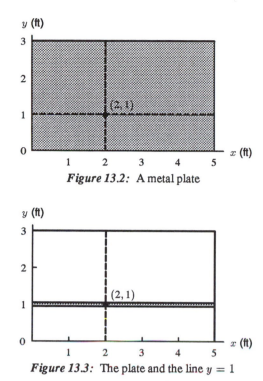

Figure 13.2: A metal plate

Table 13.2: *The temperature of the metal plate*

y						
3	85	90	**110**	135	155	180
2	100	110	**120**	145	190	170
1	**125**	**128**	**135**	**160**	**175**	**160**
0	120	135	**155**	160	160	150
	0	1	2	3	4	5

x

Figure 13.3: The plate and the line $y = 1$

Table 13.3: *Temperature along the line $y = 1$*

x (ft)	0	1	2	3	4	5
$m(x)$ (°F)	125	128	135	160	175	160

Temperature in a Metal Plate: a Two Variable Problem

Now imagine an unevenly heated thin rectangular metal plate, 3 ft by 5 ft as in Figure 13.2. We want to measure the rate at which the temperature varies over the plate. We now have a function T of two variables: $T(x, y)$ is the temperature (in °F) of the plate at the point x feet in from the left edge of the plate and y feet above the bottom edge. A brief table of values for the function T is given in Table 13.2. How does the temperature of the plate vary near the point $(2, 1)$? Since we know how to measure the rate of change of temperature along a rod, we will consider the horizontal line $y = 1$ containing $(2, 1)$. This line is like a rod; see Figure 13.3. The temperature along the line is a function of x only, since y does not change: its value at x is $T(x, 1)$, the temperature at the point $(x, 1)$. Call this function m, so $m(x) = T(x, 1)$. Table 13.3 gives some values of m; it is the row of Table 13.2 where $y = 1$. In fact, this is the same function m we studied in Example 1, so that the rod in that example is actually part of the plate.

What is the meaning of the derivative $m'(2)$ in this context? It is the rate of change of temperature T *in the x-direction* at the point $(2, 1)$ keeping y fixed. Denoting the rate of change of T in the x-direction by $T_x(2, 1)$, we have

$$T_x(2, 1) = m'(2) = \lim_{h \to 0} \frac{m(2 + h) - m(2)}{h}$$
$$= \lim_{h \to 0} \frac{T(2 + h, 1) - T(2, 1)}{h}.$$

We call $T_x(2, 1)$ the *partial derivative of T with respect to x at the point $(2, 1)$.*

In terms of the two-variable function T, the difference quotient we calculated in Example 1 is

$$\frac{T(2 + 1, 1) - T(2, 1)}{1} = \frac{160 - 135}{1} = 25.$$

This means that the average rate of change of temperature between the points $(2, 1)$ and $(3, 1)$ is 25 degrees per foot. Also, as before this is only an approximation for the partial derivative $T_x(2, 1)$. A more precise estimate comes from the one-variable estimate $m'(2) \approx 16$, which we found from the graph in Figure 13.1. Thus,

$$T_x(2, 1) \approx 16.$$

The fact that $T_x(2, 1)$ is positive means that the temperature of the plate is increasing as you move on the plate past the point $(2, 1)$ in the direction of increasing x (that is, horizontally from left to right in Figure 13.3).

Example 2   Investigate temperature variation in the plate along the line $x = 2$. See Figure 13.4.

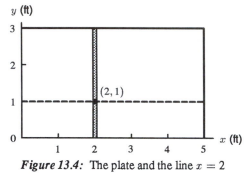

Figure 13.4: The plate and the line $x = 2$

Table 13.4: *Temperature along the line $x = 2$*

y (ft)	$n(y)$ (°F)
3	110
2	120
1	135
0	155

Solution   The temperature along this line is a function of y only, since x does not change; it is the function $n(y) = T(2, y)$ that gives the temperature at the point $(2, y)$ on the plate. Table 13.4 shows some values of n; it is the column of Table 13.2 corresponding to $x = 2$. The rate of change of temperature T in the y-direction at the point $(2, 1)$ is the same as the rate of change of n at $y = 1$, which is $n'(1)$. Denoting the rate of change of T in the y direction by $T_y(2, 1)$, we have

$$T_y(2, 1) = n'(1) = \lim_{h \to 0} \frac{n(1 + h) - n(1)}{h}$$
$$= \lim_{h \to 0} \frac{T(2, 1 + h) - T(2, 1)}{h}.$$

For example, taking $h = 1$ gives the difference quotient

$$\frac{T(2, 1 + 1) - T(2, 1)}{1} = \frac{120 - 135}{1} = -15,$$

which means that the average rate of change of temperature between the points $(2, 1)$ and $(2, 1+1) = (2, 2)$ is -15 degrees per foot. This average rate is negative because the temperature decreases as you move from $(2, 1)$ to $(2, 2)$.

The quantity $T_y(2, 1)$ is called the *partial derivative of T with respect to y at the point* $(2, 1)$. The average rate -15 is only an approximation for the partial derivative $T_y(2, 1)$. We cannot actually determine the instantaneous rate of change $T_y(2, 1)$ without more information than is given in Table 13.2.

Rates of Change and Partial Derivatives

For any function $f(x, y)$ we can use the method of the previous section to calculate the rates of change with respect to x and y. The idea is to study *separately* the influence of x and y on the value $f(x, y)$. Let us focus on values of f near a point (a, b). We hold y fixed at b and study the influence of x on f by considering the difference quotient

$$\text{Average rate of change of } f \text{ in } x\text{-direction at point } (a, b) = \frac{f(a + h, b) - f(a, b)}{h}.$$

The numerator of the difference quotient measures how much f changes in moving from the point (a, b) to the point $(a + h, b)$, which is on the same horizontal line as (a, b). Dividing by h produces the average rate of change of f between (a, b) and $(a + h, b)$, that is, the average rate of change in the x-direction.

Similarly, we hold x fixed at a, and study the influence of y on f by considering the difference quotient

$$\text{Average rate of change of } f \text{ in } y\text{-direction at point } (a, b) = \frac{f(a, b + h) - f(a, b)}{h}.$$

This time, the numerator of the difference quotient measures how much f changes in moving from the point (a, b) to the point $(a, b + h)$, which is on the same vertical line as (a, b).

To get the instantaneous rates of change we take the limit of these difference quotients as h approaches 0. This gives us the *partial derivatives of f at the point (a, b)*.

Partial derivatives of f with respect to x and y at (a, b)

$$f_x(a, b) = \begin{array}{c} \text{Rate of change of } f \text{ with respect to } x \\ \text{at the point } (a, b) \end{array} = \lim_{h \to 0} \frac{f(a + h, b) - f(a, b)}{h}$$

$$f_y(a, b) = \begin{array}{c} \text{Rate of change of } f \text{ with respect to } y \\ \text{at the point } (a, b) \end{array} = \lim_{h \to 0} \frac{f(a, b + h) - f(a, b)}{h}.$$

Just as with ordinary derivatives, there is an alternative notation:

Alternative notation

For $z = f(x, y)$ we can write

$$f_x(a, b) = \left.\frac{\partial z}{\partial x}\right|_{(a,b)} = \left.\frac{\partial f}{\partial x}\right|_{(a,b)}$$

and

$$f_y(a, b) = \left.\frac{\partial z}{\partial y}\right|_{(a,b)} = \left.\frac{\partial f}{\partial y}\right|_{(a,b)}.$$

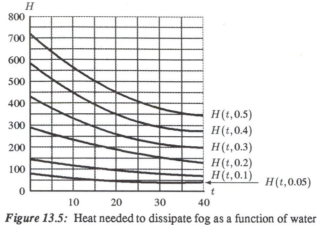

Figure 13.5: Heat needed to dissipate fog as a function of water content and temperature

We use the symbol ∂ to distinguish partial derivatives from ordinary derivatives of one variable functions. In cases where the independent variables have names different from x and y, we adjust the notation accordingly. For example, the partial derivatives of $f(u, v)$ are denoted by f_u and f_v.

Example 3   An airport can be cleared of fog by heating the air. The amount of heat required to do the job depends on the air temperature and the wetness of the fog. The graph in Figure 13.5 shows the heat $H(t, w)$ required (in calories per cubic meter of fog) as a function of the temperature t (in degrees Celsius) and the water content w (in grams per cubic meter of fog). Note that Figure 13.5 is not a contour diagram, but a set of graphs of $H(t, w)$ against t with w fixed at $0.05, 0.1, 0.2, 0.3, 0.4$, and 0.5.

Use this information to find an approximate value for $H_t(t, w)$ for $t = 10$, and $w = 0.1$. Interpret the partial derivative in practical terms.

Solution   The quantity $H_t(10, 0.1)$ is approximated by a difference quotient. This is the best that can be done with the information in Figure 13.5. The first partial derivative with respect to t is approximated by

$$H_t(10, 0.1) \approx \frac{H(10 + h, 0.1) - H(10, 0.1)}{h} \quad \text{for small } h.$$

We are free to choose h. We choose $h = 10$ because we can find a value for $H(10 + 10, 0.1) = H(20, 0.1)$ from Figure 13.5. If we take $H(10, 0.1) = 110$ and $H(20, 0.1) = 85$, we get the approximation

$$H_t(10, 0.1) \approx \frac{H(10 + 10, 0.1) - H(10, 0.1)}{10} = \frac{85 - 110}{10} = -2.5.$$

(Note that you may get a different answer if you read different values from the graph.) The geometric meaning of the partial derivative $H_t(10, 0.01)$ that we just approximated is the slope of the curve in Figure 13.5 corresponding to $w = 0.1$ at the point where $t = 10$. In practical terms, we have found that for fog at $10°$ C containing 0.1 g/m^3 of water, a $1°$ C increase in temperature will reduce the heat requirement for dissipating the fog by about 2.5 calories per cubic meter of fog. Certainly, early morning fogs tend to thin as the day warms up.

Example 4   Let

$$f(x, y) = \frac{x^2}{y + 1}.$$

Estimate $f_x(3, 2)$ and $f_y(3, 2)$ using difference quotients.

Solution   If h is small, then

$$f_x(3, 2) \approx \frac{f(3 + h, 2) - f(3, 2)}{h}.$$

With $h = 0.01$, we find

$$f_x(3, 2) \approx \frac{f(3.01, 2) - f(3, 2)}{0.01}$$

$$= \frac{\frac{3.01^2}{(2+1)} - \frac{3^2}{(2+1)}}{0.01} = 2.00333.$$

Let us improve the estimate by choosing a smaller value for h. With $h = 0.0001$, we get

$$f_x(3, 2) \approx \frac{f(3.0001, 2) - f(3, 2)}{0.0001}$$

$$= \frac{\frac{3.0001^2}{(2+1)} - \frac{3^2}{(2+1)}}{0.0001} = 2.0000333.$$

Since the difference quotient seems to be approaching 2 as h gets smaller, we conclude

$$f_x(3, 2) \approx 2.$$

Generally, as the value of h decreases, the value of the difference quotient gets closer to the true value of the partial derivative. However, for very small values of h, round-off error may cause the value of the difference quotient to move away again.

To estimate $f_y(3, 2)$, we use

$$f_y(3, 2) \approx \frac{f(3, 2 + h) - f(3, 2)}{h}.$$

With $h = 0.01$, we get

$$f_y(3, 2) \approx \frac{f(3, 2.01) - f(3, 2)}{0.01}$$

$$= \frac{\frac{3^2}{(2.01+1)} - \frac{3^2}{(2+1)}}{0.01} = -0.99668.$$

With $h = 0.0001$, we get

$$f_y(3, 2) \approx \frac{f(3, 2.0001) - f(3, 2)}{0.0001}$$

$$= \frac{\frac{3^2}{(2.0001+1)} - \frac{3^2}{(2+1)}}{0.0001} = -0.9999667.$$

Thus, it seems that the difference quotient is approaching -1, so we estimate

$$f_y(3, 2) \approx -1.$$

Example 5   Let

$$f(x, y) = \frac{x^2}{y+1}.$$

Find an expression for the difference quotient between the points $(x, y+h)$ and (x, y) and, by taking the limit as $h \to 0$, confirm the value for $f_y(3, 2)$ obtained in the previous example.

Solution   The difference quotient between $(x, y+h)$ and (x, y) is given by

$$
\begin{aligned}
\frac{f(x, y+h) - f(x, y)}{h} &= \frac{1}{h}\left(\frac{x^2}{y+h+1} - \frac{x^2}{y+1}\right) \\
&= \frac{x^2}{h}\left(\frac{1}{y+h+1} - \frac{1}{y+1}\right) \\
&= \frac{x^2}{h}\left(\frac{-h}{(y+h+1)(y+1)}\right) \\
&= -\frac{x^2}{(y+h+1)(y+1)}.
\end{aligned}
$$

Taking the limit as $h \to 0$ gives the value $f_y(x, y)$:

$$f_y(x, y) = \lim_{h \to 0}\left(-\frac{x^2}{(y+h+1)(y+1)}\right) = -\frac{x^2}{(y+1)^2}$$

and so

$$f_y(3, 2) = -\frac{3^2}{(2+1)^2} = -1.$$

Just like the derivative of a one-variable function, the partial derivative of a function is a rate of change. The difference is that in the case of a function of many variables, you have a rate of change with respect to each independent variable. The partial derivative of a function with respect to one of its variables is its rate of change with respect to that variable when you hold all the other variables fixed.

Visualizing the Partial Derivative on a Graph

The ordinary derivative of a one-variable function is the slope of its graph. Is there a way of visualizing the partial derivative of a two-variable function $f(x, y)$ as a slope? We have already seen

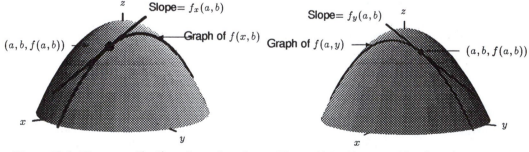

Figure 13.6: The curve $f(x, b)$ on the surface f ***Figure 13.7:*** The curve $f(a, y)$ on the surface f

that $f_x(a, b)$, the partial derivative with respect to x at the point (a, b), is just the derivative at $x = a$ of the one-variable function $f(x, b)$ obtained by holding y equal to b. The graph of this one-variable function is the curve where the vertical plane $y = b$ cuts through the graph of $f(x, y)$. See Figure 13.6. Thus, $f_x(a, b)$ is the slope of the tangent line to this curve at $x = a$.

Similarly, the graph of the one-variable function $f(a, y)$ is the curve where the vertical plane $x = a$ cuts through the graph of f, and the partial derivative $f_y(a, b)$ is the slope of this curve at $y = b$. See Figure 13.7.

Picture standing on the hillside $z = f(x, y)$, facing eastward (say) in the direction of the x-axis, there is one slope, f_x, whereas facing northward along the y-axis, there is another slope, f_y. In fact we will see shortly that there is a different slope in every direction.

Example 6   At each point labeled on the graph of the surface $z = f(x, y)$ in Figure 13.8, say whether each partial derivative is positive or negative.

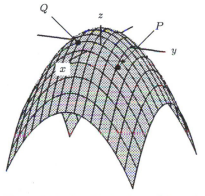

Figure 13.8: Decide the signs of f_x and f_y at P and Q

Solution   The positive x-axis points out of the page. Imagine heading off in this direction from the point marked P; you would find yourself descending quite steeply. So the partial derivative with respect to x is negative there, with quite a large absolute value. The same is true for the partial derivative with respect to y, since you would also have quite a steep descent in the positive y-direction.

At the point marked Q, heading in the positive x-direction will result in a gentle descent, whereas heading in the positive y-direction will result in a gentle ascent, so the partial derivative f_x is negative but small, and the partial derivative f_y is positive but small.

Visualizing Partial Derivatives on a Contour Diagram

We have just seen how to visualize a partial derivative by considering the slope of the curve in which a vertical plane cuts the surface generated by the function. We can also visualize a partial derivative on a contour diagram. If we move parallel to one of the axes, the partial derivative is the rate of change of the value of the function on the level curves. For example, if the values of the level curves are increasing, then the partial derivative must be positive.

Example 7   Figure 13.9 shows the contour diagram for the temperature $H(x, t)$ (in °F) in a room as a function of distance x (in feet) from a heater and time t (in minutes). What are the signs of $H_x(10, 20)$ and $H_t(10, 20)$? Estimate these partial derivatives and explain your answer in practical terms.

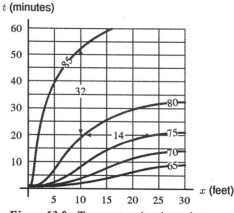

Figure 13.9: Temperature in a heated room

Solution   The point $(10, 20)$ is on the $H = 80$ contour. As x increases you move towards the $H = 75$ contour, so H is decreasing and $H_x(10, 20)$ is negative. This makes sense because as you move further from the heater, the temperature drops. On the other hand, as t increases you move towards the $H = 85$ contour, so H is increasing and $H_t(10, 20)$ is positive. This also makes sense, because it says that as time goes on, the room warms up.

To estimate the partial derivatives, we use a difference quotient. Looking at the contour diagram, we see there is a point on the $H = 75$ contour about 14 units to the right of $(10, 20)$. Hence, H decreases by 5 when x increases by 14, so the rate of change of H with respect to x is about $-5/14 \approx -0.4$. Thus, we find

$$H_x(10, 20) \approx -0.4.$$

This means that the temperature drops about half a degree for each foot you move away from the heater.

To estimate $H_t(10, 20)$, we look again at the contour diagram and notice that the $H = 85$ contour is about 32 units directly above $(10, 20)$. So H increases by 5 when t increases by 32. Hence,

$$H_t(10, 20) \approx \frac{5}{32} = 0.16.$$

This means that on the average the temperature goes up about 1/6 of a degree each minute.

Problems for Section 13.1

1. In Example 3 on page 114 we computed $H_t(10, 0.1)$ using Figure 13.5. Make a table of values for $H(t, w)$ and use it to estimate $H_t(t, w)$ for $t = 10, 20,$ and 30 and $w = 0.1, 0.2,$ and 0.3.

2. In Problem 1 we used values of a function, $H(t, w)$, to estimate $H_t(t, w)$ at several points. Estimate $H_w(t, w)$ for $t = 10, 20,$ and 30 and $w = 0.1, 0.2,$ and 0.3. What is the practical meaning of these partial derivatives?

3. The monthly mortgage payment in dollars, P, for a house is a function of three variables

$$P = f(A, r, N)$$

   where A is the amount borrowed in dollars, r is the interest rate, and N is the number of years before the mortgage is paid off.

   (a) Suppose $P(92000, 14, 30) = 1090.08$. What does this tell you, in financial terms?

   (b) Suppose $\dfrac{\partial P}{\partial r}(92000, 14, 30) = 72.82$. What is the financial significance of the number 72.82?

   (c) Would you expect $\partial P / \partial A$ to be positive or negative? Why?

   (d) Would you expect $\partial P / \partial N$ to be positive or negative? Why?

4. Suppose you borrow $\$A$ at an interest rate of $r\%$ (per month), and pay it off over t months by making monthly payments of $\$P$. Then $P = g(A, r, t)$. In financial terms, what do the following statements tell you?

   (a) $g(8000, 1, 24) \approx 376.59$ (b) $\dfrac{\partial g}{\partial A}(8000, 1, 24) \approx 0.047$

   (c) $\dfrac{\partial g}{\partial r}(8000, 1, 24) \approx 44.83$

5. Suppose that x is the average price of a new car, and that y is the average price of a gallon of gasoline. Then q_1, the number of new cars bought in a year depends on both x and y, so $q_1 = f(x, y)$. Similarly, if q_2 is the quantity of gas bought in a year, $q_2 = g(x, y)$.

   (a) What do you expect the signs of $\partial q_1 / \partial x$ and $\partial q_2 / \partial y$ to be? Explain.

   (b) What do you expect the signs of $\partial q_1 / \partial y$ and $\partial q_2 / \partial x$ to be? Explain.

6. A drug is injected into a patient's blood vessel. The function $c = f(x, t)$ represents the concentration of the drug at a distance x in the direction of the blood flow from the point of injection and a time t since the injection. What are the units of the following partial derivatives? What are their practical interpretations? What do you expect their signs to be?

   (a) $\partial c / \partial x$.   (b) $\partial c / \partial t$.

7. Suppose $\$P$ is your monthly car payment, and $P = f(P_0, t, r)$ where $\$P_0$ is the amount you borrowed, t is the number of months it takes to pay off the loan, and $r\%$ is the interest rate. What are the units, the practical meanings (in terms of money), and the signs of $\partial P / \partial t$ and $\partial P / \partial r$?

8. Part of the surface $z = f(x, y)$ is shown in Figure 13.10. The points A and B are in the xy-plane.

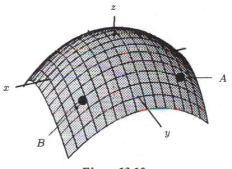

Figure 13.10

(a)   What is the sign of $f_x(A)$?   (b)   What is the sign of $f_y(A)$?

(c)   Suppose P is a point in the xy-plane which moves along a straight line from A to B. How does the sign of $f_x(P)$ change? How does the sign of $f_y(P)$ change?

9.   Given the surface $z = f(x, y)$, graphed in Figure 13.11.

(a)   What is the sign of $f_x(0, 5)$?   (b)   What is the sign of $f_y(0, 5)$?

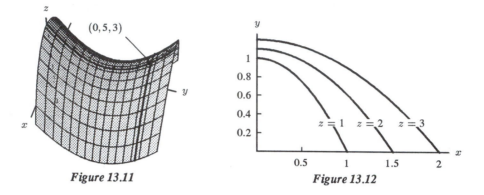

Figure 13.11   Figure 13.12

10.   Estimate $z_x(1, 0)$, $z_x(0, 1)$, $z_y(0, 1)$ from Figure 13.12.

For Problems 11 – 13 refer to Table 11.4 on page 9 giving the wind-chill factor, C, as a function $f(w, T)$ of the wind speed, w, and the temperature, T.

11.   Approximate $f_w(10, 25)$. What does your answer mean in practical terms?

12.   Approximate $f_T(5, 20)$. What does your answer mean in practical terms?

13.   From the table you can see that when the temperature is 20° F, the wind-chill factor drops by an average of about 2.6° F with every 1 mph increase in wind speed from 5 mph to 10 mph. Which partial derivative is this telling you about?

For Problems 14 and 15 refer to Table 11.5 on page 9 giving the heat index, I, as a function $f(h, T)$ of the humidity, h, and the temperature, T.

14.   Estimate $\partial I/\partial h$ and $\partial I/\partial T$ for typical weather conditions in Tucson in summer ($h = 10$, $T = 100$). What do your answers mean in practical terms for the residents of Tucson?

15.   Answer the question in Problem 14 for Boston in summer ($h = 50$, $T = 80$).

16.   Suppose that c represents the cardiac output, which is the volume of blood flowing through a person's heart and that s represents the systemic vascular resistance (SVR), which is the resistance to blood flowing through veins and arteries. Let p be a person's blood pressure. Then $p = f(c, s)$ is a function of c and s.

(a)   What does $\partial p/\partial c$ represent?

Suppose now that $p = kcs$, where k is a constant.

(b)   Sketch the level curves of p. What do they represent? Label your axes.

(c)   For a person with a weak heart, it is desirable to have the heart pumping against less resistance, while maintaining the same blood pressure. Such a person is given the drug Nitroglycerine to counter the SVR and the drug Dopamine to increase the cardiac output. Represent this on a graph showing level curves. Put a point A on the graph representing the person's state before drugs are given and a point B for after.

(d)   Right after a heart attack, a patient's cardiac output drops, thereby causing the blood pressure to drop. A common mistake made by medical residents is to get the patient's blood pressure back to normal by using drugs to increase the SVR, rather than by increasing the cardiac output. On a graph of the level curves of p, put a point D representing the patient before the heart attack, a point E representing the patient right after the heart attack, and a third point F representing the patient after the resident has given the drugs.

13.2 COMPUTING PARTIAL DERIVATIVES ALGEBRAICALLY

In computing partial derivatives, keep in mind that the partial derivative $f_x(a, b)$ is exactly the same as the ordinary derivative $m'(a)$ where m is the one-variable function $m(x) = f(x, b)$. Similarly $f_y(a, b) = n'(b)$ where $n(y) = f(a, y)$. So you can use all the techniques of differentiation from one-variable calculus to find partial derivatives.

Example 1 Let $f(x, y) = \dfrac{x^2}{y + 1}$. Find $f_x(3, 2)$ and $f_y(3, 2)$ algebraically.

Solution These are the same partial derivatives we found numerically in Example 4 on page 115. Now we will use the fact that $f_x(3, 2)$ equals the derivative of $m(x) = f(x, 2)$ at $x = 3$. Since

$$f(x, 2) = \frac{x^2}{(2 + 1)} = \frac{x^2}{3},$$

we have

$$m'(x) = \frac{2x}{3},$$

and so $f_x(3, 2) = m'(3) = 2$. Notice that the approximations we computed with difference quotients in Example 4 were quite good. Similarly, $f_y(3, 2)$ equals the derivative of $f(3, y)$ at $y = 2$. Let

$$n(y) = f(3, y) = \frac{9}{(y + 1)}.$$

Then

$$n'(y) = \frac{-9}{(y + 1)^2},$$

and so

$$f_y(3, 2) = n'(2) = -1.$$

Again, this agrees with the approximations in Example 4.

Example 2 The motion of a guitar string when vibrating can be described by a function of two variables, as described on page 5. Imagine a string 1 meter long, with the points on the string labeled by a coordinate x, $0 \le x \le 1$, as in Figure 13.13.

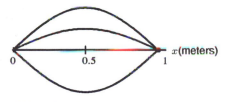

Figure 13.13: A vibrating guitar string

At time t seconds the point x has been displaced $f(x, t)$ meters from its rest position, where

$$f(x, t) = 0.003 \sin(\pi x) \sin(2765t).$$

(a) Evaluate $f_x(0.3, 1)$ and explain what it means in practical terms.

(b) Evaluate $f_t(0.3, 1)$ and explain what it means in practical terms.

Solution (a) We define

$$m(x) = f(x, 1) = 0.003 \sin(\pi x) \sin(2765)$$
$$= 0.0011 \sin(\pi x).$$

Thus, we have

$$f_x(x, 1) = m'(x) = 0.0011\pi \cos(\pi x).$$

In particular, $f_x(0.3, 1) = m'(0.3) = 0.0011\pi \cos(\pi(0.3)) \approx 0.002$. To see what the partial derivative $f_x(0.3, 1)$ means, think about the function $m(x) = f(x, 1)$ of which it is the derivative. The graph of $y = f(x, 1)$ in Figure 13.14 is a snapshot of the string at the single instant $t = 1$ second. The derivative $f_x(0.3, 1)$ is the slope of the string at the point where $x = 0.3$ (the instant when $t = 1$).

(b) Let us write

$$n(t) = f(0.3, t) = 0.003 \sin(\pi(0.3)) \sin(2765t)$$
$$= 0.0024 \sin(2765t),$$

so

$$f_t(0.3, t) = n'(t) = (0.0024)(2765) \cos(2765t) = 6.7 \cos(2765t).$$

Then

$$f_t(0.3, 1) = n'(1) = 6.7 \cos(2765(1)) \approx 6.$$

To figure out the meaning of the partial derivative $f_t(0.3, 1)$, think about the function $f(0.3, t)$ of which it is the derivative. The graph of $y = f(0.3, t)$ is a position versus time graph that

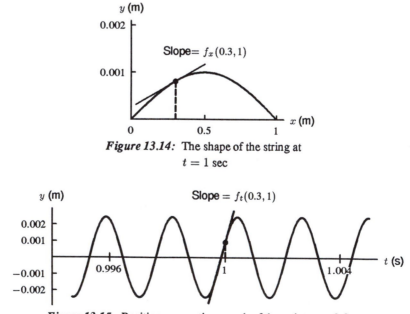

Figure 13.14: The shape of the string at
$t = 1$ sec

Figure 13.15: Position versus time graph of the point $x = 0.3$ m

tracks the movement up and down of a single point on the string, the point where $x = 0.3$. See Figure 13.15. The derivative $f_t(0.3, 1) = 6$ m/sec is thus the velocity of that point on the string at time 1 sec. The fact that $f_t(0.3, 1)$ is positive indicates that the particle is moving *upward* when $t = 1$.

The Partial Derivative Functions

So far we have focused on the partial derivatives of a function $f(x, y)$ at a fixed point (a, b). By thinking of a and b as variables, $a = x$ and $b = y$, we get the partial derivative functions $f_x(x, y)$ and $f_y(x, y)$.

Partial Derivative Functions

$$f_x(x, y) = \begin{array}{c} \text{Rate of change of } f \text{ with respect to } x \\ \text{at the point } (x, y) \end{array} = \lim_{h \to 0} \frac{f(x + h, y) - f(x, y)}{h}$$

$$f_y(x, y) = \begin{array}{c} \text{Rate of change of } f \text{ with respect to } y \\ \text{at the point } (x, y) \end{array} = \lim_{h \to 0} \frac{f(x, y + h) - f(x, y)}{h}$$

We also have the alternative notation:

For $z = f(x, y)$ we can write

$$f_x(x, y) = \frac{\partial z}{\partial x} \quad \text{and} \quad f_y(x, y) = \frac{\partial z}{\partial y}.$$

If a function $f(x, y)$ is given by a formula, then a formula for $f_x(x, y)$ can be found by regarding y as a constant and differentiating f with respect to x just as you would for a function of the single variable x. Similarly, a formula for $f_y(x, y)$ can be found by regarding x as a constant and differentiating f with respect to y.

Example 3   Compute both partial derivatives for
   (a) $f(x, y) = y^2 e^{3x}$ (b) $z = (3xy + 2x)^5$ (c) $y(x, y) = e^{x+3y} \sin(xy)$.

Solution   (a)   This is the product of a function of x (namely e^{3x}) and a function of y (namely y^2). When we differentiate with respect to x, we can think of the function of y as a constant, and vice versa.

$$f_x(x, y) = y^2 \frac{\partial}{\partial x}\left(e^{3x}\right) = 3y^2 e^{3x},$$

$$f_y(x, y) = e^{3x} \frac{\partial}{\partial y}(y^2) = 2y e^{3x}.$$

(b) Here we must use the chain rule:

$$\frac{\partial z}{\partial x} = 5(3xy + 2x)^4 \frac{\partial}{\partial x}(3xy + 2x) = 5(3xy + 2x)^4(3y + 2),$$

$$\frac{\partial z}{\partial y} = 5(3xy + 2x)^4 \frac{\partial}{\partial y}(3xy + 2x) = 5(3xy + 2x)^4 3x = 15x(3xy + 2x)^4.$$

(c) Since each function in the product is a function of both x and y, we need to use the product rule for each partial derivative:

$$g_x(x, y) = \frac{\partial(e^{x+3y})}{\partial x} \sin(xy) + e^{x+3y} \frac{\partial}{\partial x}(\sin(xy))$$
$$= e^{x+3y} \sin(xy) + e^{x+3y} y \cos(xy)$$
$$g_y(x, y) = \frac{\partial(e^{x+3y})}{\partial y} \sin(xy) + e^{x+3y} \frac{\partial}{\partial y}(\sin(xy))$$
$$= 3e^{x+3y} \sin(xy) + e^{x+3y} x \cos(xy).$$

Partial Derivatives for Functions of More Than Two Variables

For a function of two variables, $f(x, y)$, we find the partial derivatives by keeping one variable fixed and differentiating with respect to the other. Similarly, for functions of three or more variables, we find partial derivatives by differentiating with respect to one variable, regarding all the other variables as constants.

Example 4 Find all the partial derivatives of $f(x, y, z) = \frac{x^2 y^3}{z}$.

Solution To find $f_x(x, y, z)$, we consider y and z as fixed giving

$$f_x = \frac{2xy^3}{z}.$$

Similarly we have

$$f_y = \frac{3x^2 y^2}{z},$$

$$f_z = -\frac{x^2 y^3}{z^2}.$$

Problems for Section 13.2

1. Let $f(u, v) = u(u^2 + v^2)^{3/2}$.
    (a) Use a difference quotient to approximate $f_u(1, 3)$ with $h = 0.001$.
    (b) Now evaluate $f_u(1, 3)$ exactly. Was the approximation in part (a) reasonable?

2. Let $f(r, s) = r^2 s/(r^2 - s^2)$.
    (a) Use a difference quotient to approximate $f_s(2, 4)$ with $h = 0.001$.
    (b) Now evaluate $f_s(2, 4)$ exactly. Was the approximation in part (a) reasonable?

Find the indicated partial derivatives for the Problems 3–34.

3. $\dfrac{\partial A}{\partial h}$ if $A = \frac{1}{2}(a + b)h$

4. $\dfrac{\partial}{\partial m}\left(\frac{1}{2}mv^2\right)$

5. $\dfrac{\partial}{\partial B}\left(\dfrac{1}{u_0}B^2\right)$

6. $\dfrac{\partial}{\partial r}\left(\dfrac{2\pi r}{v}\right)$

7. F_v if $F = \dfrac{mv^2}{r}$

8. u_E if $u = \dfrac{1}{2}\epsilon_0 E^2 + \dfrac{1}{2\mu_0}B^2$

9. $\dfrac{\partial}{\partial v_0}(v_0 + at)$

10. $\dfrac{\partial F}{\partial m_2}$ if $F = \dfrac{Gm_1 m_2}{r^2}$

11. a_v if $a = v^2/r$

12. F_m if $F = mg$

13. $\dfrac{\partial}{\partial T}\left(\dfrac{2\pi r}{T}\right)$

14. $\dfrac{\partial}{\partial t}(v_0 t + \frac{1}{2}at^2)$

15. z_x if $z = x^2 y + 2x^5 y$

16. $\dfrac{\partial f_0}{\partial L}$ if $f_0 = \dfrac{1}{2\pi\ \sqrt{LC}}, \quad LC > 0$

17. $\dfrac{\partial y}{\partial t}$ if $y = \sin(ct - 5x)$

18. $\dfrac{\partial}{\partial y}(3x^5 y^7 - 32x^4 y^3 + 5xy)$

19. z_x if $z = \sin(5x^3 y - 3xy^2)$

20. $\dfrac{\partial}{\partial x}(a\sqrt{x}), \quad x > 0$

21. $\dfrac{\partial}{\partial M}\left(\dfrac{2\pi r^{3/2}}{\sqrt{GM}}\right), \quad GM > 0$

22. z_x if $z = \dfrac{1}{2x^2 ay} + \dfrac{3x^5 abc}{y}$

23. g_x if $g(x, y) = \ln(ye^{xy}), y > 0$

24. $\dfrac{\partial\alpha}{\partial\beta}$ if $\alpha = e^{x\beta - 3}/(2y\beta + 5)$

25. $\dfrac{\partial}{\partial T}\left(\ln\dfrac{T + 3}{V}\right), \quad \dfrac{T + 3}{V} > 0$

26. $\dfrac{\partial L}{\partial c}$ if $L = 6x^2 y^5 c^2 e^{c^2 - 3c - 7}$

27. $\dfrac{\partial}{\partial\theta}\left(3t\cos(5\theta - 1) - \tan(7t\theta^2)\right)$

28. $\dfrac{\partial}{\partial x}(xe^{\sqrt{xy}}), \quad xy > 0$

29. z_y if $z = \dfrac{3x^2 y^7 - y^2}{15xy - 8}$

30. $\dfrac{\partial}{\partial\lambda}\left(\dfrac{x^2 y\lambda - 3\lambda^5}{\sqrt{\lambda^2 - 3\lambda + 5}}\right)$

31. $\dfrac{\partial m}{\partial c}$ if $m = \dfrac{m_0}{\sqrt{1 - v^2/c^2}}$

32. $\dfrac{\partial}{\partial w}(\sqrt{2\pi xyw - 13x^7 y^3 v}), \quad 2\pi xyw - 13x^7 y^3 v > 0$

33. z_c if $z = \tan\sqrt{5wc^7 - 2c^4}, \quad 0 \le 5wc^7 - 2c^4 < \pi^2/4$

34. $\dfrac{\partial}{\partial w}\left(\dfrac{x^2 yw - xy^3 w^7}{w-1}\right)^{-7/2}$, $\quad \dfrac{x^2 yw - xy^3 w^2}{w-1} > 0$

For Problems 35–38 find the partial derivatives specified.

35. z_x and z_y for $z = x^7 + 2^y + x^y$

36. $\dfrac{\partial H}{\partial t}$ if $H = \ln(e^{t+p} + t) + \sin(p^2 t)$

37. $z_x(2,3)$ if $z = (\cos x) + y$. Estimate your answer to one decimal place using a calculator.

38. $\dfrac{\partial f}{\partial x}\bigg|_{(\pi/3,1)}$ if $f(x,y) = x\ln(y\cos x)$

39. Show that the Cobb-Douglas function

$$Q = bK^\alpha L^{1-\alpha} \quad \text{where} \quad 0 < \alpha < 1$$

satisfies Euler's theorem, which states that:

$$K\dfrac{\partial Q}{\partial K} + L\dfrac{\partial Q}{\partial L} = Q.$$

40. In order to treat patients, doctors sometimes need to know the surface area of a person's body. Since this is difficult to measure directly, an estimate is usually made from the person's height and weight. For a child of height h inches and weight w pounds, the surface area in square inches can be approximated by $S = 8.52h^{0.35}w^{0.54}$ square inches.

    (a) Find $\dfrac{\partial S}{\partial h}\bigg|_{(40,50)}$ and interpret this quantity in practical terms.

    (b) Find $\dfrac{\partial S}{\partial w}\bigg|_{(40,50)}$ and interpret this quantity in practical terms.

    (c) Which of the two partial derivatives you found is larger? Why might you have expected this?

41. Suppose you know that

$$f_x(x,y) = 4x^3 y^2 - 3y^4,$$
$$f_y(x,y) = 2x^4 y - 12xy^3.$$

    Can you find a function f which has these partial derivatives? If so, are there any others?

13.3 LOCAL LINEARITY

In Section 13.1 we learned how to study a function of two variables by varying one variable at a time. In effect, we found a way to use one-variable calculus to get some information about the rate of change of a two-variable function at a point, at least along the two lines through the point where one variable is constant. We now go off those lines, entering a truly multivariable world. Let's see how.

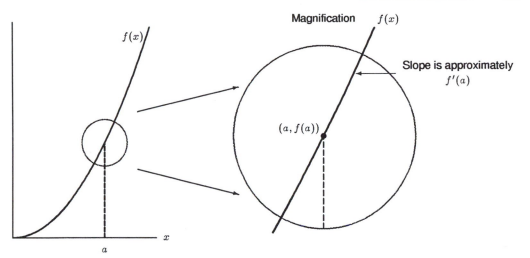

Figure 13.16: Zooming in on a portion of a function of one variable until the graph is almost straight

Review of Local Linearity for One Variable

If you zoom in on the graph of a smooth function of one variable, $y = f(x)$, around the point $x = a$, the graph looks more and more like a straight line, and thus becomes indistinguishable from its tangent line at that point. See Figure 13.16. The slope of the tangent line is the derivative $f'(a)$, and the line passes through the point $(a, f(a))$, so its equation is

$$y = f(a) + f'(a)(x - a).$$

(See Figure 13.17.) Now we approximate the values of f by the y values from the tangent line, giving the result in the following box.

The Tangent Line Approximation for values of x near a

$$f(x) \approx f(a) + f'(a)(x - a)$$

We are thinking of a as fixed, so that $f(a)$ and $f'(a)$ are constants, and the expression on the right-hand side is linear in x. The fact that f is approximately a linear function of x for x near a is expressed by saying f is *locally linear* near $x = a$.

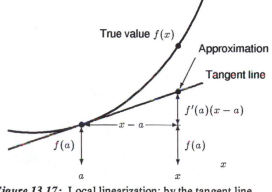

Figure 13.17: Local linearization: by the tangent line approximation

Example 1   Find the local linearization at $x = 2$ of the function m in Section 13.1.

Solution   Recall that $m(2) = 135$, and $m'(2) = 16$, so the tangent line approximation to $m(x)$ at $x = 2$ is

$$m(x) \approx m(2) + m'(2)(x - 2) = 135 + 16(x - 2) \quad \text{for } x \text{ near } 2.$$
$$= 10\,5+\ 16x$$

Figure 13.18 shows the function m and its tangent line at $x = 2$. Notice that $m(x)$ is not well approximated by values from the tangent line as x moves away from 2.

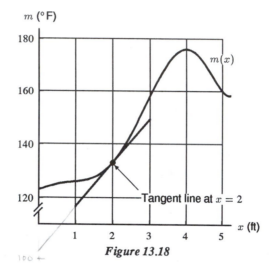

Figure 13.18

One-Variable Local Linearity Used for a Function of Two Variables

The local linearization of a function of one variable can be used to give a linear approximation to a function of two variables along a line parallel to one of the axes.

For example, in Section 13.1, we looked at the two-variable function $T(x, y)$ describing temperature as a function of position on a metal plate. We found that along the line $y = 1$ it had the same temperatures as the metal rod, that is, $m(x) = T(x, 1)$. Thus, the tangent line approximation to $m(x)$ at $x = 2$ leads to the linear approximation to $T(x, 1)$ for x near 2:

$$T(x, 1) \approx T(2, 1) + T_x(2, 1)(x - 2)$$
$$= 135 + 16(x - 2).$$

This approximation is useful if y is fixed at $y = 1$ and the value of x is changing but remains near 2.

If, on the other hand, the value of y is changing, say near 1, and x is fixed at $x = 2$, then we use a linear approximation to $T(2, y)$ for y near 1. If we know that $T_y(2, 1) = -17.5$, we have:

$$T(2, y) \approx T(2, 1) + T_y(2, 1)(y - 1) = 135 - 17.5(y - 1).$$

Thus, we can use local linearity for one-variable functions to approximate two-variable functions, as long as you keep one variable fixed. But what if we want to approximate $T(x, y)$ for (x, y) near $(2, 1)$, with neither $x = 2$ nor $y = 1$, say at the point $(2.04, 0.97)$? For this we need to develop a true two-variable linear approximation.

Local Linearity for Functions of Two Variables

Zooming in on a Surface and a Contour Diagram

We want to develop an approximation for functions of two variables that is analogous to the local linearization for functions of one variable. For a function of one variable, local linearity means that as we zoom in on the graph it looks like a straight line. So it is natural to ask: What happens if we zoom in on the graph of a two-variable function? Figure 13.19 shows three successive views of the graph of such a function, each one a closer view than the previous one. Notice that each view is flatter than the previous one: The closer we zoom in, the more the graph looks like the graph of a plane.

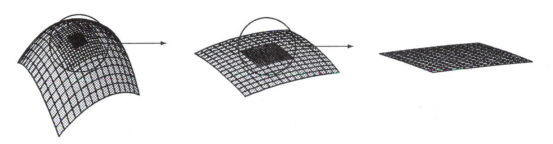

Figure 13.19: Zooming in on a function of two variables until the graph is almost flat

The same effect is seen by zooming in on the contour diagram of the function. Figure 13.20 shows three successive views of the contours near a point. Notice that as you zoom in, the contours look more like equally spaced parallel lines, that is like the contours of a plane.

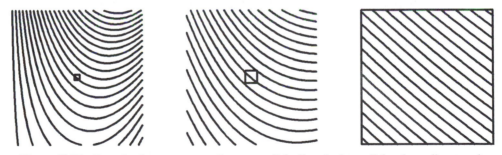

Figure 13.20: Zooming in on a contour diagram until the lines look parallel and equally spaced

The Tangent Plane

Thus, as we zoom in on a surface it generally becomes indistinguishable from a plane. This plane is called the *tangent plane* to the surface at the point. The tangent plane is to functions of two variables what the tangent line is to functions of one variable. Figure 13.21 shows the graph of a function with its tangent plane at a point.

What is the equation of the tangent plane? The x-slope of the graph of f at $(a, b, f(a, b))$ is the partial derivative $f_x(a, b)$, and the y-slope is $f_y(a, b)$. Thus, we have the following equation:

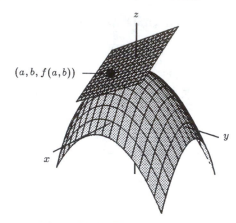

Figure 13.21: The tangent plane

The Tangent Plane

$$z = f(a,b) + f_x(a,b)(x-a) + f_y(a,b)(y-b).$$

You can see that the plane has the right slopes by checking the coefficients of x and y, and you can check that it passes through the point $(a, b, f(a,b))$ by substituting $x = a$ and $y = b$. Here we are thinking of a and b as fixed, so $f(a,b)$, $f_x(a,b)$, and $f_y(a,b)$ are constants; thus, the right side is a linear function of x and y.

Local Linearization of a Function of Two Variables

We now approximate the values of f by z values from the tangent plane, giving the following result:

The Tangent Plane Approximation for (x, y) near (a, b)

$$f(x,y) \approx f(a,b) + f_x(a,b)(x-a) + f_y(a,b)(y-b)$$

We are thinking of a and b as being fixed, so the expression on the right side is linear in x and y. This "almost equality" is called the *local linearization* of f near $x = a$, $y = b$.

Since the local linearization contains both variables x and y, it can be used to approximate f at points whether neither $x = a$ nor $y = b$.

Example 2   Consider the heated plate of Section 13.1 with temperature $T(x, y)$ at the point (x, y). Use the fact that $T(2, 1) = 135$, $T_x(2, 1) = 16$, and $T_y(2, 1) = -17.5$ to estimate the temperature at the point $(2.04, 0.97)$.

Solution   Local linearization gives us the tangent plane approximation

$$T(x, y) \approx T(2, 1) + T_x(2, 1)(x - 2) + T_y(2, 1)(y - 1)$$
$$T(x, y) \approx 135 + 16(x - 2) - 17.5(y - 1).$$

Substituting $x = 2.04$, $y = 0.97$ gives

$$T(x, y) \approx 135 + 16(2.04 - 2) - 17.5(0.97 - 1) = 136.17$$

Example 3   Find the local linearization of $f(x, y) = x^2 + y^2$ at the point $(3, 4)$. See how close an estimate of $f(2.9, 4.2)$ it gives.

Solution   The value of the function at the point $(3, 4)$ is given by

$$f(3, 4) = 3^3 + 4^2 = 25.$$

We have $f_x(x, y) = 2x$, so

$$f_x(3, 4) = 6,$$

and $f_y(x, y) = 2y$, so

$$f_y(3, 4) = 8.$$

Therefore, for (x, y) near $(3, 4)$ we have

$$f(x, y) \approx 25 + 6(x - 3) + 8(y - 4).$$

The approximate value of $f(2.9, 4.2)$ yielded by this linearization is

$$f(2.9, 4.2) \approx 25 + 6(-0.1) + 8(.2) = 26.$$

NOTE: the difference was substituted into the original equation

This compares favorably with the true value $f(2.9, 4.2) = (2.9)^2 + (4.2)^2 = 26.05$. Notice, however, that this local linearization does not give a good approximation at points farther away from $(3, 4)$. For example, if $x = 2$, $y = 2$, the local linearization gives

$$f(2, 2) \approx 25 + 6(-1) + 8(-2) = 3$$

whereas the true value of the function is

$$f(2, 2) = 2^2 + 2^2 = 8.$$

Example 4   Find the equation for the plane tangent to the surface $z = x^2 + y^2$ at the point $(x, y) = (3, 4)$.

Solution   Let $z = f(x, y) = x^2 + y^2$. In Example 3, we found the local linearization:

$$f(x, y) = x^2 + y^2 \approx 25 + 6(x - 3) + 8(y - 4)   \text{ for } (x, y) \text{ near } (3, 4).$$

Thus, the equation for the tangent plane is at $(3, 4)$

$$z = 25 + 6(x - 3) + 8(y - 4) = -25 + 6x + 8y.$$

$= 25 + 6x - 18 + 8y - 32$
$= 25 + 6x + 8y - 50$
$= -25 + 6x + 8y$

Local Linearity from a Numerical Point of View

To make a linear approximation to a function that is given in tabular form rather than as a formula, we approximate the required partial derivatives by difference quotients.

Example 5   The volume V of one pound of steam is a function $f(t, p)$ of the temperature t and pressure p of the steam. Detailed knowledge of f is critical to the safe design of boilers. Extensive tables, called steam tables, have been published, of which Table 13.5 forms a small portion.

    (a)   Use the table to give a linear approximation of $V = f(t, p)$ for t near $500°$ F and p near 24 lb/in^2.

    (b)   Estimate the volume of one pound of steam at a temperature of $510°$ F and a pressure of 25 lb/in^2.

TABLE 13.5   *Volume (in cubic feet) of one pound of steam at various temperatures and pressures*

		Pressure p (lb/in^2)			
		20	22	24	26
	480	27.85	25.31	23.19	21.39
Temperature	500	28.46	25.86	23.69	21.86
t	520	29.06	26.41	24.20	22.33
(°F)	540	29.66	26.95	24.70	22.79

Solution   (a)   We use local linearization around the point $t = 500$, $p = 24$, giving

$$f(t, p) \approx f(500, 24) + f_t(500, 24)(t - 500) + f_p(500, 24)(p - 24).$$

Directly from the table, we read the value

$$f(500, 24) = 23.69.$$

Next we approximate $f_t(500, 24)$ by a difference quotient. We look at the $p = 24$ column and compute the average rate of change between $t = 500$ and $t = 520$:

$$f_t(500, 24) \approx \frac{f(520, 24) - f(500, 24)}{520 - 500}$$
$$= \frac{24.20 - 23.69}{20} = 0.0255.$$

Note that $f_t(500, 24)$ is positive. Since steam expands, when heated, we expect the rate of change of volume with respect to temperature to be positive.

    Next we approximate $f_p(500, 24)$ by looking at the $t = 500$ row and computing the average rate of change between $p = 24$ and $p = 26$:

$$f_p(500, 24) \approx \frac{f(500, 26) - f(500, 24)}{26 - 24}$$
$$= \frac{21.86 - 23.69}{2} = -0.915.$$

Note that $f_p(500, 24)$ is negative, because subjecting steam to greater pressures confines it to smaller volumes. Thus, the rate of change of volume with respect to pressure is negative.

Using these approximations for the partial derivatives, we obtain the local linearization:

$$V = f(t, p) \approx 23.69 + 0.0255(t - 500) - 0.915(p - 24) \text{ ft}^3 \quad \begin{array}{l} \text{for } t \text{ near } 500\,^\circ\text{F} \\ \text{and } p \text{ near } 24 \text{ lb/in}^2. \end{array}$$

(b)   When $t = 510$ and $p = 25$, we have

$$V \approx 23.69 + 0.0255(510 - 500) - 0.915(25 - 24) = 23.03\text{ft}^3$$

We should be able to see the local linearity of a function from its table of values. Recall that the table of values of a linear function has all its rows linear with the same slope and all its columns linear with the same slope. (The rows and the columns do not have to have the same slope as one another, however.) As an example, let us tabulate the function $f(x, y) = x^2y$ for (x, y) near $(3, 1)$. Table 13.6 is a brief table of values, rounded to 4 decimal places.

TABLE 13.6   *Values of $f(x, y) = x^2y$ rounded to 4 decimal places*

		0.98	0.99	1.00	1.01	1.02
	2.98	8.7028	8.7915	8.8804	8.9692	9.0580
	2.99	8.7613	8.8506	8.9401	9.0295	9.1189
x	3.00	8.8200	8.9100	9.0000	9.0900	9.1800
	3.01	8.8789	8.9694	9.0601	9.1507	9.2413
	3.02	8.9380	9.0291	9.1204	9.2116	9.3028

It is clear from the table (as well as from the formula for f) that f is not linear. None of the columns in the table is linear, and the rows, while individually linear, do not have the same slope. However, let us round off all the entries in Table 13.6 to two decimal places. The results are in Table 13.7.

TABLE 13.7   *Approximate values of $f(x, y) = x^2y$ rounded to 2 decimal places*

		0.98	0.99	1.00	1.01	1.02
	2.98	8.70	8.79	8.88	8.97	9.06
	2.99	8.76	8.85	8.94	9.03	9.12
x	3.00	8.82	8.91	9.00	9.09	9.18
	3.01	8.88	8.97	9.06	9.15	9.24
	3.02	8.94	9.03	9.12	9.21	9.30

Observe that going down any column, as x increases by 0.01, the value of the function increases by 0.06 every time. Similarly, as y increases by 0.01 going across a row, the value of the function increases by 0.09. Thus, the data in Table 13.7 appears to be linear. The x-slope equals $0.06/0.01 = 6$ and the y-slope equals $0.09/0.01 = 9$. The entries in Table 13.7 are therefore given by the linear formula $9 + 6(x - 3) + 9(y - 1)$. Since Table 13.7 gives approximate values for f, we can write

$$f(x, y) \approx 9 + 6(x - 3) + 9(y - 1) \quad \text{for } (x, y) \text{ near } (3, 1).$$

In fact, this linear function *is* the local linearization of f (see Problem 1 at the end of this section).

What If There are Three or More Variables?

Partial derivatives and linear approximations for functions of three or more variables follow the same pattern as linear approximations for functions of two variables. As for functions of two variables, we write the linear approximation of $f(x, y, z)$ at (a, b, c) as

$$f(x, y, z) \approx f(a, b, c) + f_x(a, b, c)(x - a) + f_y(a, b, c)(y - b) + f_z(a, b, c)(z - c).$$

Example 6   Find a linear approximation for

$$f(u, v, w) = \frac{u^2}{2v + e^w}$$

that is valid for (u, v, w) near $(3, 1, 0)$. Use it to approximate $f(3.03, 1.02, -0.02)$.

Solution   The local linearization for f around the point $(3, 1, 0)$ is of the form

$$f(u, v, w) \approx f(3, 1, 0) + f_u(3, 1, 0)(x - 3) + f_v(3, 1, 0)(y - 1) + f_w(3, 1, 0)z.$$

We compute the values of the function and its partial derivatives:

$$f(3, 1, 0) = \frac{3^2}{2(1) + e^0} = 3,$$

$$f_u(3, 1, 0) = \frac{2u}{2v + e^w}\bigg|_{u=3, v=1, w=0} = 2,$$

$$f_v(3, 1, 0) = \frac{-2u^2}{(2v + e^w)^2}\bigg|_{u=3, v=1, w=0} = -2,$$

$$f_w(3, 1, 0) = \frac{-e^w u^2}{(2v + e^w)^2}\bigg|_{u=3, v=1, w=0} = -1.$$

Therefore,

$$f(u, v, w) \approx 3 + 2(u - 3) - 2(v - 1) - w \quad \text{for } (u, v, w) \text{ near } (3, 1, 0).$$

The approximation gives

$$f(3.03, 1.02, -0.02) \approx 3 + 2(0.03) - 2(0.02) - (-0.02) = 3.04,$$

which compares well with the exact value

$$f(3.03, 1.02, -0.02) = \frac{(3.03)^2}{2(1.02) + e^{-0.02}} = 3.03983 \cdots.$$

It is important to remember that the local approximation is precisely that. It is only valid near the point at which it is evaluated. For example, in the last example, $f(4, 2, 1) = 2.38156$ but the local approximation produces a value of 2.

Problems for Section 13.3

1. Find the local linearization of the function $f(x, y) = x^2 y$ at the point $(3, 1)$.

2. Find the equation of the tangent plane to $z = e^y + x + x^2 + 6$ at the point $(1, 0, 9)$.

3. Find the equation of the tangent plane to $z = y e^{x/y}$ at the point $(1, 1, e)$.

4. Find the equation of the tangent plane to $z = \frac{1}{2}(x^2 + 4y^2)$ at the point $(2, 1, 4)$.

5. Consider the surface given by the equation $z = 1 + x^2 - y^2$. Where is the tangent plane parallel to the plane $3x + 2y + 3z = 4$?

6. At what point on the surface $z = 1 + x^2 + y^2$ is its tangent plane parallel to the following planes? (a) $z = 5$ (b) $z = 5 + 6x - 10y$

7. A student was asked to find the equation of the tangent plane to the surface $z = x^3 - y^2$ at the point where $(x, y) = (2, 3)$. His answer was

$$z = 3x^2(x - 2) - 2y(y - 3) - 1.$$

  (a) At a glance, how do you know this is wrong?
  (b) What mistake did the student make?
  (c) Answer the question correctly.

8. Consider the tangent plane to the surface $z = x^2 + 2y^2$ at the point where $(x, y) = (1, 2)$. As x and y decrease to 0, your height on both the surface and the tangent plane also decreases to 0. Do you get to 0 faster on the surface or on the plane? How do you know?

9. In Example 5 we found a linear approximation for $V = f(t, p)$ near $(500, 24)$. Now find a linear approximation near $(480, 20)$.

10. In Example 5 we found a linear approximation for $V = f(t, p)$ near $(500, 24)$.

  (a) Test the accuracy of this approximation by comparing its predicted value with the nearby values in the table. Do you notice anything? Explain your answer.
  (b) Can you suggest a better approximation near $(485, 21)$?

11. Using Figure 11.58 on page 39, develop a linear function that gives you approximately the monthly payment, m, on a 5-year car loan if you borrow P dollars at r percent interest. What is the practical significance of the constants in your formula?

12. The productive capacity P of the United States as a function of capital K and labor L is modeled by the Cobb-Douglas production function

$$P(K, L) = AK^{0.21}L^{0.79},$$

where A is a constant. Suppose that capital increases by 1% but that labor use remains unchanged. Use local linearization to estimate the percentage increase in production.

13.   The schematic diagram for a transistor in common-emitter configuration is shown in Figure 13.22. The transistor would be wired into an electrical circuit at the three points marked with the letters B (base), C (collector), and E (emitter). What the transistor is doing at any point in time can be summarized by giving the measurements of the three currents i_b, i_c, and i_e in the wires leading to the transistor and the two voltages v_b and v_c between pairs of terminals. All five of these measurements can be determined from measurements of i_b and v_c alone, because there are functions f and g (called the *characteristics* of the transistor) such that $i_c = f(i_b, v_c)$ and $v_b = g(i_b, v_c)$, and $i_e = -i_b - i_c$. The units are microamps (μA) for i_b, volts (V) for v_c, and milliamps (mA) for i_c.

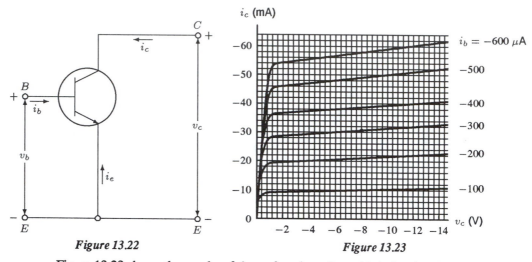

Figure 13.22   **Figure 13.23**

Figure 13.23 shows the graphs of f as a function of v_c with i_b fixed at the values -100, -200, -300, -400, -500, and -600. Find a linear approximation for f that is valid when the transistor is made to operate with i_b near $-300\,\mu A$ and v_c near $-8V$.

13.4   DIRECTIONAL DERIVATIVES

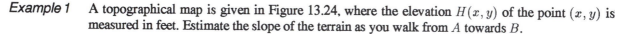

The partial derivatives of a multivariable function f tell you the rates of change of f in the directions of each of the coordinate axes. In this section we will see how to compute the rate of change of f in an arbitrary direction.

Example 1   A topographical map is given in Figure 13.24, where the elevation $H(x, y)$ of the point (x, y) is measured in feet. Estimate the slope of the terrain as you walk from A towards B.

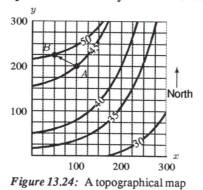

Figure 13.24: A topographical map

Solution Initially you are on the $H = 45$ contour. By the time you reach B you are on the $H = 50$ contour. The displacement vector from A to B has x component -50 and y component 25, so its length is $\sqrt{(-50)^2 + 25^2} \approx 56$, so you have risen 5 feet for a run of 56 feet, so the slope of the terrain in that direction is $5/56 \approx 0.09$.

Directional Derivatives

For a function $z = f(x, y)$, the *directional derivative* at a point (a, b) in a certain direction in the xy plane is the rate of change of f as you head in that direction. We will give the proper definition of this below in terms of difference quotients, but first we look at another example.

Example 2 For each of the functions f, g, and h, in Figure 13.25, decide whether the directional derivative at the indicated point is positive, negative, or zero, in each of the directions shown.
(a) the vector $\vec{i} + 2\vec{j}$,   (b) the vector $2\vec{i} + \vec{j}$.

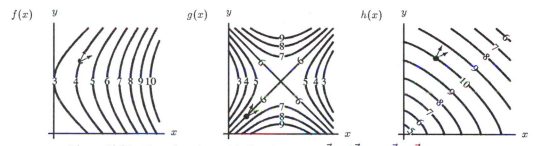

Figure 13.25: Three functions with direction vectors $\vec{i} + 2\vec{j}$ and $2\vec{i} + \vec{j}$ marked.

Solution On the contour diagram for f, the vector $\vec{i} + 2\vec{j}$ points in the direction tangent to the contour, so the function is not changing in that direction. Thus, the directional derivative is zero in that direction. The vector $2\vec{i} + \vec{j}$ points from the contour marked 4 towards the contour marked 5, so the values of the function are increasing, and thus the directional derivative is positive in that direction.

On the contour diagram for g, the vector $\vec{i} + 2\vec{j}$ points from the contour marked 6 towards the contour marked 5, so the function is decreasing in that direction. Thus, the directional derivative is negative. On the other hand, the vector $2\vec{i} + \vec{j}$ points in the direction of the contour marked 7, and hence the directional derivative is positive.

Finally, on the contour diagram for h, both vectors point from the $h = 10$ contour to the $h = 9$ contour, so both directional derivatives are negative.

Now we define the directional derivative at any point (a, b) using a difference quotient. In particular we determine the derivative in the direction of a unit vector $\vec{u} = u_1\vec{i} + u_2\vec{j}$. This derivative is written as

$$f_{\vec{u}}(a, b).$$

We use a unit vector because this will make the computation of the directional derivative much neater.

We define the **directional derivative** as

$$f_{\vec{u}}(a, b) = \lim_{h \to 0} \frac{f(a + hu_1, b + hu_2) - f(a, b)}{h}.$$

where $\vec{u} = u_1\vec{i} + u_2\vec{j}$ is a unit vector.

The numerator gives the change in f between the point (a, b) and a neighboring point which is displaced by an amount $h\vec{u}$ from (a, b). See Figure 13.26. Since $\|\vec{u}\| = 1$, the displacement $h\vec{u}$ corresponds to moving a distance h (positive or negative) in the direction of $\vec{u}$. Thus, dividing by h gives the average rate of change, and taking the limits gives the instantaneous rate of change in the direction of $\vec{u}$.

Notice that if $\vec{u}$ is in the direction of $\vec{i}$, that is, $u_1 = 1$, $u_2 = 0$, then the directional derivative reduces to f_x, since

$$f_{\vec{i}}(a, b) = \lim_{h \to 0} \frac{f(a + h, b) - f(a, b)}{h} = f_x(a, b).$$

Similarly, if $\vec{u}$ is in the direction $\vec{j}$ then the directional derivative $f_{\vec{j}} = f_y$.

If $\vec{u}$ is not a unit vector, we can still talk about the directional derivative in the $\vec{u}$ direction (provided $\|\vec{u}\| \neq 0$). In that case we are moving a distance $h\|\vec{u}\|$ in the direction of $\vec{u}$ (again, h negative means moving in the opposite direction), so we divide by $\|h\vec{u}\|$:

$$f_{\vec{u}}(a, b) = \lim_{h \to 0} \frac{f(a + hu_1, b + hu_2) - f(a, b)}{h\|\vec{u}\|}.$$

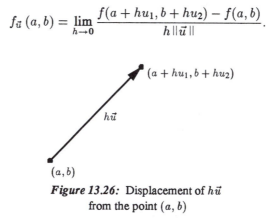

Figure 13.26: Displacement of $h\vec{u}$
from the point (a, b)

Example 3   Find the directional derivative of $f(x, y, z) = x^2 + y^2 + z^2$ in the direction of the vector $\vec{i} + \vec{j} + \vec{k}$ at the point $(1, 0, 0)$.

Solution   First we have to find a unit vector in the same direction as the vector $\vec{i} + \vec{j} + \vec{k}$. Since this vector has magnitude $\sqrt{3}$, a unit vector is

$$\vec{u} = \frac{1}{\sqrt{3}}(\vec{i} + \vec{j} + \vec{k}) = \left(\frac{1}{\sqrt{3}}\vec{i} + \frac{1}{\sqrt{3}}\vec{j} + \frac{1}{\sqrt{3}}\vec{k}\right).$$

Thus,

$$
\begin{aligned}
f_{\vec{u}}\,(1,0,0) &= \lim_{h \to 0} \frac{f((1,0,0) + h\vec{u}\,) - f(1,0,0)}{h} \\
&= \lim_{h \to 0} \frac{f(1 + \frac{h}{\sqrt{3}}, \frac{h}{\sqrt{3}}, \frac{h}{\sqrt{3}}) - f(1,0,0)}{h} \\
&= \lim_{h \to 0} \frac{(1 + \frac{h}{\sqrt{3}})^2 + (\frac{h}{\sqrt{3}})^2 + (\frac{h}{\sqrt{3}})^2 - 1}{h} \\
&= \lim_{h \to 0} \frac{h^2 + (\frac{2}{\sqrt{3}})h}{h} \\
&= \lim_{h \to 0} (h + \frac{2}{\sqrt{3}}) \\
&= \frac{2}{\sqrt{3}}.
\end{aligned}
$$

Problems for Section 13.4

1. True or false? $f_{\vec{u}}\,(1,2)$ is a number. Explain your answer.
2. Suppose $f(x, y) = x + \ln y$. Using difference quotients, estimate the following.
   (a) The rate of change of f as you leave the point $(1, 4)$ going towards the point $(3, 5)$.
   (b) The rate of change of f as you arrive at the point $(3, 5)$.

For Problems 3–8 use Figure 13.27, showing level curves of $f(x, y)$, to estimate the directional derivatives.

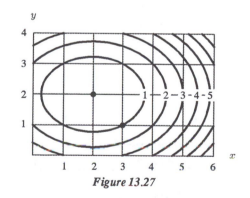

Figure 13.27

3. $f_{\vec{i}}\,(3, 1)$

4. $f_{\vec{j}}\,(3, 1)$

5. $f_{\vec{u}}\,(1, 3)$ where $\vec{u} = (-\vec{i} + \vec{j}\,)/\sqrt{2}$

6. $f_{\vec{u}}\,(3, 1)$ where $\vec{u} = (\vec{i} - \vec{j}\,)/\sqrt{2}$

7. For what part of the rectangular region shown in Figure 13.27 is $f_{\vec{i}}$ positive?

8. For what part of the rectangular region shown in Figure 13.27 is $f_{\vec{j}}$ negative?

9. The surface $z = g(x, y)$ is shown in Figure 13.28. What is the sign of each of the following directional derivatives?

    (a) $g_{\vec{u}}(2, 5)$ where $\vec{u} = (\vec{i} - \vec{j})/\sqrt{2}$.

    (b) $g_{\vec{u}}(2, 5)$ where $\vec{u} = (\vec{i} + \vec{j})/\sqrt{2}$.

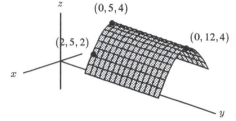

Figure 13.28

10. Use a difference quotient to estimate the rate of change of $f(x, y) = 2x^2 + y^2$ at the point $(2, 1)$ as you move in the direction of the vector $\vec{u} = (\vec{i} + \vec{j})/\sqrt{2}$.

11. Using a difference quotient, estimate the rate of change of $f(x, y) = xe^y$ at the point $(1, 1)$ as you move in the direction of the vector $\vec{i} + 2\vec{j}$.

12. Using the limit of a difference quotient, compute the rate of change of $f(x, y) = 2x^2 + y^2$ at the point $(2, 1)$ as you move in the direction of the vector $\vec{u} = (\vec{i} + \vec{j})/\sqrt{2}$.

13.5   THE GRADIENT

The rate of change of a one-variable function $f(x)$ at $x = a$ can be described by a single number, its derivative, $f'(a)$. We have seen that the rate of change of a two-variable function $f(x, y)$ at a point $(x, y) = (a, b)$ cannot be described by a single number, because before you can talk about the rate of change you have to specify a direction. The rate of change in the x-direction is $f_x(a, b)$, the rate of change in the y-direction is $f_y(a, b)$. In general, the rate of change in the direction of a unit vector $\vec{u}$ is the directional derivative $f_{\vec{u}}(a, b)$. This suggests that a vector quantity might be better than a scalar quantity for giving complete information about a function's rate of change.

In this section we will define a vector called the *gradient* which describes how the function changes near a point. We use geometric properties to define the gradient vector. Then we show how to compute the components of the gradient vector algebraically.

Geometric Definition of the Gradient Vector

> The **gradient vector** of f at the point (a, b) is defined to be the vector with the following properties
> - The direction of the gradient vector is the direction in which f is increasing at the greatest rate.
> - The magnitude of the gradient vector is the rate of change of f in that direction.
>
> The gradient of f at (a, b) is written grad $f(a, b)$.

Example 1   Figure 13.29 shows the topographical map from Example 1 on page 136. It is the contour diagram of the function $H(x, y)$ which gives height above the sea level at the point (x, y). What is the practical significance of the gradient vectors at A and B?

Solution   The gradient vector points in the direction of greatest increase; on a topographical map, that means it points in the direction that will take you directly uphill. The magnitude of the gradient vector measures the steepness of the slope. The length of the gradient vector at A is longer than that of the gradient vector at B. That is because the contours are closer together at A, so the slope is steeper at A.

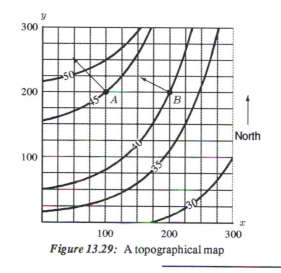

Figure 13.29: A topographical map

Notice that the gradient vector at any point is perpendicular to the contour through that point. This is always the case. To see why, look at Figure 13.30. It shows a close-up of the contours of f around the point (a, b). The contours appear straight, parallel, and equally spaced, because f is

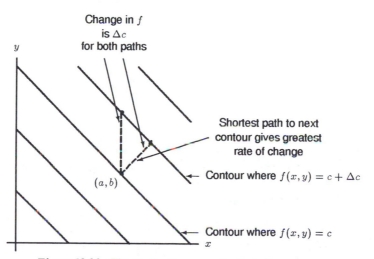

Figure 13.30: The gradient is perpendicular to the contours

locally linear at (a, b). The directional derivative in any direction is approximated by the difference quotient

$$\frac{\text{Change in } f}{\text{Distance traveled}}.$$

Hence, the largest directional derivative will be obtained by moving in the direction that takes you to the next contour in the shortest possible distance, which is the direction perpendicular to the contour.

Properties of the Gradient Vector

- The gradient vector of f at the point (a, b) is perpendicular to the contour of f through (a, b), and in the direction of increasing f.
- The magnitude of the gradient vector, $\|\text{grad } f\|$, is large when the contours are close together and small when they are far apart.

The Gradient Vector of a Function of More Than Two Variables

The definition of the gradient vector applies to functions of any number of variables.

The **gradient vector** of f at $(a_1, a_2, \ldots, a_n)$ is a vector in n-space defined as follows:
- The direction of the gradient vector is the direction in which f is increasing at the greatest rate.
- The magnitude of the gradient vector is the rate of change of f in that direction.

Just as the gradient vector of a function of two variables is perpendicular to the contours, or level curves, so the gradient of a function of three variables is perpendicular to the level surfaces. The reason is the same; the function has the same value all over a level surface, so to change the value as quickly as possible you need to move directly away from the level surface, that is, perpendicular to the surface.

Example 2 Let $f(x, y, z) = x^2 + y^2$ and $g(x, y, z) = -x^2 - y^2 - z^2$. What can you say about the direction of the following vectors?

(a) grad $f(0, 1, 1)$ (b) grad $f(1, 0, 1)$ (c) grad $g(0, 1, 1)$ (d) grad $g(1, 0, 1)$?

Solution Consider the cylinder $x^2 + y^2 = 1$ in Figure 13.31. This is a level surface of f and contains both the points $(0, 1, 1)$ and $(1, 0, 1)$. Since the value of f does not change at all in the z-direction, the gradient vectors are horizontal. They are perpendicular to the cylinder and point outward because the value of f increases as you move out.

Similarly, both points also lie on the same level surface of g, namely the level surface $-x^2 - y^2 - z^2 = -2$, which may also be written $x^2 + y^2 + z^2 = 2$. Part of this level surface is shown in Figure 13.32. Notice that this time the gradient vectors point inward, since the negative signs mean that the function increases (from large negative values to small negative values) as you move inwards.

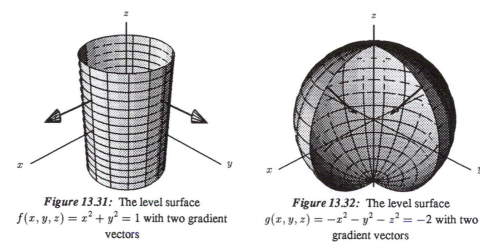

Figure 13.31: The level surface
$f(x, y, z) = x^2 + y^2 = 1$ with two gradient
vectors

Figure 13.32: The level surface
$g(x, y, z) = -x^2 - y^2 - z^2 = -2$ with two
gradient vectors

The Relationship Between the Gradient Vector and Directional Derivatives

The gradient vector at a point gives us the direction and magnitude of the greatest rate of change of f at that point; the directional derivative at that point in some direction gives the rate of change of f in that direction. We will now see that there is a close relationship between the gradient and the directional derivative.

Imagine zooming in on a function $f(x, y)$ at a point (a, b). By local linearity, the contours of f around (a, b) will look like the contours of a linear function. Figure 13.33 shows such contours and the gradient vector at the point (a, b). Suppose you want to find the directional derivative $f_{\vec{u}}(a, b)$ in the direction of $\vec{u}$, the unit vector shown in Figure 13.33. If you move a small distance h in the direction of $\vec{u}$, the directional derivative is approximated by the difference quotient

$$\frac{\text{Change in } f}{h}.$$

To estimate the change in f, we use the gradient vector. Because the contours are approximately parallel, moving h units in the direction of $\vec{u}$ takes you to the same contour as moving $h \cos \theta$ units in the direction of grad $f(a, b)$, where θ is the angle between $\vec{u}$ and grad $f(a, b)$. Thus,

$$\text{Change in } f = (\text{Rate of change})(\text{Distance traveled}) \approx \| \text{grad } f \|(h \cos \theta).$$

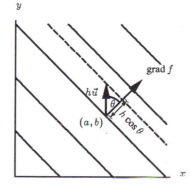

Figure 13.33: Computing the
directional derivative in the direction
of $\vec{u}$ from grad f by zooming in on
the contour diagram of f

Hence, since $\vec{u}$ is a unit vector, we have

$$f_{\vec{u}}(a,b) \approx \frac{\text{Change in } f}{h} \approx \frac{\|\operatorname{grad} f(a,b)\| h \cos\theta}{h}$$
$$= \|\operatorname{grad} f(a,b)\| \cos\theta = \|\operatorname{grad} f\| \|\vec{u}\| \cos\theta = \operatorname{grad} f(a,b) \cdot \vec{u}.$$

This approximation gets better as we choose h smaller and smaller, and in the limit we get the following relationship.

Directional Derivatives and the Gradient Vector

If $\vec{u}$ is a unit vector making an angle of θ with the gradient vector:

$$f_{\vec{u}}(a,b) = \operatorname{grad} f(a,b) \cdot \vec{u} = \|\operatorname{grad} f(a,b)\| \cos\theta$$

In words:
The directional derivative at the point (a,b) in the direction $\vec{u}$ is the dot product of the gradient vector at (a,b) with the vector $\vec{u}$ (provided $\vec{u}$ is a unit vector).

This relationship between the directional derivative and the gradient vector tells us a good deal about how a function changes near a point. Imagine the point, and hence grad $f(a,b)$, is fixed, but that the unit vector $\vec{u}$ can rotate. The dot product of grad $f_{\vec{u}}(a,b)$ and $\vec{u}$ is maximum when $\vec{u}$ is in the same direction as grad $f(a,b)$. Then $\theta = 0$ and the maximum rate of change of f at this point is given by

$$f_{\vec{u}}(a,b) = \|\operatorname{grad} f(a,b)\| \cos 0 = \|\operatorname{grad} f(a,b)\|,$$

as we know from the definition of the gradient. Similarly, the minimum rate of change of f at this point is obtained by taking $\vec{u}$ in the opposite direction to grad f, so that $\theta = \pi$. Thus, the minimum rate of change of f is given by

$$f_{\vec{u}}(a,b) = \|\operatorname{grad} f\| \cos\pi = -\|\operatorname{grad} f(a,b)\|.$$

Whenever $\vec{u}$ is perpendicular to the gradient vector, so $\theta = \pi/2$, the rate of change of f in that direction is zero:

$$f_{\vec{u}}(a,b) = \|\operatorname{grad} f(a,b)\| \cos\frac{\pi}{2} = 0.$$

If $\vec{u}$ is tangent to a level curve or level surface, the rate of change of f in this direction is zero, and therefore, provided the gradient is not zero, $\vec{u}$ must be perpendicular to the gradient vector.

In summary, we have the following results.

Rates of Change and the Gradient Vector at a Point

- The maximum rate of change of f at a point is in the direction of the gradient vector at that point, and has magnitude $\|\operatorname{grad} f\|$
- The minimum rate of change of f is in the direction opposite to the gradient, and has value $-\|\operatorname{grad} f\|$
- The rate of change of f in any direction perpendicular to the gradient vector is 0
- The gradient vector is perpendicular to the level curves or level surfaces at every point

---

These results hold for functions of any number of variables although the discussion has been for two variables.

Components of the Gradient Vector

As for any vector, the x-component of grad f is the dot product of grad f with the unit vector $\vec{i}$ in the x-direction. However we now know that this dot product is the directional derivative in the x-direction, which is just the partial derivative with respect to x. That is,

$$x\text{-component of grad } f(a, b) = \text{grad } f(a, b) \cdot \vec{i} = f_{\vec{i}}(a, b) = f_x(a, b).$$

Similarly,

$$y\text{-component of grad } f(a, b) = \text{grad } f(a, b) \cdot \vec{j} = f_{\vec{j}}(a, b) = f_y(a, b).$$

Similar results hold for the gradient of a function of three or more variables. If we now consider a general point, we get the following formulas for the components of the gradient vector.

Components of the Gradient Vector in Cartesian Coordinates.

In two dimensions

$$\text{grad } f = f_x \vec{i} + f_y \vec{j}.$$

In three dimensions

$$\text{grad } f = f_x \vec{i} + f_y \vec{j} + f_z \vec{k}.$$

In n dimensions the components of the gradient vector are

$$\text{grad } f(a_1, a_2, \ldots, a_n) = (f_{x_1}(a_1, a_2, \ldots, a_n), f_{x_2}(a_1, a_2, \ldots, a_n), \ldots, f_{x_n}(a_1, a_2, \ldots, a_n)).$$

Alternative Notation for the Gradient

You can think of $\dfrac{\partial f}{\partial x}\vec{i} + \dfrac{\partial f}{\partial y}\vec{j}$ as the result of applying the vector operator

$$\nabla = \frac{\partial}{\partial x}\vec{i} + \frac{\partial}{\partial y}\vec{j}$$

to the function f. Thus, we get the alternative notation

$$\text{grad } f = \nabla f.$$

Similarly, for three variables, we have grad $f = \nabla f$ with

$$\nabla = \frac{\partial}{\partial x}\vec{i} + \frac{\partial}{\partial y}\vec{j} + \frac{\partial}{\partial z}\vec{k}.$$

Example 3

(a) Find the components of the gradient vector of $f(x, y) = x + e^y$ at the point $(1, 1)$.

(b) Use the gradient to find the rate of change of f in the direction of the vector $\vec{i} - \vec{j}$.

Solution   (a)   We have

$$\text{grad } f = f_x \vec{i} + f_y \vec{j} = \vec{i} + e^y \vec{j},$$

so at the point $(1, 1)$,

$$\text{grad } f(1, 1) = \vec{i} + e\vec{j}.$$

(b)   A unit vector in the direction of $\vec{i} - \vec{j}$ is $\vec{u} = (1/\sqrt{2})\vec{i} - (1/\sqrt{2})\vec{j}$, so

$$f_{\vec{u}}(1, 1) = \text{grad } f(1, 1) \cdot \vec{u} = (\vec{i} + e\vec{j}) \cdot \left(\frac{\vec{i}}{\sqrt{2}} - \frac{\vec{j}}{\sqrt{2}} \right) = \frac{1 - e}{\sqrt{2}} \approx -1.2.$$

Example 4   Suppose the function $\rho = f(x, y, z)$ gives the density ρ mg/m^3 at point (x, y, z) of an impurity in a cloud of gas. Suppose that x, y, z are in meters and that the gas is contained in the region $-3 \leq x \leq 3, -3 \leq y \leq 3, 0 \leq z \leq 2$ and the function is given by

$$\rho = f(x, y, z) = -2x^2 - 3y^2 + \frac{1}{z + 1}.$$

(a)   What is the physical significance of the level surface

$$\rho = f(x, y, z) = 1 ?$$

(b)   In what direction should we move from the point $(2, 1, 1)$ to achieve the greatest rate of change in density of the impurity? What is that greatest rate of change?

(c)   In what direction should we move from the point $(2, 1, 1)$ to achieve the least (that is, the most negative) rate of change of density? What is that least rate of change?

(d)   Find a vector such that the rate of change of density at $(2, 1, 1)$ in that direction is zero.

Solution   (a)   The surface $\rho = f(x, y, z) = 1$ contains all the points where the density of the impurity is 1.
(b)   The directional derivative in the direction of a unit vector $\vec{u}$ at the point $(2, 1, 1)$ is given by

$$f_{\vec{u}}(2, 1, 1) = \text{grad } f(2, 1, 1) \cdot \vec{u}$$

Therefore, if $\vec{u}$ is in the direction of grad ρ we get the greatest rate of change. Thus, the direction we want is given by the vector

$$\text{grad } \rho = f_x \vec{i} + f_y \vec{j} + f_z \vec{k} = -4x\vec{i} - 6y\vec{j} - \frac{\vec{k}}{(z + 1)^2},$$

evaluated at the point $(2, 1, 1)$, giving

$$\text{grad } \rho = -8\vec{i} - 6\vec{j} - \frac{\vec{k}}{4}.$$

The greatest rate of change of ρ, which is in the direction of grad ρ, is given by

$$\| \text{grad } \rho \| = \sqrt{(-8)^2 + (-6)^2 + \left(\frac{-1}{4} \right)^2} \approx 10 \frac{\text{mg/m}^3}{\text{m}}.$$

(c)   Since the directional derivative in the direction of $\vec{u}$ is found by taking the dot product of $\vec{u}$ and grad ρ at the point $(2, 1, 1)$, the minimum (most negative) directional derivative occurs in the direction opposite to grad ρ at the point. Thus, the direction we want is given by the vector

$$- \text{grad}\, \rho = 8\vec{i} + 6\vec{j} + \frac{\vec{k}}{4}.$$

The rate of change of density in this direction is

$$-\| \text{grad}\, \rho \| \approx -10\frac{\text{mg/m}^3}{\text{m}}.$$

(d)   Moving a direction $\vec{u}$ which is perpendicular to grad ρ will make the rate of change of ρ zero (because the dot product of grad ρ and $\vec{u}$ will be zero). Thus, any vector $\vec{u}$ satisfying

$$\text{grad}\, \rho \cdot \vec{u} = (-8\vec{i} - 6\vec{j} - \frac{\vec{k}}{4}) \cdot (u_1\vec{i} + u_2\vec{j} + u_3\vec{k}) = -8u_1 - 6u_2 - \frac{u_3}{4} = 0$$

will do. Since we have only one equation to determine three quantities u_1, u_2, u_3, we can pick two arbitrarily, say $u_2 = 8$, $u_3 = 0$, giving $u_1 = -6$. Thus, the rate of change of ρ in the direction of the vector

$$\vec{u} = -6\vec{i} + 8\vec{j}$$

will be zero. There are many of other possible vectors obtained by choosing other values for u_2 and u_3.

Problems for Section 13.5

1.   True or false? $f_{\vec{u}}(x_0, y_0)$ is a scalar. Explain your answer.

2.   True or false? Explain your answer. Suppose that f has continuous partial derivatives. For any point P in the domain of f, there is always a direction in which the rate of change of f is 0.

3.   If $f(x, y) = x^2 + \ln y$, find (a) ∇f,   (b) ∇f at $(4, 1)$.

In Problems 4–13 compute the gradient of the given function.

4. $f(m, n) = m^2 + n^2$

5. $f(x, y) = \frac{3}{2}x^5 - \frac{4}{7}y^6$

6. $f(s, t) = \frac{1}{\sqrt{s}}(t^2 - 2t + 4)$

7. $f(x, y) = \sin(xy) + \cos(xy)$

8. $f(s, t) = t \cos^2 s$

9. $f(x, y) = y \ln x$

10. $f(\alpha, \beta) = \sqrt{5\alpha^2 + \beta}$

11. $f(\alpha, \beta) = \frac{2\alpha + 3\beta}{2\alpha - 3\beta}$

12. $f(x, y) = xye^{\sqrt{x}}$

13. $f(u, v) = \frac{u^2 - \sqrt{v}}{u + v}$

In Problems 14–21 find ∇z.

14. $z = \sin(x/y)$ 15. $z = xe^y$

16. $z = (x + y)e^y$ 17. $z = \tan^{-1}(x/y)$

18. $z = \sin(x^2 + y^2)$ 19. $z = xe^y/(x + y)$

20. $z = \sqrt{x^2 + y^2}$ 21. $z = \tan^{-1}(x + y)$

In Problems 22–26 compute the gradient at the specified point.

22. $f(m, n) = 5m^2 + 3n^4$, at $(5, 2)$ 23. $f(x, y) = x^2y + 7xy^3$, at $(1, 2)$

24. $f(x, y) = \sqrt{\tan x + y}$, at $(0, 1)$ 25. $f(x, y) = \sin(x^2) + \cos y$, at $(\frac{\sqrt{\pi}}{2}, 0)$

26. $f(\alpha, \beta) = \dfrac{5\alpha + 3\beta}{\sqrt{\beta}}$, at $(0, 1)$

27. If $f(x, y) = x^2y$ and $\vec{v} = 4\vec{i} - 3\vec{j}$, find the directional derivative at the point $(2, 6)$ in the direction of $\vec{v}$.

28. If $f(x, y, z) = x^2 + 3xy + 2z$, find the directional derivative at the point $(2, 0, -1)$ in the direction of $2\vec{i} + \vec{j} - 2\vec{k}$.

29. Find the directional derivative of $f(x, y, z) = 3x^2y^2 + 2yz$ at the point $(-1, 0, 4)$ in the directions (a) $\vec{i} - \vec{k}$ and   (b) $-\vec{i} + 3\vec{j} + 3\vec{k}$.

30. Find the directional derivative of $f(x, y) = e^x \tan(y) + 2x^2y$ at the point $(0, \frac{\pi}{4})$ in the directions (a) $\vec{i} - \vec{j}$ and   (b) $\vec{i} + \sqrt{3}\vec{j}$.

31. Find the directional derivative of $z = x^2y$ at the point $(1, 2)$ in the direction $\theta = 5\pi/4$. In which direction is the directional derivative the largest?

32. Find the rate of change of $f(x, y) = x^2 + y^2$ at the point $(1, 2)$ in the direction of the vector $\vec{u} = 0.6\vec{i} + 0.8\vec{j}$.

33. Let $f(x, y) = (x + y)/(1 + x^2)$ and let $P = (1, -2)$. Find the directional derivative at P in the direction of the vectors
    (a) $\vec{v} = 3\vec{i} - 2\vec{j}$.   (b) $\vec{v} = -\vec{i} + 4\vec{j}$.
    (c) What is the direction of greatest increase at P ?

34. A student was asked to find the directional derivative of $f(x, y) = x^2e^y$ at the point $(1, 0)$ in the direction of $\vec{v} = 4\vec{i} + 3\vec{j}$. His answer was

$$f_{\vec{u}}(1, 0) = \nabla f(1, 0) \cdot \vec{u} = \frac{8}{5}\vec{i} + \frac{3}{5}\vec{j}.$$

    (a) At a glance, how do you know this is wrong?
    (b) What is the correct answer?

35. Find the rate of change of $f(x, y) = xe^y$ at the point $(1, 1)$ in the direction of the vector $\vec{i} + 2\vec{j}$.

36. The directional derivative of $z = f(x, y)$ at the point $A = (2, 1)$ in the direction towards the point $B = (1, 3)$ is $-2/\sqrt{5}$ and the directional derivative in the direction towards the point $C = (5, 5)$ is 1. Compute $\partial f/\partial x$ and $\partial f/\partial y$ at the point A.

37. (a) Sketch the surface $z = y^2$ in three dimensions.
    (b) Sketch the level curves on this surface in the xy plane.
    (c) If you are standing on the surface at the point $(2, 3, 9)$, in which direction should you move to climb the fastest? (Give your answer as a 2-vector.)

38. Given $z = y - \sin x$,
    (a) Sketch the contours for $z = -1, 0, 1, 2$.
    (b) A bug starts on the surface at the point $(\pi/2, 1, 0)$ and walks on the surface in the direction of the vector $(0, 1, 0)$. Is the bug walking in a valley or on top of a ridge? Explain.
    (c) On the contour for $z = 0$ above, draw the gradients of z at $x = 0$, $x = \pi/2$, and $x = \pi$.

39. Suppose that $F(x, y, z) = x^2 + y^4 + x^2 z^2$ gives the concentration of salt in a fluid at the point (x, y, z), and that you are at the point $(-1, 1, 1)$.
    (a) In which direction should you move if you want the concentration to increase the fastest?
    (b) Suppose you start to move in the direction you found in part (a) at a speed of 4 units/sec. How fast is the concentration changing? Explain your answer.

40. Consider the function $f(x, y)$. If you start at the point $(4, 5)$ and move towards the point $(5, 6)$, the directional derivative is 2. Starting at the point $(4, 5)$ and moving towards the point $(6, 6)$ gives a directional derivative of 3. Find ∇f at the point $(4, 5)$.

41. The temperature at any point in the plane is given by the function

$$T(x, y) = \frac{100}{x^2 + y^2 + 1}.$$

    (a) Where on the plane is it hottest? What is the temperature at that point?
    (b) Find the direction of the greatest increase in temperature at the point $(3, 2)$. What is the magnitude of that greatest increase?
    (c) Find the direction of the greatest decrease in temperature at the point $(3, 2)$.
    (d) Does the vector you found in part (b) point towards the origin?
    (e) Find a direction at the point $(3, 2)$ in which the temperature does not increase or decrease.
    (f) What shape are the level curves of T?

42. Consider S to be the surface represented by the equation $F = 0$, where

$$F(x, y, z) = x^2 - \left(\frac{y}{z^2} \right).$$

    (a) Find all points on S where a normal vector is parallel to the xy-plane.
    (b) Find the tangent plane to S at the points $(0, 0, 1)$ and $(1, 1, 1)$.
    (c) Find the unit vectors $\vec{u}_1$ and $\vec{u}_2$ pointing in the direction of maximum increase of F at the points $(0, 0, 1)$ and $(1, 1, 1)$ respectively.

43. Consider the functions $f(x, y) = x - y^2$ and $g(x, y) = 2x + \ln y$. Sketch three level curves for each function. Show that the level curves of g always intersect the level curves of f at right angles.

44. The sketch in Figure 13.34 shows the level curves of a function $z = f(x, y)$. At the points $(1, 1)$ and $(1, 4)$ on the sketch, draw a vector representing grad f. Explain how you decided the approximate direction and length of each vector.

45. Figure 13.35 represents the level curves $f(x, y) = c$; the values of f on each curve are marked. In each of the following parts, decide whether the given quantity is positive, negative or zero. Explain your answer.

    (a) The value of $\nabla f \cdot \vec{i}$ at P.   (b) The value of $\nabla f \cdot \vec{j}$ at P.   (c) $\partial f / \partial x$ at Q.
    (d) $\partial f / \partial y$ at Q.

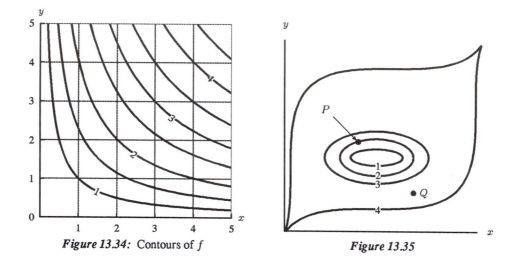

Figure 13.34: Contours of f *Figure 13.35*

46. In Figure 13.35, which is larger: $\|\nabla f\|$ at P or $\|\nabla f\|$ at Q? Explain how you know.

47. In Problem 33 on page 44, we considered the temperature distribution given by the function

$$T(x, y) = 100 - x^2 - y^2$$

and asked in which direction a heat-seeking bug should move from any point in the plane to increase its temperature fastest. Now answer this question using the gradient.

13.6   THE DIFFERENTIAL OF A FUNCTION

Although the gradient vector can be very useful, there are situations in which it is not appropriate. For instance, in Example 7 on page 118, we looked at a function $H(x, t)$ that gave temperature in a room as a function of distance from a heater and time. Then grad $H = H_x \vec{i} + H_t \vec{j}$ is a vector whose first component is in units of degrees/foot and whose second component is in units of degrees/minute. In situations like this in which the variables (here, x and t) have different units, the components of the gradient have different units. Then it is usually impossible to assign units to the magnitude of the gradient vector. Similarly, directional derivatives are only physically meaningful in such situations along the axes — that is, the partial derivatives are meaningful, but directional derivatives in other directions are not.

Such situations are better understood by shifting the focus from the rate of change of a function at a point to the increment, or change, in the value of a function between two points. We will

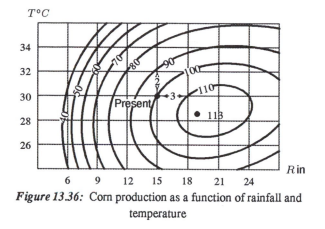

Figure 13.36: Corn production as a function of rainfall and temperature

approximate the increment using local linearity. Because of our new point of view, we will adopt a different notation for local linearity.

Example 1 Figure 13.36 shows the corn production function $C = f(R, T)$ discussed in Section 11.4, page 30. To estimate the effect of climatic change on agriculture, we are interested in the effect on the function of small deviations from the current climate, which is represented by the point $(15, 30)$. Use local linearization to study this effect.

Solution Since we are interested in what happens near the point $(15, 30)$ which represents the current climate, we will use the local linearization

$$f(R, T) \approx f(15, 30) + f_R(15, 30)(R - 15) + f_T(15, 30)(T - 30).$$

It is convenient to rewrite this in terms of small changes in the variables. Since R will stay close to 15, we let ΔR stand for a small change in R from 15, that is, $\Delta R = R - 15$. Similarly, we write $\Delta T = T - 30$, and $\Delta f = f(R, T) - f(15, 30)$. Using this notation, we rewrite the local linearization as

$$\Delta f \approx f_R(15, 30)\,\Delta R + f_T(15, 30)\,\Delta T.$$

We estimate the partial derivatives from the contour diagram:

$$f_R(15, 30) \approx \frac{f(18, 30) - f(15, 30)}{3} = \frac{110 - 100}{3} = \frac{10}{3} = 3.3$$

$$f_T(15, 30) \approx \frac{f(15, 32) - f(15, 30)}{2} = \frac{90 - 100}{2} = \frac{-10}{2} = -5.$$

Thus, the change in f can be expressed

$$\Delta f \approx 3.3\Delta R - 5\Delta T.$$

This approximation can be used to estimate the effect of small climatic changes. For example, if rainfall increases by half an inch and temperature decreases by one degree, then $\Delta R = 0.5$ and $\Delta T = -1$, so

$$\Delta f \approx (3.3)(0.5) - (5)(-1) = 6.65.$$

So corn production will increase by about 6.65%.

The Differential

The way we wrote local linearity above is useful in general. We let $\Delta x = x - a$, $\Delta y = y - b$, and $\Delta f = f(x, y) - f(a, b)$. Then the local linearization

$$f(x, y) \approx f(a, b) + f_x(a, b)(x - a) + f_y(a, b)(y - b)$$

becomes

$$\Delta f \approx f_x(a, b)\Delta x + f_y(a, b)\Delta y.$$

It is useful to have an idealized version of this equation, where the small changes Δf, Δx, and Δy are replaced by df, dx, and dy:

The Differential of a Function $f(x, y)$

The differential is

$$df = f_x \, dx + f_y \, dy.$$

The differential at a point (a, b) is

$$df(a, b) = f_x(a, b) \, dx + f_y(a, b) \, dy.$$

Example 2   Compute the differentials of the following functions.

     (a) $f(x, y) = x^2 e^{5y}$.   (b) $z = x \sin(xy)$.   (c) $f(x, y) = x \cos(2x)$.

Solution   (a)   Since $f_x(x, y) = 2xe^{5y}$ and $f_y(x, y) = 5x^2 e^{5y}$, we have

$$df = 2xe^{5y} \, dx + 5x^2 e^{5y} \, dy.$$

     (b)   Since $\partial z / \partial x = \sin(xy) + xy \cos(xy)$ and $\partial z / \partial y = x^2 \cos(xy)$,

$$dz = (\sin(xy) + xy \cos(xy)) \, dx + x^2 \cos(xy) \, dy.$$

     (c)   Since $f_x(x, y) = \cos(2x) - 2x \sin(2x)$ and $f_y(x, y) = 0$,

$$df = (\cos(2x) - 2x \sin(2x)) \, dx + 0 \, dy = (\cos(2x) - 2x \sin(2x)) \, dx.$$

You may find the differential confusing because we haven't said what df, dx, and dy really are. In practice you can almost always think of them as infinitesimally small changes in f, x, and y. However, in standard mathematical treatments of calculus there is no place for infinitesimal quantities; if you think about them for a while, you will see that it is difficult to say exactly what is meant by an infinitesimal. Nonetheless, differentials are useful because when you replace the infinitesimal changes by small (but real) changes, they give the same approximations you get from local linearity. In addition, they are easier to manipulate than the local linearization.

Example 3   The density ρ of carbon dioxide gas CO_2 depends upon its temperature T and pressure P. The ideal gas model for CO_2 gives what is called the state equation

$$\rho = \frac{0.5363 P}{T + 273.15},$$

where ρ is in g/cm^3, T in degrees centigrade, and P in atmospheres.

     (a)   Compute the differential $d\rho$.

(b)   Use the differential to estimate the change in density that the gas would undergo if its temperature and pressure were changed from 50°C and 5 atm to 45°C and each of the three pressures 5.1 atm, 5.2 atm, and 5.3 atm.

Solution   (a)   The definition of the differential tells us that for $\rho = f(T, P)$

$$d\rho = f_T(T, P)\, dT + f_P(T, P)dP$$
$$= \frac{-0.5363P}{(T + 273.15)^2}\, dT + \frac{0.5363}{T + 273.15}\, dP.$$

(b)   Since we will make all three approximations using the differential $d\rho$ at the same point $(T, P) = (50, 5)$, we substitute $T = 50$ and $P = 5$ into the differential:

$$d\rho = f_T(50, 5)\, dT + f_P(50, 5)dP$$
$$= \frac{-0.5363(5)}{(50 + 273.15)^2}\, dT + \frac{0.5363}{50 + 273.15}\, dP$$
$$= -2.57 \times 10^{-5}\, dT + 1.66 \times 10^{-3}\, dP.$$

Notice that the coefficient of dT is negative because increasing the temperature will expand the gas (if the pressure is kept constant) and therefore decrease its density. The coefficient of dP is positive, because increasing the pressure will compress the gas (if the temperature is kept constant) and therefore increase its density.

Now we compute the three estimates asked using the approximation $\Delta P \approx dP$ and $\Delta T \approx dT$.

(i)   If (T, P) changes to $(45, 5.1)$, then $\Delta T = -5$ and $\Delta P = 0.1$. So

$$\Delta\rho \approx -2.57 \times 10^{-5}(-5) + 1.66 \times 10^{-3}(0.1) \approx 0.0003 \text{ gm/cm}^3.$$

(ii)   If (T, P) changes to $(45, 5.2)$, then $\Delta T = -5$ and $\Delta P = 0.2$. So

$$\Delta\rho \approx -2.57 \times 10^{-5}(-5) + 1.66 \times 10^{-3}(0.2) \approx 0.0005 \text{ gm/cm}^3.$$

(iii)   If (T, P) changes to $(45, 5.3)$, then $\Delta T = -5$ and $\Delta P = 0.3$. So

$$\Delta\rho \approx -2.57 \times 10^{-5}(-5) + 1.66 \times 10^{-3}(0.3) \approx 0.0006 \text{ gm/cm}^3.$$

The differential for functions of one variable, or for functions of more than two variables, is defined in an analogous way.

The differential of a one-variable function $f(x)$ is

$$df = f'\, dx.$$

The differential of an n-variable function $f(x_1, x_2, \ldots, x_n)$ is

$$df = f_{x_1}\, dx_1 + f_{x_2}\, dx_2 + \cdots + f_{x_n}\, dx_n.$$

Problems for Section 13.6

1. If $z = e^{-x}\cos(y)$ find dz.

Find the differentials of the functions in Problems 2– 4.

2. $f(x, y) = \sin(xy)$.

3. $h(x, t) = e^{-3t}\sin(x+5t)$.

4. $g(u, v) = u^2 + uv$.

Find differentials of the functions in Problems 5–8 at the given point.

5. $f(x, y) = xe^{-y}$ at $(1, 0)$

6. $g(x, t) = x^2\sin(2t)$ at $(2, \frac{\pi}{4})$

7. $P(K, L) = 1.01K^{0.75}L^{0.25}$ at $(1, 100)$

8. $F(m, r) = Gm/r^2$ at $(100, 10)$

9. Find the differential of $f(x, y) = \sqrt{x^2 + y^3}$ at the point $(1, 2)$. Use it to estimate $f(1.04, 1.98)$.

10. For the Cobb-Douglas production function $P = 40L^{0.25}K^{0.75}$, find the differential dP when $L = 2$ and $K = 16$.

11. The area of a rectangle with sides x and y is given by $z = xy$. Find the approximate maximum error dz if the measured length of the sides are $x = 5$ feet and $y = 6$ feet with maximum errors of 0.01 feet each.

12. One mole of ammonia gas is contained in a vessel which is capable of changing its volume (imagine a piston, for example). The total energy, U (in Joules), of the ammonia, is a function of the volume, V (in cubic meters), of the container, and the temperature, T (in degrees Kelvin), of the gas. The differential dU is given by

$$dU = 840\,dV + 27.32\,dT.$$

(a) How does the energy change if the volume is held constant and the temperature is increased slightly?

(b) How does the energy change if the temperature is held constant and the volume is increased slightly?

(c) Find the change in energy if the gas is compressed by 100 cm³ and heated by 2° K.

13. The radius of a right circular cone is 4 inches and its altitude is 9 inches. If the maximum error in the radius is 0.01 inches and in the altitude is 0.02 inches, estimate the error in the computed volume.

14. The value of π is to be determined empirically by measuring the perimeter, L, and the area A of a circle. Show that if the measured value of L and A are in error by λ and μ respectively then the resulting error in π is $2\lambda - \mu$.

15. The area of a triangle can be calculated from the formula $S = \frac{1}{2}ab\sin(C)$. Show that if an error of $\frac{\pi}{1800}$ radians is made in measuring C then the error in S is approximately $\frac{\pi}{1080}S\cot C$.

16. The period of oscillation of a pendulum clock is $2\pi\sqrt{\frac{l}{g}}$, where l depends on the temperature according to $l = l_0(1 + \alpha t)$ for some value of α which characterizes the clock. The clock is set to the correct period for temperature $= t_0$. How many seconds a day does the clock lose when the temperature is slightly greater than t_0 and show that this loss is independent of l_0.

17.   In a pendulum experiment, g was determined using the formula

$$g = \frac{4\pi^2 l}{T^2}$$

where the length l is given by $l = s + \frac{k^2}{s}$, $k < s$. If the measurements of k and s are accurate to within 1% find the maximum percentage error in l. If the measurement of T is accurate to $\frac{1}{2}$% find the maximum percentage error in the computed value of g.

18.   The differential dU of the energy of a body of gas is given by

$$dU = \alpha\, dT + \beta\, dV,$$

where α and β are constants that depend on the gas, T and V are the temperature and volume of the gas. The heat capacity of a gas is a measure of how much heat it can absorb. It turns out that the heat capacity is one of the constants α, β. Which one is it and why?

19.   The coefficient of thermal expansion of a liquid, β, relates the change in its volume V (in m^3) to an increase in its temperature T (in degrees Celsius):

$$dV = \beta V\, dT.$$

(a)   Let ρ be the density (in kg/m^3) of 1 kg of water as a function of temperature. Write an expression for $d\rho$ in terms of ρ and dT.

(b)   The graph below shows density of water as a function of temperature. Use it to estimate β when $T = 20$ and when $T = 80$.

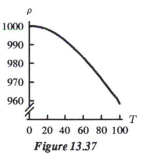

Figure 13.37

20.   The gas equation for one mole of oxygen relates its pressure in atmospheres, its temperature in degrees Kelvin, and its volume in cubic decimeters:

$$T = 16.574\frac{1}{V} - 0.52754\frac{1}{V^2} + 0.3879P + 12.187VP.$$

(a)   Find the temperature T and differential dT if the volume of gas is 25 dm^3 and the pressure is 1 atmosphere.

(b)   Use your answer to part (a) to estimate how much the volume of oxygen would have to change if the pressure increased by 0.1 atmosphere and the temperature remained constant.

13.7   THE CHAIN RULE

How Do We Compose Functions of Many Variables?

The purpose of the chain rule is to enable us to differentiate composite functions, so we start by thinking about what it means to compose functions. To compose functions of one variable we substitute one function into another. A function of more than one variable has more than one place you can substitute a function; for example, if $z = f(x, y)$, then we could substitute two functions, $x = g(t)$, and $y = h(t)$. Then the composite function is $f(g(t), h(t))$. Although it looks complicated, this is now a function of just one variable, t. The next example shows where this sort of substitution arises in practice, and how it leads to the chain rule.

Example 1   Figure 13.38 shows the contour diagram of the corn production, C, as a function of temperature, T, and rainfall, R (Example 1 on page 151). Figures 13.39 and 13.40 show how rainfall and temperature are predicted to vary with time according to a certain model of global warming. Estimate the change in corn production between the year 2000 and the year 2001. Hence, estimate dC/dt when $t = 2000$.

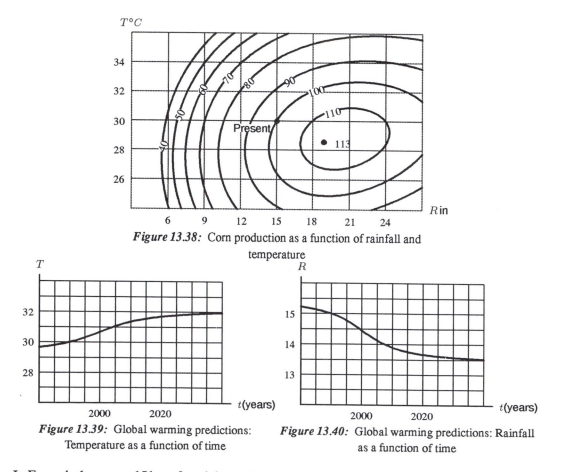

Figure 13.38: Corn production as a function of rainfall and temperature

Figure 13.39: Global warming predictions: Temperature as a function of time

Figure 13.40: Global warming predictions: Rainfall as a function of time

Solution   In Example 1 on page 151 we found that, as long as the temperature and rainfall stay close to their current values of $R = 15$ and $T = 30$, a change ΔR in rainfall and a change ΔT in temperature will produce a change in corn production given by

$$\Delta C \approx 3.3 \Delta R - 5 \Delta T.$$

Now we have a situation where both R and T are functions of time t (in years from 1990), and we need to find the effect of a small change in time, Δt, on R and T. Looking at Figure 13.40, we see that the slope of the graph for R versus t is about $-1/15 \approx -0.07$ when $t = 2000$. Similarly, the slope of the graph of T is about $1/15 \approx 0.07$ when $t = 2000$. Thus, around the year 2000,

$$\Delta R \approx -0.07 \Delta t, \quad \text{and} \quad \Delta T \approx 0.07 \Delta t.$$

Substituting these into the equation for ΔC, we get

$$\Delta C \approx (3.3)(-0.07)\Delta t - (5)(0.07)\Delta t \approx -0.6 \Delta t.$$

Since at present $C = 100$, corn production will decline by about 0.6 % between the years 2000 and 2001.

Now $\Delta C \approx -0.6 \Delta t$ tells us that when $t = 2000$,

$$\frac{\Delta C}{\Delta t} \approx -0.6 \quad \text{and therefore that} \quad \frac{dC}{dt} \approx -0.6.$$

The Chain Rule

In general, given $z = f(x, y)$, where each of x and y is itself a function of t, that is, if $x = g(t)$ and $y = h(t)$, the local linearizations for Δx and Δy in terms of Δt may be substituted into the local linearization for Δz in terms of Δx and Δy, to obtain the local linearization for Δz in terms of Δt. We will formulate this in terms of differentials in the following example and thus get one version of the multivariable chain rule.

Example 2   Suppose that $z = f(x, y)$, $x = m(t)$, and $y = n(t)$ are three differentiable functions. Find a general expression for dz/dt.

Solution   On one hand, the derivative dz/dt that we want to compute is the coefficient of dt in the equation

$$dz = \frac{dz}{dt} \, dt.$$

On the other hand, we have the local linearization

$$dz = \frac{\partial z}{\partial x} \, dx + \frac{\partial z}{\partial y} \, dy,$$

into which we can substitute

$$dx = \frac{dx}{dt} \, dt \quad \text{and} \quad dy = \frac{dy}{dt} \, dt,$$

yielding

$$dz = \frac{\partial z}{\partial x}\frac{dx}{dt} \, dt + \frac{\partial z}{\partial y}\frac{dy}{dt} \, dt$$

$$= \left(\frac{\partial z}{\partial x}\frac{dx}{dt} + \frac{\partial z}{\partial y}\frac{dy}{dt} \right) dt.$$

Comparison of the coefficients of dt in the first and last formula for dz yields one version of the multivariable chain rule.

A Chain Rule Identity

$$\frac{dz}{dt} = \frac{\partial z}{\partial x}\frac{dx}{dt} + \frac{\partial z}{\partial y}\frac{dy}{dt}$$

The next example shows how this identity can be used.

Example 3   Suppose that $z = f(x, y) = x \sin y$, where $x = t^2$ and $y = 2t + 1$. Let $z = g(t)$. Compute $g'(t)$ by two different methods.

Solution   Since $z = g(t) = f(t^2, 2t + 1) = t^2 \sin(2t + 1)$, it is possible to compute $g'(t)$ directly with one-variable methods:

$$g'(t) = t^2 \frac{d}{dt}(\sin(2t + 1)) + (\frac{d}{dt}(t^2))\sin(2t + 1)$$
$$= 2t^2 \cos(2t + 1) + 2t \sin(2t + 1).$$

The chain rule provides an alternative route to the same answer. We have

$$\frac{dz}{dt} = \frac{\partial z}{\partial x}\frac{dx}{dt} + \frac{\partial z}{\partial y}\frac{dy}{dt} = (\sin y)(2t) + (x \cos y)(2)$$
$$= 2t \sin(2t + 1) + 2t^2 \cos(2t + 1).$$

For functions of one variable, there is only one version of the chain rule. For functions of several variables, however, there are several different versions of the chain rule. We now consider the case when $z = f(x, y)$ and x and y are both functions of two new variables.

Example 4   Suppose that $z = f(x, y)$, $x = m(u, v)$, and $y = n(u, v)$ are three differentiable functions. Find general expressions for $\partial z/\partial u$ and $\partial z/\partial v$.

Solution   The partial derivatives $\partial z/\partial u$ and $\partial z/\partial v$ that we are interested in are the coefficients of du and dv in the equation

$$dz = \frac{\partial z}{\partial u}du + \frac{\partial z}{\partial v}dv.$$

The chain rule will give us a second formula for dz. Start with the linearizations

$$dz = \frac{\partial z}{\partial x}dx + \frac{\partial z}{\partial y}dy,$$
$$dx = \frac{\partial x}{\partial u}du + \frac{\partial x}{\partial v}dv,$$
$$dy = \frac{\partial y}{\partial u}du + \frac{\partial y}{\partial v}dv.$$

Substituting the formulas for dx and dy into the formula for dz gives

$$dz = \frac{\partial z}{\partial x}\left(\frac{\partial x}{\partial u}du + \frac{\partial x}{\partial v}dv\right) + \frac{\partial z}{\partial y}\left(\frac{\partial y}{\partial u}du + \frac{\partial y}{\partial v}dv\right)$$

$$= \left(\frac{\partial z}{\partial x}\frac{\partial x}{\partial u} + \frac{\partial z}{\partial y}\frac{\partial y}{\partial u}\right)du + \left(\frac{\partial z}{\partial x}\frac{\partial x}{\partial v} + \frac{\partial z}{\partial y}\frac{\partial y}{\partial v}\right)dv.$$

Comparison of coefficients of du and dv in the two formulas for dz yields another version of the multivariable chain rule:

A Chain Rule Identity

$$\frac{\partial z}{\partial u} = \frac{\partial z}{\partial x}\frac{\partial x}{\partial u} + \frac{\partial z}{\partial y}\frac{\partial y}{\partial u}$$

$$\frac{\partial z}{\partial v} = \frac{\partial z}{\partial x}\frac{\partial x}{\partial v} + \frac{\partial z}{\partial y}\frac{\partial y}{\partial v}$$

Example 5   Let $w = x^2 e^y$, $x = 4u$, and $y = 3u^2 - 2v$. Compute $\partial w/\partial u$ and $\partial w/\partial v$ using the chain rule.

Solution   We have

$$\frac{\partial w}{\partial u} = \frac{\partial w}{\partial x}\frac{\partial x}{\partial u} + \frac{\partial w}{\partial y}\frac{\partial y}{\partial u}$$

$$= 2xe^y(4) + x^2 e^y(6u)$$

$$= (8x + 6x^2 u)e^y$$

$$= (32u + 96u^3)e^{3u^2 - 2v}.$$

and

$$\frac{\partial w}{\partial v} = \frac{\partial w}{\partial x}\frac{\partial x}{\partial v} + \frac{\partial w}{\partial y}\frac{\partial y}{\partial v}$$

$$= 2xe^y(0) + x^2 e^y(-2)$$

$$= -2x^2 e^y$$

$$= -32u^2 e^{3u^2 - 2v}.$$

A Thermodynamic Example

A chemist investigating the properties of carbon dioxide may want to know how the internal energy U of 1 kilogram of CO_2 depends on its temperature, T, pressure, P, and volume, V. Since any one of the three quantities T, P, and V is determined by the other two, the internal energy U can be thought of as a function of T and P, or of T and V, or of P and V.

The chemist will write, for example, $(\frac{\partial U}{\partial T})_P$ to indicate the partial derivative of U with respect to T *holding P constant*, meaning that for this computation U is viewed as a function of T and P. If U is to be viewed as a function of T and V, then the chemist would write $(\frac{\partial U}{\partial T})_V$ for the partial derivative of U with respect to T holding V constant.

The notation can be summarized by three formulas for the differential dU.

$$dU = \left(\frac{\partial U}{\partial T}\right)_P dT + \left(\frac{\partial U}{\partial P}\right)_T dP,$$

$$dU = \left(\frac{\partial U}{\partial T}\right)_V dT + \left(\frac{\partial U}{\partial V}\right)_T dV,$$

$$dU = \left(\frac{\partial U}{\partial P}\right)_V dP + \left(\frac{\partial U}{\partial V}\right)_P dV.$$

All of the six partial derivatives of U appearing in the above formulas have physical meaning, but they are not all equally easy to measure experimentally in the laboratory. An identity among the partial derivatives may make it possible to evaluate one of the partials in terms of others that are more easily measured. Here is a typical example.

Example 6   Find a way to evaluate $\left(\dfrac{\partial U}{\partial T}\right)_P$ experimentally.

Solution   Since V is a function of P and T, we have

$$dV = \left(\frac{\partial V}{\partial T}\right)_P dT + \left(\frac{\partial V}{\partial P}\right)_T dP.$$

Substitution into the second formula above for dU gives

$$dU = \left(\frac{\partial U}{\partial T}\right)_V dT + \left(\frac{\partial U}{\partial V}\right)_T \left(\left(\frac{\partial V}{\partial T}\right)_P dT + \left(\frac{\partial V}{\partial P}\right)_T dP\right)$$

$$= \left(\left(\frac{\partial U}{\partial T}\right)_V + \left(\frac{\partial U}{\partial V}\right)_T \left(\frac{\partial V}{\partial T}\right)_P\right) dT + \left(\frac{\partial U}{\partial V}\right)_T \left(\frac{\partial V}{\partial P}\right)_T dP.$$

Comparison with the first formula above for dU yields the two identities

$$\left(\frac{\partial U}{\partial T}\right)_P = \left(\frac{\partial U}{\partial T}\right)_V + \left(\frac{\partial U}{\partial V}\right)_T \left(\frac{\partial V}{\partial T}\right)_P$$

$$\left(\frac{\partial U}{\partial P}\right)_T = \left(\frac{\partial U}{\partial V}\right)_T \left(\frac{\partial V}{\partial P}\right)_T$$

The first identity expresses $(\frac{\partial U}{\partial T})_P$, the partial we are interested in, in terms of three others. It happens that two of them, namely $(\frac{\partial U}{\partial T})_V$, called the heat capacity, and $(\frac{\partial V}{\partial T})_P$, called the expansion coefficient, can be easily measured experimentally. The third, the internal pressure $(\frac{\partial U}{\partial V})_T$, cannot easily be measured directly, but it can be related to $(\frac{\partial P}{\partial T})_V$, which is measurable. (The second law of thermodynamics, which is not a mathematical but a physical principle, shows that $(\frac{\partial U}{\partial V})_T = T(\frac{\partial P}{\partial T})_V - P$.) Thus, $(\frac{\partial U}{\partial T})_P$ can be determined indirectly using the identity we have derived for it.

Problems for Section 13.7

For Problems 1–7, find dz/dt using the chain rule.

1. $z = xy^2$, $x = e^{-t}$, $y = \sin t$.

2. $z = x \sin y + y \sin x$, $x = t^2$, $y = \ln t$.

3. $z = \ln(x^2 + y^2)$, $x = 1/t$, $y = \sqrt{t}$.

4. $z = \sin(x/y)$, $x = 2t$, $y = 1 - t^2$.

5. $z = xe^y$, $x = 2t$, $y = 1 - t^2$.

6. $z = (x + y)e^y$, $x = 2t$, $y = 1 - t^2$.

7. $z = \tan^{-1}(x/y)$, $x = 2t$, $y = 1 - t^2$.

For Problems 8–17, find $\partial z/\partial u$ and $\partial z/\partial v$.

8. $z = xe^{-y} + ye^{-x}$, $x = u \sin v$, $y = v \cos u$.

9. $z = \cos(x^2 + y^2)$, $x = u \cos v$, $y = u \sin v$.

10. $z = xe^y$, $x = \ln u$, $y = u$.

11. $z = (x + y)e^y$, $x = \ln u$, $y = u$.

12. $z = \tan^{-1}(x/y)$, $x = \ln u$, $y = u$.

13. $z = \sin(x/y)$, $x = u^2 + v^2$, $y = u^2 - v^2$.

14. $z = xe^y$, $x = u^2 + v^2$, $y = u^2 - v^2$.

15. $z = (x+y)e^y$, $x = u^2 + v^2$, $y = u^2 - v^2$.

16. $z = \tan^{-1}(x/y)$, $x = u^2 + v^2$, $y = u^2 - v^2$.

17. $z = \sin(x/y)$, $x = \ln u$, $y = u$.

18. When you are diving in the ocean, the pressure, P, that you experience is a function of ρ, the density of water, and h, your depth below the surface. Suppose you are descending into water of variable density in such a way that, at one moment, $d\rho/dt = 5$ gm/cc/sec and $dh/dt = 2$ cm/sec. At the same moment, suppose $\partial P/\partial \rho = 1$ and $\partial P/\partial h = 2$. Find how fast the pressure is changing at that moment. Explain your reasoning.

19. Suppose $w = f(x, y, z) = 3xy + yz$ and that x, y, z are functions of u and v such that

$$x = \ln u + \cos v, \quad y = 1 + u \sin v, \quad z = uv.$$

(a) Find $\partial w/\partial u$ and $\partial w/\partial v$ at $(u, v) = (1, \pi)$.

(b) Suppose now that u and v are also functions of t such that

$$u = 1 + \sin(\pi t), \quad v = \pi t^2.$$

Use your answer to part (a) to find dw/dt at $t = 1$.

20. Suppose $f(x, y)$ is a function of x and y and define

$$g(u, v) = f(e^u + \cos v, e^u + \sin v).$$

Find $g_u(0, 0)$ given that

$f(0, 0) = \pi$,	$f_x(0, 0) = 1$,	$f_x(1, 2) = 3$,	$f_x(2, 1) = 5$,	$f(2, 1) = e$,
$g(0, 0) = e$,	$f_y(0, 0) = 2$,	$f_y(1, 2) = 4$,	$f_y(2, 1) = 6$,	$f(1, 2) = \pi^2$.

For Problems 21–22, suppose the quantity z can be expressed either as a function of Cartesian coordinates (x, y) or as a function of polar coordinates (r, θ), so that $z = f(x, y) = g(x, y)$.

21. (a) Find $\partial z/\partial r$ and $\partial z/\partial \theta$ in terms of $\partial z/\partial x$ and $\partial z/\partial y$.
     (b) Solve the equations you have just written down for $\partial z/\partial x$ and $\partial z/\partial y$.
     (c) Show that the expressions you get in part (b) are the same as you would get by using the chain rule directly to write down $\partial z/\partial x$ and $\partial z/\partial y$.

22. Show that

$$\left(\frac{\partial z}{\partial x}\right)^2 + \left(\frac{\partial z}{\partial y}\right)^2 = \left(\frac{\partial z}{\partial r}\right)^2 + \frac{1}{r^2}\left(\frac{\partial z}{\partial \theta}\right)^2.$$

23. Let $F(x, y, z)$ be a function and define a function $z = f(x, y)$ implicitly by letting $F(x, y, f(x, y)) = 0$. Use the chain rule to show that

$$\frac{\partial z}{\partial x} = -\frac{\partial F/\partial x}{\partial F/\partial z} \quad \text{and} \quad \frac{\partial z}{\partial y} = -\frac{\partial F/\partial y}{\partial F/\partial z}.$$

13.8 SECOND-ORDER PARTIAL DERIVATIVES

Since the partial derivatives of a function are themselves functions, we can calculate their partial derivatives, called *second-order partial derivatives*. For a function $f(x, y)$, we can differentiate f_x with respect to either x or y, and we can differentiate f_y with respect to either x or y, so there are four second-order partial derivatives: the two partial derivatives $(f_x)_x$ and $(f_x)_y$ of f_x and the two partial derivatives $(f_y)_x$ and $(f_y)_y$ of f_y. It is convenient to omit the parentheses from the notation, writing, for example, f_{xy} instead of $(f_x)_y$. We therefore have f_{xx}, f_{xy}, f_{yx}, and f_{yy}. Similarly, if $z = f(x, y)$, we will usually shorten $\frac{\partial}{\partial x}\left(\frac{\partial z}{\partial x}\right)$ to $\frac{\partial^2 z}{\partial x^2}$ and $\frac{\partial}{\partial y}\left(\frac{\partial z}{\partial x}\right)$ to $\frac{\partial^2 z}{\partial y \partial x}$.

The Second-Order Partial Derivatives of $z = f(x, y)$ are as follows

$$\frac{\partial^2 z}{\partial x^2} = f_{xx} = (f_x)_x \qquad \frac{\partial^2 z}{\partial x \partial y} = f_{yx} = (f_y)_x$$

$$\frac{\partial^2 z}{\partial y \partial x} = f_{xy} = (f_x)_y \qquad \frac{\partial^2 z}{\partial y^2} = f_{yy} = (f_y)_y$$

Example 1   Compute the four second-order partial derivatives of $f(x, y) = y \cos x + 3x^2 e^y$.

Solution   From $f_x(x, y) = -y \sin x + 6x e^y$ we get

$$f_{xx}(x, y) = \frac{\partial}{\partial x}(-y \sin x + 6x e^y) = -y \cos x + 6e^y$$

and

$$f_{xy}(x, y) = \frac{\partial}{\partial y}(-y \sin x + 6x e^y) = -\sin x + 6x e^y.$$

From $f_y(x, y) = \cos x + 3x^2 e^y$ we get

$$f_{yx}(x, y) = \frac{\partial}{\partial x}(\cos x + 3x^2 e^y) = -\sin x + 6xe^y$$

and

$$f_{yy}(x, y) = \frac{\partial}{\partial y}(\cos x + 3x^2 e^y) = 3x^2 e^y.$$

Observe that in this example we have $f_{xy} = f_{yx}$.

What Do the Second-Order Partial Derivatives Tell Us?

A second-order partial derivative is the derivative of a derivative. Since all derivatives can be interpreted as rates of change, a second-order partial derivative is the rate of change of a first derivative with respect to some quantity.

Example 2   Let us return to the guitar string of Example 2, page 121. The string is 1 meter long and at time t seconds, the point x meters from one end is displaced $f(x, t)$ meters from its rest position, where

$$f(x, t) = 0.003 \sin(\pi x) \sin(2765t).$$

Compute the four second-order partial derivatives of f at the point $(x, t) = (0.3, 1)$ and describe their meaning in practical terms.

Solution   First we compute $f_x(x, t) = 0.003\pi \cos(\pi x) \sin(2765t)$, from which we get

$$f_{xx}(x, t) = \frac{\partial}{\partial x}(f_x(x, t)) = -0.003\pi^2 \sin(\pi x) \sin(2765t),$$
$$f_{xx}(0.3, 1) \approx -0.01;$$

and

$$f_{xt}(x, t) = \frac{\partial}{\partial t}(f_x(x, t)) = (0.003)(2765)\pi \cos(\pi x) \cos(2765t),$$
$$f_{xt}(0.3, 1) \approx 14.$$

On page 121 we saw that $f_x(0.3, 1) = 0.002$ is the slope of the guitar string at the point $x = 0.3$ at the instant $t = 1$. More generally, $f_x(x, t)$ is the slope of the string at the point x at the time t. Therefore, $f_{xx}(x, t)$ is the concavity of the string at that point and time. The fact that $f_{xx}(0.3, 1) = -0.01$ indicates that the shape of the string at $t = 1$ is concave down. See Figure 13.41.

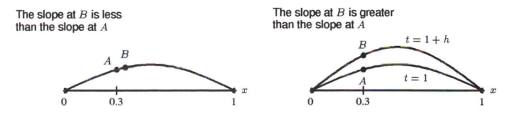

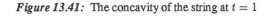

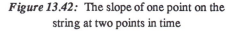

Figure 13.41: The concavity of the string at $t = 1$ **Figure 13.42:** The slope of one point on the string at two points in time

On the other hand, $f_{xt}(0.3, 1)$ is the rate of change of the slope of the string with respect to time; that $f_{xt}(0.3, 1) = 14$ indicates that at time $t = 1$ the slope at the point $x = 0.3$ is increasing at a rate of 14 units/sec. See Figure 13.42.

Now we compute

$$f_t(x, t) = (0.003)(2765) \sin(\pi x) \cos(2765t),$$

from which we get

$$f_{tx}(x, t) = \frac{\partial}{\partial x}(f_t(x, t)) = (0.003)(2765)\pi \cos(\pi x) \cos(2765t),$$
$$f_{tx}(0.3, 1) \approx 14;$$

and

$$f_{tt}(x, t) = \frac{\partial}{\partial t}(f_t(x, t)) = -(0.003)(2765)^2 \sin(\pi x) \sin(2765t),$$
$$f_{tt}(0.3, 1) \approx -7200.$$

On page 121 we saw that $f_t(0.3, 1) \approx 6$ m/sec is the velocity of the guitar string at the point $x = 0.3$ at the instant $t = 1$. More generally, $f_t(x, t)$ is the velocity of the string at the point x at the time t. Therefore, $f_{tx}(x, t)$ and $f_{tt}(x, t)$ will both be rates of change of velocity at the point x and the time t. That $f_{tx}(0.3, 1) = 14$ indicates that at the time $t = 1$ the velocities of the points on the string just to the right of $x = 0.3$ are greater than the velocity of the point $x = 0.3$. The velocity as you move past the point $x = 0.3$ at time $t = 1$ increases at a rate of 14 (m/sec)/m. See Figure 13.43. That $f_{tt}(0.3, 1) = -7200$ indicates that the velocity of the point $x = 0.3$ is decreasing at time $t = 1$ at a rate of 7200(m/sec)/sec. Thus, $f_{tt}(0.3, 1) = -7200$ m/sec^2 is the acceleration of the point $x = 0.3$ at time $t = 1$. See Figure 13.44.

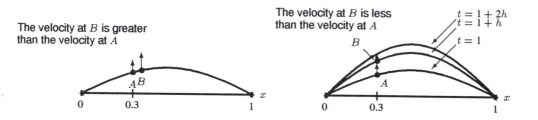

Figure 13.43: The velocity of different points on the string at $t = 1$

Figure 13.44: The velocity of one point on the string at two points in time

The Mixed Partials are Equal

Notice that in Example 1 and Example 2 the mixed partials f_{xy} and f_{yx} or f_{xt} and f_{tx} are equal. This is no accident. For almost all functions $z = f(x, y)$ that we will encounter, it will be true that

$$f_{xy}(x, y) = f_{yx}(x, y)$$

or, in the other notation, that

$$\frac{\partial^2 z}{\partial y \partial x} = \frac{\partial^2 z}{\partial x \partial y}.$$

In words, if we differentiate first with respect to x, getting f_x and then with respect to y, getting f_{xy}, the result will be the same as first differentiating with respect to y, getting f_y, and then differentiating with respect to x, getting f_{yx}. The order in which we take the derivatives does not affect the final outcome.

There are some exceptions to the rule that $f_{xy} = f_{yx}$. However, if f_{xy} and f_{yx} are both continuous, then it can be proved that they are equal.

Problems for Section 13.8

In Problems 1–8, calculate all four second-order partial derivatives and show that $z_{xy} = z_{yx}$.

1. $z = \sin(\frac{x}{y})$.

2. $z = xe^y$.

3. $z = (x + y)e^y$.

4. $z = \tan^{-1}(\frac{x}{y})$.

5. $z = \sin(x^2 + y^2)$.

6. $z = \frac{xe^y}{x+y}$.

7. $z = \sqrt{x^2 + y^2}$.

8. $z = \tan^{-1}(x + y)$.

In Problems 9–12, calculate all four second-order partial derivatives and show that $f_{xy} = f_{yx}$.

9. $f(x, y) = x^y$

10. $f(x, y) = \ln(xy)$

11. $f(x, y) = (x + y)^2$

12. $f(x, y) = (x + y)^3$

13. If $z = f(x) + yg(x)$, what can you say about z_{yy}? Why?

14. If $z_{xy} = 4y$, what can you say about (a) z_{yx}? (b) z_{xyx}? (c) z_{xyy}?

In Problems 15–25, use the level curves of the function $z = f(x, y)$ to decide the sign (positive, negative, or zero) of each of the following partial derivatives at the point P. Assume the x- and y-axes are in the usual positions.

(a) $f_x(P)$ (b) $f_y(P)$ (c) $f_{xx}(P)$ (d) $f_{yy}(P)$ (e) $f_{xy}(P)$

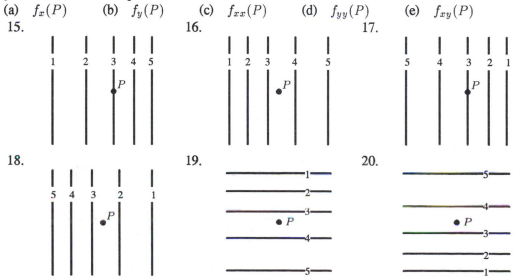

15.

16.

17.

18.

19.

20.

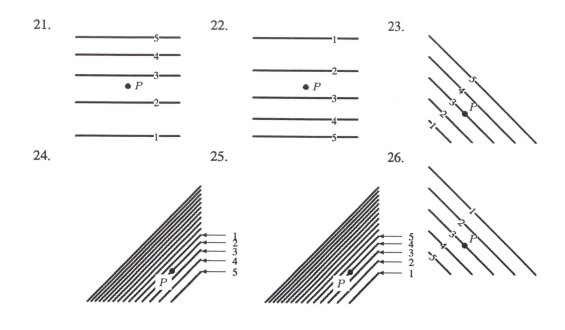

13.9 QUADRATIC APPROXIMATIONS

Just as a function of one variable can usually be better approximated by a quadratic function than by a linear function, so can a function of several variables. In Section 13.3, we saw how to approximate $f(x, y)$ by a linear function (its local linearization). In this section, we see how to improve this approximation of $f(x, y)$ using a quadratic function.

Quadratic Approximations Near (0,0)

The local linearization of a function is the best linear approximation to the function at a point. For a function of a single variable, local linearity tells us that

$$f(x) \approx f(a) + f'(a)(x - a) \quad \text{for } x \text{ near } a.$$

We can improve on this by taking a second-order Taylor polynomial

$$f(x) \approx f(a) + f'(a)(x - a) + \frac{f''(a)}{2}(x - a)^2.$$

The local linearization of a function of two variables near (a, b) is

$$f(x, y) \approx f(a, b) + f_x(a, b)(x - a) + f_y(a, b)(y - b),$$

or, if $(a, b) = (0, 0)$,

$$f(x, y) \approx f(0, 0) + f_x(0, 0)x + f_y(0, 0)y.$$

To get a better approximation to a function $f(x, y)$ we use a quadratic polynomial $P(x, y)$

$$f(x, y) \approx P(x, y) = A + Bx + Cy + Dx^2 + Exy + Fy^2.$$

5. $z = \sin(x^2 + y^2)$

$z_x = 2x\cos(x^2 + y^2)$

$z_y = 2y\cos(x^2 + y^2)$

$z_{xy} = 4xy\cos(x^2 + y^2)$

$z_{xx} = 2(\cos(x^2 + y^2)) - 4x^2(\sin(x^2 + y^2))$

$z_{yy} = 2(\cos(x^2 + y^2)) - 4x^2(\sin(x^2 + y^2))$

(?)
7. $z = \sqrt{x^2 + y^2} = (x^2 + y^2)^{1/2}$

$z_x = \frac{1}{2}(x^2 + y^2)^{-1/2} \cdot 2x = \frac{x}{\sqrt{x^2 + y^2}}$

$z_y = \frac{1}{2}(x^2 + y^2)^{-1/2} \cdot 2y = \frac{y}{\sqrt{x^2 + y^2}}$

$z_{xy} = z_{yx} =$

$z_{xx} =$

$z_{yy} =$

13.8

5-25 odd

(?)
9. $f(x,y) = x^y$

$f_x = yx^{y-1}$

$f_y = x^y \ln x$

$f_{xx} = y(y-1)x^{(y-2)}$

$f_{yy} = x^y (\ln x)^2$

$f_{xy} =$

$f_{yx} =$

(?)
11. $f(x,y) = (x+y)^2$

$f_x = 2(x+y)(1) = 2y(x+y)$

$f_y = 2(x+y)(1) = 2x(x+y)$

$f_{xx} =$

$f_{yy} =$

$f_{xy} =$

13. $z_{yy} = 0$; because g is only a function of x

15 a. pos.   17 a. neg.   19 a. zero

b. zero   b. zero   b. neg.

c. pos.   c. neg.   c. zero

d. zero   d. zero   d. neg.

e. zero   e. zero   e. zero

21 a. zero   23 a. pos.   25 a. neg.

b. pos.   b. pos.   b. pos.

c. zero   c. zero   c. pos.

d. pos.   d. zero   d. pos.

e. zero   e. zero   e. neg.

What quadratic polynomial should we choose? We would certainly want the approximation to be exact at $(0,0)$, which means that

$$f(0,0) = P(x,y) = A$$

If we arrange for the first partials of f and P to agree at $(0,0)$, that is

$$f_x(0,0) = P_x(0,0) = B \quad \text{and} \quad f_y(0,0) = P_y(0,0) = C,$$

then f and P will stay close near $(0,0)$ because they will have the same local linearizations at $(0,0)$. The natural next step is to insist that the second partials of f and P be equal at $(0,0)$. In other words, we ask that $P_{xx}(0,0) = f_{xx}(0,0)$, $P_{xy}(0,0) = f_{xy}(0,0)$, and $P_{yy}(0,0) = f_{yy}(0,0)$. Then

$$f_{xx}(0,0) = P_{xx}(0,0) = 2D \quad \text{and} \quad f_{xy}(0,0) = P_{xy}(0,0) = E \quad \text{and} \quad f_{yy}(0,0) = P_{yy}(0,0) = 2F$$

Hence, we can find the coefficients A, B,..., F and thus the best quadratic approximation for f near $(0,0)$ is given by the following polynomial:

Taylor Polynomial of Degree 2 Approximating $f(x,y)$ for (x,y) near (0,0)

$$f(x,y) \approx P(x,y)$$
$$= f(0,0) + f_x(0,0)x + f_y(0,0)y + \frac{f_{xx}(0,0)}{2}x^2 + f_{xy}(0,0)xy + \frac{f_{yy}(0,0)}{2}y^2.$$

Example 1   Let $f(x,y) = \sqrt{x + 2y + 1}$.

    (a)   Compute the local linearization of f at $(0,0)$.

    (b)   Compute the Taylor polynomial of degree 2 for f at $(0,0)$.

    (c)   Compare the linear and quadratic approximations in part (a) and part (b) with the true values for $f(x,y)$ at the points $(0.1, 0.1)$, $(-0.1, 0.1)$, $(0.1, -0.1)$, $(-0.1, -0.1)$.

Solution   Let us first collect the computations that we will need.

$$\begin{array}{lll}
f(x,y) = & (x + 2y + 1)^{1/2} & f(0,0) = 1 \\
f_x(x,y) = & \frac{1}{2}(x + 2y + 1)^{-1/2} & f_x(0,0) = 1/2 \\
f_y(x,y) = & (x + 2y + 1)^{-1/2} & f_y(0,0) = 1 \\
f_{xx}(x,y) = & -\frac{1}{4}(x + 2y + 1)^{-3/2} & f_{xx}(0,0) = -1/4 \\
f_{xy}(x,y) = & -\frac{1}{2}(x + 2y + 1)^{-3/2} & f_{xy}(0,0) = -1/2 \\
f_{yy}(x,y) = & -(x + 2y + 1)^{-3/2} & f_{yy}(0,0) = -1.
\end{array}$$

    (a)   The local linearization $L(x,y)$ of f at $(0,0)$ is given by

$$f(x,y) \approx L(x,y) = f(0,0) + f_x(0,0)x + f_y(0,0)y = 1 + \frac{1}{2}x + y.$$

(b) The second-degree Taylor polynomial, $P(x, y)$, for f at $(0, 0)$ is given by

$$f(x, y) \approx P(x, y)$$

$$= f(0, 0) + f_x(0, 0)x + f_y(0, 0)y + \frac{f_{xx}(0, 0)}{2}x^2 + f_{xy}(0, 0)xy + \frac{f_{yy}(0, 0)}{2}y^2$$

$$= 1 + \frac{1}{2}x + y - \frac{1}{8}x^2 - \frac{1}{2}xy - \frac{1}{2}y^2.$$

Notice that the local linearization of f is the same as the linear part of the Taylor polynomial of degree 2 for f. The extra terms in the Taylor polynomial of degree 2 can be thought of as "correction terms" to the linear approximation.

(c) Table 13.8 records the values of $f(x, y)$, $L(x, y)$, and $P(x, y)$. Observe that the quadratic approximations $P(x, y)$ are closer to the true values $f(x, y)$ than are the linear approximations $L(x, y)$. Of course both approximations are exact at $(0, 0)$.

TABLE 13.8 *Linear and quadratic approximations to f near $(0, 0)$*

Point	Linear	Quadratic	True
(x, y)	$L(x, y)$	$P(x, y)$	$f(x, y)$
$(0, 0)$	1	1	1
$(0.1, 0.1)$	1.15	1.13875	1.140175
$(-0.1, 0.1)$	1.05	1.04875	1.048809
$(0.1, -0.1)$	0.95	0.94875	0.948683
$(-0.1, -0.1)$	0.85	0.83875	0.836660

Quadratic Approximations near (a, b)

Recall that the local linearization for a function $f(x, y)$ at a point (a, b) is

$$f(x, y) \approx f(a, b) + f_x(a, b)(x - a) + f_y(a, b)(y - b).$$

This suggests that a quadratic polynomial approximation $P(x, y)$ for $f(x, y)$ near a point (a, b) should be written in terms of $(x - a)$ and $(y - b)$ instead of x and y. If we require that $P(a, b) = f(a, b)$ and that the first- and second-order partial derivatives of P and f at (a, b) be equal, then after some algebra, we get the following polynomial:

Taylor Polynomial of Degree 2 Approximating $f(x, y)$ for (x, y) near (a, b)

$$f(x, y) \approx P(x, y)$$

$$= f(a, b) + f_x(a, b)(x - a) + f_y(a, b)(y - b)$$

$$+ \frac{f_{xx}(a, b)}{2}(x - a)^2 + f_{xy}(a, b)(x - a)(y - b) + \frac{f_{yy}(a, b)}{2}(y - b)^2.$$

(?)
3.

$f(x,y)$	$\ln(1+x^2-y)$
f_x	$\dfrac{2x}{(1+x^2-y)}$
f_y	$\dfrac{-1}{(1+x^2-y)}$
f_{xx}	$\dfrac{2(1+x^2-y)-4x^2}{(1+x^2-y)^2}$
f_{yy}	$\dfrac{-1}{(1+x^2-y)^2}$
f_{xy}	$\dfrac{2x}{(1+x^2-y)^2}$

$$P(x,y) = f(0,0) + f_x(0,0)x + f_y(0,0)y + \tfrac{1}{2}f_{xx}(0,0)x^2 + f_{xy}(0,0)xy + \tfrac{1}{2}f_{yy}(0,0)y^2 \quad |3.9$$

$$= 0 + 0x - 1y + \tfrac{1}{2}(2)x^2 + \tfrac{1}{2}(-1)y^2 \qquad\qquad 3,5,7,11$$

$$= -y + x^2 - \tfrac{y^2}{2}$$

(?)
5.

z	xe^y	2.72
z_x	e^y	2.72
z_y	yxe^{y-1}	1
z_{xx}	e^y	2.72
z_{xy}	ye^{y-1}	1
z_{yy}	$y(y-1)xe^{y-2}$	0

$$P(x,y) = f(1,1) + f_x(1,1)x + f_y(1,1)y + \tfrac{1}{2}f_{xx}(1,1)x^2 + f_{xy}(1,1)xy + \tfrac{1}{2}f_{yy}(1,1)y^2$$

$$= 2.72 + 2.72x + y + \tfrac{1}{2}(2.72)x^2 + xy$$

These coefficients are derived in the same way as for $(a, b) = (0, 0)$.

Example 2   Find the Taylor polynomial of degree 2 at the point $(1, 2)$ for the function

$$f(x, y) = \frac{1}{xy}.$$

Solution   First we calculate all the necessary partial derivatives and their values at $(1, 2)$.

Derivative	Formula	Value at $(1, 2)$
$f(x, y)$	$1/xy$	$1/2$
$f_x(x, y)$	$-1/x^2 y$	$-1/2$
$f_y(x, y)$	$-1/xy^2$	$-1/4$
$f_{xx}(x, y)$	$2/x^3 y$	1
$f_{xy}(x, y)$	$1/x^2 y^2$	$1/4$
$f_{yy}(x, y)$	$2/xy^3$	$1/4$

So that the Taylor polynomial at (1,2) is

$$\frac{1}{xy} \approx P(x, y)$$

$$= \frac{1}{2} - \frac{1}{2}(x - 1) - \frac{1}{4}(y - 2) + \frac{1}{2}(1)(x - 1)^2 + \frac{1}{4}(x - 1)(y - 2) + \left(\frac{1}{2}\right)\left(\frac{1}{4}\right)(y - 2)^2$$

$$= \frac{1}{2} - \frac{x - 1}{2} - \frac{y - 2}{4} + \frac{(x - 1)^2}{2} + \frac{(x - 1)(y - 2)}{4} + \frac{(y - 2)^2}{8}.$$

Problems for Section 13.9

Find the quadratic Taylor polynomials about $(0, 0)$ for the functions in Problems 1–3.

1. $e^{-2x^2 - y^2}$
2. $\sin 2x + \cos y$
3. $\ln(1 + x^2 - y)$

For each of the functions in Problems 4–11, find a quadratic approximation valid near $(1, 1)$. Evaluate the approximation at $(1.1, 1.1)$ and compare it with the exact value of the function.

4. $z = \sin(x/y)$
5. $z = xe^y$
6. $z = (x + y)e^y$
7. $z = \sin(x^2 + y^2)$
8. $z = \sqrt{x^2 + y^2}$
9. $z = \tan^{-1}(x + y)$
10. $z = \tan^{-1}(x/y)$
11. $z = \dfrac{xe^y}{x + y}$

13.10 PARTIAL DIFFERENTIAL EQUATIONS

Heat Flow

Imagine a room heated by a radiator along one wall. What will happen after a window on the opposite wall is opened on a cold day? The temperature in the room will begin to drop, quickly near the window, more slowly near the heater. Eventually the temperature will stabilize, with the room temperature near the window close to the outside temperature, the temperature near the radiator close to what it was before the window was opened, and the temperature in the middle of the room somewhere in between. The temperature $T = u(x,t)$ at any point in the room is a function of the distance x in meters from the heated wall and the time t in minutes since the window was opened. Figure 13.45 shows how the temperature, H, might look as a function of x at various values of t.

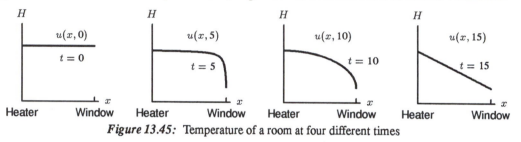

Figure 13.45: Temperature of a room at four different times

The partial derivative $u_t(x,t)$ gives the rate of change of temperature with respect to time at any point and time in the room. The temperature in a region of the room will be increasing if the heat flow into it from the nearby hotter parts of the room is greater than the heat flow out of it to the nearby colder parts and it will be decreasing if the inflow is less than the outflow.

The heat flow between two points will depend on the temperature difference between them: the greater the temperature difference, the greater the rate of flow. A more precise formulation of this statement is Newton's Law of Cooling, which predicts that the rate of heat flow past a point x at time t is proportional to the partial derivative $u_x(x,t)$. This makes sense, since $u_x(x,t)$ measures the rate of change of u with respect to x at a fixed time t; in other words, it tells us how large the temperature difference between neighboring points is.

Example 1   Figure 13.46 gives the graph of temperature $H = u(x,t)$ at a fixed time t. Use Newton's Law of Cooling to determine whether the temperature is increasing or decreasing
(a)   at the point p (b)   at the point q.

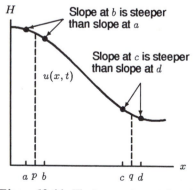

Figure 13.46: The temperature at time t

Solution (a) The question is whether the small region $a < x < b$ that contains p is gaining or losing energy. (See Figure 13.46.) Since energy flows from hotter regions to colder regions, the energy flow is from left to right on the graph. We might say that energy flows downhill on a temperature graph; Newton's Law of Cooling asserts that the steeper the slope, which equals $u_x(x, t)$, the greater the rate of flow. Energy flows into the region $a < x < b$ at the point a, and it flows out at b. The outflow is greater than the inflow because the tangent line at b is steeper than the tangent line at a. The region $a < x < b$ is losing energy and so its temperature is decreasing. Thus, $u_t(p, t) < 0$.

(b) Energy flows into the small region $c < x < d$ at the point c and flows out at d, because the slopes $u_x(c, t)$ and $u_x(d, t)$ are negative. But the inflow is greater than the outflow, by Newton's Law of Cooling, because the tangent line at c is steeper than the tangent line at d. The region $c < x < d$ is gaining energy and so its temperature is increasing. Thus, $u_t(q, t) > 0$.

The Heat Equation

In a problem about heat flow, whether the temperature is increasing or decreasing at some point is determined by the sign of the partial derivative u_t. Example 1 shows that the sign of u_t can be found by studying the shape of the graph of $u(x, t)$ against x for fixed t. The relevant feature of the graph is the way in which the slope $u_x(x, t)$ changes when x varies, and that is determined by the concavity of the graph, which is measured by the second derivative $u_{xx}(x, t)$. In Example 1 the graph of u is concave down at $x = p$, so $u_{xx}(p, t) < 0$, and we found that $u_t(p, t) < 0$. At the point q we have $u_{xx}(q, t) > 0$ and $u_t(q, t) > 0$. In a heat flow problem the two derivatives u_t and u_{xx} are definitely related, for they always have the same sign. In many situations the two derivatives u_t and u_{xx} are actually proportional.

So the function $u(x, t)$ satisfies the following equation:

The One-Dimensional Heat Equation (or Diffusion Equation)

$$u_t(x, t) = A u_{xx}(x, t) \quad \text{where } A \text{ is a positive constant.}$$

The heat equation is an example of a partial differential equation (PDE), that is, an equation involving one or more of the partial derivatives of a function. The unknown quantity in a PDE is the function itself. The value of the constant of proportionality depends on the situation.

Example 2 Which of the following two functions satisfies the heat equation $u_t = u_{xx}$?
(a) $u(x, t) = e^{-4t} \sin(2x)$ (b) $u(x, t) = \sin(x + t)$

Solution (a) $u_t = -4e^{-4t} \sin(2x)$, $u_x = 2e^{-4t} \cos(2x)$, $u_{xx} = -4e^{-4t} \sin(2x)$, so $u_t = u_{xx}$.
Thus, $u(x, t) = e^{-4t} \sin(2x)$ is a solution.

(b) $u_t = \cos(x + t)$, $u_x = \cos(x + t)$, $u_{xx} = -\sin(x + t)$, so $u_t \neq u_{xx}$.
Thus, $u(x, t) = \sin(x + t)$ is not a solution.

Example 3   A 10 cm metal rod is wrapped with insulation so that heat can flow along the rod but cannot radiate into the air except at the ends. The temperature H (°C) at x cm from one end and at time t seconds is a function $H = u(x,t)$ that satisfies the heat equation $u_t(x,t) = 0.1u_{xx}(x,t)$. At $t = 0$ the temperature at several points are given in Table 13.9.

TABLE 13.9   *Temperature in metal rod at time $t = 0$*

x (cm)	0	2	4	6	8	10
$u(x,0)$ (°C)	50	52	56	62	70	80

   (a)   Is the temperature at the point $x = 6$ increasing or decreasing at $t = 0$?

   (b)   Make an estimate of the temperature $T = u(6,1)$ at the point $x = 6$ at time $t = 1$.

Solution   (a)   The graph of $u(x,0)$ in Figure 13.47 shows that $u(x,0)$ is concave up, so $u_{xx}(6,0) > 0$. Since $u_t(6,0) = 0.1u_{xx}(6,0)$ we have $u_t(6,0) > 0$ also. Since $u_t(6,0)$ gives the rate of change of temperature at $x = 6$ with respect to time, the fact that it is positive indicates that the temperature at $x = 6$ is increasing. You might have guessed this intuitively from the observation that the temperature of 62 at $x = 6$ is below the average $(56+70)/2 = 63$ of the temperatures of the neighboring points at $x = 4$ and $x = 8$ (which is because $u_{xx}(6,0) > 0$). The average effect of the neighboring points is to heat up the point at $x = 6$.

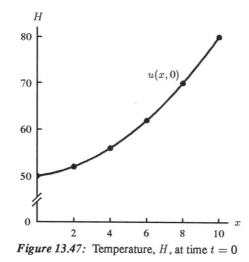

Figure 13.47: Temperature, H, at time $t = 0$

   (b)   To estimate the temperature at $t = 1$ from the temperature at $t = 0$ we must first estimate the rate of change of temperature with respect to time, $u_t(6,0)$. Since $u_t(6,0) = 0.1u_{xx}(6,0)$, we will first approximate $u_{xx}(6,0)$, which we can do from Table 13.9. Since $u_{xx}(6,0)$ is approximated by a difference quotient of $u_x(x,0)$, we first approximate u_x at two points near $x = 6$.

$$u_x(5,0) \approx \frac{u(6,0) - u(4,0)}{6-4} = \frac{62-56}{2} = 3, \quad \text{and} \quad u_x(7,0) \approx \frac{u(8,0) - u(6,0)}{8-6} = 4.$$

Therefore,

$$u_{xx}(6,0) \approx \frac{u_x(7,0) - u_x(5,0)}{7-5} \approx \frac{4-3}{2} = 0.5.$$

Thus,

$$u_t(6,0) \approx 0.1u_{xx}(6,0) \approx (0.1)(0.5) = 0.05 \text{ deg/sec}.$$

Finally we can make the local linear approximation for the temperature at $t = 1$ that we want:

$$u(6,1) \approx u(6,0) + u_t(6,0)(1) \approx 62 + (0.05)(1) = 62.05°\text{C}.$$

Warning!

Getting accurate numerical approximations of solutions to PDEs is generally quite difficult. Example 3 was given to show that quantitative information can be extracted from a PDE, not as a practical way to get accurate approximations to solutions.

Boundary Conditions

The heat equation $u_t = u_{xx}$ has many, many solutions. More information than this equation alone is required in order to specify a single solution, information analogous to the initial conditions we are familiar with from the study of ordinary differential equations. In Example 1, for example, we would need to know the temperature in the room at the time the window was opened, the outside temperature at all times, and the temperature very near the heater (which tells what the heater is doing). This information is collectively referred to as the boundary condition. The problems at the end of this section contain examples of how the boundary conditions can be used.

A Traveling Wave

Think about a bottle on a quiet sea that floats up then down as a wave rolls through. The vertical velocity of the bottle will depend on the shape and horizontal velocity of the wave.

Example 4   Suppose that the height y of the sea (and hence of a floating bottle) above normal at time t seconds and distance x meters from a reference point is given by the function $y = u(x,t)$, which is graphed at $t = 0$ in Figure 13.48. Suppose that the wave is moving in the direction of increasing x.

(a)   Determine whether $u_x(x,0)$ and $u_t(x,0)$ are positive or negative at the following points:

   (i) $x = p$,   (ii) $x = q$,   (iii) $x = r$.

(b)   Would $u_t(p,0)$ be greater or smaller if the wave were traveling faster?

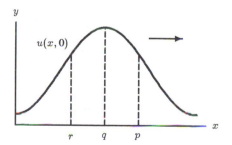

Figure 13.48:   The height of the wave at $t = 0$; wave moving to the right

Solution (a)
  i The partial derivative $u_x(p, 0)$ gives the slope of the tangent line to the graph of $u(x, 0)$ at $x = p$. From Figure 13.48 it is clear that the slope is negative, so $u_x(p, 0) < 0$. On the other hand, $u_t(p, 0)$ equals the vertical velocity of a bottle floating at the point $x = p$ at the time $t = 0$. Since the bottle is rising as the wave rolls by, that velocity is positive. Thus, $u_t(p, 0) > 0$.

  ii We have $u_x(q, 0) = 0$, because the tangent to the wave at $x = q$ at the time 0 is horizontal, slope 0. A bottle at $x = q$ would at time 0 be exactly at the crest of the wave, momentarily stopped as it reverses direction from rising to falling, so its velocity would be zero. Thus, $u_t(q, 0) = 0$.

  iii At $t = 0$ the point r is on the back side of the wave, so $u_x(r, 0) > 0$. A bottle at $x = r$ would be falling at $t = 0$, so $u_t(r, 0) < 0$.

(b) If the wave were moving faster, then a floating bottle would rise and fall much faster, so the positive velocity $u_t(p, 0)$ would be greater than for the original, slower wave.

The Traveling Wave Equation

Observe that in Example 4 the derivatives $u_x(x, 0)$ and $u_t(x, 0)$ are of opposite sign or are both zero together, for all x. In this example there is no x for which $u_x(x, 0)$ and $u_t(x, 0)$ are both positive or both negative, suggesting that the two derivatives u_x and u_t are not completely independent of each other.

To investigate the relationship between the derivatives, let us suppose that a wave is moving to the right with velocity c, and that its positions at two nearby times t and $t + \Delta t$ are as shown in Figure 13.49. If Δt is small enough, $u_x(p, t)$, the slope of the graph at B, is well approximated by the slope of the secant line between the points A and B. Note that during time interval Δt the wave moves horizontally a distance of $c\Delta t$. Thus, we have

$$
\begin{aligned}
u_x(p, t) &= \text{Slope of tangent at } B \\
&\approx \text{Slope of secant between } A \text{ and } B \\
&= \frac{u(p, t) - u(p, t + \Delta t)}{c\Delta t} \\
&= -\frac{1}{c} \frac{u(p, t + \Delta t) - u(p, t)}{\Delta t} \\
&\approx -\frac{1}{c} u_t(p, t)
\end{aligned}
$$

If $u(x, t)$ represents a wave moving in the positive x-direction with speed c then it satisfies the following PDE:

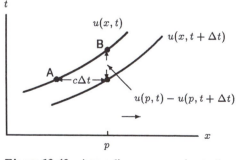

Figure 13.49: A traveling wave moving to the right at speed c

> **The Traveling Wave Equation**
>
> $$u_t(x, t) = -cu_x(x, t) \quad \text{where } c \text{ is a positive constant.}$$

Analytic Representation of Traveling Waves

Suppose that a traveling wave is moving in the positive x-direction with speed c and that the wave has a known shape $y = f(x)$ at time $t = 0$, which means that $u(x, 0) = f(x)$. By time t the wave will have moved a distance ct (distance = rate $\times$ time) to the right, so at time t it will have shape $y = f(x - ct)$. In other words, $u(x, t) = f(x - ct)$.

Example 5

    (a) Write an equation for $u(x, t)$ that describes a wave whose shape at time $t = 0$ is $y = \sin x$ and that is moving in the positive direction with speed 0.5.

    (b) Show that the function $u(x, t)$ found in part (a) satisfies the traveling wave equation.

    (c) Sketch the graphs of $u(x, t)$ against x for $t = 0, 1, 2$.

Solution   (a) $u(x, t) = \sin(x - ct) = \sin(x - 0.5t)$.

    (b) Since $u_t(x, t) = -0.5\cos(x - 0.5t)$ and $u_x(x, t) = \cos(x - 0.5t)$, the function $u(x, t)$ satisfies the traveling wave equation with $c = 0.5$:

$$u_t(x, t) = -0.5u_x(x, t).$$

    (c) The graphs are in Figure 13.50. Notice that the forward motion of the wave is clearly visible.

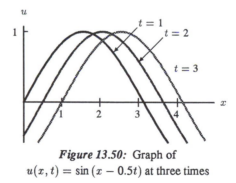

Figure 13.50: Graph of
$u(x, t) = \sin(x - 0.5t)$ at three times

A String Under Tension

What is the mathematics of a vibrating string under tension, such as a plucked guitar string? Let $y = u(x, t)$ be the displacement at time t of the point on the string x units from one end, where displacement is measured from the equilibrium position on the straight string. Then it can be shown that u satisfies the following equation.

The One-Dimensional Wave Equation:

$$u_{tt} = c^2 u_{xx} \quad \text{where } c \text{ is a positive constant.}$$

Let's see what this equation says. The function $u(x, t)$ describes the motion of a mass (the string) under the influence of a force (tension). Thus, Newton's Second Law of Motion must apply, so we expect an equation of the type $F = ma$, where F is force, m is mass, and a is acceleration. The term $u_{tt}(x, t)$ in the wave equation is the acceleration of the point x at time t. The term $u_{xx}(x, t)$ is closely related to the force, for it measures the concavity of the string at the point x at time t. The greater the concavity, the stronger the force back towards equilibrium, as anyone knows who has shot an arrow from a bow.

Problems for Section 13.10

1. (a)   Find an estimate for $u(4, 1)$ and $u(8, 1)$ where $u(x, t)$ is as in Example 3.
    (b)   Use your answers to part (a) and the estimate for $u(6, 1)$ worked out in Example 3 to make an estimate for $u(6, 2)$.

2. Sketch the graph of $u(x, t) = 1 - (x - 2t)^2$ for $t = 0, 1, 2$. Do you see a traveling wave? What is its speed?

3. We will model the spread of an epidemic through a region. Suppose $I(x, y, t)$ represents the density of sick people per unit area at the point (x, y) in the plane at time t. Suppose I satisfies the diffusion equation:

    $$\frac{\partial I}{\partial t} = D\left(\frac{\partial^2 I}{\partial x^2} + \frac{\partial^2 I}{\partial y^2}\right)$$

    where D is a constant. Suppose we know that for some particular epidemic

    $$I = e^{ax + by + ct}.$$

    What can you say about the relationship between a, b, and c?

4. For what values of the constants a and b will the function $f(x, y) = (x + y)e^{ax + by}$ satisfy the differential equation

    $$\frac{\partial^2 f}{\partial x \partial y} - \frac{\partial f}{\partial x} - \frac{\partial f}{\partial y} + f = 0?$$

Show that the functions in Problems 5–8 satisfy Laplace's equation:

$$F_{xx} + F_{yy} = 0.$$

5. $F(x, y) = e^{-x} \sin y.$

6. $F(x, y) = \tan^{-1} \frac{y}{x}.$

7. $F(x, y) = e^x \sin(y) + e^y \sin(x).$

8. $F(x, y) = e^{ax + by}$, if $a^2 + b^2 = 0.$

9. The function $f(x, t) = 0.003 \sin(\pi x) \sin(2765t)$ was used in Example 2 of Section 13.8 to describe a vibrating guitar string. Show that it is a solution of the wave equation. $f_{tt} = c^2 f_{xx}$ for some c.

10. Suppose that f is any differentiable function of one variable. Define V, a function of two variables by

$$V(x, t) = f(x + ct).$$

Show that V satisfies the partial differential equation

$$\frac{\partial V}{\partial t} = c\frac{\partial V}{\partial x}.$$

11. Show that $u = x^2 y - \frac{1}{2}xy^2 + 2\sin x + 3y^4 - 5$ satisfies the second-order partial differential equation

$$\frac{\partial^2 u}{\partial x \partial y} = 2x - y.$$

12. If $V = f(u, v, w)$, and if $u^3 = x^3 + y^3 + z^3$, $v^2 = x^2 + y^2 + z^2$, $w = x + y + z$, show that

$$xV_x + yV_y + zV_z = uV_u + vV_v + wV_w.$$

13. Show that the partial differential equation

$$z_{xx} + 2kz_{xy} + f^2 z_{yy} = 0, \qquad f \neq 1,$$

can be written as $z_{vv} = 0$ by setting $u = y + \alpha x$, $v = y - x$ and choosing a suitable value for α.

14. Consider the *wave equation*:

$$\frac{\partial^2 y}{\partial t^2} = a^2\frac{\partial^2 y}{\partial x^2}$$

where a^2 is a constant, and y is the displacement of the wave at any point x and at any time t (so y is a function of x and t). The analogue of initial conditions for an ordinary differential equation are *boundary conditions*, or values of the function, or a derivative, with one or other of the variables held constant. A partial differential equation with boundary conditions is called a *boundary value problem*.

Write the boundary conditions for a vibrating string of length L for which:

(a) The ends (at $x = 0$ and $x = L$) are fixed.
(b) The initial shape is given by $f(x)$.
(c) The initial velocity distribution is given by $g(x)$.

15. Figure 13.51 is the graph of the temperature versus position at one instant in a metal rod, where temperature $T = u(x, t)$ satisfies the heat equation $u_t = u_{xx}$. Determine for which x the temperature is increasing, for which it is decreasing, and use this information to sketch a graph of temperature at a slightly later time.

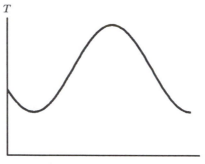

Figure 13.51: Temperature at time t

16. The gravitational potential, $V(x, y, z)$, at any point (x, y, z) outside a mass m located at the point (x_0, y_0, z_0), is defined as m/r, where r is the distance from (x, y, z) to (x_0, y_0, z_0). Show that V satisfies Laplace's equation:

$$\frac{\partial^2 V}{\partial x^2} + \frac{\partial^2 V}{\partial y^2} + \frac{\partial^2 V}{\partial z^2} = 0$$

17. Assuming the solution to the wave equation

$$\frac{\partial^2 y}{\partial t^2} = 4 \frac{\partial^2 y}{\partial x^2}$$

is of the form $y = F(x + 2t) + G(x - 2t)$, find a solution satisfying the boundary conditions:

$$y(0, t) = y(5, t) = 0, \quad y(x, 0) = 0, \quad \left.\frac{\partial y}{\partial t}\right|_{t=0} = 5 \sin \pi x, \quad 0 < x < 5, \quad t > 0.$$

18. Find the relationship between a and b that must hold for $u(x, t) = e^{at} \sin(bx)$ to satisfy the heat equation $u_t = u_{xx}$.

19. (a) Suppose you wish to study heat conduction in a 1 meter metal rod, $0 \le x \le 1$, wrapped in insulation, whose ends are maintained at $0°$ C at all times (for instance because they are stuck into ice baths). The conditions at the ends of the rod represent a boundary condition on the possible functions $u(x, t)$ that could describe the temperature in the rod. The boundary condition must hold in addition to the PDE $u_t = u_{xx}$. State the boundary condition as a pair of equations.
    (b) Determine all possible values of a and b such that $u(x, t) = e^{at} \sin(bx)$ satisfies both the PDE and the boundary conditions of part (a).

20. (a) Show that $u(x, t) = ax + b$ satisfies the heat equation.
    (b) The solution in part (a) is called a steady-state solution of the heat equation because it represents a temperature that does not depend on time. Find and graph the steady state solution such that $u(0, t) = 20$ and $u(10, t) = 50$ for all t.

21. (a) Verify that the function

$$u(x, t) = \frac{1}{2\sqrt{\pi t}} e^{-x^2/(4t)}$$

satisfies the heat equation $u_t = u_{xx}$ for $t > 0$ and all x.

(b) Sketch the graphs of $u(x, t)$ against x for $t = 0.01, 0.1, 1, 10$. You are observing heat conduction in an infinitely long insulated rod that at $t = 0$ is $0°$ everywhere except at the origin $x = 0$, and that is infinitely hot at $t = 0$ at the origin.

22. The temperature T of a metal plate can be described by a function $T = u(x, y, t)$ of three variables, the two space variables x and y and the time variable t. In many situations this function will satisfy the two-dimensional heat equation:

$$u_t = A(u_{xx} + u_{yy}) \quad \text{where } A \text{ is a positive constant.}$$

Find conditions on a, b, and c such that

$$u(x, y, t) = e^{-at} \sin(bx) \sin(cy)$$

satisfies this equation.

23. (a) Show that if a function $u(x, t)$ represents a wave that is traveling in the negative x-direction with speed c, then $u_t(x, t) = cu_x(x, t)$.

(b) Show that a wave traveling in the negative x-direction with speed c can be represented by the formula $u(x, t) = f(x + ct)$, where $f(x) = u(x, 0)$.

24. By focusing on the difference between the traveling wave equation $u_t = -cu_x$ and the modified equation $u_t = -uu_x$, explain why the modified equation might be used to describe a breaking wave.

25. (a) Verify that $u(x, t) = \sin(ax) \sin(at)$ satisfies the wave equation $u_{tt} = u_{xx}$. This solution represents a vibration with period $2\pi/a$, since $u(x, t + 2\pi/a) = u(x, t)$.

(b) Suppose that you wish to study the vibrations of a 1-meter string with fixed ends (such as a guitar string), so that $u(0, t) = 0$ and $u(1, t) = 0$ for all t, and such that $u_{tt} = u_{xx}$. The condition on the ends is a boundary condition. Find all $a > 0$ such that $u(x, t) = \sin(ax) \sin(at)$ satisfies both the PDE and the boundary condition. (It has been known to musicians since at least the time of Pythagoras that stretched strings can be made to vibrate at only special frequencies.)

26. You can generate a traveling wave on a string under tension by giving one end a snap. This suggests that there should be traveling wave solutions of the wave equation.

(a) Show that if f is an arbitrary function, then $u(x, t) = f(x - ct)$ is a solution of the wave equation $u_{tt} = c^2 u_{xx}$.

(b) Show that if g is an arbitrary function, then $u(x, t) = g(x + ct)$ is a solution of the wave equation $u_{tt} = c^2 u_{xx}$.

(c) Show that if f and g are arbitrary functions, then $u(x, t) = f(x - ct) + g(x + ct)$ is also a solution of the wave equation. In fact, all solutions of the wave equation can be written in this form, as the sum of a forward traveling wave and a backward traveling wave.

27. (a) Show that $u(x, t) = (\sin t)(\sin x)$ is a solution of the wave equation $u_{tt} = u_{xx}$.

(b) Use the identities

$$\cos(x + t) = (\cos x)(\cos t) - (\sin x)(\sin t)$$
$$\cos(x - t) = (\cos x)(\cos t) + (\sin x)(\sin t)$$

to express the function $u(x, t)$ of part (a) as the sum of a forward traveling wave and a backward traveling wave.

28. The vibration of a two dimensional object under tension, such as a drum head, is described by a function $u(x, y, t)$ of three variables, two space variables x and y and one time variable t. Such a function often satisfies the two dimensional wave equation: $u_{tt} = c^2(u_{xx} + u_{yy})$. Find conditions on a, b, and e such that $u(x, y, t) = \sin(ax)\sin(by)\sin(et)$ satisfies this equation.

29. Let u satisfy Burger's equation $u_t = u_{xx} + u_x^2$ and set $\omega = e^u$. Show that ω satisfies the heat equation $\omega_t = \omega_{xx}$. This change of variables is known as the Hopf-Cole transformation.

REVIEW PROBLEMS FOR CHAPTER THIRTEEN

For Problems 1–4, find the indicated partial derivatives.

1. $\dfrac{\partial z}{\partial x}$ and $\dfrac{\partial z}{\partial y}$ if $z = (x^2 + x - y)^6$.

2. $\dfrac{\partial F}{\partial L}$ if $F(L, K) = 3\sqrt{LK}$.

3. $\dfrac{\partial f}{\partial p}$ and $\dfrac{\partial f}{\partial q}$ if $f(p, q) = e^{p/q}$.

4. $\dfrac{\partial f}{\partial x}$ if $f(x, y) = (\ln y)e^{xy}$.

Find both partial derivatives for the functions in Problems 5–8.

5. $z = x^4 - x^7y^3 + 5xy^2$ 6. $z = \tan\theta/r$ 7. $w = s\ln(s + t)$

8. $w = \arctan(ue^{-v})$

9. Let $f(w, z) = e^{w \ln z}$.
    (a) Use difference quotients, with $h = 0.01$ to approximate $f_w(2, 2)$ and $f_z(2, 2)$.
    (b) Now evaluate $f_w(2, 2)$ and $f_z(2, 2)$ exactly.

10. Find the directional derivative of $z = x^2 - y^2$ at the point $(3, -1)$ in the direction $\theta = \frac{\pi}{4}$. In which direction is the directional derivative the largest?

11. Find the quadratic Taylor polynomial about $(0, 0)$ for $f(x, y) = \cos(x + 2y)\sin(x - y)$.

Are the statements in Problems 12–16 true or false? Explain your answer.

12. $f_{\vec{u}}(a, b) = \|\nabla f(a, b)\|$

13. $\operatorname{grad} f \cdot \vec{u} = 0$ where $\vec{u}$ is tangent to the level curves of f at the point where $\operatorname{grad} f$ is computed.

14. There is a point where $\|\operatorname{grad} f\| = 0$ and where there is a directional derivative which is not zero.

15. There is a function with $\|\operatorname{grad} f\| = 4$ and $f_{\vec{i}} f = 5$ at some point.

16. There is a function with $\|\operatorname{grad} f\| = 4$ and $f_{\vec{j}} = -3$ at some point.

17. The quantity Q (in pounds) of beef that a certain community buys during a week is a function $Q = f(b, c)$ of the prices of beef, b, and chicken, c, during the week. Is $\partial Q/\partial b$ positive or negative? What about $\partial Q/\partial c$?

18. Suppose the cost of producing one unit of a certain product is given by

$$c = a + bx + ky$$

where x is the amount of labor used (in man hours) and y is the amount of raw material used (by weight) and a and b are constants. What does $\partial c/\partial x = b$ mean? What is the practical interpretation of b?

19. People commuting to a city can choose to go either by bus or by train. The number of people who choose either method depends in part upon the price of each. Let $f(P_1, P_2)$ be the number of people who take the bus when P_1 is the price of a bus ride and P_2 is the price of a train ride. What can you say about the signs of $\partial f/\partial P_1$ and $\partial f/\partial P_2$? Explain the reasons for your answers.

20. A soft drink company is interested in seeing how the demand for its products is affected by prices. The company believes that the quantity, q, of its soft drinks sold depends on p_1, the average price of the company's soft drinks, p_2, the average price of competing soft drinks, and p_3, the average amount of money spent by the company on advertising. Suppose also that

$$q = C - 8 \cdot 10^6 p_1 + 4 \cdot 10^6 p_2 + p_3.$$

(a) What does the constant C represent?

(b) Find the marginal demand for soft drinks with respect to changes in p_1, p_2, and p_3. Explain why the signs and relative magnitudes of your answers are reasonable.[Hint: The marginal demand is the rate of change of quantity demanded with price.]

21. Imagine a box-shaped building with a flat roof. Suppose it has length x feet, width y feet, height z feet and is oriented as shown in Figure 13.52.

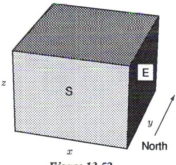

Figure 13.52

Each of the walls, the roof and the floor loses heat at a different rate because they are made of different materials and have different exposures. (For example, the southern exposure loses less heat because it receives more of the sun's rays.) The heat loss (measured in BTUs per square foot per hour, where a BTU is a unit of heat) is given in the following table. Let $H(x, y, z)$ represent the total heat loss per hour from the house.

Exposure	N	S	E	W	Floor	Roof
Heat loss per ft^2 per hour (BTU)	8	3	6	7	1	12

(a) Find a formula for H in terms of x, y, z.

(b) Find $\partial H/\partial x$.

(c) Interpret $\partial H/\partial x$ in terms of heat loss. Explain the presence of each term in $\partial H/\partial x$ as a contribution from some parts of the house. [Hint: Think geometrically about how the house changes as x varies.]

22.   Suppose that x is the price of one brand of gasoline and y is the price of a competing brand. Then q_1, the quantity of the first brand sold in a fixed time period, depends on both x and y, so $q_1 = f(x, y)$. Similarly, if q_2 is the quantity of the second brand sold in the same time, $q_2 = g(x, y)$.

   (a)   What do you expect the signs of $\partial q_1/\partial x$ and $\partial q_2/\partial y$ to be? Explain.

   (b)   What do you expect the signs of $\partial q_1/\partial y$ and $\partial q_2/\partial x$ to be? Explain.

23.   In the 1940s the quantity, q, of beer sold each year in Britain was found to depend on I (the aggregate personal income, adjusted for taxes and inflation), p_1 (the average price of beer) and p_2 (the average price of all other goods and services). Would you expect $\partial q/\partial I, \partial q/\partial p_1, \partial q/\partial p_2$ to be positive or negative? Give reasons for your answers.

24.   The acceleration due to gravity, g, at a distance r from the center of a planet of mass m is given by

$$g = \frac{Gm}{r^2}$$

   where G is the universal gravitational constant.

   (a)   Find $\partial g/\partial m$ and $\partial g/\partial r$.

   (b)   Interpret each of the partial derivatives you found in part (a) as the slope of a graph in the plane and sketch the graph.

25.   Suppose that the function $P = f(K, L)$ expresses the production of a firm as a function of the capital invested, K, and its labor costs, L.

   (a)   Suppose

$$f(K, L) = 60K^{1/3}L^{2/3}.$$

   Find the relationship between K and L if the marginal productivity of capital (that is, the rate of change of production with capital) equals the marginal productivity of labor cost (that is, the rate of change of production with labor cost). Put your answer in the simplest form possible.

   (b)   Now suppose

$$f(K, L) = cK^a L^b$$

   (a, b, c positive constants). What must be true of a, b and c if the relationship between K and L is to be the same as you found in part (a)?

26.   In analyzing a factory and deciding whether or not to hire more workers, it is useful to know under what circumstances productivity increases. Suppose $P = f(x_1, x_2, x_3)$ is the total quantity produced as a function of x_1, the number of workers, and any other variables x_2, x_3. We define the average productivity of a worker as P/x_1. Show that the average productivity increases as x_1 increases when marginal production, $\partial P/\partial x_1$, is greater than the average productivity, P/x_1.

27.   The total amount, P, of a certain product produced by a factory can be represented by the function

$$P = 40L^{1/4}K^{3/4}$$

   where L is the number of workers (in 100's) and K is the amount of capital used (in millions of dollars).

   (a)   Find and interpret $\partial P/\partial L$ at $L = 1, K = 16$.

   (b)   Suppose a factory is currently operating with $L = 1, K = 16$ and the factory wants to increase production by increasing L and K. What is the best ratio for the increase in L to the increase in K?

28. Show that if F is any differentiable function of one variable, then $V(x, y) = x F(2x + y)$ satisfies

$$x\frac{\partial V}{\partial x} - 2x\frac{\partial V}{\partial y} = V.$$

29. Find a particular solution to the differential equation in Problem 28 satisfying

$$V(1, y) = y^2.$$

30. The level curves representing the height of a surface above the xy-plane are shown in Figure 13.53. Sketch the paths that would be followed by particles starting at P and at Q.

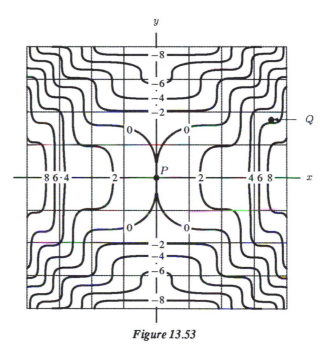

Figure 13.53

31. Figure 13.54 shows the level curves of a function $f(x, y)$. Give the approximate value of $f_{\vec{u}}(3, 1)$ with $\vec{u} = (-2\vec{i} + \vec{j})/\sqrt{5}$. Explain your answer.

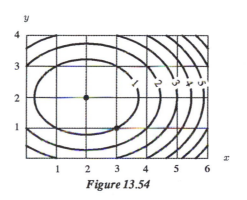

Figure 13.54

32. Suppose that the values of the function $f(x, y)$ near the point $x = 2$, $y = 3$ are given in Table 13.10. Estimate the following.

    (a) $\left.\dfrac{\partial f}{\partial x}\right|_{(2,3)}$, and $\left.\dfrac{\partial f}{\partial y}\right|_{(2,3)}$.

    (b) The rate of change of f at $(2, 3)$ in the direction of the vector $\vec{i} + 3\vec{j}$.

    (c) The maximum possible rate of change of f as you move away from the point $(2, 3)$. In which direction should you move to obtain this rate of change?

    (d) Find a vector tangent to the level curve of f through the point $(2, 3)$. What is the equation of the level curve?

    (e) Find the differential of f at the point $(2, 3)$. If $dx = 0.03$, $dy = 0.04$, find df. What does df represent?

TABLE 13.10

		x	
		2.00	2.01
	3.00	7.56	7.42
y	3.02	7.61	7.47

33. Find the equation of the tangent plane to $z = \sqrt{17 - x^2 - y^2}$ at the point $(3, 2, 2)$.

34. Find the equation of the tangent plane to $z = 8/(xy)$ at the point $(1, 2, 4)$.

35. Show that a normal to the surface $z = F(x, y)$ at a point (x_0, y_0, z_0) is given by

$$(F_x(x_0, y_0))\vec{i} + (F_y(x_0, y_0))\vec{j} - \vec{k}$$

36. Find an equation of the tangent plane and of a normal vector to the surface $x = y^3 z^7$ at the point $(1, -1, -1)$.

37. Find the point(s) on $x^2 + y^2 + z^2 = 8$ where the tangent plane is parallel to the plane $x - y + 3z = 0$.

38. Show that the tangent plane to

$$z = \frac{x^2}{a^2} + \frac{y^2}{b^2}$$

at the point (x_1, y_1, z_1) is given by

$$z + z_1 = \frac{2xx_1}{a^2} + \frac{2yy_1}{b^2}.$$

39. Two surfaces are said to be tangential at a point P if they have the same tangent plane. Show that the surfaces $z = \sqrt{2x^2 + 2y^2 - 25}$ and $z = \frac{1}{5}(x^2 + y^2)$ are tangential at the point $(4, 3, 5)$.

40. Two surfaces are said to be orthogonal to each other at a point P if the normals to their tangent planes are perpendicular at P. Show that $z = \frac{1}{2}(x^2 + y^2 - 1)$ and $z = \frac{1}{2}(1 - x^2 - y^2)$ are orthogonal at all points of intersection.

41. Each diagram (I) – (IV) in Figure 13.55 represents the level curves of a function $f(x, y)$. For each function f, consider the point above P on the surface $z = f(x, y)$ and choose from the lists which follow:

(a) A vector which could be the normal to the surface at that point;

(b) An equation which could be the equation of the tangent plane to the surface at that point

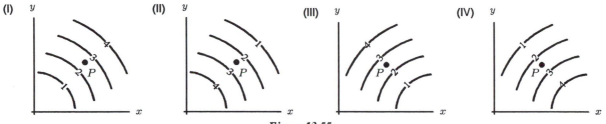

Figure 13.55

Vectors

(E) $2\vec{i} + 2\vec{j} - 2\vec{k}$
(F) $2\vec{i} + 2\vec{j} + 2\vec{k}$
(G) $2\vec{i} - 2\vec{j} + 2\vec{k}$
(H) $-2\vec{i} + 2\vec{j} + 2\vec{k}$

Equations

(I) $x + y + z = 4$
(J) $2x - 2y - 2z = 2$
(K) $-3x - 3y + 3z = 6$
(L) $-\dfrac{x}{2} + \dfrac{y}{2} - \dfrac{z}{2} = -7$

CHAPTER FOURTEEN

OPTIMIZATION

In one-variable calculus you learned how to find the maximum and minimum values of a function of one variable. In real life, you often want to consider more than one variable in an optimization problem. For example, you may have $10,000 to invest in new equipment and advertising for your business. What combination of equipment and advertising will yield the greatest extra income? Or, what combination of drugs will lower a patient's temperature the most? In this chapter we will consider optimization problems, both where the variables are completely free to vary (unconstrained optimization), and where there is a constraint on the variables (for example, a budget constraint).

14.1 LOCAL AND GLOBAL EXTREMA

Review of the One-Variable Case

A function of one variable is said to have a *local maximum* at $x = a$ if $f(a) \geq f(x)$ for x near a, and if a is not an endpoint of the domain of f. It has a *global maximum* at $x = a$ if $f(a) \geq f(x)$ for *all* values of x in the domain of f (including endpoints). The term global maximum is also used when we are considering a restricted domain; for example, if we are maximizing over a closed interval $a \leq x \leq b$, a function could have a global maximum at an endpoint. See Figure 14.1. Functions need not have local or global extrema — it depends on the function and on the domain under consideration. For example, $f(x) = x^2$ has a local minimum at $x = 0$, and this local minimum is also the global minimum, but it has no local or global maxima (see Figure 14.2). If, on the other hand we look at the same function on the domain $1 \leq x \leq 2$ then f has a global minimum at $x = 1$ and a global maximum at $x = 2$ (see Figure 14.3). Local maxima or minima occur at *critical points*. These are points where the derivative is either 0 or undefined. If the derivative is neither of these, then it must be a positive or negative number, so the graph must be sloping up or down: and you can't have a local maximum or minimum at such a point.

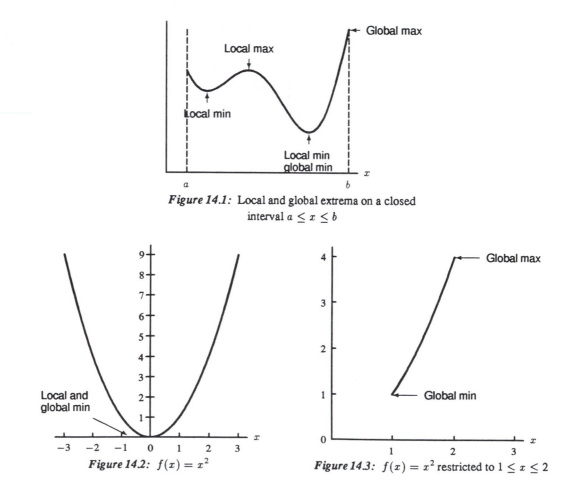

Figure 14.1: Local and global extrema on a closed
interval $a \leq x \leq b$

Figure 14.2: $f(x) = x^2$

Figure 14.3: $f(x) = x^2$ restricted to $1 \leq x \leq 2$

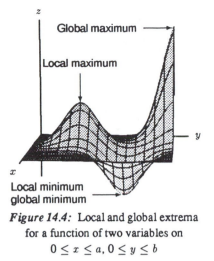

Figure 14.4: Local and global extrema
for a function of two variables on
$0 \leq x \leq a, 0 \leq y \leq b$

Generalization to the Multivariable Case

In the multivariable case we similarly have to distinguish between local and global extrema. Picture a surface in space, as in Figure 14.4. A function $f(x, y)$ has a *local maximum* at a point $(x, y) = (a, b)$ which is not on the boundary of its domain if $f(a, b) \geq (x, y)$ for all points (x, y) near (a, b). It has a global maximum at (a, b) if $f(a, b)$ is the largest value that f attains in its domain. We can also consider the problem of optimizing f over a particular region, in which case the idea of a global maximum or minimum applies to that region.

How do you Detect a Local Maximum or Minimum?

If it is defined at a point and is not zero, the gradient vector of a function points in a direction where the function is increasing. Since at a local maximum there is no direction in which the function is increasing, the gradient vector must be zero or undefined there. Similarly, the negative of the gradient vector points in a direction where the function is decreasing, and so it must be zero or undefined at a local minimum as well. For functions of two variables, you can see that the gradient vector must be zero at a local maximum by looking at the contour diagram. Figure 14.5 shows the contour diagram of a function with a local maximum, and Figure 14.6 shows its gradient vectors at various points. Around the maximum the vectors are all pointing in, perpendicularly to the contours, but at the maximum there is nowhere to point, and the gradient vector is zero.

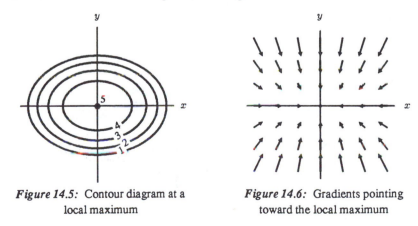

Figure 14.5: Contour diagram at a
local maximum

Figure 14.6: Gradients pointing
toward the local maximum

For functions of three or more variables, it is harder to draw pictures, but the principle remains the same:

> If P is a local maximum or a local minimum, then grad $f(P)$ is zero or undefined.

Thus, points where the gradient vector is zero or undefined play the same role as points where the derivative is zero or undefined in one-variable calculus, and we will give them the same name.

> **Critical points**
>
> A critical point of a function f of many variables is a point P where grad $f(P)$ is zero or undefined.

How Do You Find Critical Points?

To find critical points you set grad f equal to zero, which is the same thing as simultaneously setting each component equal to zero. Since the components of grad f are the partial derivatives of f, we have:

> To find critical points, set all partial derivatives equal to zero (and also look for points where one or more of them is undefined).

Since there is more than one partial derivative, finding the critical points of a multivariable function will usually involve setting more than one equation equal to zero.

Example 1 Find and analyze the critical points of $f(x, y) = x^2 - 2x + y^2 - 4y + 5$.

Solution Since
$$\text{grad } f = (2x - 2)\vec{i} + (2y - 4)\vec{j},$$
grad f is zero when
$$2x - 2 = 0, \quad \Rightarrow 2x = 2 \Rightarrow x = 1$$
$$2y - 4 = 0, \quad \Rightarrow 2y = 4 \Rightarrow y = 2$$

that is, when $(x, y) = (1, 2)$. Hence, f has only one critical point, namely the point $(1, 2)$. What is the behavior of f near $(1, 2)$? Look at the values of the function in Table 14.1.

TABLE 14.1 *Values of $f(x, y)$ near the point* $(1, 2)$

		x				
		0.8	0.9	1.0	1.1	1.2
	1.8	0.08	0.05	0.04	0.05	0.08
	1.9	0.05	0.02	0.01	0.02	0.05
y	2.0	0.04	0.01	0.00	0.01	0.04
	2.1	0.05	0.02	0.01	0.02	0.05
	2.2	0.08	0.05	0.04	0.05	0.08

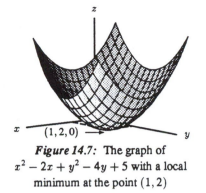

Figure 14.7: The graph of
$x^2 - 2x + y^2 - 4y + 5$ with a local
minimum at the point $(1, 2)$

From the table, we see that the function seems to have a minimum value of 0 at $(1, 2)$. We can verify this by completing the square:

$$f(x, y) = (x - 1)^2 + (y - 2)^2.$$

The graph of this is like the graph of $x^2 + y^2$, shifted to $(1, 2)$. The graph of $x^2 + y^2$ is shown in Figure 11.16 on page 19; it looks like a parabolic bowl with its vertex at $(0, 0)$. The graph of f, shown in Figure 14.7, is similar, except that the vertex has been shifted to $(1, 2)$. So $(1, 2)$ must be a local minimum of f (as well as a global minimum).

Example 2   Find any critical points of $f(x, y) = -\sqrt{x^2 + y^2}$.

Solution   Here we look again for places where grad $f = 0$ or is undefined. The partial derivatives are given by

$$\frac{\partial f}{\partial x} = -\frac{x}{\sqrt{x^2 + y^2}},$$

$$\frac{\partial f}{\partial y} = -\frac{y}{\sqrt{x^2 + y^2}}.$$

These are *never* both zero; but they are both undefined at $(x, y) = (0, 0)$. Thus, $(0,0)$ is a critical point, and a possible extreme point. In fact, the graph of f (Figure 14.8) shows that it is an upward pointing cone, with vertex at $(0,0)$.

So here is an example where a maximum (both local and global) occurs at a place where no tangent plane exists.

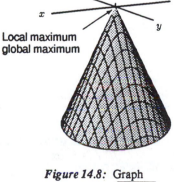

Local maximum
global maximum

Figure 14.8: Graph
of $f(x, y) = -\sqrt{x^2 + y^2}$

Example 3   Find the local extrema of the function $f(x, y) = 8y^3 + 12x^2 - 24xy$.

Solution   Begin by looking for critical points:

$$\frac{\partial f}{\partial x} = 24x - 24y,$$

$$\frac{\partial f}{\partial y} = 24y^2 - 24x.$$

Setting these equations equal to zero gives the system

$$x = y,$$
$$x = y^2,$$

which has two solutions $(0,0)$ and $(1, 1)$. But are these maxima, minima or neither? Let's look at the contours near the points: Figure 14.9 shows the contour diagram of this function. Notice that $f(1, 1) = -4$ and that there is no other -4 contour. The contours near $P = (1, 1)$ appear oval in shape, and show that f is increasing in value no matter in which direction you move away from P. This suggests that f has a local minimum at the point $(1, 1)$. The level curves near $Q = (0, 0)$ show a very different behavior. While $f(0, 0) = 0$, we see that f takes on both positive and negative values at nearby points. Thus, it appears that $(0,0)$ is a *saddle point*, that is, a critical point which is neither a local maximum or minimum.

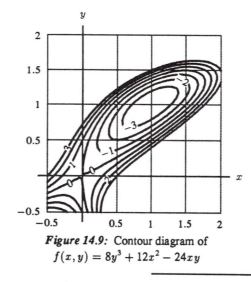

Figure 14.9: Contour diagram of
$f(x, y) = 8y^3 + 12x^2 - 24xy$

Saddle Points

In one-variable calculus, critical points can indicate local maxima, minima, or inflection points of the function. The situation is similar in more than one variable: critical points can occur at local maxima or minima, or at places which are neither – the function can be greater in some directions and less in others. A simple example can be found by looking at the function $f(x, y) = x^2 - y^2$, shown in Figure 14.10. At $(0,0)$ this function has a flat tangent plane, and value 0. Moving along the y-axis in either direction from $(0,0)$ gives values for f that are negative, while moving along the x-axis in either direction gives values of f that are positive. Since f has the appearance of a saddle near $(0,0)$ we call any point with these properties a *saddle point*.

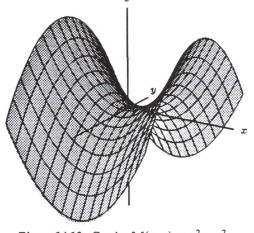

Figure 14.10: Graph of $f(x, y) = x^2 - y^2$,
showing a saddle point

Let's look at the level curves of f near the saddle point (0,0) (Figure 14.11). They are hyperbolas; the ones intersecting the x-axis correspond to positive values of f, and the ones intersecting the y-axis correspond to negative values of f. You can see that near (0,0) there are always values of f which are above and below the value of f at (0,0).

Contrast this with the appearance of level curves near a local maximum or minimum, for example, near (0,0) of $g(x, y) = x^2 + y^2$ (Figure 14.12). These are circles which surround the point, so that near (0,0) the function is increasing or decreasing (in this case, increasing) no matter in which direction you look.

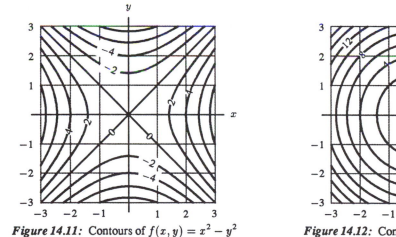

Figure 14.11: Contours of $f(x, y) = x^2 - y^2$

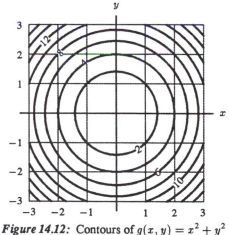

Figure 14.12: Contours of $g(x, y) = x^2 + y^2$

The Second Derivative Test

We have seen how to recognize whether a critical point is a maximum, minimum, or saddle point, by looking at the contour diagram. Now we will develop an analytic method for making the distinction. In one-variable calculus, we had the second derivative test: If $f'' > 0$ then the graph is concave up, so the point is a minimum; if $f'' < 0$ then the graph is concave down, hence a maximum; if $f'' = 0$ then anything can happen (maximum, minimum or inflection point).

There is a generalization of the second derivative test to multivariable functions – we shall describe it here only for functions of two variables.

Quadratic Functions

Let's begin by looking at what can happen at critical points of functions which are of the form $f(x, y) = ax^2 + bxy + cy^2$, where a, b and c are constants. You may think that these are very special functions, but using the Taylor series expansion we can approximate most functions by quadratic functions, and the approximation is usually good enough so that the function and its quadratic Taylor polynomial have the same classification of the critical point into maximum, minimum, saddle point, or none of these. So understanding quadratic functions is the key.

Example 4   Find and analyze the local extrema of the function $f(x, y) = x^2 + xy + y^2$.

Solution   The partial derivatives are

$$\frac{\partial f}{\partial x} = 2x + y,$$
$$\frac{\partial f}{\partial y} = x + 2y.$$

The only critical point is $(0, 0)$, and the value of the function there is $f(0, 0) = 0$. So, if f is always positive near $(0, 0)$, then $(0, 0)$ is a local minimum, and if f is always negative near $(0, 0)$, it is a local maximum, and if f takes both positive and negative values it is neither. One way to find out is to look at the graph in Figure 14.13, which shows that $(0, 0)$ is a local minimum.

How could we have known the graph would look like this without drawing it? There is an algebraic way to determine if a quadratic function is always negative, always positive, or neither, and that is to complete the square. If we write

$$f(x, y) = x^2 + xy + y^2 = (x + \frac{1}{2}y)^2 + \frac{3}{4}y^2,$$

then we can see that $f(x, y)$ is a sum of two squares, so it must always be ≥ 0; thus, the critical point must be a minimum.

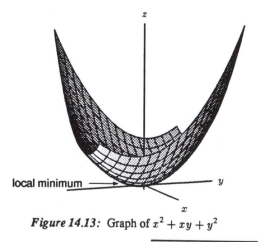

Figure 14.13: Graph of $x^2 + xy + y^2$

In general, to complete the square of a quadratic function $f(x, y) = ax^2 + bxy + cy^2$, you write (assuming $a \neq 0$)

$$ax^2 + bxy + cy^2 = a[x^2 + \frac{b}{a}xy + \frac{c}{a}y^2]$$

$$= a\left[\left(x + \frac{b}{2a}y\right)^2 + \left(\frac{c}{a} - \frac{b^2}{4a^2}\right)y^2\right]$$

$$= a\left[\left(x + \frac{b}{2a}y\right)^2 + \left(\frac{4ac - b^2}{4a^2}\right)y^2\right].$$

The basic shape of the graph of f depends on whether the coefficient of y^2 is positive, negative, or zero. The sign of $4ac - b^2$ determines the sign of this coefficient.

- If $4ac - b^2 > 0$, then both squares inside the brackets have positive coefficients, so the function has a maximum or a minimum. It has a minimum if a is positive, since then it will be a right-side-up paraboloid, and it has a maximum if a is negative, since then the whole thing is turned upside down.

- If $4ac - b^2 < 0$ then the two squares have different signs; this means the function sometimes goes up and sometimes goes down, like $x^2 - y^2$, and hence it has a saddle point at $(0, 0)$.

- If $4ac - b^2 = 0$, then the quadratic function is $a(x + by/2a)^2$, whose graph is a parabolic cylinder.

We assumed that $a \neq 0$ above; it turns out that rules above also cover the case where $a = 0$. Thus, we have:

Shape of the graph of $f(x, y) = ax^2 + bxy + cy^2$.

Let

$$D = 4ac - b^2.$$

(a) If $D > 0$ and $a > 0$, the graph is a parabolic bowl.

(b) If $D > 0$ and $a < 0$, the graph is an inverted parabolic bowl.

(c) If $D < 0$, the graph is a saddle.

(d) If $D = 0$, the graph is a parabolic cylinder.

The quantity $D = 4ac - b^2$ is called the *discriminant* of the quadratic function.

The General Case

The reason we spent so much time studying functions of the form $ax^2 + bxy + cy^2$ is that, near a critical point, most functions of two variables behave like such quadratic functions. This means that we can apply the discriminant analysis to almost *any* function to classify its critical points as local maxima, minima or saddle points.

To see why this is so, we look at the quadratic Taylor polynomial of f described on page 168:

$$f(x, y) \approx f(a, b) + f_x(a, b)(x - a) + f_y(a, b)(y - b)$$
$$+ \frac{1}{2}f_{xx}(a, b)(x - a)^2 + f_{xy}(a, b)(x - a)(y - b) + \frac{1}{2}f_{yy}(a, b)(y - b)^2.$$

At critical points, where $f_x(a, b) = f_y(a, b) = 0$, the expansion simplifies to

$$f(x, y) \approx f(a, b) + \frac{1}{2}f_{xx}(a, b)(x - a)^2 + f_{xy}(a, b)(x - a)(y - b) + \frac{1}{2}f_{yy}(a, b)(y - b)^2,$$

or

$$f(x, y) - f(a, b) \approx \frac{1}{2}f_{xx}(a, b)(x - a)^2 + f_{xy}(a, b)(x - a)(y - b) + \frac{1}{2}f_{yy}(a, b)(y - b)^2.$$

In words, this says that the value of f at points near (a, b) differs from $f(a, b)$ by an amount that behaves just like a quadratic functions. In this case, the discriminant is

$$4(\frac{1}{2}f_{xx})(\frac{1}{2}f_{yy}) - f_{xy}{}^2,$$

which simplifies to

$$f_{xx}f_{yy} - f_{xy}{}^2,$$

where the partials are evaluated at (a, b). We then have the following result:

Second Derivative Test for Functions of Two Variables

Suppose (a, b) is a critical point of f. Let

$$D = f_{xx}(a, b)f_{yy}(a, b) - f_{xy}(a, b)^2.$$

(a) If $D > 0$ and $f_{xx}(a, b) > 0$, (a, b) is a minimum.
(b) If $D > 0$ and $f_{xx}(a, b) < 0$, (a, b) is a maximum.
(c) If $D < 0$, (a, b) is a saddle point.
(d) If $D = 0$, anything can happen.

Example 5   Find the local maxima, minima, and saddle points of the function

$$f(x, y) = \frac{x^2}{2} + 3y^3 + 9y^2 - 3xy + 9y - 9x.$$

Solution   The partial derivatives are $f_x = x - 3y - 9$ and $f_y = 9y^2 + 18y - 3x + 9$. Setting $f_x = 0$ gives $x = 3y + 9$, and setting $f_y = 0$ gives

$$9y^2 + 18y + 9 = 3x$$
$$= 9y + 27,$$
$$9y^2 + 9y - 18 = 0,$$
$$9(y + 2)(y - 1) = 0, \quad \text{so} \quad y = -2 \text{ or } 1.$$

Therefore, the critical points are $(3, -2)$ and $(12, 1)$. The discriminant is

$$D(x, y) = f_{xx}f_{yy} - f_{xy}^2 = (1)(18y + 18) - (-3)^2$$
$$= 18y + 9.$$

Since $D(3, -2) = -36 + 9 < 0$, $(3, -2)$ is a saddle point, and since $D(12, 1) = 18 + 9 > 0$, $(12, 1)$ is a local minimum.

1. MISSISSIPPI: 58; 83

ALABAMA: 88, 85

PENNSYLVANIA: 89; 78

NEW YORK: 85; 75

CALIFORNIA: 100; 65

ARIZONA: 111; 85

MASSACHUSETTS: 82; 74

5. $z = -x^2 - y^2$

$z_x = -2x \Rightarrow -2x = 0 \Rightarrow x = 0$

$z_y = -2y \Rightarrow -2y = 0 \Rightarrow y = 0$

MAX: $(0,0)$

MIN: $(1,1)(1,-1)(-1,-1)(-1,1)$

3. $z = x^2 + y^2$

$z_x = 2x \Rightarrow 2x = 0 \Rightarrow x = 0$

$z_y = 2y \Rightarrow 2y = 0 \Rightarrow y = 0$

MIN: $(0,0)$

MAX: $(1,1)(-1,1)(-1,-1)(1,-1)$

7. a. max   c. max

    b. saddle   d. none

9. (See Sol. Man.)

(?)

11. $P = 400 - 3x^2 - 4x + 2xy - 5y^2 + 48y$

$P_x = -6x - 4 + 2y$

$6x = -4 + 2y$

$x = -\frac{2}{3} + \frac{1}{3}y$

$P_y = 2x - 10y + 48$

$10y - 48 = 2x$

$10y - 48 = 2\left(-\frac{2}{3} + \frac{1}{3}y\right)$

$10y - 48 = -1\frac{1}{3} + \frac{2}{3}y$

$11\frac{1}{3}y - 48\frac{2}{3} = 0$

(?)

13. $f = (x+y)(xy+1) = x^2y + x + xy^2 + y$

$f_x = (xy+1) + (x+y)(y) = (xy+1) + (xy + y^2) = y^2 + 2xy + 1 = 0$

$f_y = (xy+1) + (x+y)(x) = (xy+1) + (xy + x^2) = x^2 + 2xy + 1 = 0 \Rightarrow x^2 = y^2 \Rightarrow x = \pm y$

for $x = y \Rightarrow x^2 + 2x(x) + 1 = 0 \Rightarrow x^2 + 2x^2 + 1 = 0 \Rightarrow 3x^2 = -1 \Rightarrow x = 0$

for $x = -y \Rightarrow x^2 + 2x$

15. $f = x^2 + y^3 - 3xy$

$f_x = 2x - 3y \Rightarrow 2x - 3y = 0 \Rightarrow 2x = 3y$

$f_y = 3y - 3x \Rightarrow 3y - 3x = 0 \Rightarrow 3y = 3x$

17. $E = 1 - \cos x + \frac{y^2}{2}$

$E_x = \sin x = 0$ when $x = 0, \pm\pi, \pm 2\pi \cdots$

$E_y = 2\left(\frac{y}{2}\right) = y = 0$ when $y = 0$

Critical Points: $(-\pi, 0), (0,0), (\pi, 0), (2\pi, 0) \cdots$

$D = E_{xx} E_{yy} - E_{xy}^2 =$

Problems for Section 14.1

1. By looking at the weather map in Figure 11.1 on page 2, find the maximum and minimum daily temperatures in the states of Mississippi, Alabama, Pennsylvania, New York, California, Arizona, and Massachusetts.

2. By looking at the table of UV exposure as a function of latitude and year in Table 22 on page 67, find the latitude and year that will have the worst UV exposure between now and the year 2010.

In Problems 3–5, find the global maximum and minimum of the given function over the square $-1 \le x \le 1, -1 \le y \le 1$, and say whether it occurs on the boundary of the square. (Hint: consider the graph of the function.)

3. $z = x^2 + y^2$
        4. $z = x^2 - y^2$
        5. $z = -x^2 - y^2$

6. Suppose $f(x, y) = A - (x^2 + Bx + y^2 + Cy)$. What values of A, B, and C give $f(x, y)$ a maximum value of 15 at the point $(-2, 1)$?

For Problems 7–9, use Figure 14.14, which shows level curves of some function $f(x, y)$.

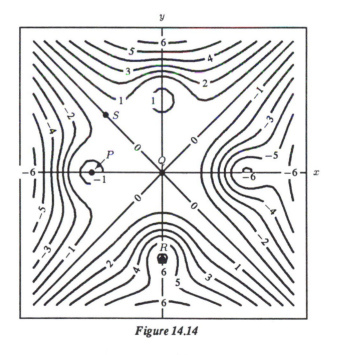

Figure 14.14

7. Decide whether each of the points is a local max, local min, saddle point, or none of these:
   (a) P
           (b) Q
           (c) R
           (d) S

8. Sketch ∇f around P and Q. Your sketch should make clear the direction of ∇f; don't worry about the magnitude.

9. Put arrows showing the direction of ∇f at the points where $\|\nabla f\|$ is largest.

10. (a) Find all the critical points of the function

$$f(x, y) = 8xy - \frac{1}{4}(x + y)^4.$$

   (b) Are these local maxima, local minima or saddle points? Justify your answer.

11. Find the values of x and y which maximize

$$P(x, y) = 400 - 3x^2 - 4x + 2xy - 5y^2 + 48y.$$

For Problems 12–14, find the local maxima, minima and saddle points of the given function.

12. $f(x, y) = \frac{x^2}{2} + 3y^3 + 9y^2 - 3xy + 9y - 9x.$ 13. $f(x, y) = (x + y)(xy + 1)$.

14. $f(x, y) = \sin x + \sin y + \sin(x + y), \quad 0 < x < \pi, 0 < y < \pi.$

For Problems 15 and 16, find the local maxima, local minima, and saddle points of the functions given. Dicide if the local maxima or minima are global maxima or minima. Explain.

15. $f(x, y) = x^2 + y^3 - 3xy$ 16. $f(x, y) = xy + \ln x + y^2 - 10$ $(x > 0)$

17. Find and classify the critical points of the function

$$E = 1 - \cos x + \frac{y^2}{2}.$$

18. (a) Find all the critical points of the function

$$f(x, y) = \sin x \sin y.$$

   (b) Are these local maxima, local minima or saddle points? Justify your answer.

19. Suppose $f_x = f_y = 0$ at $(1, 3)$ and $f_{xx} > 0$, $f_{yy} > 0$, $f_{xy} = 0$.

   (a) What can you conclude about the behavior of the function near the point $(1, 3)$?
   (b) Draw a possible sketch for the contour diagram.

20. Suppose that for some function $f(x, y)$ near the point (a, b), we have $f_x = f_y = 0$, $f_{xx} > 0$, $f_{yy} = 0$, $f_{xy} > 0$.

   (a) What can you conclude about the shape of the function near (a, b)?
   (b) Draw a possible contour diagram.

14.2 UNCONSTRAINED OPTIMIZATION

In this section we give some examples of the theory of optimization applied to practical problems.

Monopolies, Oligopolies, and Collusion

A *monopoly* is a market in which there is only one supplier for a particular good. For example, your local electric company is a monopoly, as was Bell Telephone before its break-up. ("Mono" means one and "polein" means sell in Greek.) A market in which a small number of companies compete, so that the behavior of any one firm can influence the rest, is called an *oligopoly* ("oligo" means few). In this section, we will model the behavior of an oligopoly consisting of two companies, and see why they might choose to work together as a monopoly rather than compete.[1]

----

[1] Adapted from *An Application of Calculus in Economics: Oligopolistic Competition*, Donald R. Sherbert (COMAP, Inc., Lexington, MA, 1982)

Assumptions about Price, Quantity, and Revenue

The first person to study the behavior of an oligopoly was the French economist Cournot in 1838. He imagined two owners of mineral water springs, who sold the water from their springs to people who brought their own containers. Thus, there were no costs of production. The question he considered was how much water each would choose to sell, and how much revenue (which in this case is the same as profit) each would make.

Suppose the two spring owners sell q_1 and q_2 liters of water a day, respectively. We assume that they charge the same price of \$$p$ per liter. Then the revenue that the owners make are, respectively,

$$R_1 = pq_1,$$
$$R_2 = pq_2.$$

In order to see how the revenues behave as the quantities sold change, we need to know how the price depends on the quantity sold. In general, the more of a good that is available, the less consumers are willing to pay, per unit. Thus, the price paid for a good is usually assumed to decrease as the quantity sold increases. Since the total quantity sold is $q_1 + q_2$, we will assume that if

$$p = f(q_1 + q_2),$$

then f is a decreasing function. In particular, we will assume that f is a linear function of $q_1 + q_2$, so that we can write

$$p = b - m \cdot (q_1 + q_2)$$

where b and m are positive constants.

We will now model and compare ways the spring owners might behave.

Owners Acting Separately: The Cournot Hypothesis

We know that the two owners make revenues of

$$R_1 = pq_1 = (b - m \cdot (q_1 + q_2))q_1 = bq_1 - mq_1^2 - mq_1q_2,$$
$$R_2 = pq_2 = (b - m \cdot (q_1 + q_2))q_2 = bq_2 - mq_2^2 - mq_1q_2.$$

Cournot decided to assume that each owner would look at the other owner's output as given (a constant) and maximize his own output. The first owner, therefore, regards q_2 as a constant and wants to maximize R_1 as a function of q_1. Thus, the first owner wants to produce q_1 to satisfy

$$\frac{\partial R_1}{\partial q_1} = b - 2mq_1 - mq_2 = 0.$$

Similarly, the second owner regards q_1 as a constant and wants to maximize R_2 by producing q_2 to satisfy

$$\frac{\partial R_2}{\partial q_2} = b - 2mq_2 - mq_1 = 0.$$

Solving these equations simultaneously gives

$$q_1 = q_2 = \frac{b}{3m}.$$

At this production level, the price per liter is

$$p = b - m \cdot (q_1 + q_2) = b - m\left(\frac{2b}{3m}\right) = \frac{b}{3},$$

and therefore each owner receives a revenue of

$$R_1 = R_2 = \frac{b^2}{9m}.$$

Since, viewed as a function of one variable, each revenue is a quadratic whose graph opens downwards, these critical points must both be local and global maxima.

Owners Acting in Collusion

Suppose, instead of looking at the production of the other as fixed, the owners get together and decide that they will choose q_1 and q_2 to maximize total revenue, $R = R_1 + R_2$. The joint revenue is

$$R = R_1 + R_2 = pq_1 + pq_2 = p(q_1 + q_2),$$

thus,

$$R = [b - m \cdot (q_1 + q_2)](q_1 + q_2) = b(q_1 + q_2) - m \cdot (q_1 + q_2)^2.$$

At the maximum of R, as a function of the two variables q_1 and q_2, we have

$$\frac{\partial R}{\partial q_1} = b - 2m \cdot (q_1 + q_2) = 0$$

$$\frac{\partial R}{\partial q_2} = b - 2m \cdot (q_1 + q_2) = 0.$$

These two equations are identical, and there are infinitely many solutions to them, corresponding to any pair (q_1, q_2) satisfying $q_1 + q_2 = b/2m$. Any such combination gives the maximum revenue because the surface representing R is a parabolic cylinder opening downwards, so all its critical points are both local and global maxima. If, for example, the owners decide to sell equal quantities, then

$$q_1 = q_2 = \frac{b}{4m},$$

and for these quantities,

$$p = b - m \cdot (q_1 + q_2) = b - m\left(\frac{2b}{4m}\right) = \frac{b}{2}.$$

Therefore, the total revenue is

$$R = p(q_1 + q_2) = \frac{b^2}{4m}$$

and so each owner receives

$$R_1 = R_2 = \frac{b^2}{8m}.$$

The important thing to notice is that each owner does *better* when cooperating with the other than separately. (Each owner gets a revenue of $b^2/8m$ instead of $b^2/9m$.) Thus, there is an incentive for the owners to cooperate and share the revenues; this is called *collusion*. Oligopolies are often observed to work cooperatively in practice.

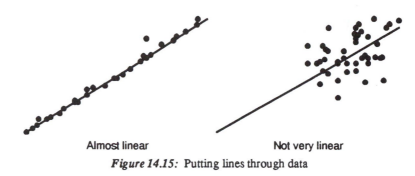

Figure 14.15: Putting lines through data

Curve Fitting

Another simple application of optimization is to the problem of linear approximation by the method of least squares. This is a technique used by many disciplines for taking discrete experimental data and fitting a line to it. By "fitting a line" we mean finding a line that in some sense comes as close as possible to the data points. If the data points appear to be roughly linear, this can be a useful procedure, as it allows the person conducting the experiment to quantify, by means of a formula, the behavior of the data. If the data is not nearly linear, then fitting a line to the data has dubious value. See Figure 14.15.

The *least squares* method measures the distance from a line to the data points by adding together all of the squares of the vertical distances from each point to the line. The line with a minimal sum of square distances is called the *least squares line*, or sometimes the *regression* line. (The reason we use the squares of the distances is to prevent the contributions from points above the line and below the line from canceling.)

Example 1   Find a least squares line for the following data points: $(1, 1)$, $(2, 1)$, and $(3, 3)$.

Solution   A line may be specified by an equation $y = b + mx$. If we can find b and m then we'll have found the line. So, for this problem, b and m are the two variables. What is the function that we want to minimize? It is the sum of the three squared vertical distances from the points to the line (see Figure 14.16).

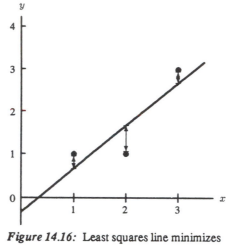

Figure 14.16: Least squares line minimizes
the sum of the squares of these vertical
distances

The y- coordinate of a point on the line has values $b + m$, $b + 2m$ and $b + 3m$ at x values 1, 2 and 3 respectively. Then the vertical distance of each data point to the line is $1 - (b + m)$, $1 - (b + 2m)$, and $3 - (b + 3m)$, so the sum of squares is

$$S = f(m, b) = (1 - (b + m))^2 + (1 - (b + 2m))^2 + (3 - (b + 3m))^2.$$

Let's look for critical points. First we differentiate S with respect to m:

$$\begin{aligned}
\frac{\partial S}{\partial m} &= 2(1 - (b + m))(-1) + 2(1 - (b + 2m))(-2) + 2(3 - (b + 3m))(-3) \\
&= -2 + 2b + 2m - 4 + 4b + 8m - 18 + 6b + 18m \\
&= -24 + 12b + 28m.
\end{aligned}$$

Now we'll compute the partial derivative with respect to b:

$$\begin{aligned}
\frac{\partial S}{\partial b} &= -2(1 - (b + m)) - 2(1 - (b + 2m)) - 2(3 - (b + 3m)) \\
&= -2 + 2b + 2m - 2 + 2b + 4m - 6 + 2b + 6m \\
&= -10 + 6b + 12m.
\end{aligned}$$

Setting both partials equal to zero yields a system of two linear equations in two unknowns, which is easy to solve:

$$\begin{aligned}
0 &= -24 + 12b + 28m = -4(6 - 3b - 7m) \\
0 &= -10 + 6b + 12m = -2(5 - 3b - 6m).
\end{aligned}$$

The solution to this pair of equations is $m = 1$ and $b = -1/3$. Since S is a sum of squares, it is evident that this is a local minimum. So the regression line is $y = x - 1/3$. Do these results make sense? Without the point (2,1) the line that comes as close as possible to the remaining points (1,1) and (3,3) is a line that passes through both of them: it would have slope 1 and y intercept 0. Introducing the point (2,1) moves the y intercept down (from 0 to $-1/3$).

There are general formulas for finding the slope and y-intercept of a least squares line, and their derivation will be explored in the exercises at the end of the section. Many calculators have these formulas built in, so that all you need to do is enter the data points and out pops the line (i.e. m and b). There are other measures of interest that are typically computed at the same time. For example, the *correlation coefficient*, which is a measure of how close the data points actually come to fitting the least squares line.

Gradient Search

So far we have searched for values that maximize or minimize a function $f(x, y)$ by first finding the critical points of f. Finding the critical points amounts to solving the equation grad $f = 0$, which is really a pair of simultaneous equations for x and y:

$$\frac{\partial f}{\partial x}(x, y) = 0 \quad \text{and} \quad \frac{\partial f}{\partial y}(x, y) = 0.$$

But solving such equations can be a very difficult problem in its own right! In many cases even equations in a single-variable must be solved numerically, say by Newton's method, rather than explicitly. You should not be surprised if you cannot find explicit values for critical points. In practice, most optimization problems are solved approximately by numerical methods. And in the real world, approximations are often all there are.

One of the simplest numerical methods of optimization can be explained by analogy with a mountain climber who wishes to maximize his elevation by getting to the top of the highest mountain. All he has to do is keep going up, and eventually he will get to the top of some mountain. If he started near enough to the highest mountain, that is probably the mountain he will conquer. If not, there is a chance that he will go up a lower mountain, winding up at a local rather than a global maximum, but if lucky, he will still be pretty high.

The method of optimization this suggests is illustrated in the next example. It is a minimization problem, so you should imagine a hiker who seeks the lowest valley by always going down.

Example 2   Four hundred cubic meters of gravel are to be delivered to a landfill by a trucker. She plans to purchase an open-top box in which to transport the gravel in numerous trips. She figures that the cost to her will be the cost of the box plus $0.10 per trip. The box must have height 0.5 m, but she can choose the length and width. The cost of the box will be $20/m^2 for the ends and $10/m^2 for the bottom and sides. Notice the tradeoff she faces: A smaller box is cheaper to buy but will require more trips. What size box should she buy to minimize her over-all costs? [2]

Solution   We will first get an algebraic expression for the trucker's cost. Let the length and width of the box be x meters and y meters, so that the box measures x m $\times$ y m $\times$ 0.5 m as in Figure 14.17. The

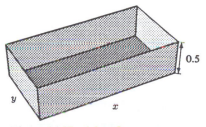

Figure 14.17: A box for transporting gravel

[2]adapted from Claude McMillan, Jr., *Mathematical Programming*, 2nd ed., Wiley, New York, 1975, p156-157

volume of the box is $0.5xy$ m^3, so delivery of 400 m^3 of gravel will require $400/0.5xy$ trips. The trucker's cost can be itemized as follows:

TABLE 14.2   *Trucker's itemized cost*

400/(0.5xy) at $0.10/trip	80/xy
2 ends at $20/m^2 × 0.5 y m^2	20y
2 sides at $10/m^2 × 0.5 x m^2	10x
1 bottom at $10/m^2 × xy m^2	10 xy
total cost	80/(xy) + 10x + 10xy + 20y

Our problem is to choose x and y to minimize the function

$$f(x,y) = \frac{80}{xy} + 10x + 10xy + 20y.$$

The critical points are given by solving the system:

$$\frac{\partial f}{\partial x} = \frac{-80}{x^2 y} + 10y + 10 = 0,$$

$$\frac{\partial f}{\partial y} = \frac{-80}{xy^2} + 10x + 20 = 0.$$

It does not appear easy to solve this system of equations, so we will take another approach to minimizing the function f. We will pick a starting point (x_0, y_0) that may not minimize f but which we hope might not be too far from a minimizing point. In this example we will start with $(x_0, y_0) = (5, 5)$, which is definitely not a critical point of f because

$$\text{grad } f(5,5) = 59.4\vec{i} + 69.4\vec{j} \neq 0\vec{i} + 0\vec{j}\,.$$

We plan to move from (x_0, y_0) to a new point (x_1, y_1) such that $f(x_1, y_1) < f(x_0, y_0)$. Motion in the direction of grad $f(x_0, y_0)$ will increase f as rapidly as possible, so we decrease f in the opposite direction, the direction of steepest descent, stopping just as f values would begin to increase again. Figure 14.18 gives the graph of

$$f[(x_0, y_0) - t \text{ grad } f(x_0, y_0)] = f(5 - 59.4t, 5 - 69.4t)$$

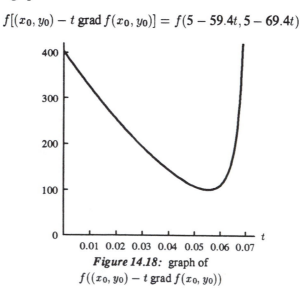

Figure 14.18: graph of
$f((x_0, y_0) - t \text{ grad } f(x_0, y_0))$

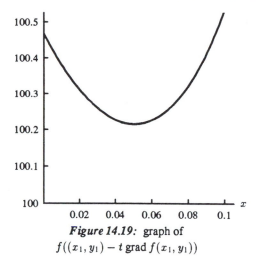

Figure 14.19: graph of
$$f((x_1, y_1) - t \operatorname{grad} f(x_1, y_1))$$

for positive t. As t increases from 0, $f(5 - 59.4t, 5 - 69.4t)$ decreases, attaining a local minimum at $t \approx 0.0554$. So we take

$$(x_1, y_1) = (5 - (59.4)(0.0554), 5 - (69.4)(0.0554)) = (1.71, 1.16).$$

Notice that $f(1.71, 1.16) \approx 100.47 < f(5, 5) = 403.2$, so we have decreased the cost function f by quite a lot.

We can further decrease f by moving away from $(1.71, 1.16)$ in the direction opposite to $\operatorname{grad} f(1.71, 1.16) = -1.99\vec{i} + 2.33\vec{j}$. Figure 14.19 gives the graph of

$$f[(x_1, y_1) - t \operatorname{grad} f(x_1, y_1)] = f(1.71 + 1.99t, 1.16 - 2.33t)$$

which achieves a local minimum at $t = 0.050$. So we take

$$(x_2, y_2) = (1.71 + (1.99)(0.050), 1.16 - (2.33)(0.050))$$
$$= (1.81, 1.04).$$

Notice that $f(1.81, 1.04) \approx 100.22 < f(1.71, 1.16) \approx 100.47$. The move from (x_1, y_1) to (x_2, y_2) decreased the cost f by a rather small amount, only \$0.25, so while we have not actually achieved a minimum, we may feel that for practical purposes we are close enough. The trucker will round off, buying a box of dimension 1.8m $\times$ 1m $\times$ 0.5m.

Problems for Section 14.2

1. Assume that two products are produced in quantities q_1 and q_2 and sold at prices of p_1 and p_2 respectively, and that the cost of producing them is given by

$$C = 2q_1^2 + 2q_2^2 + 10.$$

   (a) Find the maximum profit that can be made, assuming the prices are fixed.
   (b) Find the rate of change of that maximum profit as p_1 increases.

2. A cruise missile has a remote guidance device which is sensitive to both temperature and humidity. Army engineers have worked out a formula to show the range at which the missile can be controlled:

$$\text{Max range in miles} = 12{,}000 - t^2 - 2ht - 2h^2 + 200t + 260h,$$

where t is the temperature in $^\circ$F and h is percent humidity. What are the optimal atmospheric conditions for controlling the missile?

3. It is a familiar fact that some items are sold at different prices to different groups of people. For example, there are sometimes discounts for senior citizens or for children. The reason is that these groups may be more sensitive to price and so a small discount will have greater impact on their purchasing decisions. The seller faces an optimization problem: How large a discount to offer in order to maximize profits?

A theater believes that it will sell q_c child tickets and q_a adult tickets at prices p_c and p_a, according to the following demand functions:

$$q_c = r{p_c}^{-4},$$
$$q_a = s{p_a}^{-2},$$

and that its operating costs are proportional to the total number of tickets sold. What should be the relative price of children's and adults' tickets?

4. Show that the function $f(x, y)$ in Example 2 achieves a minimum at the point $(2, 1)$.

5. Design a cube-shaped milk carton box of width w, length l, and height h which holds 512cm^3 of milk. The sides of the box cost 1 cent/cm^2 and the bottoms cost 2 cent/cm^2. Find the dimensions of the box that minimize the total cost.

6. An international airline has a regulation that each passenger can carry, at no charge, up to two suitcases each having the sum of its width and length and height less than or equal to 54 inches. Find the dimensions of the suitcase of maximum volume that the passenger may carry under this airline regulation.

7. An irrigation canal has a trapezoidal cross-section of area 50 square feet, as in Figure 14.20.

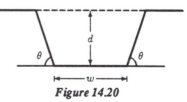

Figure 14.20

The average flow velocity in the canal is inversely proportional to the wetted perimeter, p, of the canal, that is, to the perimeter of the trapezoid in Figure 14.20, excluding the top. Thus, to maximize the flow we must minimize p. Find the depth d, base width w, and angle θ that give the maximum flow rate.[3]

8. This exercise will lead you to a derivation of general formulas for the slope and y-intercept of a least squares line. Assume that you have n data points $(x_1, y_1), (x_2, y_2), \cdots, (x_n, y_n)$. Let the equation of the sought-after least squares line be $y = b + mx$.

    (a) For each data point (x_i, y_i), show that the corresponding point directly above or below it on the least squares line has y-coordinate $b + mx_i$.

---

[3] Adapted from *Mathematical Foundations of Design: Civil Engineering Systems*, Robert M. Stark and Robert L. Nichols, McGraw-Hill, New York, 1972.

(?)

3. $q_a = r p_a^{-4}$

$q_b = s p_b^{-2}$

5. $V = lwh = 512 \Rightarrow l = \dfrac{512}{wh}$

$C = 2wh + 2lh + 4lw$

$C = 2wh + 2\left(\dfrac{512}{wh}\right)h + 4\left(\dfrac{512}{wh}\right)w$

$f_h = 2w - \dfrac{1024}{h^2} = 0 \Rightarrow h = \dfrac{512}{w^2}$

$f_w = 2h - \dfrac{2048}{w^2} = 0 \Rightarrow w = \dfrac{1024}{h^2}$

$\dfrac{h}{w} = \dfrac{h^2}{2w^2} \Rightarrow \dfrac{h}{w} - \dfrac{1}{2}\left(\dfrac{h}{w}\right)^2 = 0 \Rightarrow \dfrac{h}{w}\left(1 - \dfrac{1}{2}\dfrac{h}{w}\right) = 0 \Rightarrow \dfrac{h}{w} = 2 \Rightarrow h = 2w$

$2w = \dfrac{512}{w^2} \Rightarrow 2w^3 = 512 \Rightarrow w^3 = 256$

$w = \sqrt[3]{256} \; ; \; h = 2\sqrt[3]{256} \; ; \; l = \dfrac{512}{2\sqrt[3]{256}}$

14.2

3-11 odd

(?)

7. $P = w + 2l$

$P = w + 2\left(\dfrac{d}{\sin\theta}\right)$

$A = 50 = \tfrac{1}{2}d\,(b_1 + b_2)$

$= b_1 d + 2\left(\tfrac{1}{2}d \cdot \tfrac{1}{2}(b_1 + b_2)\right)$

$= d\left[b_1 + \tfrac{1}{2}b_2 - \tfrac{1}{2}b_1\right]$

$= d\,\dfrac{(b_1 + b_2)}{2}$

9. Find the Regression Line for $(-1,2)(0,-1)(1,1)$

$f(m,b) = (2-(b-m))^2 + (-1-(b))^2 + (1-(b-m))^2$

$f_m = 2(2-b+m) + 0 + 2(1-b-m)(-1)$

$\qquad = 4 - 2b + 2m - 2 + 2b + 2m$

$\qquad = 2 + 4m \Rightarrow 4m = -2 \Rightarrow m = -\tfrac{1}{2}$

$f_b = 2(2-b+m)(-1) + 2(-1-b)(-1) + 2(1-b-m)(-1)$

$\qquad = -4 + 2b - 2m + 2 + 2b - 2 + 2b + 2m$

$\qquad = -4 + 6b \Rightarrow 6b = 4 \Rightarrow b = \tfrac{2}{3}$

$y = \tfrac{2}{3} - \tfrac{1}{2}x$

11.

(b) For each data point (x_i, y_i), show that the square of the vertical distance from it to the point found in part (a) is $(y_i - (b + mx_i))^2$.

(c) Now form the function $f(b, m)$ which is the sum of all of the n squared distances found in part (b). That is,

$$f(b, m) = \sum_{i=1}^{n}(y_i - (b + mx_i))^2$$

Show that the partial derivatives $\dfrac{\partial f}{\partial b}$ and $\dfrac{\partial f}{\partial m}$ are given by

$$\frac{\partial f}{\partial b} = -2\sum_{i=1}^{n}(y_i - (b + mx_i))$$

and

$$\frac{\partial f}{\partial m} = -2\sum_{i=1}^{n}(y_i - (b + mx_i)) \cdot x_i.$$

(d) Now set the partial derivatives equal to zero and solve for m and b. This is easier than it looks; you can simplify the appearance of the equations by temporarily substituting other symbols for the sums. For example, write SY for $\sum y_i$, SX for $\sum x_i$, SYX for $\sum y_i x_i$ and SXX for $\sum x_i^2$. Remember, the x_i and y_i are all constants. You should get a pair of simultaneous linear equations in m and b; solving for m and b will give you formulas in terms of SX, SY, SXY, and SXX.

(e) Apply these formulas to the data points $(1, 1), (2, 1), (3, 3)$ to verify that you get the same result as in Example 1.

9. Compute the regression line for the point $(-1, 2), (0, -1), (1, 1)$ using least squares.

10. The following data indicates the increase in the cost of a first class stamp in the United States over the last 70 years.

TABLE 14.3 *Increase in cost of stamps*

Year	1920	1932	1958	1963	1968	1971	1974	1975	1978	1981	1985	1988	1991
Postage	0.02	0.03	0.04	0.05	0.06	0.08	0.10	0.13	0.15	0.20	0.22	0.25	0.29

(a) Find the line of best fit through the data. Using this line, predict the cost of a postage stamp in the year 2000.

(b) Plot the data. Does it look linear?

(c) Plot the year against the natural logarithm of the price: does this look linear? If it is linear, what does that tell you about the price of a stamp as a function of time? Find the line of best fit through this data, and use your answer to again predict the cost of a postage stamp in the year 2000.

11. A biological rule of thumb states that when the area A of an island increases tenfold, the number of species, N, living on it doubles. Consider the following table on some of the islands in the West Indies showing the area of each island and the number of species of reptiles and amphibians living on it.

TABLE 14.4   *Number of species on various islands*

Island	Area	Number
Redonda	1	5
Saba	8	9
Montserrat	75	15
Puerto Rico	3460	75
Jamaica	4240	70
Hispaniola (Haiti and Dominican Rep.)	29520	130
Cuba	44420	125

(a)   What sort of function of A is N, according to this rule of thumb?

(b)   What sort of function of $\ln A$ is $\ln N$, according to this rule of thumb?

(c)   Tabulate $\ln N$ against $\ln A$ and find the line of best fit. Does your answer agree with the biological rule?

12.   The government wants to build a pipe that will pump water up from a dam to a reservoir, as in Figure 14.21.

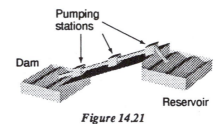

Figure 14.21

The cost, C, (in millions of dollars) will depend on the diameter, d, of the pipe (in inches) and the number, n, of pumping stations, according to the following formula[4]:

$$C = 0.15n + 3\left(\frac{d}{50}\right)^{-4.87} + \left(\frac{d}{50}\right)^{1.8} + 3\left(\frac{d}{50}\right)^{1.8} n^{-1}.$$

Using the gradient search method, find the optimal number of pumping stations and pipe diameter.

14.3   CONSTRAINED OPTIMIZATION

Many, perhaps most, real optimization problems are constrained by external circumstances. For example, a city wanting to build or improve a public transportation system has only a limited number of tax dollars it can spend on the project. Any nation trying to maintain its balance of trade cannot spend more on imports than it earns on exports. (The US violated this constraint throughout the '80's.) In this section, we will look at a mathematical way to find an optimization value under such constraints.

---

[4]from *Globally Optimal Design*, Douglass J. Wilde, John Wiley & Sons, New York, 1978

A Graphical Example

Let's consider the example of trying to maximize the production of a firm under a budget constraint. Suppose production P is a function of two variables, x_1 and x_2, which could be two raw materials, or labor and capital, or the number of two different types of workers (doctors and nurses, for example, or engineers and secretaries). We will suppose that

$$P = x_1^{2/3} x_2^{1/3}.$$

Suppose that x_1 and x_2 must be purchased at prices of p_1 and p_2 per unit. What is the maximum value of P that can be produced with a budget of $\$B$?

If we want to maximize P without regard to the budget, we should simply increase x_1 and x_2 as far as we can. However, the budget will prevent us from increasing x_1 and x_2 beyond a certain point. Exactly how does the budget constrain us? With prices of p_1 and p_2, the amount spent on x_1 is $p_1 x_1$ and the amount spent on x_2 is $p_2 x_2$, so we must have

$$p_1 x_1 + p_2 x_2 \leq B.$$

Let's look at the case when $p_1 = p_2 = 100$ and $B = 378$. Then

$$100 x_1 + 100 x_2 \leq 378$$

or

$$x_1 + x_2 \leq 3.78.$$

Graphically, this constraint is represented by the line in Figure 14.22. Any point on or below the line represents a pair of values of x_1 and x_2 that we can afford. A point on the line completely exhausts the budget, a point below the line represents values of x_1 and x_2 which can be bought without using up the budget. Any point above the line represents a pair of values that breaks the budget — we cannot afford such a combination.

Figure 14.22 also shows some contours of P. Since we want to maximize P, we want to find the point which lies on the level curve with the largest possible P value *and* which lies within the budget. The point we are looking for must lie on the budget line because we should clearly spend all the available money if we want to maximize P. If we move along the line representing the budget constraint in Figure 14.22, we can always increase P *unless* we are at the point where the budget constraint is tangent to the contour $P = 2$. For example, if we are to the left of the point of tangency, moving right will increase P; if we are to the right of the point of tangency, moving left will increase

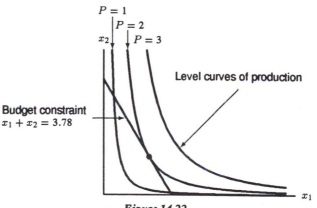

Figure 14.22

P. Thus, the maximum value of P on the budget constraint occurs at the point where the budget constraint is tangent to the contour $P = 2$.

In theory, we could find the values of x_1 and x_2 giving this maximum by reading them off the graph. In practice, however, we often use the method of Lagrange multipliers described in this section.

Analytical Solutions for Constrained Optimization

Substitution

The previous example asked us to maximize

$$P = f(x_1, x_2) = x_1^{2/3} x_2^{1/3}$$

subject to the constraint

$$x_1 + x_2 = 3.78.$$

You may wonder how we could solve this problem algebraically, without drawing a graph. One method is *substitution*: solve the constraint for one of the variables, say x_2:

$$x_2 = 3.78 - x_1,$$

substitute into P:

$$P = x_1^{2/3}(3.78 - x_1)^{1/3},$$

and now maximize P as you would any function of one variable.

However, the method of substitution does not always work, for example if the constraint cannot be solved for one variable. The following method avoids this problem.

Lagrange Multipliers

Recall in the previous section that the maximum production was achieved at the point where the budget constraint was tangent to a level curve of the production function. The method of Lagrange multipliers uses this fact in algebraic form. Figure 14.23 shows that at the optimum point, the gradient of P and the normal $\vec{n}$ to the budget plane are parallel, so

$$\text{grad } P = \lambda \vec{n}$$

for some scalar λ, called the *Lagrange multiplier*. Since

$$\text{grad } P = \left(\frac{2}{3} x_1^{-1/3} x_2^{1/3}\right) \vec{i} + \left(\frac{1}{3} x_1^{2/3} x_2^{-2/3}\right) \vec{j} \quad \text{and} \quad \vec{n} = \vec{i} + \vec{j},$$

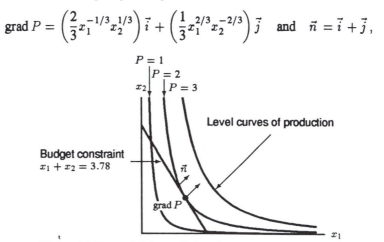

Figure 14.23: grad P is parallel to $\vec{n}$ at maximum production

we have, after equating the components,

$$\frac{2}{3}x_1^{-1/3}x_2^{1/3} = \lambda,$$

$$\frac{1}{3}x_1^{2/3}x_2^{-2/3} = \lambda.$$

Eliminating λ gives $\frac{2}{3}x_1^{-1/3}x_2^{1/3} = \frac{1}{3}x_1^{2/3}x_2^{-2/3}$, so

$$2x_2 = x_1.$$

Since we must also satisfy the constraint

$$x_1 + x_2 = 3.78$$

we have $x_1 = 2.52$ and $x_2 = 1.26$. For these values,

$$P = f(2.52, 1.26) = (2.52)^{2/3}(1.26)^{1/3} = 2.$$

Thus, as before, we see that the maximum value of P is 2; we also learn that this maximum occurs with $x_1 = 2.52$ and $x_2 = 1.26$.

Lagrange Multipliers in General

Suppose we want to optimize an *objective function* $f(x, y)$ subject to the *constraint* $g(x, y) = c$. We look for the points on the constraint where the level curves of f are parallel to the level curves of g. See Figure 14.24. Since grad f points in the direction in which f increases fastest, if there is a component of grad f parallel to the constraint, moving along the constraint in that direction will increase f, and moving in the opposite direction will decrease f. Thus, if we are on the constraint at

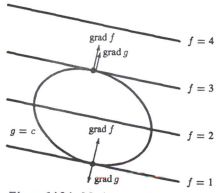

Figure 14.24: Maximum and minimum values of $f(x, y)$ on $g(x, y)$ at points where grad f is parallel to grad g

a point where grad f has a component parallel to the constraint, we cannot be at a point where f has a maximum or minimum. The only points where f may have a maximum or minimum are points where grad f is perpendicular to the constraint, which means parallel to grad g. Thus:

> To optimize $f(x, y)$ subject to $g(x, y) = 0$ using Lagrange multipliers, solve the equations
>
> $$\operatorname{grad} f = \lambda \operatorname{grad} g,$$
>
> subject to
>
> $$g(x, y) = c$$

Notice that if f and g are functions of two variables, $\operatorname{grad} f = \lambda \operatorname{grad} g$ leads to two equations, so together with the constraint $g(x, y) = c$, we have three equations for the three unknowns x, y, λ.

Example 1   Find the maximum and minimum values of $x + y$ on the circle $x^2 + y^2 = 4$.

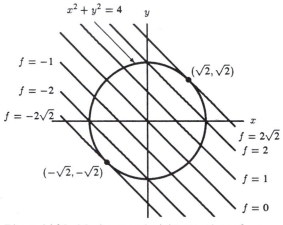

Figure 14.25: Maximum and minimum values of $x + y$
on $x^2 + y^2 = 4$

Solution   The objective function is

$$f(x, y) = x + y,$$

and the constraint is

$$g(x, y) = x^2 + y^2 = 4.$$

Since $\operatorname{grad} f = f_x \vec{i} + f_y \vec{j} = \vec{i} + \vec{j}$ and $\operatorname{grad} g = g_x \vec{i} + g_y \vec{j} = 2x\vec{i} + 2y\vec{j}$, then $\operatorname{grad} f = \lambda \operatorname{grad} g$ gives:

$$1 = 2\lambda x,$$
$$1 = 2\lambda y,$$

so $x = y$. We also know that $x^2 + y^2 = 4$, giving $x = y = \sqrt{2}$ or $x = y = -\sqrt{2}$.

Since $f(x, y) = x + y$, we see that the maximum value of $x + y$ is $2\sqrt{2}$, and occurs when $x = y = \sqrt{2}$; the minimum value of $x + y$ is $-2\sqrt{2}$, and occurs when $x = y = -\sqrt{2}$. See Figure 14.25.

How to Distinguish Maxima from Minima

There is no easy test to distinguish among maxima, minima, and critical points which are neither. However, as you can see from the examples, a graph of the constraint and some level curves can usually make it clear which points are maxima, which points are minima and which are neither.

Equality versus Inequality

The production problem that we looked at first was to maximize

$$P = f(x_1, x_2)$$

subject to a budget constraint

$$p_1 x_1 + p_2 x_2 \leq B.$$

This budget constraint is an example of an inequality constraint, because it contains an inequality sign rather than an equals sign. The fact that there is an inequality means that the constraint restricts us to a region of the plane rather than a curve in the plane. For example, in the case of a budget, the line is

$$p_1 x_1 + p_2 x_2 = B.$$

(This represents the case when the budget is exhausted.) The inequality $p_1 x_1 + p_2 x_2 < B$ determines a triangular region in a plane.

To maximize or minimize a function

$$f(x, y)$$

subject to an inequality constraint

$$g(x, y) \leq c,$$

use Lagrange multipliers to find the critical values on the boundary

$$g(x, y) = c,$$

and use the regular methods for unconstrained optimization to find critical values on the interior

$$g(x, y) < c.$$

Test the values of f at all the critical points (including the end points of the boundary, if any) and pick out the maximum and minimum.

The Meaning of λ

In our previous examples, we never found (or needed) the value of λ. However, λ does have a practical interpretation. Let's look back at the production problem where we wanted to maximize

$$P = f(x_1, x_2) = x_1^{2/3} x_2^{1/3}$$

subject to the constraint

$$g(x_1, x_2) = x_1 + x_2 = 3.78.$$

We solve the equations

$$\frac{2}{3} x_1^{-1/3} x_2^{1/3} = \lambda,$$

$$\frac{1}{3} x_1^{2/3} x_2^{-2/3} = \lambda,$$

$$x_1 + x_2 = 3.78,$$

to get $x_1 = 2.52$, $x_2 = 1.26$. Continuing to find λ gives us

$$\lambda = 0.53.$$

Suppose now we did another, apparently unrelated calculation. Suppose our budget is increased slightly, from 3.78 to 4.78, giving a new budget constraint of

$$x_1 + x_2 = 4.78.$$

Then the corresponding solution is at $x_1 = 3.19$ and $x_2 = 1.59$ and the new maximum value (instead of $P = 2$) is

$$P = (3.19)^{2/3}(1.59)^{1/3} \approx 2.53.$$

Notice that the amount by which P has increased is 0.53, the value of λ. Thus, in this example, the value of λ represent the extra production achieved by increasing the budget by one – in other words, "the extra bang you get for an extra buck of budget." In summary:

Interpretation of λ:

- The value of λ is approximately the increase in the optimum value of P when the budget is increased by 1 unit.

More precisely:

- The value of λ represents the rate of change of the optimum value of P as the budget increases.

Justification of Interpretation of λ

To interpret λ in general, we look at how the optimum value of the objective function changes as the constraint is altered. Suppose we have found a point (x_0, y_0) which gives an optimum value of $f(x_0, y_0)$ and satisfies the constraint

$$g(x_0, y_0) = c.$$

Suppose now the constraint is relaxed

$$g(x, y) = c + \Delta c.$$

How does the optimum value of f change? Suppose the new optimal point is $(x_0 + \Delta x, y_0 + \Delta y)$, then

$$g(x_0 + \Delta x, y_0 + \Delta y) = c + \Delta c.$$

Thus, using the local linearization of g, we have

$$g(x_0, y_0) + \frac{\partial g}{\partial x}\Delta x + \frac{\partial g}{\partial y}\Delta y \approx c + \Delta c,$$

giving

$$\frac{\partial g}{\partial x}\Delta x + \frac{\partial g}{\partial x}\Delta y \approx \Delta c.$$

Now we want to estimate the change in the optimum value of f:

$$f(x_0 + \Delta x, y_0 + \Delta y) - f(x_0, y_0) \approx \frac{\partial f}{\partial x}\Delta x + \frac{\partial f}{\partial y}\Delta y.$$

But at the point (x_0, y_0), we know that grad $f = \lambda$ grad g, so $\frac{\partial f}{\partial x} = \lambda\frac{\partial g}{\partial x}$, and $\frac{\partial f}{\partial y} = \lambda\frac{\partial g}{\partial y}$. Thus,

$$f(x_0 + \Delta x, y_0 + \Delta y) - f(x_0, y_0) \approx \frac{\partial f}{\partial x}\Delta x + \frac{\partial f}{\partial y}\Delta y = \lambda\frac{\partial g}{\partial x}\Delta x + \lambda\frac{\partial g}{\partial y}\Delta y \approx \lambda\Delta c.$$

Therefore, the optimum value has changed by approximately λ times the change in the constraint. More precisely:

The value of λ is the rate of change of the optimum value of f as c increases (where $g(x, y) = c$). If the optimum value of f is written as $f(x_0, y_0, c)$ to emphasize its dependence on the parameter c, then we have

$$\frac{\partial f}{\partial c}(x_0, y_0, c) = \lambda.$$

Example 2   Suppose the quantity of goods produced according to the function $P(x_1, x_2) = x_1^{2/3} x_2^{1/3}$ is maximized subject to the constraint $x_1 + x_2 \le 3.78$. What price must the product sell for if it is to be worth an increased budget for its production?

Solution   On page 213 we found that $\lambda = 0.53$, and therefore increasing the budget by \$1 increases production by about 0.53 unit. In order to make the increase in budget profitable, the extra goods produced must sell for more than \$1. Thus, the price per unit must be at least $1/0.53 = \$1.89$.

Problems for Section 14.3

1. An industry produces a product from two raw materials. The quantity produced, Q, can be given by the Cobb-Douglas function:

$$Q = cx^a y^b,$$

where x and y are quantities of each of the two raw materials used and a, b, and c are positive constants. Suppose the first raw material costs $\$P_1$ per unit and the second costs $\$P_2$ per unit. Find the maximum production possible if no more than $\$K$ can be spent on raw materials.

2. Each person tries to balance his or her time between leisure and work. The tradeoff is, as we all know, that as you work less your income falls. Therefore each person has indifference curves which connect the number of hours of leisure, l, and income, s. If, for example, you are indifferent between 0 hours of leisure and an income of \$450 a week on the one hand, and 10 hours of leisure and an income of \$300 a week on the other hand, then the points $l = 0$, $s = 450$, and $l = 10$, $s = 300$ both lie in the same indifference curve. Table 14.5 below gives information on three of your indifference curves, I, II, and III.

TABLE 14.5

Weekly Income			Weekly Leisure Hours		
I	II	III	I	II	III
450	500	550	0	20	40
300	350	400	10	30	50
200	250	300	20	40	60
150	200	250	30	50	70
100	150	200	50	70	90

(a) Sketch the three indifference curves on squared paper.

(b) Now suppose you have only 100 hours a week available for work and leisure combined, and that you earn \$4/hour. Write an equation in terms of l and s which represents this constraint.

(c) On the same squared paper, sketch a graph of this constraint.

(d) Estimate from the graph what combination of leisure hours and income you would choose under these circumstances. Give the corresponding number of hours per week you would work. Make very clear on your drawing, or explain in words, how you made this estimate.

3. Figure 14.26 shows ∇f for a function $f(x, y)$ and two curves $g(x, y) = 1$ and $g(x, y) = 2$. Notice that $g = 1$ is the inside curve and $g = 2$ is the outside curve. Mark the following points on a copy of the figure.

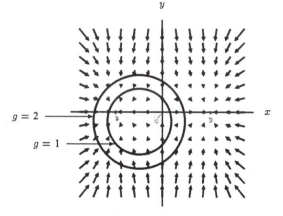

Figure 14.26

(a) The point(s) A where f has a local maximum.

(b) The point(s) B where f has a saddle point.

(c) The point C where f has a maximum on $g = 1$.

(d) The point D where f has a minimum on $g = 1$.

(e) If you had used Lagrange multipliers to find C, what would the sign of λ have been? Why?

4. Design a closed cylindrical container which holds 100 cm^3 and has the minimal possible surface area. What should its dimensions be?

5. The quantity, Q, of a good produced by a company is given by

$$Q = aK^{0.6}L^{0.4},$$

where a is a positive constant, K is the quantity of capital and L is the quantity of labor used. Capital costs are \$20 per unit, labor costs are \$10 per unit, and the company wants to keep costs for capital and labor combined to be no higher than \$150. Suppose you are asked to consult for the company, and learn that 5 units each of capital and labor are being used.

(a) What do you advise? Should the plant use more or less labor? More or less capital? If so, by how much?

(b) Write a one sentence summary that could be used to "sell" your advice to the board of directors.

3. (See p. 216)   5. $Q = 3K^{.6}L^{.4}$ $\nabla Q = k \nabla C$

$C = 20K + 10L = 150$

$Q_K = .6 \cdot 3K^{-.4}L^{.4} = 20k$

$Q_L = .4 \cdot 3K^{.6}L^{-.6} = 10k$

a. $\dfrac{.6 \cdot 3K^{-.4}L^{.4}}{.4 \cdot 3K^{.6}L^{-.6}} = 1.5\dfrac{L}{K} = \dfrac{20k}{10k} = 2 \Rightarrow L = \dfrac{4}{3}K$

$20K + 10L = 150$ $20\left(\dfrac{9}{2}\right) + 10L = 150$

$20K + 10\left(\dfrac{4}{3}K\right) = 150$ $10L = 150 - 90$

$27.5K = 150 \Rightarrow K = \dfrac{9}{2}$ $10L = 60 \Rightarrow L = 6$

Reduce capital by $\dfrac{1}{2}$ units; $\left(\dfrac{9}{2} - 5 = -\dfrac{1}{2}\right)$, and increase labor by 1 unit $(6-5=1)$

b. (See sol. man.)

(?)
7. $q = 6W^{3/4}K^{1/4}$ $\nabla q = k \nabla B$

$B = 10W + 20K = 3000$

$q_W = 4.5W^{-1/4}K^{1/4} = 10k$

$q_{1K} = 1.5W^{3/4}K^{-3/4} = 20k$

a. $\dfrac{4.5W^{-1/4}K^{1/4}}{1.5W^{3/4}K^{-3/4}} = 3\dfrac{K}{W} = \dfrac{20k}{10k} \Rightarrow \dfrac{3K}{W} = 2 \Rightarrow 3K = 2W \Rightarrow W = \dfrac{3}{2}K$

$10W + 20K = 3000$ $10W + 20(85.7) = 3000$

$10\left(\dfrac{3}{2}K\right) + 20K = 3000$ $10W = 3000 - 171.4$

$35K = 3000 \Rightarrow K = 85.7$ $W = 282.6$

(?)

9a. $V(D,N) = 1000 D^{.6} N^{.3}$

$40000 D + 10000 N \leq 600000$

b. $B = 40 \ D + 10 \ N = 600$

$\nabla V = k \nabla B$

$V_D = B_D k = 40 k$

$V_N = B_N k = 10 k$

$\dfrac{V_D}{V_N} = 4$; At the optimum value of V, the increase in the number of visits to the number of doctors is 4 times the corresponding rate for nurses.

c. $\nabla V = \nabla B k$

$600 D^{-.4} N^{.3} = 40 k$

$300 D^{.6} N^{-.7} = 10 k$

$\dfrac{600 D^{-.4} N^{.3}}{40000} = k = \dfrac{300 D^{.6} N^{-.7}}{10000}$

$D^{-.4} N^{.3} = 2 D^{.6} N^{-.7}$

$N^{.3}(N^{.7}) = 2 D^{.6}(D^{.4})$

$N = 2D$

$600 - 40D - 10N = 0$

$600 - 40D - 10(2D) = 0$

$600 = 60D \Rightarrow D = 10 \ ; N = 20$

$k = \dfrac{600(10)^{-.4}(20)^{.3}}{40} = 14.7$

The clinic should have 10 doctors and 20 nurses

$V = 1000(10)^{.6}(20)^{.3} = 9780$

d. $k = 14.7$; for $\$1000$, 14.7 more visits can be added

e.

6. Suppose that the quantity, Q, produced of a certain good depends on the number of units of labor, L, and of capital, K, according to the function

$$Q = 900L^{1/2}K^{2/3}.$$

Suppose also that labor costs \$100 per unit and that capital costs \$200 per unit. What combination of labor and capital should be used to produce 36,000 units of the goods at minimum cost? What is that minimum cost?

7. Suppose the quantity, q, of a product produced depends on the number of workers, W, and the amount of capital invested, K, and is represented by the Cobb-Douglas function

$$q = 6W^{3/4}K^{1/4}.$$

In addition, labor costs are \$10 per worker and capital costs are \$20 per unit, and the budget is \$3000.

(a) What are the optimum number of workers and the optimum number of units of capital?

(b) Check that at the optimum values of W and K, the ratio of the marginal productivity of labor to the marginal productivity of capital is the same as the ratio of the cost of a unit of labor to the cost of a unit of capital.

(c) Recompute the optimum values of W and K when the budget is increased by one dollar. Check that increasing the budget by \$1 allows the production of λ extra units of the good, where λ is the Lagrange multiplier.

8. Consider a manufacturing process which produces x units of one product and y units of another. Suppose the total cost in dollars, C, of producing these two products can be approximated by the function

$$C = 5x^2 + 2xy + 3y^2 + 800.$$

(a) If the production quota for the total number of products (both types combined) is 39, find the minimum production cost.

(b) Estimate the additional production cost or savings if the production quota is raised to 40 or lowered to 38.

9. The director of a neighborhood health clinic has an annual budget of \$600,000. He wants to allocate his budget so as to maximize the number of patient visits, V, which is given as a function of the number of doctors, D, and the number of nurses, N, by:

$$V = 1000D^{0.6}N^{0.3}.$$

Doctors receive a salary of \$40,000, while nurses get \$10,000.

(a) Set up the director's constrained optimization problem.

(b) Describe, in words, the conditions which must be satisfied by $\partial V/\partial D$ and $\partial V/\partial N$ for V to have an optimum value.

(c) Solve the problem formulated in part (a).

(d) Find the value of the Lagrange multiplier and interpret its meaning in this problem.

(e) What is the marginal cost of a patient visit at the optimum point? Will that marginal cost rise or fall with the number of visits? Why?

10. Consider a firm which manufactures a commodity at two different factories. The total cost of manufacturing depends on the quantities, q_1 and q_2, supplied by each factory, and is expressed by the *joint cost function*, $C = f(q_1, q_2)$. Suppose the joint cost function is approximated by

$$f(q_1, q_2) = 2q_1^2 + q_1q_2 + q_2^2 + 500$$

and that the company's objective is to produce 200 units, at the same time minimizing production costs. How many units should be supplied by each factory?

11. Minimize
$$f(x, y, z) = \sqrt{(x-a)^2 + (y-b)^2 + (z-c)^2},$$
subject to the constraint $Ax + By + Cz + D = 0$.

12. The energy required to compress a gas from pressure p_1 to pressure p_{N+1} in N stages is proportional to
$$E = \left(\frac{p_2}{p_1}\right)^2 + \left(\frac{p_3}{p_2}\right)^2 + \cdots + \left(\frac{p_{N+1}}{p_N}\right)^2 - N.$$

Show how to choose the intermediate pressures $p_2, \ldots, p_N$ so as to minimize the energy requirement. [5]

REVIEW PROBLEMS FOR CHAPTER FOURTEEN

1. Consider $f(x, y) = x + y + \dfrac{1}{x} + \dfrac{4}{y}$.
   (a) Find and classify the local maxima, minima, and saddle points.
   (b) What are the global maxima and minima? Explain.

2. Suppose $f_x = f_y = 0$ at $(1, 3)$ and $f_{xx} < 0$, $f_{yy} < 0$, $f_{xy} = 0$. Draw a possible contour diagram.

3. The quantity of a product demanded by consumers is affected by its price; the *demand function* gives the dependence of the quantity demanded based on price. In some cases the quantity of one product demanded depends on the price of other products. For example, the demand for tea may be affected by the price of coffee; the demand for cars may be affected by the price of gas. Suppose the quantities demanded, q_1 and q_2, of two products depend on their prices p_1 and p_2 as follows
$$q_1 = 150 - 2p_1 - p_2$$
$$q_2 = 200 - p_1 - 3p_2$$
   (a) What does the fact that q_1 is a function of p_1 and p_2 (instead of p_1 alone) tell you?
   (b) What does the fact that the coefficients of p_1 and p_2 are negative tell you? Give an example of two products that might be related this way.
   (c) Suppose one manufacturer sells both of these products. How should the manufacturer set prices to earn the maximum possible revenue? What is that maximum possible revenue?

4. Find the least squares line for the data points $(0, 4)$, $(1, 3)$, $(2, 1)$.

5. Find the minimum and maximum of the function $z = 4x^2 - xy + 4y^2$ over the closed disc $x^2 + y^2 \leq 2$.

6. An international organization must decide how to spend the $2000 they have been allotted for famine relief in a remote area. They expect to divide the money between buying rice at $5/sack and beans at $10/sack. The number, P, of people who would be fed if they buy x

---

[5] Rutherford Aris, *Discrete Dynamic Programming*, Blaisdell, New York, 1964, p.35.

sacks of rice and y sacks of beans is given by

$$P = x + 2y + \frac{x^2 y^2}{2 \cdot 10^8}.$$

What is the maximum number of people that can be fed, and how should the organization allocate its money?

7. The Cobb-Douglas equation models the total quantity, q, of a commodity produced as a function of the number of workers, W, and the amount of capital invested, K, by the production function

$$q = cW^{1-a}K^a$$

where a and c are positive constants. Assume labor costs are $\$p_1$ per worker, capital costs are $\$p_2$ per unit, and there is a fixed budget of $\$b$. Show that when W and K are at their optimal levels, the ratio of marginal productivity of labor to marginal productivity of capital equals the ratio of the cost of one unit of labor to one unit of capital.

8. What are the maximum and minimum values of $f(x, y) = -3x^2 - 2y^2 + 20xy$ on the line $x + y = 100$?

9. A family wants to move to a house at a point that is better situated with respect to the children's school and both the parents' places of work. See Figure 14.27.

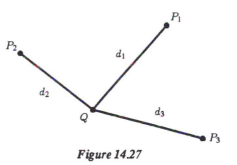

Figure 14.27

Currently they live at Q, the school is at P_1, the mother's work is at P_2, and the father's work is at P_3. They want to minimize $d = d_1 + d_2 + d_3$, where

$$d_1 = \text{distance to school,}$$
$$d_2 = \text{distance to mother's work,}$$
$$d_3 = \text{distance to father's work.}$$

(a) Show that grad d_i is a unit vector pointing directly away from P_i, for $i = 1, 2, 3$. [Hint: Draw contours of d_i, paying careful attention to the spacing of the contours.]
(b) Use your answer to part (a) to draw grad d at the point Q in Figure 14.27. In what direction should the family move to decrease d?
(c) Find as best you can the point on the diagram where grad $d = 0$. What geometric condition characterizes this location?

CHAPTER FIFTEEN

INTEGRATING FUNCTIONS OF MANY VARIABLES

To find the total population of a region, given the population density as a function of position, we need a definite integral. The definite integral represents the limit of a sum, where each term in the sum represents the contribution to the total population from a small subregion. In one-variable calculus we considered problems where the population density only depended on one variable; for example, distance along a turnpike, or distance from the center of the city. If the density depends on more than one variable, we need a multivariable integral to calculate it. In this chapter we will develop the multivariable integral and its interpretations.

15.1 THE DEFINITE INTEGRAL OF A FUNCTION OF TWO VARIABLES

In single-variable calculus, to calculate the total population from a variable population density requires a definite integral. Thus, to investigate integrals of functions of two variables we will start by considering a population density in the plane and how to calculate the total population from that.

Population Density of Foxes in England

England has been largely free of the disease rabies, which is spread by animals. In predicting the spread of an epidemic that might be introduced from Europe at one of the southern ports, it is useful to know the population of animals such as foxes that would be responsible for spreading the disease. The contour diagram in Figure 15.1 shows the population density $D = f(x, y)$ of foxes in the southwestern corner of England, where x and y are in kilometers from the southwest corner of the map, and D is in foxes per square kilometer.[1] It was obtained from estimates of population densities in various parts of England. The bold contour is the coastline (approximately), and may be thought of as the $D = 0$ contour; obviously the density is zero outside it.

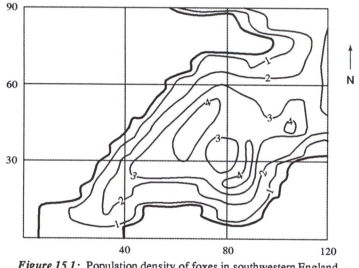

Figure 15.1: Population density of foxes in southwestern England

Example 1 Estimate the total fox population in the part of England represented by the map in Figure 15.1.

Solution We will subdivide the area into the nine areas shown in Figure 15.1 and estimate the population in each subdivision. We will be particularly concerned with finding upper and lower bounds for the population. Our first attempt at an upper bound is to find an upper bound for the population density in each rectangle, multiply it by the area of the rectangle to get an upper bound for the population in that rectangle, then add up all these upper bounds to get an upper bound for the total population. The top left rectangle is in the ocean, so the population is 0 everywhere in that rectangle. The next rectangle to the right just touches the $D = 3$ contour, so the density in that region is at most 3 foxes per square kilometer. For the rectangle immediately below the top left one, we notice that its bottom right corner is halfway between the $D = 2$ and $D = 3$ contours, so we estimate that the maximum

[1] adapted from J. D. Murray, *Mathematical Biology*, Springer-Verlag, 1989

population density in that rectangle is 2.5 foxes per square kilometer. In the middle rectangle we see the density goes above 4 inside the $D = 4$ contour, but not above 5, since there is no $D = 5$ contour. Hence, we choose a maximum of 5 for that rectangle. Continuing in this way, we find upper bounds as tabulated below.

TABLE 15.1 *Upper bounds for the population densities*

0	3	3
2.5	5	5
2.5	4	5

Each region has an area of $30 \times 40 = 1200$ km^2, so we obtain an upper bound on the total fox population of

$$(0 + 3 + 3 + 2.5 + 5 + 5 + 2.5 + 4 + 5)(1200) = 36,000 \text{ foxes}.$$

What about a lower bound? Since each region contains part of the sea, the minimum density on each region is zero, so our lower bound is 0. So all we know is that the fox population is between 0 and 36,000; taking the average, we estimate it to be 18,000.

Notice that there is a big difference between our lower and upper bounds in the previous example. To do better, we need a finer subdivision.

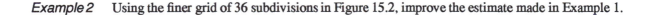

Example 2   Using the finer grid of 36 subdivisions in Figure 15.2, improve the estimate made in Example 1.

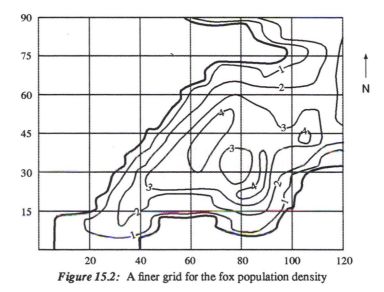

Figure 15.2: A finer grid for the fox population density

Solution   Proceeding in the same way as before, we obtain the following upper and lower estimates.

TABLE 15.2   *Lower estimates of the fox population using the finer grid*

0	0	0	0	0	0.25
0	0	0	0	0	0.25
0	0	0	1	2.25	2.25
0	0	0	2	2	0
0	0	1	1	0	0
0	0	0	0	0	0

TABLE 15.3   *Upper estimates of the fox population using the finer grid*

0	0	0	1.5	3	3
0	0	1	3	3	3
0	0	3	5	3.75	4.5
0	2.5	4	5	5	4.5
0	2.5	4	4.25	5	1
1.25	2.25	2	2	2	0

This time the area of each subdivision is $15 \times 20 = 300 \text{ km}^2$, so the total lower estimate is

$$(0.25 + 0.25 + 1 + 2.25 + 2.25 + 2 + 2 + 1 + 1)(300) = 3600 \text{ foxes},$$

and the total upper estimate is

$$(1.5 + 3 + 3 + 1 + 3 + 3 + 3 + 3 + 5 + 3.75 + 4.5 + 2.5 + 4 + 5$$
$$+5 + 4.5 + 2.5 + 4 + 4.25 + 5 + 1 + 1.25 + 2.25 + 2 + 2 + 2)(300) = 24,300 \text{ foxes}.$$

Even with 36 subdivisions, there is still a wide disparity between the upper and lower estimates, but we have narrowed the gap; by taking finer and finer subdivisions we could make the upper and lower estimates closer and closer. The average of our two estimates is about 14,000, so we will take that as our estimate for now.

Definition of the Definite Integral

We now give a working definition of a Riemann integral for a real valued function f of two variables defined on a rectangular region. A review of the definite integral is given in Appendix C. Given a function $f(x, y)$ defined on a rectangular region $a \leq x \leq b$ and $c \leq y \leq d$, we construct a Riemann sum by subdividing the region into smaller rectangles. We can do this by subdividing each of the intervals $a \leq x \leq b$ and $c \leq y \leq d$ into n and m equal subintervals, which yields a total of nm subrectangles. See Figure 15.3.

To compute the Riemann sum, we multiply the area of each rectangle by the value of the function at a point in the rectangle, and then add all the resulting numbers together. We will call

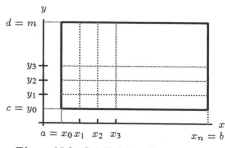

Figure 15.3: Subdivision of a rectangle

the area of each subrectangle ΔA. Thus, $\Delta A = \Delta x \Delta y$, where $\Delta x = (b - a)/n$ is the width of each subdivision along the x-axis, and $\Delta y = (d - c)/m$ is the width of each subdivision along the y-axis. In the fox density example, we constructed upper and lower sums by taking the maximum or minimum value of the function in each subrectangle. Since this value may be hard to find, it is sometimes more convenient to define a Riemann sum by choosing a definite point at which to evaluate the Riemann sum in each subrectangle; for example, the bottom left hand corner. Then the Riemann sum is

$$\sum_{i,j} f(x_i, y_j) \Delta x \Delta y.$$

Taking the limit as Δx and Δy tend to 0 we get

The **definite integral** of f on the rectangle R ($a \leq x \leq b, c \leq y \leq d$), is defined to be

$$\int_R f \, dA = \lim_{\Delta x, \Delta y \to 0} \sum_{i,j} f(x_i, y_j) \Delta x \Delta y.$$

Such an integral is called a **two-variable integral**.

Sometimes we use the notation

$$\int_R f \, dA = \int_R f(x, y) \, dx \, dy.$$

Here it is useful to think of dA as being the area of an infinitesimal rectangle of length dx and height dy, so that $dA = dx \, dy$.

The Riemann sum used in the definition is just one type of Riemann sum. For a general Riemann sum the function can be evaluated anywhere in each subdivision; in fact, the subdivisions don't have to be rectangular, they can be regions of any shape. The main thing is to approximate the definite integral by making the subdivisions smaller and smaller.

The Region R

In our definition of the definite integral $\int_R f(x, y) \, dA$, the region R is a rectangle. However, the definite integral can be defined for many other shaped regions which includes triangles, circles, and those bounded by the graphs of other functions.

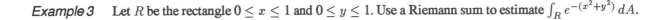

Example 3   Let R be the rectangle $0 \leq x \leq 1$ and $0 \leq y \leq 1$. Use a Riemann sum to estimate $\int_R e^{-(x^2+y^2)} \, dA$.

Solution   We will divide R into 16 subrectangles by dividing each edge into 4. We want to construct lower and upper sums for the integral, by getting lower and upper estimates of f on each subdivision. From Figure 15.4 we see that f decreases as we move away from the origin. Thus, to get an upper sum we evaluate f on each sub-rectangle at the corner nearest the origin. For example, in the rectangle $0 \leq x \leq 0.25, 0 \leq y \leq 0.25$, we will evaluate f at $(0,0)$.

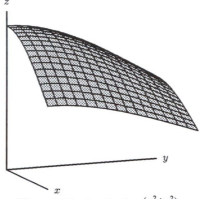

Figure 15.4: Graph of $e^{-(x^2+y^2)}$

TABLE 15.4   *Values of $e^{-(x^2+y^2)}$*

				y		
		0.0	0.25	0.50	0.75	1.00
	0.0	1	0.9394	0.7788	0.5698	0.3679
	0.25	0.9394	0.8825	0.7316	0.5353	0.3456
x	0.50	0.7788	0.7316	0.6065	0.4437	0.2865
	0.75	0.5698	0.5353	0.4437	0.3247	0.2096
	1.00	0.3679	0.3456	0.2865	0.2096	0.1353

To calculate the Riemann sum, we multiply function values by the area of each subdivision, which is $0.25^2 = 0.0625$, and add them all up. In fact it's easier to add up first and then multiply. Using Table 15.4 we find that the resulting upper sum is

$$[(\ 1 + 0.9394 + 0.7788 + 0.5698) +$$
$$(0.9394 + 0.8825 + 0.7316 + 0.5353)$$
$$+(0.7788 + 0.7316 + 0.6065 + 0.4437) +$$
$$(0.5698 + 0.5353 + 0.4437 + 0.3247)](0.0625) = 0.68.$$

To get a lower sum, we must evaluate f at the opposite corner of each rectangle because the surface slopes down in both the x and y directions. This yields a lower estimate of 0.44. Thus,

$$0.44 \leq \int_R e^{-(x^2+y^2)}\, dA \leq 0.68.$$

As in the one-variable case, computing Riemann sums is tedious to do by hand, and is best done by a computer or calculator. In the previous example, if we keep $m = n$ and let

$$S_n = \sum_{i,j} e^{-(x_i^2 + y_j^2)} \Delta x \Delta y$$

then for $n = 8, \ 16, \ 32,$ and 64, we get from a calculator:

$$S_8 = 0.61, \quad S_{16} = 0.59, \quad S_{32} = 0.57, \quad \text{and } S_{64} = 0.565.$$

Each S_n as calculated here is an upper estimate for $\int_R e^{-(x^2+y^2)} \, dA$. Note that as n increases the sums S_n are decreasing. Even though this will continue, to find S_n requires n^2 computations. So as n increases the time needed to compute S_n increases by a factor of n^2. Hence, we should look for better ways to approximate the integral.

Interpretations of the Two-Variable Integral

Interpretation as a Volume

Just as the definite integral of a positive one-variable function can be interpreted as an area under the graph of the function, so the definite integral of a two-variable function can be interpreted as a volume under its graph. In the one-variable case you can see this by visualizing the Riemann sums as the total area of rectangles above the subdivisions. In the two-variable case you get solid bars instead of rectangles, because each value of f is multiplied by an area, not a length. As the number of subdivisions grows, the tops of the bars approximate the surface better and better, and the volume of the bars gets closer and closer to the volume under the surface, above the rectangle R. Figure 11.14 on page 18 illustrates this process.

Example 4   Find the volume under the graph of $f(x, y) = 2 - x^2 - y^2$ lying above the rectangle $-1 \leq x \leq 1$ and $-1 \leq y \leq 1$. (See Figure 15.5.)

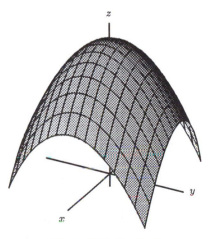

Figure 15.5: Graph of $f(x, y) = 2 - x^2 - y^2$ above $-1 \leq x \leq 1$ and $-1 \leq y \leq 1$

Solution   The volume we want is

$$\int_R (2 - x^2 - y^2)\, dA$$

where R is the rectangle $-1 \leq x \leq 1$, $-1 \leq y \leq 1$.

Using a computer program that calculates Riemann sums, we find the values in Table 15.5. The sum S_n was calculated by subdividing the rectangle into n^2 subrectangles and evaluating f at the point with minimum x and y values. Thus, the value of the integral seems to be about 5.3.

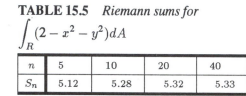

TABLE 15.5   *Riemann sums for*

$$\int_R (2 - x^2 - y^2)\, dA$$

n	5	10	20	40
S_n	5.12	5.28	5.32	5.33

Interpretation of the Definite Integral as an Average Value

As in the one-variable case, the definite integral can be used to compute the average value of a function:

$$\begin{array}{ll} \text{Average value of } f \\ \text{on the rectangle } R \end{array} = \frac{1}{\text{Area of } R} \int_R f\, dA$$

We can rewrite this as Integral = (Area of R)(Average value). Thus, if we interpret the integral as the volume under the graph of f, then we can think of the average value of f as the height of the box with the same volume that is on the same rectangular base. (See Figure 15.6.) One way to think of this is to imagine that the volume under the graph is made out of wax; if the wax melted and was allowed to level out within walls erected on the perimeter of R, then it would end up looking like the rectangular box, and having a constant height equal to the average value of f.

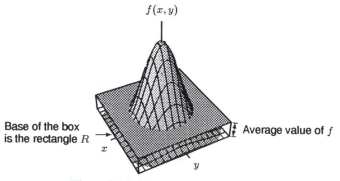

Figure 15.6: Volume and average value

Interpretation of the Definite Integral When f is a Density Function

A function of two variables can represent a density per unit area, for example, the fox population density (in foxes per unit area), or the density of a thin metal plate. Then the integral $\int_R f\, dA$ represents the total population or total mass in the rectangle R.

What About the Fundamental Theorem of Calculus?

The Fundamental Theorem of Calculus was a very important tool in one-variable calculus; it enabled us to calculate definite integrals algebraically by finding antiderivatives of the integrand, instead of using Riemann sums. However, there is no exact analog for the Fundamental Theorem of Calculus in many dimensions. In one-variable calculus, the Fundamental Theorem relied on interpreting the integrand as the rate of change of something. Now we have a situation where the integrand is a function of two variables. In Chapter 18 we will see how such a function can arise (as the rate of divergence of a vector field per unit area), and we eventually get something like the Fundamental Theorem; but it is not as useful as the one-variable Fundamental Theorem for evaluating integrals. We will have to look elsewhere to find a way of evaluating two-variable integrals algebraically. The answer is in the next section, which shows how the single-variable Fundamental Theorem of Calculus is adapted to the multivariable case.

Problems for Section 15.1

1. A function $f(x, y)$ has the values in the table below. Let R be the rectangle $1 \le x \le 1.2$, $2 \le y \le 2.4$. Find the Riemann sums which are reasonable over and underestimates for $\int_R f(x, y)\, dA$ with $\Delta x = 0.1$ and $\Delta y = 0.2$.

TABLE 15.6

		x		
		1.0	1.1	1.2
	2.0	5	7	10
y	2.2	4	6	8
	2.4	3	5	4

2. A solid is formed above the rectangle R with $0 \le x \le 2$, $0 \le y \le 4$ by the graph of $f(x, y) = 2 + xy$. Using Riemann sums with four subdivisions, find upper and lower bounds for the volume of this solid.

3. Let R be the rectangle with vertices $(0, 0)$, $(4, 0)$, $(4, 4)$, and $(0, 4)$ and let $f(x, y) = \sqrt{xy}$.

   (a) Find reasonable upper and lower bounds for $\int_R f\, dA$ without subdividing R.

   (b) Estimate $\int_R f\, dA$ by partitioning R into four subrectangles and evaluating f at its maximum and minimum values on each subrectangle.

4. Figure 15.7 shows the distribution of temperature, in °F, in a 10-foot by 10-foot heated room. Using Riemann sums, estimate the average temperature in the room.

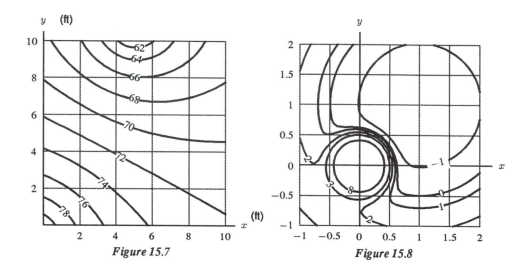

Figure 15.7   Figure 15.8

5. Figure 15.8 shows the contour diagram of a function $z = f(x, y)$. Contours are drawn and labeled for $z = 0, \pm 1, \pm 2, \ldots$. Note that the contours for 4, 5, 6, and 7 have been omitted. Let R be the rectangle $-0.5 \le x \le 1$, $-0.5 \le y \le 1$. Is the integral

$$\int_R f \, dA$$

positive or negative? Explain your reasoning.

6. A biologist studying insect populations measures the population density of flies and mosquitos at various points in a rectangular study region. The graphs of the two population densities for the region are shown in Figures 15.9 and 15.10. Assuming that the units along the corresponding axes are the same in the two graphs, are there more flies or more mosquitos in the region?

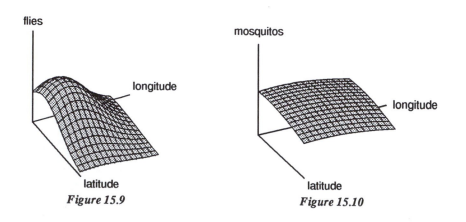

Figure 15.9   Figure 15.10

(3)

1. Lower Sum $= \sum\sum f(x_i, y_j)\Delta x \Delta y$ Upper Sum $= \sum\sum f(x_i, y_j)\Delta x \Delta y$ 15.1

        $= (4+6+3+4)\Delta x \Delta y$ $= (7+10+6+8)\Delta x \Delta y$ 1-13 odd

        $= 17(.1)(.2)$ $= 31(.1)(.2)$

        $= .34$ $= .62$

3.

5. More of the surface is greater than zero so the integral is positive

7.

In Problems 7 and 8, use a computer or calculator program that finds two-dimensional Riemann sums to estimate the given integral over the given region.

7. If R is the rectangle $1 \leq x \leq 2$, $1 \leq y \leq 3$, find $\int_R x^2 + y^2 \, dA$.

8. If R is the rectangle $-\pi \leq x \leq 0$, $0 \leq y \leq \pi/2$, find $\int_R \sin(xy) \, dA$.

9. The hull of a certain boat has width $w(x, y)$ feet at a point x feet from the front and y feet below the water line. A table of values of w follows. Set up a definite integral that gives the volume of the hull below the waterline, and then estimate the value of the integral.

Front of boat $\longrightarrow$ Back of boat

		0	10	20	30	40	50	60
	0	2	8	13	16	17	16	10
Depth	2	1	4	8	10	11	10	8
below	4	0	3	4	6	7	6	4
waterline								
(in feet)	6	0	1	2	3	4	3	2
	8	0	0	1	1	1	1	1

10. Let $f(x, y)$ be a function of x and y which is independent of y, that is, $f(x, y) = g(x)$ for some one-variable function g.

    (a) What does the graph of f look like?
    (b) Let R be the rectangle $a \leq x \leq b$, $c \leq y \leq d$. By interpreting the integral as a volume, and using your answer to part (a), express $\int_R f \, dA$ in terms of a one-variable integral.

11. Let $f(x, y)$ be a function of x and y which is independent of x, that is, $f(x, y) = g(y)$ for some one-variable function g.

    (a) What does the graph of f look like?
    (b) Let R be the rectangle $a \leq x \leq b$, $c \leq y \leq d$. By interpreting the integral as a volume, and using your answer to part (a), express $\int_R f \, dA$ in terms of a one-variable integral.

12. A health insurance company wants to know what proportion of its policies are going to cost them a lot of money because the insured people are over 65 and sick. In order to compute this proportion, the company defines a *disability index*, x, with $0 \leq x \leq 1$, where $x = 0$ represents perfect health and $x = 1$ represents total disability. In addition, the company uses a density function, $f(x, y)$, defined in such a way that the quantity

$$f(x, y) \, \Delta x \, \Delta y$$

represents the fraction of the population with disability index between x and $x + \Delta x$, and aged between y and $y + \Delta y$. The company knows from experience that a policy no longer covers its costs if the insured person is over 65 and has a disability index exceeding 0.8. Write an expression for the fraction of the company's policies held by people meeting these criteria.

13. Figure 15.11 shows contours for the annual frequency of tornados per 10,000 square miles in the U.S.[2] Each grid square is 100 miles on a side. Use the map to estimate the total number of tornados per year in (a) Texas   (b) Florida   (c) Arizona.

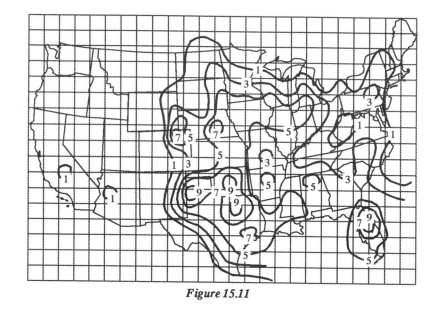

Figure 15.11

14. Figure 15.12 shows contours of annual rainfall (in centimeters) in Oregon.[3] Use it to estimate how much rain falls in Oregon in one year. Each grid square is 100 kilometers on a side.

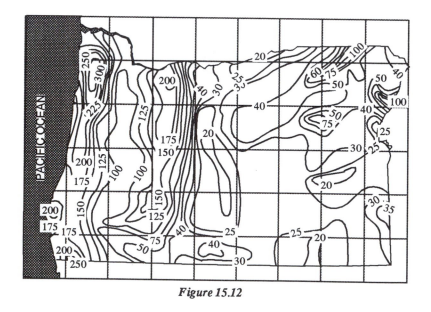

Figure 15.12

---

[2]From *Modern Physical Geography*, Alan H. Strahler and Arther H. Strahler, Fourth Edition, John Wiley & Sons, New York, 1992, p. 128

[3]from *Physical Geography of the Global Environment*, H. J. de Blij and Peter O. Muller, John Wiley & Sons, New York, 1993, p. 133

15.2   ITERATED INTEGRALS

Although we can approximate a two-variable definite integral as closely as we like using Riemann sums, we don't yet have any way of computing one exactly. In this section we will find a way of doing this by expressing a two-variable integral in terms of one-variable integrals.

The Fox Population Again

When we estimated the fox population from the population density function $D = f(x, y)$, we found that we did not get very good estimates, even when we subdivided the area into 36 pieces. We now look at a different method of estimating the integral. We could do better by subdividing even further, but that would be a lot of work. Here is another way of estimating the population. The idea is to break the area up into horizontal strips, estimate the population in each strip, and then add up all these estimates. Figure 15.13 shows three strips.

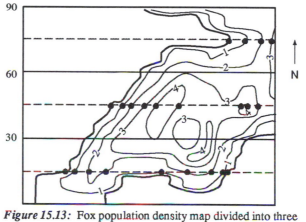

Figure 15.13: Fox population density map divided into three strips

---

Example 1   Estimate the fox population by estimating the population in each horizontal strip.

Solution   To estimate the population in each strip, we take a cross section of the density function through the middle of each strip. For example, look at the bottom strip in Figure 15.13. We have put a line through the middle of the strip, at $y = 15$, and marked every point where that line crosses a contour. In Figure 15.14, we have plotted the fox densities D against horizontal distance x along this cross section; the data points we obtained from the contour map are marked on the graph. Thus, Figure 15.14 is a graph of the one-variable function $D = f(x, 15)$.

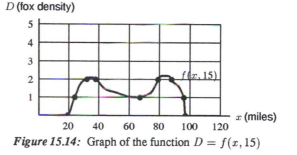

Figure 15.14: Graph of the function $D = f(x, 15)$

Now we approximate the population in this strip by supposing that the density does not vary across the width of the strip. Then the population is the width of the strip, which is 30 miles, times the integral of the one-variable density function $f(x, 15)$. Thus, we have

$$\text{Population in bottom strip} \approx 30 \int_0^{120} f(x, 15) \, dx.$$

We can estimate the integral by counting grid squares in Figure 15.14; there are about 6 grid squares, and each has a width of 20, so the integral is about 120. Thus, the fox population in the bottom strip is estimated to be about $30 \times 120 = 3,600$. To estimate the population in the other strips, we follow the same procedure; Figure 15.15 shows the graphs of the functions $D = f(x, 45)$ and $D = f(x, 75)$, which were obtained by taking data points from the cross sections $y = 45$ and $y = 75$ on the contour map.

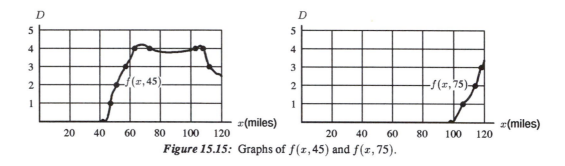

Figure 15.15: Graphs of $f(x, 45)$ and $f(x, 75)$.

From these graphs we estimate by counting grid squares that

$$\int_0^{120} f(x, 45) \, dx \approx 260 \quad \text{and} \quad \int_0^{120} f(x, 75) \, dx \approx 30.$$

Thus, the population estimates for these strips are $30 \times 260 = 7,800$, and $30 \times 30 = 900$. Combining all three estimates we get an estimate for the total population of $3,600 + 7,800 + 900 = 12,300$, or about 12,000 foxes. This is consistent with the upper and lower bounds we found before.

If we want a better estimate by this method, we can take more strips. In general, the method is to divide the diagram into horizontal strips of width Δy. If y_i is the midpoint on the i-th subdivision on the y-axis, then a cross-section through the middle of the i-th strip gives us the one-variable function of $D = f(x, y_i)$; this is a function of x, with y fixed at y_i. Integrating with respect to x and multiplying by the width Δy gives us the estimate

$$\text{Population of } i\text{-th strip} \approx \left(\int_0^{120} f(x, y_i) \, dx \right) \cdot \Delta y,$$

and adding these all up we get:

$$\text{Total population} \approx \sum_{i=1}^n \left(\int_0^{120} f(x, y_i) \, dx \right) \cdot \Delta y.$$

This sum is itself a Riemann sum for another integral, this time with respect to y; it is the integral from $y = 0$ to $y = 90$ of the function whose value at y is the integral with respect to x

$$\int_0^{120} f(x, y)\, dx.$$

We have expressed the integral of $f(x, y)$ as an *iterated integral*

$$\int_R f\, dA = \int_0^{90} \left(\int_0^{120} f(x, y)\, dx \right) dy.$$

The inside integral is the integral of f with respect to x, with y held fixed; the answer depends on y. The outside integral is with respect to y. We have discovered a way of expressing two-variable integrals in terms of nested one-variable integrals:

If R is the rectangle $a \le x \le b, c \le y \le d$, and if $f(x, y)$ is a function of two variables, then

$$\int_R f\, dA = \int_c^d \left(\int_a^b f(x, y)\, dx \right) dy.$$

The inside integral is performed with respect to x, holding y constant, and then the result is integrated with respect to y. The expression $\int_c^d \left(\int_a^b f(x, y)\, dx \right) dy$, or simply $\int_c^d \int_a^b f(x, y)\, dx\, dy$ is called an iterated integral or a double integral.

Here is another way to see that a two-variable integral can be computed as an iterated integral. This time we look at Riemann sums and a table of values.

Example 2   Compute $\int_R f(x, y) dA$ where R is the rectangle $0 \le x \le 1, 0 \le y \le 3$ and $f(x, y) = x + y^2$.

Solution   We compute the Riemann sum

$$\sum_{i,j} f(x_i, y_j) \Delta x \Delta y$$

and see what happens as Δx and Δy approach zero. Suppose $\Delta x = 0.25$ and $\Delta y = 0.5$. If we choose x_i and y_j each to have the two least values in each subrectangle, we get $x_i = 0, 0.25, 0.5, 0.75$ and $y_j = 0, 0.5, 1, 1.5, 2, 2.5$. The values of $f(x, y) = x + y^2$ that we need are given in Table 15.7.

TABLE 15.7   *Values for* $f(x, y) = x + y^2$

		0.0	0.5	1.0	1.5	2.0	2.5
	0.0	0	0.25	1.00	2.25	4.00	6.25
x	0.25	0.25	0.50	1.25	2.50	4.25	6.50
	0.50	0.50	0.75	1.50	2.75	4.50	6.75
	0.75	0.75	1.00	1.75	3.00	4.75	7.00

The Riemann sum we want is the sum of all entries in Table 15.7 times $\Delta x \Delta y$. We organize the sum by first adding all the values in each column, multiplying by Δx, then adding these results and multiplying by Δy. The sum for a typical column, say the column with $y = 1.5$, looks like this:

$$\sum_i (x_i + 1.5^2)\Delta x = (2.25 + 2.50 + 2.75 + 3.00)(0.25)$$

This is a Riemann sum for the section of f with $y = 1.5$, so as Δx approaches 0 the sum approaches the usual one-variable integral for the section of f with $y = 1.5$:

$$\int_0^1 (x + 1.5^2)dx$$

When we add up all the column sums to compute the full Riemann sum we get

$$\left(\sum_i (x_i + 0^2)\Delta x + \sum_i (x_i + 0.5^2)\Delta x + \cdots + \sum_i (x_i + 2.5^2)\Delta x \right) \Delta y$$

$$= \sum_j \left(\sum_i (x_i + y_j^2)\Delta x \right) \Delta y$$

Just as each column sum (the inner sum) approaches an integral in x as Δx approaches 0, the outer sum also approaches an integral in y as Δy approaches zero. Thus, the sum of sums, where y is fixed in each inner sum, approaches an integral of integrals, where y is fixed in each inner integral:

$$\int_0^3 \left(\int_0^1 (x + y^2)dx \right) dy$$

To evaluate this integral we first compute the inner integral, remembering to treat y as a constant, and then integrate the resulting expression in y:

$$\int_R (x + y^2)dA = \int_0^3 \left(\int_0^1 (x + y^2)dx \right) dy$$

$$= \int_0^3 \left(\frac{x^2}{2} + y^2 x \right) \Big|_0^1 dy = \int_0^3 \left(\frac{1}{2} + y^2 \right) dy$$

$$= \left(\frac{1}{2}y + \frac{1}{3}y^3 \right) \Big|_0^3 = \frac{21}{2}.$$

The Parallel Between Repeated Summation and Repeated Integration

It is helpful to notice how the summation and integral notation in Example 2 parallel each other. Viewing a Riemann sum as a sum of sums:

$$\sum_{i,j} f(x_i, y_j)\Delta x \Delta y = \sum_j \left(\sum_i f(x_i, y_j)\Delta x \right) \Delta y$$

leads us to see a two-variable integral as an integral of integrals:

$$\int_R f(x,y)dA = \int_c^d \left(\int_a^b f(x,y)dx \right) dy,$$

where R is the rectangle $a \le x \le b$ and $c \le y \le d$.

Example 3   A building is 8 feet wide and 16 feet long. It has a flat roof that is 12 feet high at one corner, and 10 feet high at each of the adjacent corners. What is the volume of the building?

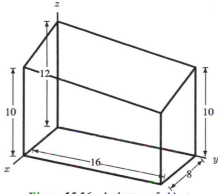

Figure 15.16: A slant-roofed hut

Solution   If we put the high corner on the z-axis, the long side along the y-axis, and the short side along the x-axis, as in Figure 15.16, then the roof is a plane with z-intercept 12, and x slope $(-2)/8 = -1/4$, and y slope $(-2)/16 = -1/8$. Hence, the equation of the roof is

$$z = 12 - \frac{1}{4}x - \frac{1}{8}y.$$

To calculate the volume, we will integrate over the rectangle $0 \le x \le 8, 0 \le y \le 16$. Setting the integral up as an iterated integral, we get

$$\int_0^8 \int_0^{16} (12 - \frac{1}{4}x - \frac{1}{8}y) \, dy \, dx.$$

The inside integral is

$$\int_0^{16} (12 - \frac{1}{4}x - \frac{1}{8}y) \, dy = \left(12y - \frac{1}{4}xy - \frac{1}{8}\frac{y^2}{2} \right) \Big|_0^{16} = 176 - 4x.$$

Then the outside integral is

$$\int_0^8 (176 - 4x) \, dx = (176x - 2x^2) \Big|_0^8 = 1280.$$

So the volume of the building is $1,280$ cubic feet.

The Order of Integration

In Example 2 we could clearly have chosen to add up the rows (fixed y) first instead of the columns. Similarly, in Example 1, we could have taken vertical strips (fixed x) instead of horizontal strips. This leads to an iterated integral where x is constant in the inner integral instead of y. Thus,

$$\int_R f(x, y)\, dA = \int_a^b \left(\int_c^d f(x, y)dy \right)\, dx$$

where R is the rectangle $a \le x \le b$ and $c \le y \le d$.

For any function you are likely to meet, it doesn't matter in which order you integrate over a rectangle region R; you get the same two-variable definite integral either way.

$$\int_R f\, dA = \int_c^d \left(\int_a^b f(x, y)dx \right)\, dy = \int_a^b \left(\int_c^d f(x, y)dy \right)\, dx$$

Example 4   Compute the integral of Example 2 as an iterated integral with x fixed in the inner integral.

Solution

$$\int_R (x + y^2)dA = \int_0^1 \left(\int_0^3 (x + y^2)dy \right)\, dx = \int_0^1 \left. \left(xy + \frac{y^3}{3}\right) \right|_0^3\, dx$$

$$= \int_0^1 (3x + 9)dx = \left. \frac{3}{2}x^2 + 9x \right|_0^1 = \frac{21}{2}.$$

It is a little surprising to see the same final answers in Example 2 and 4 are the same when the intermediate results are so different.

Iterated Integrals Over Odd-Shaped Regions

Example 5   The density at the point (x, y) of a right triangular metal plate, as shown in Figure 15.17, is $\rho(x, y)$. Express its mass as an iterated integral.

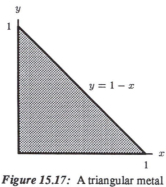

Figure 15.17: A triangular metal plate

Solution   The mass is

$$\int_R \rho(x, y) \, dA,$$

where R is the triangle. The sloping edge of the triangle is the line $y = 1 - x$. We want to compute this integral using an iterated integral. Think about how an iterated integral over a rectangle, such as

$$\int_a^b \int_c^d f(x, y) \, dy \, dx,$$

works. This integral is over the rectangle $a \le x \le b$, $c \le y \le d$. The inside integral with respect to y is an integral along vertical strips from $y = c$ to $y = d$. There is one such integral for each x strip between $x = a$ and $x = b$. Thus, the value of the inside integral depends on x. After computing the inside integral with respect to y, we compute the outside integral with respect to x, which means putting together all the contributions from the individual vertical strips that make up the rectangle. See Figure 15.18.

For the triangular region in Figure 15.17, the idea is the same. The only difference is that the individual vertical strips no longer all go from $y = c$ to $y = d$. The vertical strip that enters the triangle at the point $(x, 0)$ leaves it at the point $(x, 1 - x)$, because the top edge of the triangle is the line $y = 1 - x$. Thus, on this vertical strip, y goes from 0 to $1 - x$. Hence, the inside integral is

$$\int_0^{1-x} \rho(x, y) \, dy.$$

Finally, since there is one of these integrals for each x strip between 0 and 1, the outside integral goes from 0 to 1. Thus, the iterated integral we want is

$$\text{Mass} = \int_0^1 \int_0^{1-x} \rho(x, y) \, dy \, dx.$$

See Figure 15.19. We could have chosen to integrate in the opposite order keeping y fixed in the inner integral instead of x. The limits are formed by looking at horizontal strips instead of vertical ones, and expressing the x-values at the end points in terms of y. A typical horizontal strip goes from $x = 0$ to $x = y - 1$, and since y values overall range between 0 and 1, the iterated integral is

$$\text{Mass} = \int_0^1 \int_0^{y-1} \rho(x, y) \, dx \, dy.$$

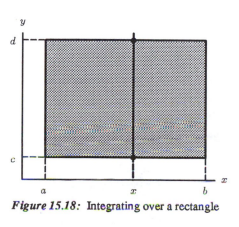

Figure 15.18: Integrating over a rectangle

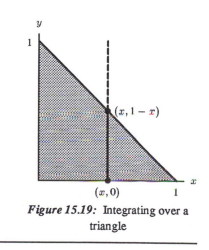

Figure 15.19: Integrating over a triangle

Iterated integrals with variable limits on the inner integral

There are two things to remember
- The limits on the outer integral must be constants.
- If the inner integral is with respect to x, its limits should be constants or expressions in terms of y, and vice versa.

Example 6   Find the mass M of a metal plate R bounded by $y = x$ and $y = x^2$ with density given by $\rho(x, y) = 1 + xy$ kg/meter^2. (See Figure 15.20.

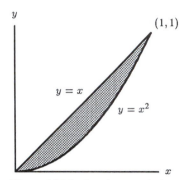

Figure 15.20: Another metal plate

Solution   The mass is

$$M = \int_R \rho(x, y) \, dA.$$

We will integrate along vertical strips first; this means we will do the y integral first, which goes from the bottom boundary $y = x^2$ to the top boundary $y = x$. Thus, the inside integral is

$$\int_{x^2}^{x} \rho(x, y) \, dy = \int_{x^2}^{x} (1 + xy) \, dy = \left. \left(y + x\frac{y^2}{2} \right) \right|_{y=x^2}^{y=x}$$

$$= \left(x + \frac{x^3}{2} \right) - \left(x^2 + \frac{x^5}{2} \right) = x - x^2 + \frac{x^3}{2} - \frac{x^5}{2}.$$

The left edge of the region is at $x = 0$ and the right edge is at the intersection point of $y = x$ and $y = x^2$, which is $(1, 1)$. Thus, the x-coordinate of the vertical strips can vary from $x = 0$ to $x = 1$, and so the mass is

$$M = \int_0^1 \left(x - x^2 + \frac{x^3}{2} - \frac{x^5}{2} \right) dx = \left. \left(\frac{x^2}{2} - \frac{x^3}{3} + \frac{x^4}{8} - \frac{x^6}{12} \right) \right|_0^1 = \frac{5}{24} \text{ kg}.$$

Example 7   A city is in the form of a semicircle region of radius 3 miles bordering on the ocean. Find the average distance from any point in the city to the ocean.

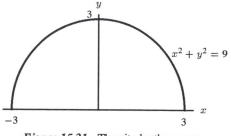

Figure 15.21: The city by the ocean

Solution   Think of the ocean as everything below the x-axis in the xy-plane and think of the city as the upper half of the circular disk of radius 3 bounded by $x^2 + y^2 = 9$. (See Figure 15.21). The distance from any point (x, y) in the city to the ocean is the vertical distance to the x-axis, namely y. Thus, we want to compute

$$\text{Average distance} = \frac{\int_R y\, dA}{\text{area}(R)}$$

where R is the upper half of the circle $x^2 + y^2 = 9$ and the x-axis. The area of R is $\pi 3^2/2$. To compute the integral, let's try making the inner integral with respect to y. Then a typical vertical strip goes from the x-axis, namely $y = 0$, to the semicircle. The limits must be expressed in x so we solve $x^2 + y^2 = 9$ for y in terms of x to get $y = \sqrt{9 - x^2}$. Since x varies from -3 to 3 throughout the region, the integral is:

$$\int_R y\, dA = \int_{-3}^{3} \left(\int_0^{\sqrt{9-x^2}} y\, dy \right) dx = \int_{-3}^{3} \left(\frac{y^2}{2} \Big|_0^{\sqrt{9-x^2}} \right) dx$$

$$= \int_{-3}^{3} \frac{1}{2}(9 - x^2)dx = \frac{1}{2}\left(9x - \frac{x^3}{3}\right)\Big|_{-3}^{3} = \frac{1}{2}(18 - (-18)) = 18.$$

Therefore, the average distance is $\frac{18}{9\pi/2} = \frac{4}{\pi}$ miles.

What if we had chosen the inner integral to be with respect to x instead ? Then the limits would have been obtained by looking at horizontal strips, not vertical, and we should solve $x^2 + y^2 = 9$ for x in terms of y, rather than y in terms of x. We get $x = -\sqrt{9 - y^2}$ at the left end of the strip and $x = \sqrt{9 - y^2}$ at the right. Now y varies from 0 to 3 so the integral becomes:

$$\int_0^3 \left(\int_{-\sqrt{9-y^2}}^{\sqrt{9-y^2}} y\, dx \right) dy = \int_0^3 \left(yx \Big|_{x=-\sqrt{9-y^2}}^{x=\sqrt{9-y^2}} \right) dy = \int_0^3 2y\sqrt{9 - y^2}dy$$

$$= -\frac{2}{3}(9 - y^2)^{3/2}\Big|_0^3 = -\frac{2}{3}(0 - 27) = 18.$$

Needless to say, we get the same average distance as before.

In the examples so far, the region was given and the problem was to determine the limits for an iterated integral. Sometimes the limits are given and we want to determine the region.

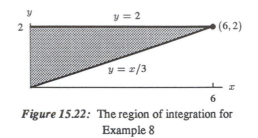

Figure 15.22: The region of integration for
Example 8

Example 8   Sketch the region of integration for the iterated integral $\int_0^6 \int_{x/3}^2 x\sqrt{y^3 + 1}\, dy dx$.

Solution   The inner integral is with respect to y, so we should imagine vertical strips crossing the region of integration. The bottom of each strip is $y = x/3$, a line through the origin, and the top is $y = 2$, a horizontal line. Since the outer integral's limits are 0 and 6, the whole region is contained between the vertical lines $x = 0$ and $x = 6$. Notice the lines $y = 2$ and $y = x/3$ meet where $x = 6$. The region is shown in Figure 15.22 .

Reversing the Order of Integration

It can sometimes be helpful to reverse the order of integration in a double integral. Surprisingly enough, an integral which is difficult or impossible with the limits in one order can be quite straightforward in the other. The next example is such a case.

Example 9   Evaluate $\int_0^6 \int_{x/3}^2 x\sqrt{y^3 + 1}\, dy\, dx$.

Solution   Since many elementary functions have no elementary antiderivatives, and since $\sqrt{y^3 + 1}$ is such a function, we have no algebraic way to do the inner integral. We will try a method which might surprise you: We interchange the order of integration. From the region we sketched, we see that horizontal strips go from $x = 0$ to $x = 3y$. For the whole region, y varies from 0 to 2. Thus, when we change the order of integration we get:

$$\int_0^2 \int_0^{3y} x\sqrt{y^3 + 1}\, dx dy.$$

Now we can at least do the inner integral. What about for the outer integral?

$$\int_0^2 \int_0^{3y} x\sqrt{y^3 + 1}\, dx dy = \int_0^2 \left(\frac{x^2}{2}\sqrt{y^3 + 1}\right)\Bigg|_{x=0}^{x=3y} dy$$

$$= \int_0^2 \frac{9y^2}{2}(y^3 + 1)^{1/2} dy$$

$$= (y^3 + 1)^{3/2}\Big|_0^2 = 27 - 1 = 26.$$

---

1. $\int_R \sqrt{x+y} \, dA \Rightarrow 0 \leq x \leq 1; \ 0 \leq y \leq 2$

$= \int_0^1 \int_0^2 \sqrt{x+y} \, dy \, dx$

$= \dfrac{16-\sqrt{2}}{15} - \dfrac{8}{15}$

3. $\int_R (2x+3y)^2 \, dA$

$= \int_0^1 \int_{y-1}^{-y+1} (2x+3y)^2 \, dx \, dy$

$= \dfrac{13}{6}$

5. $\int_{-1}^3 \int_{-2}^{\frac{1}{4}-\frac{3}{4}x} f(x,y) \, dy \, dx$

7. $\int_1^3 \int_{\frac{x+1}{2}}^{\frac{4-5}{2}} f(x,y) \, dy \, dx$

9.   11.   13.   15

17   19   b. $\int_0^2 \int_0^{-2x+4} q(x,y) \, dy \, dx$

(?)

b. $\int_0^1 \int_{-\sqrt{1-x^2}}^{\sqrt{1-x^2}} f(x,y) \, dy \, dx$ b. $\int_0^8 \int_{\frac{8-x}{2}}^{\frac{8-x}{2}} f(x,y) \, dx \, dy$ (?) 21.

(?)

23. (See Man.)

(?)

25.

Thus, reversing the order of integration made the integral in the previous problem much easier. Notice that to reverse the order of integration, it is essential first to sketch the region over which the integration is being performed.

Problems for Section 15.2

For Problems 1–3, evaluate the given integral.

1. $\int_R \sqrt{x+y} \, dA$, where R is the rectangle $0 \le x \le 1, 0 \le y \le 2$.

2. $\int_R (5x^2 + 1) \sin 3y \, dA$, where R is the rectangle $-1 \le x \le 1, 0 \le y \le \pi/3$.

3. $\int_R (2x + 3y)^2 \, dA$, where R is the triangle with vertices at $(-1, 0), (0, 1)$, and $(1, 0)$.

For each of the regions R in Problems 4–7, set up $\int_R f \, dA$ as an iterated integral.

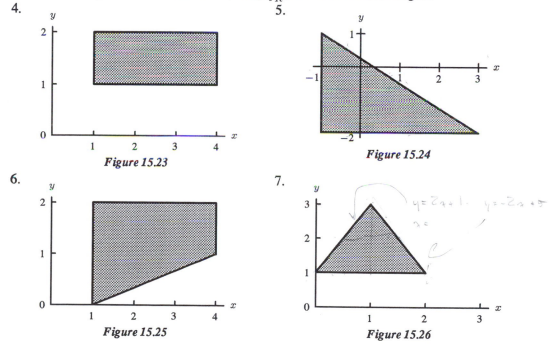

Figure 15.23

Figure 15.24

Figure 15.25

Figure 15.26

8. (a) Let R be the rectangle $0 \le x \le 1, 0 \le y \le 2$. Compute $\int_R \sqrt{x+y} \, dA$.

   (b) Do the integral in part (a) in reverse order.

Sketch the region over which the integral in Problems 9–14 are computed.

9. $\int_1^3 \int_0^4 f(x, y) \, dy \, dx$

10. $\int_0^2 \int_0^x f(x, y) \, dy \, dx$

11. $\int_0^1 \int_{x^2}^x f(x, y) \, dy \, dx$

12. $\int_1^5 \int_x^{2x} f(x, y) \, dy \, dx$

13. $\int_1^4 \int_{\sqrt{y}}^y f(x, y) \, dx \, dy$

14. $\int_{-2}^0 \int_{-\sqrt{9-x^2}}^0 f(x, y) \, dy \, dx$

15. Consider the integral $\int_0^4 \int_0^{-(y-4)/2} g(x, y)\, dx\, dy$.

    (a) Sketch the region over which the integration is being performed.
    (b) Write the integral with the order of the integration reversed.

Sketch the region of integration in Problems 16–19 and then reverse the order of integration.

16. $\int_0^1 \int_y^{\sqrt{y}} f(x, y)\, dx\, dy$ 17. $\int_{-1}^1 \int_0^{\sqrt{1-y^2}} f(x, y)\, dx\, dy$ 18. $\int_0^{1/2} \int_{2x}^1 f(x, y)\, dy\, dx$

19. $\int_{-4}^0 \int_0^{2x+8} f(x, y)\, dy\, dx + \int_0^4 \int_0^{-2x+8} f(x, y)\, dy\, dx$

For Problems 20–23 set up, but do not evaluate, an iterated integral for the volume of the solid.

20. Between the graph of $f(x, y) = 25 - x^2 - y^2$ and the xy-plane.
21. Between the graph of $f(x, y) = 25 - x^2 - y^2$ and the plane $z = 16$.
22. The three-sided pyramid whose base is the xy-plane and whose three sides are the vertical planes $y = 0$ and $y - x = 4$ and the slanted plane $2x + y + z = 4$.
23. Same as Problem 22 with the base in the plane $z = -6$.

For Problems 24–26, find the volumes of the given regions.

24. Under the graph of $f(x, y) = xy$ and above the square $0 \le x \le 2, 0 \le y \le 2$ in the xy-plane.
25. The solid between the planes $z = 3x + 2y + 1$ and $z = x + y$ and above the triangle with vertices $(1, 0, 0), (2, 2, 0)$, and $(0, 1, 0)$ in the xy-plane. See Figure 15.27.

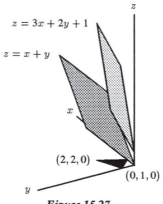

Figure 15.27

26. The region R bounded by the graph of $ax + by + cz = 1$ and the coordinate planes. Assume $a, b,$ and c are positive.
27. Find the average distance to the y-axis for points in the region bounded by the x-axis and the graph of $y = x - x^2$.
28. Prove that for a right triangle the average distance from any point in the triangle to one of the legs is one-half the length of the other leg.

29. Evaluate $\int_0^1 \int_y^1 \sin(x^2)\, dx\, dy$ 30. Evaluate $\int_0^1 \int_{e^y}^e \frac{x}{\ln x}\, dx\, dy$

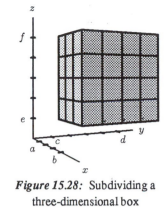

Figure 15.28: Subdividing a
three-dimensional box

15.3   THREE-VARIABLE INTEGRALS

A function of three variables can be integrated over a region R in 3-space in the same way as a function of two variables is integrated over a region in 2-space. Again, we start with a Riemann sum. First we subdivide R into smaller regions, then we multiply the volume of each region by a value of the function in that region, and then we add the results. For example, if R is the box $a \leq x \leq b$, $c \leq y \leq d$, $e \leq z \leq f$, then we subdivide each side into l, m, and n pieces, thereby chopping R into lmn smaller boxes, as shown in Figure 15.28. The volume of each smaller box is

$$\Delta V = \Delta x \Delta y \Delta z,$$

where

$$\Delta x = \frac{a - b}{l}, \quad \Delta y = \frac{c - d}{m}, \quad \Delta z = \frac{e - f}{n}.$$

Using this subdivision, we construct a Riemann sum

$$\sum_{i,j,k} f(x_i, y_j, z_k) \, \Delta V.$$

As Δx, Δy, and Δz approach 0, this Riemann sum approaches the definite integral

$$\int_R f \, dV.$$

Just as in the case of two-variable integrals, we can evaluate this integral as an iterated integral:

Three-variable integral as an iterated integral or triple integral

$$\int_R f \, dV = \int_e^f \left(\int_c^d \left(\int_a^b f(x, y, z) \, dx \right) dy \right) dz.$$

where y and z are treated as constants in the innermost (dx) integral, and z is treated as a constant in the middle (dy) integral. Any other order of integration will give the same answer.

Example 1   A cube C is 4 centimeters on a side and is made of a material of variable density. If one corner is at the origin and the adjacent corners are on the positive x, y, and z axes, then the function giving its density (in gm/cm³) at the point (x, y, z) is $\rho(x, y, z) = 1 + xyz$. Find the mass of the cube.

Solution   Consider a small piece ΔV of the cube, small enough so that the density remains close to constant over the piece. The mass of ΔV is then the density times the volume, i.e., it is $\rho(x, y, z) \, \Delta V$. To get the total mass, we need to put all the small pieces together in a Riemann sum and take a limit. Thus, the mass is the volume integral

$$
\begin{aligned}
M = \int_C \rho \, dV &= \int_0^4 \int_0^4 \int_0^4 (1 + xyz) \, dx \, dy \, dz = \int_0^4 \int_0^4 \left[x + \frac{1}{2} x^2 yz \right]_{x=0}^{x=4} dy \, dz \\
&= \int_0^4 \int_0^4 (4 + 8yz) \, dy \, dz = \int_0^4 \left[4y + 4y^2 z \right]_{y=0}^{y=4} dz \\
&= \int_0^4 (16 + 64z) \, dz = 16z + 32z^2 \bigg|_{z=0}^{z=4} = 576 \, \text{gm}.
\end{aligned}
$$

Example 2   Express the volume of the building described in example 3 on page 237 as a triple integral in several different ways.

Solution   We want the volume of the solid region W given by $0 \le x \le 8$, $0 \le y \le 16$, and $0 \le z \le 12 - x/4 - y/8$.

First if we keep x, y fixed in this region W, then z may vary from 0 to $12 - x/4 - y/8$. The idea is to first add up the volumes of a stack of cubes, each of which has volume $\Delta x \Delta y \Delta z$, starting at $(x, y, 0)$ and going up to $(x, y, 12 - \frac{x}{4} - \frac{y}{8})$. The sum of these volumes is given by

$$
\left(\sum_k \Delta z \right) \Delta x \Delta y
$$

where the line from $(x, y, 0)$ to $(x, y, 12 - \frac{x}{4} - \frac{y}{8})$ is divided into n equal intervals of length Δz. Now adding each of these volume stacks along a line parallel to the y-axis and going through the fixed point $(x, y, 0)$ gives a slice of volume from the region W. This slice of volume is given by

$$
\left(\sum_j \sum_k \Delta z \Delta y \right) \Delta x
$$

where the line, in Figure 15.29, parallel to the y-axis is divided into m equal intervals of length Δy.

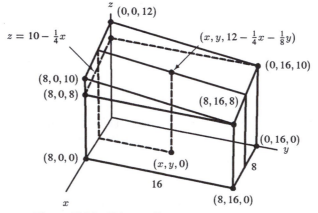

Figure 15.29: Volume of building as triple integral

Now dividing the interval from 0 to 8 into l equal intervals length Δx and adding each of the volume slices given above we obtain an approximation for the volume of W given by

$$\sum_i \sum_j \sum_k \Delta z \Delta y \Delta x$$

which approximates

$$\text{Vol}(W) = \int_0^8 \int_0^{16} \int_0^{12 - x/4 - y/8} dz \, dy \, dx.$$

If, in Figure 15.29, instead of taking a line through $(x, y, 0)$ parallel to the y-axis we had taken one parallel to the x-axis, then we would have obtained the iterated integral given by

$$\text{Vol}(W) = \int_0^{16} \int_0^8 \int_0^{12 - x/4 - y/8} dz \, dx \, dy.$$

Example 3   Set up an iterated integral to compute the mass of the cone bounded by $z = \sqrt{x^2 + y^2}$ and $z = 3$, if the density is given by $\rho(x, y, z) = z$.

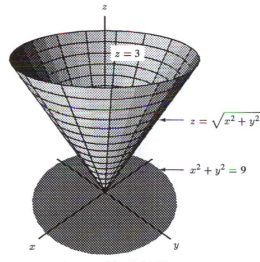

Figure 15.30

Solution   The cone is shown in Figure 15.30. If we integrate with respect to z first, the inner integral is

$$\int_{\sqrt{x^2+y^2}}^3 z \, dz.$$

To compute the limits for the outer x and y integral, we need to find the intersection of the cone $z = \sqrt{x^2 + y^2}$ and the horizontal plane $z = 3$. It is the circle $x^2 + y^2 = 9$. Setting up the limits for this xy-region is just like setting up the limits of a double integral. If we integrate next with respect to y, the limits are obtained by solving $x^2 + y^2 = 9$ for y in term of x. The limits on the outer integral are obtained by the overall range of x from -3 to 3. The final answer is

$$\text{Mass} = \int_{-3}^3 \int_{-\sqrt{9-x^2}}^{\sqrt{9-x^2}} \int_{\sqrt{x^2+y^2}}^3 z \, dz \, dy \, dx.$$

There are three things to remember for a triple integral:
- The limits for the outer integral are constants.
- The limits for the middle integral can involve only one variable (that in the outer integral).
- The limits for the inner integral can involve two variables (those on the two outer integrals).

Problems for Section 15.3

In Problems 1–4, find the three-variable integrals of the given functions over the given regions.

1. $f(x, y, z) = x^2 + 5y^2 - z$, W is the rectangular box $0 \le x \le 2, -1 \le y \le 1, 2 \le z \le 3$.
2. $f(x, y, z) = \sin x \cos(y + z)$, W is the cube $0 \le x \le \pi, 0 \le y \le \pi, 0 \le z \le \pi$.
3. $h(x, y, z) = ax + by + cz$, W is the rectangular box $0 \le x \le 1, 0 \le y \le 1, 0 \le z \le 2$.
4. $f(x, y, z) = e^{-x-y-z}$, W is the rectangular box with corners at $(0, 0, 0)$, $(a, 0, 0)$, $(0, b, 0)$, and $(0, 0, c)$.

For Problems 5–11 describe or sketch the region of integration for the triple integrals. If the limits do not make sense, say why.

5. $\displaystyle\int_0^6 \int_0^{6-x} \int_0^{6-x-2y} f(x, y, z)\, dz\, dy\, dx$

6. $\displaystyle\int_0^6 \int_0^{6-x} \int_0^{6-x-2y} f(x, y, z)\, dz\, dy\, dx$

7. $\displaystyle\int_0^1 \int_0^z \int_0^x f(x, y, z)\, dz\, dy\, dx$

8. $\displaystyle\int_0^3 \int_{-\sqrt{9-y^2}}^0 \int_{\sqrt{x^2+y^2}}^3 f(x, y, z)\, dz\, dx\, dy$

9. $\displaystyle\int_1^3 \int_1^{x+y} \int_0^y f(x, y, z)\, dz\, dx\, dy$

10. $\displaystyle\int_0^1 \int_0^{2-x} \int_0^3 f(x, y, z)\, dz\, dy\, dx$

11. $\displaystyle\int_{-1}^1 \int_0^{\sqrt{1-x^2}} \int_0^{\sqrt{2-x^2-y^2}} f(x, y, z)\, dz\, dy\, dx$

12. Set up $\int_R f\, dV$ as an iterated integral in all six possible orders of integration, where R is the hemisphere bounded by the upper half of $x^2 + y^2 + z^2 = 1$ and the xy-plane.

13. Find the mass of the solid bounded by the xy-plane, yz-plane, xz-plane, and the plane $\frac{x}{3} + \frac{y}{2} + \frac{z}{6} = 1$, if the density of the solid is given by $\rho(x, y, z) = x + y$.

14. Find the average value of the sum of the squares of three numbers x, y, z where each number is between 0 and 2.

15. Find the volume of the solid formed by the intersections of the cylinders $x^2 + z^2 = 1$ and $y^2 + z^2 = 1$.

15.4   NUMERICAL INTEGRATION: THE MONTE CARLO METHOD

There are many one-variable definite integrals in which that the integrand has no elementary antiderivative. A familiar example is $\int_0^1 e^{-x^2} dx$. To evaluate such an integral, we must use some

3. $\int_0^1 \int_0^1 \int_0^2 (ax + by + cz)\, dz\, dy\, dx = a + b + 2c$ 15.3

5.   (?)   7. can't do   9. can't do   11.   3-15 odd

13. $\dfrac{x}{3} + \dfrac{y}{2} + \dfrac{z}{6} = 1$

$\dfrac{6x}{3} + \dfrac{6y}{2} + z = 1$

$-2x - 3y + 1 = z$

$m = \int_R \rho\, dV$

$= \int_0^3 \int_0^{-\frac{2}{3}x + 2} \int_0^{-2x - 3y + 1} (x + y)\, dz\, dy\, dx$

$= \left(\text{using } \text{DERIVE}\right) = \dfrac{15}{6}$

15.

numerical technique such as the trapezoid rule or Simpson's rule. For a two or three-variable integral, we are just as likely to encounter intractable integrals. One can always use Riemann sums to approximate the integral. One can get greater accuracy by using a variant of Simpson's rule (see Problem 7). In this section, we give an alternative method called the Monte Carlo method (after the gambling casino).

A One-Variable Example

Let us consider the integral $\int_0^1 x^2 dx$, which we know to have the value $1/3$. We now approach it probabilistically. We graph the function $y = x^2$ in the square $0 \leq x \leq 1, 0 \leq y \leq 1$ and throw darts at the square. We expect that some darts will land above the curve, some below. The ratio of the number of darts above to those below gives a reasonable estimate of the ratio of these two areas. Since the total area plotted is the unit square of area 1, the ratio of the number of darts which land below the curve to the total area gives an estimate for the area under the curve. This is the basis of the Monte Carlo method.

Example 1   Approximate the value of the integral $\displaystyle\int_0^1 x^2 dx$ using the Monte Carlo method.

Solution   We choose points from the unit square of Figure 15.31 at random, and we expect that the ratio of the number of points in region R, say N_R, to that of all of those picked from the square, say N, will approximate the integral:

$$\frac{N_R}{N} \approx \frac{\int_0^1 x^2 dx}{1} = \int_0^1 x^2 dx.$$

Since we are selecting the points at random we cannot expect to get the same ratio every time, but as the number of points increase, the approximation should improve. Since we won't get the same answer every time, we try the same computation several times and compare the answers to get some idea of the accuracy. Table 15.8 shows the values of N_R/N for six different trials with $N = 50$ points.

TABLE 15.8   *Six trials for $N = 50$ points*

$N = 50$	1	2	3	4	5	6
N_R/N	0.2	0.24	0.52	0.36	0.38	0.28

These approximations are not particularly good. We may improve the accuracy by taking their average, giving 0.33, which has two digit accuracy.

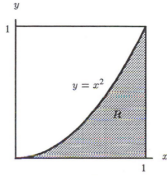

Figure 15.31: Region whose area
is $\int_0^1 x^2 \, dx$
as fraction of the unit square

Repeating this process again for $N = 50$ gives the results found in Table 15.9.

TABLE 15.9 *Six more trials for $N = 50$ points*

$N = 50$	1	2	3	4	5	6
N_R/N	0.44	0.42	0.28	0.34	0.28	0.32

Notice that the average of these is 0.347. This is not as close to the true value of 1/3 as before, but remember that this is a random process. Each time it is repeated we expect a different result.

To continue with this example we now approximate $\int_0^1 x^2 dx$ by taking increasingly large values of N, say $N = 10, 100, 1000$, and $10,000$. The results of one experiment are given in Table 15.10, but if you carry out a similar experiment your results will probably be slightly different.

TABLE 15.10 *Value of N_R/N as N increases*

N	10	100	1000	10000
N_R/N	0.2000	0.3400	0.3250	0.3343

It appears that as we take more points from the unit square, the ratio gets closer to the exact value of 1/3.

The basis of the Monte Carlo method is the generation of random numbers. Fortunately, nearly all programming languages have a random number generator built in. In the previous example we used a random number generator that gives numbers between 0 and 1. Given two random numbers, say x and y, we have a point (x, y) in the unit square. We need only check to see if $y \leq x^2$. If this is true, we know the point is in the region under the parabola and count it. If this is not true, then we pick another pair of random numbers. As Table 15.10 shows, as more points are considered, our approximation usually gets better. However, it is possible for the approximation to get poorer.

When we use a Monte Carlo method, we assume that when we pick points from a region R, no point is any more likely to be chosen than any other. That is, the points that we can choose are all equally likely to be found in R. If we know the area of R, say $A(R)$, and if we choose N points from R and find that M of these N points are in S, where S is a region contained in R, then we expect, for N large, that

$$\frac{A(S)}{A(R)} \approx \frac{M}{N}$$

or

$$A(S) \approx A(R) \cdot \frac{M}{N}.$$

A Two-Variable Example

We can extend the idea of Monte Carlo method to evaluating integrals of more than one variable.

Example 2 Use a Monte Carlo method to approximate the two-variable integral

$$\text{Vol} = \int_0^1 \int_0^1 e^{-(x^2+y^2)} dx\, dy.$$

Solution   This integral gives the volume of the region W above the unit square and below the graph of $z = e^{-(x^2+y^2)}$. Since the volume we are considering is contained in the cube C given by $0 \leq x \leq 1$, $0 \leq y \leq 1$, and $0 \leq z \leq 1$, we count only points of the form (x, y, z) which lie in the cube and which satisfy the condition

$$0 \leq z \leq e^{-(x^2+y^2)}.$$

If N_R of the N points chosen satisfy this condition, then, since the volume of the bounding cube is 1, we have

$$\text{Vol}(W) \approx \frac{N_R}{N}$$

So

$$\int_0^1 \int_0^1 e^{-(x^2+y^2)} dx\, dy = \text{Vol}(W) \approx \frac{N_R}{N}.$$

Table 15.11 shows the value of N_R/N for ten trials of $N = 100$ points each. The average of the ten values N_R/N is 0.563. We take this as an approximate value of the integral $\int_0^1 \int_0^1 e^{-(x^2+y^2)} dx\, dy$.

TABLE 15.11   *Ten trials with $N = 100$*

$N = 100$	1	2	3	4	5	6	7	8	9	10
N_R/N	0.54	0.60	0.57	0.60	0.51	0.53	0.59	0.56	0.56	0.57

Taking $N = 10,000$ gives

$$\int_0^1 \int_0^1 e^{-(x^2+y^2)} dx\, dy \approx N_R/N \approx 0.5654.$$

Here we chose the cube C to be the smallest cube containing the volume W. This is important. Intuitively, the better "the fit" between the two volumes, the fewer random points are needed to obtain a reasonable approximation. In fact, the biggest problem with the Monte Carlo method is finding sufficiently small rectangular box which encloses the volume.

Problems for Section 15.4

For Problems 1–3 use several runs of random numbers to decide on an approximate answer.

1. Use a Monte Carlo method to approximate the integral

$$\int_0^1 \sqrt{1 - x^2}\, dx$$

and explain why this gives an approximation for $\pi/4$.

2. Use a Monte Carlo method to approximate the integral

$$\int_0^1 e^{-x^2} dx.$$

3. Use a Monte Carlo method to approximate the integral

$$\int_0^1 \frac{1}{x^4 - 3x^2 + 2} dx.$$

We now give another Monte Carlo method for approximating an integral. Consider the integral $\int_R f(x, y) dx\, dy$ and the average value of the function given by $\frac{1}{\text{Area}(R)} \int_R f(x, y) dx\, dy$. The motivation for this is that when we divide the region R into subregions R_i, and take a point (x_i, y_i) in R_i for $i = 1, 2, \cdots, n$, then

$$\sum_i f(x_i, y_i) \text{Area}(R_i) \approx \int_R f(x, y) dx dy$$

If all the Area (R_i)s are equal, then

$$n\, \text{Area}(R_i) = \text{Area}(R)$$

and

$$\frac{\text{Area}(R)}{n} \sum_{i=1}^n f(x_i, y_i) \approx \int_R f(x, y) dx dy$$

so

$$\frac{1}{n} \sum_{i=1}^n f(x_i, y_i) \approx \frac{1}{\text{Area}(R)} \int_R f(x, y) dx\, dy,$$

where $\frac{1}{n} \sum_{i=1}^n f(x_i, y_i)$ is the average of the n values $f(x_i, y_i)$.
By choosing n points (x_i, y_i) at random from R, we can use the approximation:

$$\text{Area}(R) \cdot \frac{1}{n} \sum_{i=1}^n f(x_i, y_i) \approx \int_R f(x, y) dx dy.$$

Use this Monte Carlo method to approximate the integrals in Problems 4–6:

4. $\int_0^1 \int_0^1 e^{-xy} dx\, dy$ 5. $\int_0^1 \int_0^1 xy^{xy} dx\, dy$ 6. $\int_0^1 \int_0^1 x^{-y} dy\, dx$

7. Here is a way to use Simpson's rule twice to approximate a definite integral. Suppose the integral is $\int_1^5 \int_2^6 \sqrt{x^2 + y^2}\, dy\, dx$. Use Simpson's rule with $\Delta y = 1$ to approximate the inner integral when x is fixed at 1. Repeat for $x = 1.5, 2, 2.5, 3, 3.5, 4, 4.5, 5$. You now have approximations for the inner integral at nine different values of x. Now use Simpson's rule again with $\Delta x = 1$, using the nine different values for the inner integral, to approximate the outer integral (and therefore the whole double integral).

15.5 TWO-VARIABLE INTEGRALS IN POLAR COORDINATES

Integration in Polar Coordinates

We started this chapter by putting a rectangular grid on a population density map, in order to construct a Riemann sum that gave an estimate for the fox population. However, there is no particular reason to always use a rectangular grid; sometimes a polar grid is more appropriate. Riemann sums can be constructed from any subdivision of a region, not just a rectangular subdivision.

Example 1 A biologist studying insect populations around a circular lake divides the area into sectors as in Figure 15.32, and measures the population density in each sector, with the results shown (in millions per square mile). Estimate the total insect population around the lake.

A review of polar coordinates is in Appendix A.

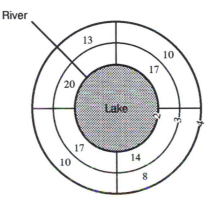

Figure 15.32: An insect infested lake

Solution To get our estimate we will multiply the population density in each sector by the area of that sector. Unlike the rectangles in a rectangular grid, the sectors in a polar grid do not all have the same area. The inner sectors have area

$$\frac{1}{4}(\pi 3^2 - \pi 2^2) = \frac{5\pi}{4} \approx 3.93,$$

and the outer sectors have area

$$\frac{1}{4}(\pi 4^2 - \pi 3^2) = \frac{7\pi}{4} \approx 5.50,$$

so the estimated population is

$$(20)(3.93) + (17)(3.93) + (14)(3.93) + (17)(3.93) +$$
$$(13)(5.50) + (10)(5.50) + (8)(5.50) + (10)(5.50) \approx 493 \text{ million insects.}$$

What is dA In Polar Coordinates?

The subdivision in this example is the sort you get when you use a polar grid rather than a rectangular grid. A rectangular grid is constructed by putting down vertical and horizontal lines; these correspond to taking $x = k$ (a constant) and $y = l$ (another constant), respectively. In polar coordinates, setting $r = k$ gives you a circle around the origin (of radius k), and setting $\theta = k$ gives you a ray emanating from the origin (at angle k with the x-axis). A polar grid is built up out of these circles and rays, just as a rectangular grid is built up out of horizontal and vertical lines.

Figure 15.33 shows a subdivision of the polar region $a \leq r \leq b$, $\alpha \leq \theta \leq \beta$, using n subdivisions each way. This is the sort of region that is naturally represented in polar coordinates; a sort of circular rectangle bent around the origin.

In general, if you divide R as shown in Figure 15.33, you will get a Riemann sum similar to the ones coming from rectangular grids:

$$\sum_{i,j} f(r_i, \theta_j) \, \Delta A.$$

However, calculating the area ΔA is more complicated. Figure 15.34 shows ΔA. For Δr, $\Delta \theta$ small, the shaded region is approximately rectangular, with sides $r\Delta\theta$ and Δr, so

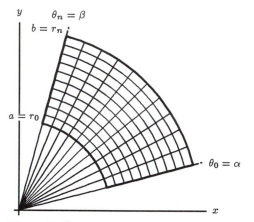

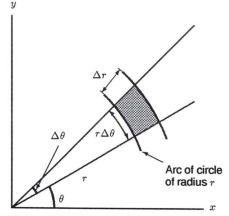

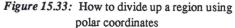

Figure 15.33: How to divide up a region using polar coordinates

Figure 15.34: Area ΔA in polar coordinates

$$\Delta A \approx r\Delta\theta\Delta r.$$

Thus, the Riemann sum is approximately

$$\sum_{i,j} f(r_i, \theta_j) \, r_i \Delta\theta \Delta r.$$

If we take the limit as Δr and $\Delta \theta$ approach 0, we end up with

$$\int_R f \, dA = \int_\alpha^\beta \int_a^b f(r, \theta) \, r \, dr \, d\theta.$$

When computing integrals in polar coordinates, put $dA = r \, dr \, d\theta$ or $dA = r \, d\theta \, dr$.

Example 2   Compute the integral of $f(x, y) = \frac{1}{(x^2 + y^2)^{3/2}}$ over the region R shown in Figure 15.35.

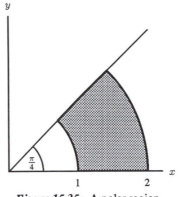

Figure 15.35: A polar region

Solution The region can be described in polar coordinates by the inequalities $1 \leq r \leq 2, 0 \leq \theta \leq \pi/4$. In polar coordinates, since $r = \sqrt{x^2 + y^2}$, we can write f as

$$f(r, \theta) = \frac{1}{r^3},$$

Then

$$\int_R f \, dA = \int_0^{\pi/4} \int_1^2 \frac{1}{r^3} r \, dr \, d\theta = \int_0^{\pi/4} \left(\int_1^2 r^{-2} \, dr \right) d\theta$$

$$= \int_0^{\pi/4} \left[-\frac{1}{r} \right]_{r=1}^{r=2} dr = \int_0^{\pi/4} \frac{1}{2} \, d\theta = \frac{\pi}{8}.$$

Example 3 On the basis of the shape of each region in Figure 15.36, decide whether to integrate using polar or Cartesian coordinates. Set up the integral of an arbitrary function $z = f(x, y)$ over the region.

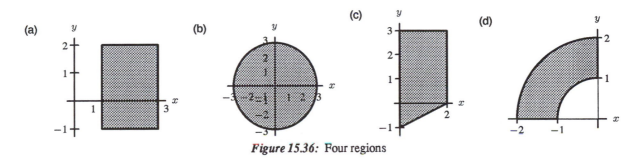

Figure 15.36: Four regions

Solution (a) Since this is a rectangular region, Cartesian coordinates are the right choice. The rectangle is described by the inequalities $1 \leq x \leq 3$ and $-1 \leq y \leq 2$, so the integral is

$$\int_{-1}^2 \int_1^3 f(x, y) \, dx \, dy.$$

(b)   A circle is best described in polar coordinates. The radius is 3, so r goes from 0 to 3, and the range of θ is 0 to 2π to describe the whole circle, so the integral is

$$\int_0^{2\pi} \int_0^3 f(r\cos\theta, r\sin\theta)\, r\, dr\, d\theta.$$

(c)   The bottom boundary of this trapezoid is the line $y = (x/2) - 1$, and the top is the line $y = 3$, so use Cartesian coordinates. If we integrate with respect to y first we get $(x/2) - 1$ for the lower limit of the integral and 3 for the upper limit. After that we integrate with respect to x from $x = 0$ to $x = 2$. So the integral is

$$\int_0^2 \int_{(x/2)-1}^3 f(x,y)\, dy\, dx.$$

(d)   This is another polar region: it is a piece of a ring with inner radius 1 and outer radius 2, so r goes from 1 to 2. Its position in the second quadrant shows that θ goes from $\pi/2$ to π. The integral is

$$\int_{\pi/2}^{\pi} \int_1^2 f(r\cos\theta, r\sin\theta)\, r\, dr\, d\theta.$$

Problems for Section 15.5

For each of the regions R in Problems 1-4, set up $\displaystyle\int_R f\, dA$ as an iterated integral in polar coordinates.

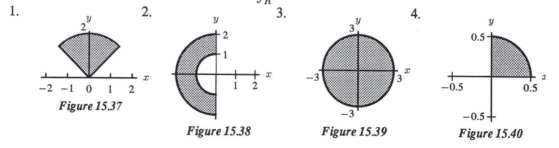

1.

Figure 15.37

2.

Figure 15.38

3.

Figure 15.39

4.

Figure 15.40

Sketch the region over which the integrals in Problems 5–11 are computed.

5. $\displaystyle\int_0^{2\pi} \int_1^2 f(r,\theta) r\, dr\, d\theta.$

6. $\displaystyle\int_{\pi/2}^{\pi} \int_0^1 f(r,\theta) r\, dr\, d\theta.$

7. $\displaystyle\int_{\pi/6}^{\pi/3} \int_0^1 f(r,\theta) r\, dr\, d\theta.$

8. $\displaystyle\int_3^4 \int_{3\pi/4}^{3\pi/2} f(r,\theta) r\, d\theta\, dr.$

9. $\displaystyle\int_0^{\pi/4} \int_0^{1/\cos\theta} f(r,\theta) r\, dr\, d\theta.$

10. $\displaystyle\int_{\pi/4}^{\pi/2} \int_0^{2/\sin\theta} f(r,\theta) r\, dr\, d\theta.$

11. $\displaystyle\int_0^4 \int_{-\pi/2}^{\pi/2} f(r,\theta) r\, d\theta\, dr.$

1. $\int_{\pi/4}^{\frac{3\pi}{4}} \int_0^2 f \, r \, dr \, d\theta$ 3. $\int_0^{2\pi} \int_0^3 f \, r \, dr \, d\theta$ 5.   7.   15.5

1-21 odd

9.   11.   13. $\int_R (x^2 - y^2) \, dA = \int_0^{\pi/4} \int_1^2 r^2 (\cos^2\theta - \sin^2\theta) \, r \, dr \, d\theta$

$= \left(\text{using DERIVE} \right) = 0$

(?)

15. $\int_{-1}^0 \int_{-\sqrt{1-x^2}}^{\sqrt{1-x^2}} f(x,y) \, dy \, dx$ $y = \sqrt{1-x^2}$

$\int_{\pi/2}^{3\pi/2} \int_0^1 (r\cos\theta, r\sin\theta) \, r \, dr \, d\theta$ $y^2 = 1 - x^2$

$y^2 + x^2 = 1$

17. $\int_0^3 \int_{-x}^x f(x,y) \, dy \, dx$ 19. $z = \sqrt{8 - x^2 - y^2}$

$\int_{-\pi/4}^{\pi/4} \int_0^{3/\cos\theta} (r\cos\theta, r\sin\theta) \, r \, dr \, d\theta$

$z = \sqrt{x^2 + y^2}$

$V =$

Evaluate the integral in Problems 12–13 over the region indicated.

12. $\int_R \sin(x^2 + y^2)\, dA$, where R is the disc of radius 2 centered at the origin.

13. $\int_R (x^2 - y^2)\, dA$, where R is the first quadrant region between the circles of radius 1 and radius 2.

14. Consider the integral $\int_0^3 \int_{x/3}^1 f(x, y)\, dy\, dx$.

   (a) Sketch the region R over which the integration is being performed.
   (b) Rewrite the integral with the order of integration reversed.
   (c) Rewrite the integral in polar coordinates.

Change the integrals in Problems 15–17 to polar coordinates.

15. $\int_{-1}^0 \int_{-\sqrt{1-x^2}}^{\sqrt{1-x^2}} f(x, y)\, dy\, dx$

16. $\int_0^{\sqrt{2}} \int_y^{\sqrt{4-y^2}} f(x, y)\, dx\, dy$

17. $\int_0^3 \int_{-x}^x f(x, y)\, dy\, dx$

18. Find the volume of the region between the graph of $f(x, y) = 25 - x^2 - y^2$ and the xy plane.

19. An ice cream cone can be modeled by the region bounded by the hemisphere $z = \sqrt{8 - x^2 - y^2}$ and the cone $z = \sqrt{x^2 + y^2}$. Find its volume.

20. A disk of radius 5 cm has density 10 g/cm^2 at its center, has density 0 at its edge, and its density is a linear function of the distance from the center. Find the mass of the disk.

21. A city by the ocean surrounds a bay, and has the shape shown in Figure 15.41.

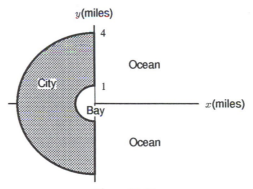

Figure 15.41

   (a) The population density of the city (in thousands of people per square mile) is given by the function $\delta(r, \theta)$, where r and θ are the usual polar coordinates with respect to the x and y axes indicated in Figure 15.41, and the distances indicated on the y-axis are in miles. Set up a double integral in polar coordinates that would give the total population of the city.
   (b) The population density decreases as you move away from the shoreline of the bay, and also decreases the further you have to drive to get to the ocean. Which of the following functions best describes this situation?
      (i) $\delta(r, \theta) = (4 - r)(2 + \cos\theta)$

(ii) $\delta(r, \theta) = (4 - r)(2 + \sin \theta)$

(iii) $\delta(r, \theta) = (r + 4)(2 + \cos \theta)$

(c)   Evaluate the integral you set up in (a) with the function you chose in (b), and give the resulting estimate for the population.

15.6   INTEGRALS IN CYLINDRICAL AND SPHERICAL COORDINATES

In our work on two-variable integrals we saw that it was easier to evaluate some integrals using polar, rather than Cartesian coordinates. Similarly, in trying to evaluate three-variable integrals, there are instances when coordinates other than rectangular ones will make the evaluation of the integral much easier.

Cylindrical Coordinates

The cylindrical coordinates of a point (x, y, z) in 3-space are obtained by representing the x and y coordinates in polar coordinates and letting the z coordinate be the z coordinate of the Cartesian coordinate system. (See Figure 15.42.)

Cartesian and cylindrical coordinates are related as follows:

$$x = r \cos \theta,$$
$$y = r \sin \theta,$$
$$z = z.$$

As with polar coordinates in the plane, note that $x^2 + y^2 = r^2$.

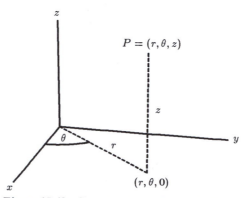

Figure 15.42: Cylindrical coordinates: (r, θ, z)

A useful way to visualize how cylindrical coordinates work is to sketch the fundamental surfaces obtained by setting one of the coordinates equal to a constant. See Figures 15.43–15.45.

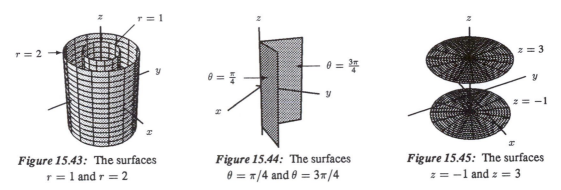

Figure 15.43: The surfaces
$r = 1$ and $r = 2$

Figure 15.44: The surfaces
$\theta = \pi/4$ and $\theta = 3\pi/4$

Figure 15.45: The surfaces
$z = -1$ and $z = 3$

Setting $r =$ constant gives a cylinder around the z-axis whose radius is equal to the constant; setting $\theta =$ constant gives a half-plane perpendicular to the xy plane, with one edge along the z-axis, making an angle θ with the x-axis. Finally, setting $z =$ constant gives a horizontal plane z units above the xy plane.

The sorts of regions that can easily be described in cylindrical coordinates are the regions whose boundaries are formed by parts of these fundamental surfaces. (For example, vertical cylinders, or wedge shaped parts of vertical cylinders.)

Example 1   Describe in cylindrical coordinates a wedge of cheese cut from a cylinder of cheese 4 cm high and 6 cm in radius if this wedge subtends an angle of $\pi/6$ at the center.

Solution   If we put the wedge in a coordinate system as shown in Figure 15.46, then it is described by the inequalities $0 \le r \le 6$, and $0 \le z \le 4$, and $0 \le \theta \le \pi/6$.

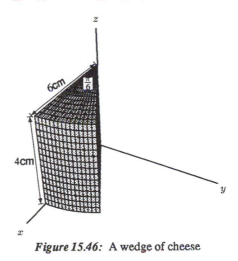

Figure 15.46: A wedge of cheese

Integration in Cylindrical Coordinates

When we integrated in polar coordinates, we had to express the area element dA in terms of polar coordinates: $dA = r \, dr \, d\theta$. To evaluate a triple integral $\int_W f \, dV$ in cylindrical coordinates, we need to express the volume element dV in cylindrical coordinates.

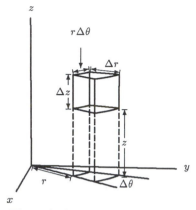

Figure 15.47: Volume element in
cylindrical coordinates

Consider the element of volume ΔV as shown in Figure 15.47. It is bounded by the fundamental surfaces described above; hence, it is a piece of a wedge out of a cylinder.

As we saw in the section on polar coordinates, the area of the base of the volume element is given approximately by $r\Delta r\Delta\theta$. Since the height is Δz, the element of volume is given approximately by $\Delta V \approx r\Delta r\Delta\theta\Delta z$.

When computing integrals in cylindrical coordinates, put $dV = r\,dr\,d\theta\,dz$. Other orders of integration are also possible.

Example 2   Find the mass of the wedge of cheese in Example 1, if its density is 0.8 grams/cm^3.

Solution   Call the wedge W. Then its mass is

$$\int_W 0.8\,dV.$$

In cylindrical coordinates this integral is

$$\int_0^4 \int_0^{\pi/6} \int_0^6 0.8\,r\,dr\,d\theta\,dz = \int_0^4 \int_0^{\pi/6} 0.4r^2\Big|_0^6 d\theta\,dz = 14.4 \int_0^4 \int_0^{\pi/6} d\theta\,dz$$

$$= 14.4(\frac{\pi}{6}) \int_0^4 dz = 14.4(\frac{\pi}{6})4 \approx 30.16\,\text{grams}.$$

Example 3   A water tank in the shape of a hemisphere has radius a; its base is its plane face. Find the volume, V, of water in the tank as a function of h, the depth of the water.

Solution   In Cartesian coordinates a sphere of radius a has the equation $x^2 + y^2 + z^2 = a^2$. Since $r^2 = x^2 + y^2$, this becomes

$$r^2 + z^2 = a^2$$

in cylindrical coordinates. Thus, if we want to describe the amount of water in the tank in cylindrical coordinates, we let r go from 0 to $\sqrt{a^2 - z^2}$, we let θ go from 0 to 2π, and we let z go from 0 to h. So its volume is

$$\int_W dV = \int_0^{2\pi} \int_0^h \int_0^{\sqrt{a^2-z^2}} r \, dr \, dz \, d\theta = \int_0^{2\pi} \int_0^h \frac{r^2}{2} \Big|_{r=0}^{r=\sqrt{a^2-z^2}} dz \, d\theta$$

$$= \int_0^{2\pi} \int_0^h \frac{1}{2}(a^2 - z^2) \, dz \, d\theta = \int_0^{2\pi} \frac{1}{2}\left(a^2 z - \frac{z^3}{3}\right) \Big|_{z=0}^{z=h} d\theta$$

$$= \int_0^{2\pi} \frac{1}{2}\left(a^2 h - \frac{h^3}{3}\right) d\theta = \pi\left(a^2 h - \frac{h^3}{3}\right).$$

Spherical Coordinates

We define another 3-dimensional system of coordinates called spherical coordinates as follows. In Figure 15.48, the point P has coordinates (x, y, z) in the rectangular coordinate system. We define spherical coordinates ρ, ϕ, and θ for P as follows: $\rho = \sqrt{x^2 + y^2 + z^2}$ is the distance of P from the origin; ϕ is the angle between the positive z axis and the line through the origin and the point P; and θ is the same as in cylindrical coordinates.

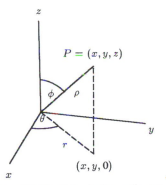

Figure 15.48: Spherical coordinates

In cylindrical coordinates,

$$x = r\cos\theta,$$
$$y = r\sin\theta,$$
$$z = z,$$

but from the figure we have $z = \rho\cos\phi$ and $r = \rho\sin\phi$, so that we have the following:

Relation between Cartesian and Spherical coordinates:

$$x = \rho\sin\phi\cos\theta,$$
$$y = \rho\sin\phi\sin\theta,$$
$$z = \rho\cos\phi.$$

Also, as noted above, $\rho^2 = x^2 + y^2 + z^2$.

Thus, each point in 3-space is represented using spherical coordinates where $0 \le \rho \le \infty, 0 \le \phi \le \pi$, and $0 \le \theta \le 2\pi$. As the name indicates, this system of coordinates is useful when there is spherical symmetry with respect to the origin, either in the region of integration or in the integrand.

You might note that the angles θ and ϕ in spherical coordinates are similar to those of longitude and latitude on the globe. The angle θ is the longitude measured east of Greenwich. The angle $\frac{\pi}{2} - \phi$ is what map-makers call the latitude, the angle measured from the equator, while ϕ itself would be measured from the north pole of the globe.

The fundamental surfaces in spherical coordinates are $\rho = k$ (a constant), which is a sphere of radius k centered at the origin, $\theta = k$ (a constant), which is the half-plane with its edge along the z-axis, and $\phi = k$ (a constant), which is a cone if $k \ne \pi/2$ and the xy-plane if $k = \frac{\pi}{2}$. (See Figures 15.49-15.51.)

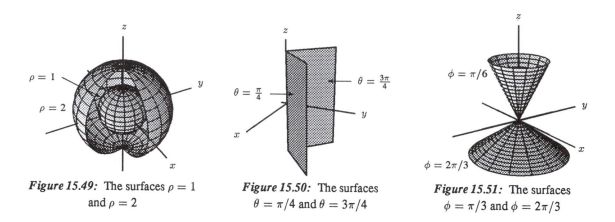

Figure 15.49: The surfaces $\rho = 1$ and $\rho = 2$

Figure 15.50: The surfaces $\theta = \pi/4$ and $\theta = 3\pi/4$

Figure 15.51: The surfaces $\phi = \pi/3$ and $\phi = 2\pi/3$

Integration in Spherical Coordinates

To use spherical coordinates in triple integrals we need to determine the volume element for $\int_W f \, dV$ where W is a region in three space and f is a function defined on W. From Figure 15.52, we see that the volume element can be approximated by a cube. One edge has length $\Delta\rho$. The edge parallel to the xy-plane is an arc of a circle made from rotating the cylindrical radius r ($= \rho \sin \phi$) through an angle $\Delta\theta$, and so has length $\rho \sin \phi \Delta\theta$. The remaining side comes from rotating the radius ρ through an angle $\Delta\phi$, and so has length $\rho\Delta\phi$. Therefore,

$$\Delta V \approx \Delta\rho(\rho\Delta\phi)(\rho \sin \phi \Delta\theta) = \rho^2 \sin \phi \Delta\rho\Delta\phi\Delta\theta$$

Thus,

When computing integrals in spherical coordinates, put $dV = \rho^2 \sin \phi \, d\rho \, d\phi \, d\theta$. Other orders of integration are also possible.

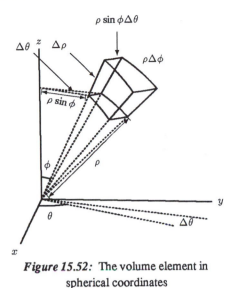

Figure 15.52: The volume element in spherical coordinates

Example 4   Use spherical coordinates to derive the formula for the volume of a solid sphere of radius a.

Solution   In spherical coordinates, a sphere of radius a is described by the inequalities $0 \leq \rho \leq a$, $0 \leq \theta \leq 2\pi$, and $0 \leq \phi \leq \pi$. Note that θ goes all the way around the circle (just like longitude), whereas ϕ only goes from 0 to π. We can find the volume by integrating the constant density function 1 over the sphere:

$$W = \int_R dV = \int_0^{2\pi} \int_0^\pi \int_0^a \rho^2 \sin \phi \, d\rho \, d\phi \, d\theta = \int_0^{2\pi} \int_0^\pi \frac{1}{3} a^3 \sin \phi \, d\phi \, d\theta$$

$$= \frac{1}{3} a^3 \int_0^{2\pi} -\cos \phi \Big|_0^\pi \, d\theta = \frac{2}{3} a^3 \int_0^{2\pi} d\theta = \frac{4\pi a^3}{3}.$$

Example 5   Find the gravitational force exerted by a solid hemisphere of radius a and constant density δ on a unit mass located at the center of the base of the hemisphere.

Solution   Assume the base of the hemisphere rests on the xy-plane with center at the origin. Because the hemisphere is symmetric about the z-axis, the force on the mass in the horizontal x or y direction is 0; the horizontal force exerted on the unit mass by the part of the hemisphere at (x, y, z) cancels with that at $(-x, -y, z)$. Thus, we need only compute the vertical z component of the force. If we use spherical coordinates, a piece of the hemisphere of volume dV located at (ρ, θ, ϕ) exerts on the unit mass a force of magnitude $G(\delta dV)/\rho^2$, since the force is $G \cdot \text{mass}/(\text{distance})^2$. The vertical component of this force is

$$(G(\delta dV)/\rho^2) \cos \phi.$$

Adding up all the contributions of all the small pieces of volume, we get a vertical force of

$$F = \int_0^{2\pi} \int_0^{\pi/2} \int_0^a (G\delta/\rho^2)(\cos \phi)\rho^2 \sin \phi \, d\rho d\phi d\theta = \int_0^{2\pi} \int_0^{\pi/2} G\delta(\cos \phi \sin \phi)\rho \Big|_{\rho=0}^{\rho=a} d\phi d\theta$$

$$= \int_0^{2\pi} \int_0^{\pi/2} G\delta a \cos\phi \sin\phi \, d\phi d\theta = \int_0^{2\pi} G\delta a \left(-\frac{(\cos\phi)^2}{2} \right) \Bigg|_{\phi=0}^{\phi=\pi/2} d\theta$$

$$= \int_0^{2\pi} G\delta a (\tfrac{1}{2}) \, d\theta = G\delta a\pi.$$

So far, the shape of the region of integration that has told us what coordinates to would be most convenient. Sometimes, however, it is the integrand that dictates what coordinates to use.

Example 6   Evaluate $\int_W \sqrt{x^2 + y^2 + z^2} \, dV$ where W is the region bounded by the cone $z = \sqrt{4x^2 + 4y^2}$ and the plane $z = 3$.

Solution   The cone and the horizontal plane suggest cylindrical coordinates. In cylindrical coordinates, the cone $z = \sqrt{4x^2 + 4y^2}$ becomes $z = 2r$. The region of integration in the xy-plane is the circular region given by the intersection of the cone and plane $z = 3$,

$$\sqrt{4x^2 + 4y^2} = 3,$$

which is a circle of radius $3/2$. Thus, in cylindrical coordinates the integral is

$$\int_W \sqrt{x^2 + y^2 + z^2} \, dV = \int_0^{2\pi} \int_0^{3/2} \int_{2r}^{3} \sqrt{r^2 + z^2} \, r \, dz dr d\theta$$

Unfortunately, integrating $\sqrt{r^2 + z^2}$ with respect to z is not easy. Let's try spherical coordinates instead because the integrand $\sqrt{x^2 + y^2 + z^2}$ is just ρ. The equation of the cone in spherical coordinates is $\phi = \arctan(1/2)$; the equation of the plane $z = 3$ is $\rho\cos\phi = 3$ or $\rho = 3/\cos\phi$. In spherical coordinates, the inner integral is usually with respect to ρ. To find the limits of integration, imagine a ray from the origin along which ρ can vary, where the ray first enters the region of integration we get the lower limit, and where it leaves the region we get the upper limit. In this case, the cone contains the region so the lower limit is 0 and upper limit is the plane $z = 3$, or in spherical coordinates, $\rho = 3/\cos\phi$. Thus, the integral is

$$\int_W \sqrt{x^2 + y^2 + z^2} dV = \int_0^{2\pi} \int_0^{\arctan(1/2)} \int_0^{3/\cos\phi} \rho\rho^2 \sin\phi \, d\rho d\phi d\theta$$

$$= \int_0^{2\pi} \int_0^{\arctan(1/2)} \left(\frac{\rho^4}{4} \Big|_0^{3/\cos\phi} \right) \sin\phi \, d\phi d\theta$$

$$= \int_0^{2\pi} \int_0^{\arctan(1/2)} \frac{3^4}{4\cos^4\phi} \sin\phi \, d\phi d\theta$$

$$= \int_0^{2\pi} \frac{3^4}{4} \frac{1}{3} \frac{1}{\cos^3\phi} \Big|_{\phi=0}^{\phi=\arctan(1/2)} d\theta = \int_0^{2\pi} \frac{3^4}{4} \frac{1}{3} \left(\frac{1}{(2/\sqrt5)^3} - \frac{1}{1} \right) d\theta$$

$$= 2\pi \frac{27}{4} \left(\frac{5\sqrt5}{8} - 1 \right)$$

1. Set Up: $\int_{-1}^{-1} \int_{\pi/4}^{3\pi} \int_0^4 (r^2 + z^2) r\, dr\, d\theta\, dz$ (Use Derive)

3.

$$F(x,y,z) = \frac{1}{\sqrt{x^2 + y^2 + z^2}}$$

$$\int_0^5 \int_0^{2\pi} \int_{\pi/2}^{\pi} \frac{1}{\rho}\, \rho \sin^2 \phi\, d\phi\, d\theta\, d\rho$$

(Use Derive)

5. Cylindrical

$$\int_0^1 \int_0^{2\pi} \int_0^4 \delta \cdot r\, dr\, d\theta\, dz$$

7. Spherical

$$\int_0^{\pi/3} \int_0^{2\pi} \int_0^3 \delta \cdot \rho^2 \sin\phi\, d\rho\, d\theta\, d\phi$$

9. Rectangular

$$\int_0^3 \int_0^1 \int_0^5 \delta \cdot dz\, dy\, dx$$

11. $\int_0^1 \int_{-\sqrt{1-x^2}}^{\sqrt{1-x^2}} \int_{-\sqrt{1-x^2-z^2}}^{\sqrt{1-x^2-z^2}} \left[\frac{1}{(x^2+y^2+z^2)} \right] dy\, dz\, dx$ (Use Derive)

13.

$$\int_{\pi/6}^{\pi/3} \int_0^2 \int_0^5 r\, dr\, dz\, d\theta$$

15. $\int_z D\, dV \Rightarrow dV = \rho^2 \sin\phi\, d\rho\, d\theta\, d\phi$

$$= \int_0^{\pi} \int_0^{2\pi} \int_0^3 (3-\rho)\rho^2 \sin\phi\, d\rho\, d\theta\, d\phi$$

$$=$$

(?)

17. Avg dist. =

(?)

19.

Problems for Section 15.6

In Problems 1 and 2, evaluate the indicated three-variable integrals in cylindrical coordinates.

1. $f(x, y, z) = x^2 + y^2 + z^2$, W is the region $0 \le r \le 4$, $\pi/4 \le \theta \le 3\pi/4$, $-1 \le z \le 1$.

2. $f(x, y, z) = \sin(x^2 + y^2)$, W is the solid cylinder with height 4 and with base of radius 1 centered on the z axis at $z = -1$.

In Problems 3 and 4, evaluate the indicated three-variable integrals in spherical coordinates.

3. $f(x, y, z) = 1/(x^2 + y^2 + z^2)^{1/2}$ over the bottom half of the sphere of radius 5 centered at the origin.

4. $f(\rho, \theta, \phi) = \sin\phi$, over the region $0 \le \theta \le 2\pi$, $0 \le \phi \le \pi/4$, $1 \le \rho \le 2$.

For Problems 5–9, choose a set of coordinate axes, and then set up the three-variable integral in an appropriate coordinate system for integrating a density function δ over the given region.

5.

6.

7. A piece of a sphere; angle at the center is $\pi/3$.

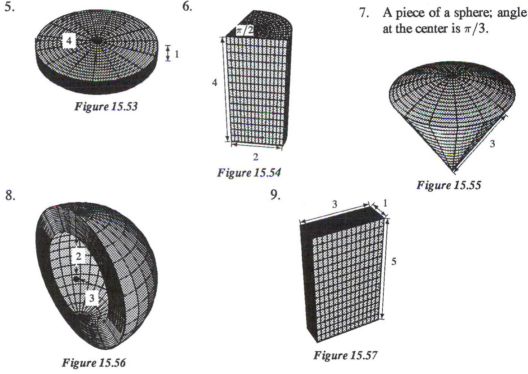

Figure 15.53

Figure 15.54

Figure 15.55

8.

9.

Figure 15.56

Figure 15.57

10. Sketch the region R over which the integration is being performed:

$$\int_0^{\pi/2} \int_{\pi/2}^{\pi} \int_0^1 f(\rho, \phi, \theta)\rho^2 \sin\phi \, d\rho \, d\phi \, d\theta.$$

11. Evaluate $\displaystyle\int_0^1 \int_{-\sqrt{1-x^2}}^{\sqrt{1-x^2}} \int_{-\sqrt{1-x^2-z^2}}^{\sqrt{1-x^2-z^2}} \frac{1}{(x^2 + y^2 + z^2)^{1/2}} \, dy \, dz \, dx$.

12. Evaluate $\int_0^1 \int_{-1}^1 \int_{-\sqrt{1-x^2}}^{\sqrt{1-x^2}} \frac{1}{(x^2+y^2)^{1/2}} \, dy \, dx \, dz$.

13. Write a triple integral representing the volume of a slice of the cylindrical cake of height 2 and radius 5 between the planes $\theta = \pi/6$ and $\theta = \pi/3$. Evaluate this integral.

14. Find the mass M of the solid region R given by

$$R = \{(\rho, \phi, \theta) : 0 \le \rho \le 3, 0 \le \theta < 2\pi, 0 \le \phi \le \pi/4\}$$

if the density at any point P is given by $\delta(P) = |P| =$ distance of P from the origin.

15. A particular spherical cloud of gas of radius 3 km is more dense at the center than towards the edge. The density, D, of the gas at a distance ρ km from the center is given by $D(\rho) = 3 - \rho$. Write an integral representing the total mass of the cloud of gas, and evaluate it.

16. Find the volume that remains after a cylindrical hole of radius R is bored through a sphere of radius a, $0 < R < a$, passing through the center of the sphere along the pole.

17. Use proper coordinates to find the average distance to the origin for points in the ice cream cone region bounded by the hemisphere $z = \sqrt{8 - x^2 - y^2}$ and the cone $z = \sqrt{x^2 + y^2}$. (Hint: The volume of this region is computed in Problem 19 on page 257.)

18. Compute the force of gravity exerted by a solid cylinder of radius R, height H, and density δ on a unit mass at the center of the base of the cylinder.

19. Let V_1 be a cylinder of height 4, radius 1, and with the origin at the center of its base. Let V_2 be the top half of the sphere of radius 5 centered at the origin. Calculate the integrals of both $f(x, y, z) = x^2 + y^2 + z^2$ and $g(x, y, z) = z$ over V_1 and over V_2.

20. Communication satellites are placed in geosynchronous orbits about the Earth — orbits of the correct altitude and velocity so that they appear to remain fixed above a particular point on the surface.

    (a) Suppose such a satellite is in orbit 4000 miles above Columbus, Ohio which is located at $40°$ latitude (north of the equator) and $83°$ longitude (west of Greenwich). Use the fact that the Earth's radius is 4000 miles to find the spherical and the rectangular coordinates of the satellite.

    (b) Suppose you live near New York City where the latitude and longitude are $41°$ north and $74°$ west and are installing a TV satellite disk. Find a unit vector giving the direction from your receiver toward the satellite. What is the distance from the disk to the satellite?

    (c) Repeat part (b) supposing you live in Los Angeles where the latitude and longitude are $34°$ north and $118°$ west.

15.7 NOTES ON CHANGE OF VARIABLES IN A DOUBLE OR TRIPLE INTEGRAL

In the previous sections, we have used some particular changes of coordinates to simplify iterated integrals. In this section, we discuss more general changes of variable. In the process, we will see where the extra factor of r comes from when we change from Cartesian to polar coordinates or the $\rho^2 \sin \phi$ when we change from Cartesian to spherical coordinates.

Polar Change of Variables Revisited

Suppose you have to evaluate the integral $\int_R (x + y)\, dA$ where R is the region in the first quadrant bounded by the circle $x^2 + y^2 = 9$ and the x and y-axes. Writing the integral in Cartesian and polar coordinates we have

$$\int_R (x + y)\, dA = \int_0^3 \int_0^{\sqrt{9 - x^2}} (x + y)\, dy dx = \int_0^{\pi/2} \int_0^3 (r \cos \theta + r \sin \theta) r\, dr d\theta$$

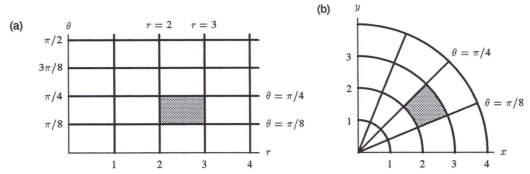

Figure 15.58: The r, θ plane with a grid and the corresponding curved grid in the x, y-plane

To convert an integral from Cartesian to polar coordinates we make three changes:

1. Substituted $x = r \cos \theta$, $y = r \sin \theta$ in the integrand,

2. Changed the limits of integration,

3. Replaced dA by $r \, dr \, d\theta$.

Any other change of variable will have three similar steps. Before going on, we will look again at the factor r in the area element $r \, dr \, d\theta$. Consider an r, θ coordinate plane with an r-axis and θ-axis at right angles just like the usual x and y-axes (See Figure 15.58). The vertical lines, say, $r = 1$, $r = 2$, $r = 3, \ldots$ and the horizontal lines, say, $\theta = \pi/8$, $2\pi/8$, $3\pi/8, \ldots$ provide a rectangular grid. For the polar coordinates, the lines $r = 1$, $r = 2$, $r = 3 \ldots$ in the $r\theta$-plane correspond to circles of radii $1, 2, 3, \ldots$ in the xy-plane, and the lines $\theta = \pi/8$, $2\pi/8$, $3\pi/8 \ldots$ in the $r\theta$-plane correspond to rays emanating from the origin. As a result, a typical rectangle (shaded) in the $r\theta$-plane with sides of length Δr and $\Delta \theta$, corresponds to a curved rectangle in the xy-plane with sides of length Δr and $r \Delta \theta$. The extra r is needed because the correspondence between r, θ and x, y not only curves the lines $r = 1, 2, 3 \ldots$ into circles, it also stretches those lines around larger and larger circles.

General Change of Variables

We now consider a general change of variable, where x, y coordinates are related to some other u, v coordinates by the equations

$$x = x(u, v) \quad y = y(u, v).$$

Just as circular regions in the xy-plane correspond to rectangular regions in the $r\theta$-plane, we assume a curvy region T in the xy-plane corresponds to a rectangular region R in the uv-plane. We want to investigate what happens when we divide R into small rectangles $R_{i,j}$. In the Figure 15.59 is a typical rectangle $R_{i,j}$ from the grid with sides of length Δu and Δv and lower left corner at (u, v).

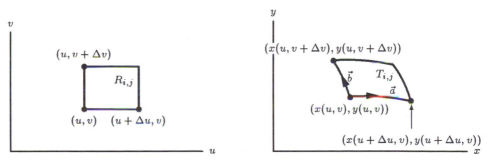

Figure 15.59: A small rectangle $R_{i,j}$ in the uv-plane and the corresponding region $T_{i,j}$ of the xy-plane

The corresponding piece $T_{i,j}$ of the xy-plane is a quadrilateral with curved sides. If we choose Δu and Δv very small, by local linearity these sides will be straight and parallel. We will not, however, get a rectangle but rather a parallelogram. Moreover, the edges will no longer have length Δu and Δv. Most of all, the area will not be $\Delta u \Delta v$.

Recall from Chapter 12 that the area of the parallelogram with sides $\vec{a}$ and $\vec{b}$ is $\|\vec{a} \times \vec{b}\|$. Thus, we need to find the sides of $T_{i,j}$ as vectors. The side of $T_{i,j}$ corresponding to the bottom side of $R_{i,j}$ has end points $(x(u,v), y(u,v))$ and $(x(u+\Delta u, v), y(u+\Delta u, v))$, so in vector form that side is

$$\vec{a} = (x(u+\Delta u, v) - x(u,v))\vec{i} + (y(u+\Delta u, v) - y(u,v))\vec{j} + 0\vec{k} \approx \left(\frac{\partial x}{\partial u}\Delta u\right)\vec{i} + \left(\frac{\partial y}{\partial u}\Delta u\right)\vec{j} + 0\vec{k}.$$

Similarly, the side of $T_{i,j}$ corresponding to the left edge of $R_{i,j}$ is given by

$$\vec{b} \approx \left(\frac{\partial x}{\partial v}\Delta v\right)\vec{i} + \left(\frac{\partial y}{\partial v}\Delta v\right)\vec{j} + 0\vec{k}.$$

Computing the cross product, we get

$$\text{Area } T_{i,j} \approx \|\vec{a} \times \vec{b}\| \approx \left|\left(\frac{\partial x}{\partial u}\Delta u\right)\left(\frac{\partial y}{\partial v}\Delta v\right) - \left(\frac{\partial x}{\partial v}\Delta v\right)\left(\frac{\partial y}{\partial u}\Delta u\right)\right|.$$

Using determinant notation, we write

$$\left(\frac{\partial x}{\partial u}\Delta u\right)\left(\frac{\partial y}{\partial v}\Delta v\right) - \left(\frac{\partial x}{\partial v}\Delta v\right)\left(\frac{\partial y}{\partial u}\Delta u\right) = \begin{vmatrix} \frac{\partial x}{\partial u} & \frac{\partial y}{\partial u} \\ \frac{\partial x}{\partial v} & \frac{\partial y}{\partial v} \end{vmatrix}\Delta u \Delta v$$

This determinant involving the partial of x and y with respect to u and v is called the *Jacobian* and is denoted $\dfrac{\partial(x,y)}{\partial(u,v)}$.

Thus, we say

$$\text{Area } T_{i,j} \approx \begin{vmatrix} \frac{\partial x}{\partial u} & \frac{\partial y}{\partial u} \\ \frac{\partial x}{\partial v} & \frac{\partial y}{\partial v} \end{vmatrix}\Delta u \Delta v.$$

Now to compute $\int_T f(x,y)\, dA$, we look at the Riemann sum obtained by dividing the region T into the little curved regions $T_{i,j}$

$$\int_T f(x,y)\, dA \approx \sum_{i,j} f(x_i, y_j) \cdot (\text{Area of } T_{i,j})$$

$$\approx \sum_{i,j} f(x_i, y_j)\left|\frac{\partial(x,y)}{\partial(u,v)}\right|\Delta u \Delta v.$$

Each point (x_i, y_j) corresponds to a point (u_i, v_j) so the sum can be written in terms of u and v

$$\sum_{i,j} f(x(u_i, v_j), y(u_i, v_j))\left|\frac{\partial(x,y)}{\partial(u,v)}\right|\Delta u \Delta v.$$

This is a Riemann sum in terms of u and v and hence as Δu and Δv approach 0, we get

$$\int_T f(x,y)\, dA = \int_R f(x(u,v), y(u,v))\left|\frac{\partial(x,y)}{\partial(u,v)}\right| du\, dv.$$

To convert an integral from Cartesian to u, v coordinates we have made three changes:
1. Substituted for x and y in the integrand in terms of u and v.
2. Changed the xy region T into a uv region R.
3. Introduced the Jacobian $\left| \dfrac{\partial(x, y)}{\partial(u, v)} \right|$.

Example 1   Verify that the Jacobian $\dfrac{\partial(x, y)}{\partial(u, v)}$ for the polar coordinate change $x = r\cos\theta$, $y = r\sin\theta$ is r.

Solution

$$\frac{\partial(x, y)}{\partial(r, \theta)} = \begin{vmatrix} \frac{\partial x}{\partial r} & \frac{\partial y}{\partial r} \\ \frac{\partial x}{\partial \theta} & \frac{\partial y}{\partial \theta} \end{vmatrix} = \begin{vmatrix} \cos\theta & \sin\theta \\ -r\sin\theta & r\cos\theta \end{vmatrix} = r\cos^2\theta + r\sin^2\theta = r.$$

Example 2   Find the area of the ellipse $\dfrac{x^2}{a^2} + \dfrac{y^2}{b^2} = 1$

Solution   Let $x = au$, $y = bv$, then the ellipse $x^2/a^2 + y^2/b^2 = 1$ in the xy-plane corresponds to the circle $u^2 + v^2 = 1$ in the uv-plane. The Jacobian is $\begin{vmatrix} a & 0 \\ 0 & b \end{vmatrix} = ab$. Thus, if we let T be the ellipse in the xy-plane and R the circle in the uv-plane, we get

$$\text{Area of } xy\text{-ellipse} = \int_T 1\, dA = \int_R 1ab\, du\, dv = ab \int_R du\, dv$$
$$= ab(\text{ Area of } uv\text{-circle}) = ab\pi.$$

Change of Variables in Three-Variable Integrals

For three variable integrals there is a similar change of variables formula. If the equations

$$x = x(u, v, w), \quad y = y(u, v, w), \quad z = z(u, v, w)$$

define a change of variables from a region S in uvw-space to a region T in xyz-space and the Jacobian of this change of variables is given by

$$\frac{\partial(x, y, z)}{\partial(u, v, w)} = \begin{vmatrix} \frac{\partial x}{\partial u} & \frac{\partial x}{\partial v} & \frac{\partial x}{\partial w} \\ \frac{\partial y}{\partial u} & \frac{\partial y}{\partial v} & \frac{\partial y}{\partial w} \\ \frac{\partial z}{\partial u} & \frac{\partial z}{\partial v} & \frac{\partial z}{\partial w} \end{vmatrix}$$

then

$$\int_T f(x, y, z)\, dx\, dy\, dz = \int_S f(x(u, v, w), y(u, v, w), z(u, v, w)) \left| \frac{\partial(x, y, z)}{\partial(u, v, w)} \right| du\, dv\, dw.$$

Problem 3 at the end of this section asks you to verify that the Jacobian for the change of variables for spherical coordinates is exactly the $\rho^2 \sin \phi$ further you already use. The next example generalizes from ellipse to ellipsoids.

Example 3   Find the volume of the ellipsoid $\dfrac{x^2}{a^2} + \dfrac{y^2}{b^2} + \dfrac{z^2}{c^2} = 1$.

Solution   Let $x = au$, $y = bv$, $z = cw$. The Jacobian is easily computed to be abc. The xyz-ellipsoid corresponds to the uvw-sphere $u^2 + v^2 + w^2 = 1$. Thus, as in Example 2,

$$\text{Volume of } xyz\text{-ellipsoid} = (abc)(\text{Volume of } uvw\text{-sphere}) = abc\frac{4}{3}\pi.$$

Problems for Section 15.7

1. Find the image, T, of the region $R = \{(u, v) \mid 0 \le u \le 3, 0 \le v \le 2\}$ under the change of variables $x = 2u - 3v$, $y = u - 2v$. Check that

$$\int_T dx\, dy = \int_R \left| \frac{\partial(x, y)}{\partial(u, v)} \right| du\, dv.$$

2. Find the image, T, of the region $R = \{(u, v) \mid 0 \le u \le 2, 0 \le v \le 2\}$ under the change of variables $x = u^2$, $y = v$. Check that

$$\int_T dx\, dy = \int_R \left| \frac{\partial(x, y)}{\partial(u, v)} \right| du\, dv.$$

3. Compute the Jacobian for the change of variables:

$$x = \rho \sin \varphi \cos \theta, \quad y = \rho \sin \varphi \sin \theta, \quad z = \rho \cos \varphi.$$

4. For the change of variables $x = 3u - 4v$, $y = 5u + 2v$, show that

$$\frac{\partial(x, y)}{\partial(u, v)} \cdot \frac{\partial(u, v)}{\partial(x, y)} = 1$$

5. Use the change of variables $x = 2u + v$, $y = u - v$ to compute the integral $\int_T (x + y)\, dA$, where T is the parallelogram formed by $(0, 0)$, $(3, -3)$, $(5, -2)$, and $(2, 1)$.

6. Use the change of variables $x = \frac{1}{2}u$, $y = \frac{1}{3}v$ to compute the integral $\int_T (x^2 + y^2)\, dA$, where T is the region bounded by the curve $4x^2 + 9y^2 = 36$.

7. Use the change of variables $u = xy$, $v = xy^2$ to compute $\int_T xy^2\, dA$, where T is the region bounded by $xy = 1$, $xy = 4$, $xy^2 = 1$, $xy^2 = 4$.

1. $x = 2u - 3v$ $y = u - 2v$ $\begin{vmatrix} x_u & y_u \\ x_v & y_v \end{vmatrix} = \begin{vmatrix} 2 & 1 \\ -3 & -2 \end{vmatrix} = 4 \times 3 = 12$ 15.7

$2u = -x - 3v$ $2v = y - u$ 1-7 odd

$u = -\frac{x}{2} - \frac{3v}{2}$ $v = \frac{y}{2} - \frac{u}{2}$

3. $\dfrac{\partial(x,y,z)}{\partial(\rho,\phi,\theta)} = \begin{vmatrix} x_\rho & x_\phi & x_\theta \\ y_\rho & y_\phi & y_\theta \\ z_\rho & z_\phi & z_\theta \end{vmatrix} = \begin{vmatrix} \sin\phi\cos\theta & \rho\cos\phi\cos\theta & \rho\sin\phi\sin\theta \\ \sin\rho\sin\theta & \rho\cos\phi\sin\theta & \rho\sin\phi\cos\phi \\ \cos\phi & -\rho\sin\phi & 0 \end{vmatrix}$

$= \cos\phi \begin{vmatrix} \rho\cos\phi\cos\theta & \rho\sin\phi\sin\theta \\ \rho\cos\phi\sin\theta & \rho\sin\phi\cos\theta \end{vmatrix} + \rho\sin\phi \begin{vmatrix} \sin\rho\cos\theta & \rho\sin\phi\sin\theta \\ \sin\rho\sin\theta & \rho\sin\phi\cos\theta \end{vmatrix}$

$= \cos\phi\left(\rho^2\cos^2\theta\cos\phi\sin\phi + \rho^2\sin^2\theta\cos\phi\sin\phi\right) + \rho\sin\phi\left(\rho\sin^2\phi\cos^2\theta + \rho\sin^2\phi\sin^2\theta\right)$

$= \rho^2\cos^2\phi\sin\phi + \rho^2\sin^3\phi$

$= \rho^2\sin\phi$

5.

7. $\begin{vmatrix} u_x & u_y \\ v_x & v_y \end{vmatrix} = \begin{vmatrix} y & x \\ y^2 & 2xy \end{vmatrix} = xy^2 - 2xy^2 = xy^2 = v$

Since $\dfrac{\partial(u,v)}{\partial(x,y)} \cdot \dfrac{\partial(x,y)}{\partial(u,v)} = 1$; $\dfrac{\partial(x,y)}{\partial(u,v)} = \dfrac{1}{v}$ so $\left|\dfrac{\partial(x,y)}{\partial(u,v)}\right| = \dfrac{1}{v}$ since $1 \leq v \leq 4$

So

8. Evaluate the integral $\int_T \cos\left(\dfrac{x-y}{x+y}\right) dx\,dy$ where T is the triangle bounded by $x + y = 1$, $x = 0$, and $y = 0$.

REVIEW PROBLEMS FOR CHAPTER FIFTEEN

1. Figure 15.60 shows contours of average annual rainfall (in inches) in South America.[4] Each grid square is 500 miles on a side. Estimate the total volume of rain that falls on the considered area in a year.

2. Figure 15.61 gives isotherms for low winter temperature in Washington, D.C.[5] The grid squares are one mile on a side. Find the average low temperature over the whole city (the city is the light shaded region).

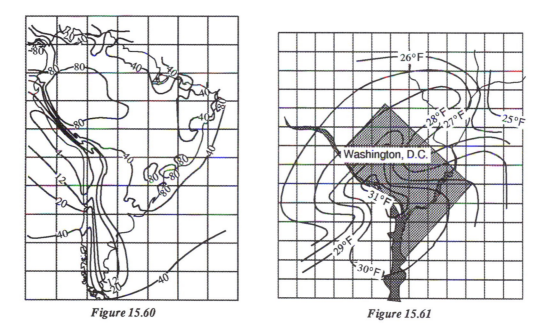

Figure 15.60   *Figure 15.61*

Sketch the regions over which the integrals in Problems 3–6 are being performed.

3. $\displaystyle\int_1^4 \int_{-\sqrt{y}}^{\sqrt{y}} f(x, y)\, dx\, dy.$

4. $\displaystyle\int_0^1 \int_0^{\sin^{-1} y} f(x, y)\, dx\, dy.$

5. $\displaystyle\int_{-1}^1 \int_{-\sqrt{1-x^2}}^{\sqrt{1-x^2}} f(x, y)\, dy\, dx.$

6. $\displaystyle\int_0^2 \int_{-\sqrt{4-y^2}}^0 f(x, y)\, dx\, dy.$

[4]From *Modern Physical Geography*, Alan H. Strahler and Arther H. Strahler, Fourth Edition, John Wiley & Sons, New York, 1992, p. 144

[5]From *Physical Geography of the Global Environment*, H. J. de Blij and Peter O. Muller, John Wiley & Sons, New York, 1993, p. 220

Calculate exactly the integrals in Problems 7–11. (Your answer may contain e, π, $\sqrt{2}$, and so on).

7. $\displaystyle\int_0^1 \int_0^z \int_0^2 (y+z)^7 \, dx \, dy \, dz.$

8. $\displaystyle\int_0^1 \int_3^4 (\sin(2-y)) \cos(3x-7) \, dx \, dy.$

9. $\displaystyle\int_0^{10} \int_0^{0.1} x e^{xy} \, dy \, dx.$

10. $\displaystyle\int_0^1 \int_0^y (\sin^3 x)(\cos x)(\cos y) \, dx \, dy.$

11. $\displaystyle\int_3^4 \int_0^1 x^2 y \cos(xy) \, dy \, dx.$

12. Write $\int_R f(x,y) \, dA$ as an iterated integral if R is the region in Figure 15.62.

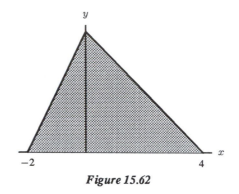

Figure 15.62

13. Evaluate $\int_R \sqrt{x^2+y^2} \, dA$ where R is the region in Figure 15.63.

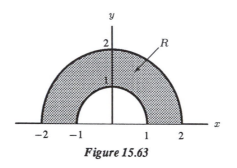

Figure 15.63

Evaluate the integrals in Problems 14–17 by changing them to cylindrical or spherical coordinates.

14. $\displaystyle\int_{-\sqrt{3}}^{\sqrt{3}} \int_{-\sqrt{3-x^2}}^{\sqrt{3-x^2}} \int_1^{4-x^2-y^2} \frac{1}{z^2} \, dz \, dy \, dx$

15. $\displaystyle\int_0^1 \int_0^{\sqrt{1-x^2}} \int_0^{\sqrt{x^2+y^2}} (z + \sqrt{x^2+y^2}) \, dz \, dy \, dx$

16. $\displaystyle\int_0^3 \int_{-\sqrt{9-z^2}}^{\sqrt{9-z^2}} \int_{-\sqrt{9-y^2-z^2}}^{\sqrt{9-y^2-z^2}} x^2 \, dx \, dy \, dz$

17. $\displaystyle\int_0^1 \int_{-\sqrt{2-x^2}}^{\sqrt{2-x^2}} \int_0^{\sqrt{2-x^2-y^2}} \sqrt{x^2+y^2+z^2} \, dz \, dy \, dx.$

18. For $R = \{(x, y, z) : 1 \leq x^2 + y^2 \leq 4, 0 \leq z \leq 4\}$ evaluate the integral

$$\int_R \frac{z}{(x^2 + y^2)^{3/2}} \, dV.$$

19. Write an integral representing the mass of a sphere of radius 3 if the density of the sphere at any point is twice the distance of that point from the center of the sphere.

20. Compute the integral $\int_0^1 \int_{-\sqrt{1-x^2}}^{\sqrt{1-x^2}} e^{-(x^2+y^2)} \, dy \, dx$.

21. A forest next to a road has the shape in Figure 15.64. The population density of rabbits is proportional to the distance from the road. It is 0 at the road, and 10 rabbits per square mile at the opposite edge of the forest. Find the total rabbit population in the forest.

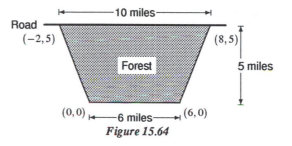

Figure 15.64

22. Consider a solid body W with an axis L passing through it, and suppose that W is spinning around on L. The *moment of inertia* of W tells you how great will be the angular acceleration of the body for a given torque (a force twisting the body). The moment of inertia may be calculated by the following triple integral:

$$\int_W S(x, y, z) d(x, y, z)^2 \, dV,$$

where $S(x, y, z)$ is the density of the body at the point (x, y, z), and $d(x, y, z)$ is the distance of the point from the axis L. Thus, for example, if L is the z-axis, then $d(x, y, z)^2 = x^2 + y^2$. Consider a rectangular brick with length 5, width 3, and height 1, and of uniform density 1. Compute the moment of inertia about each of the three axes passing through the center of the brick, perpendicular to one of the sides.

23. Compute the moment of inertia of a sphere of radius R about an axis passing through its center. Assume that the sphere has a constant density of 1.

CHAPTER SIXTEEN

PARAMETRIC CURVES AND SURFACES

In single-variable calculus, we studied the motion of a particle along a straight line. For example, if an object is thrown straight up into the air, its motion lies on a straight vertical line, going up the line for a while, reaching a high point, then moving back down the line to earth. We represent this motion by a single function $y(t)$, the height of the object above the ground at time t.

Using multivariable calculus, we can study the motion of a particle in 2-dimensional or 3-dimensional space. To do this, we must now express in terms of time t not just a single coordinate, such as $y(t)$, but all the coordinates of the particle: $x(t)$, $y(t)$, and $z(t)$ if the motion is in 3-space.

There are two important questions about motion in space that we want to be able to answer: What path the particle follows and when it reaches each point.

16.1   PARAMETRIC EQUATIONS IN TWO DIMENSIONS

What is a Parameterization?

To represent motion in the xy-plane we use two equations, one for the x-coordinate of the particle $x = f(t)$, and another for the y-coordinate $y = g(t)$. The equation for x describes the right-left motion; the equation for y describes the up-down motion. The two equations for x and y are called *parametric equations* with *parameter t*.

Example 1   Describe the motion of the particle whose coordinates at time t are

$$x = \cos t,$$
$$y = \sin t.$$

Solution   Since $(\cos t)^2 + (\sin t)^2 = 1$, we have $x^2 + y^2 = 1$. That is, at any time t the particle is at a point (x, y) on the unit circle $x^2 + y^2 = 1$. We can plot points at different times to see how the particle moves on the circle. The particle moves at a uniform speed completing one full trip counterclockwise around the circle every 2π units of time. Notice how the x-coordinate goes back and forth from 1 to 0 to -1 to 0 to 1 while the y-coordinate goes up and down from 0 to 1 to 0 to -1 to 0. The two motions combine to trace out a circle. (See Figure 16.1 and Table 16.1.)

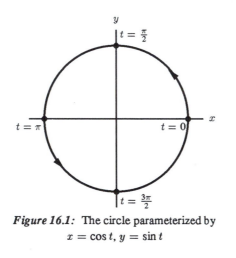

Figure 16.1: The circle parameterized by $x = \cos t, y = \sin t$

Table 16.1: *Points on the circle with $x = \cos t$, $y = \sin t$*

t	x	y
0	1	0
$\pi/2$	0	1
π	-1	0
$3\pi/2$	0	-1

The Effect of Different Parameterizations

Example 2   Describe the motion of the particle whose x and y coordinates at time t are given by the equations

$$x = \cos 3t,$$
$$y = \sin 3t.$$

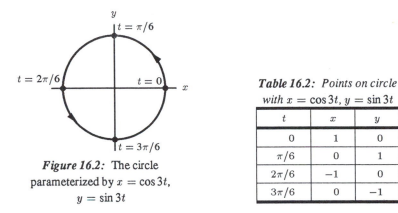

Figure 16.2: The circle parameterized by $x = \cos 3t$, $y = \sin 3t$

Table 16.2: Points on circle with $x = \cos 3t$, $y = \sin 3t$

t	x	y
0	1	0
$\pi/6$	0	1
$2\pi/6$	−1	0
$3\pi/6$	0	−1

Solution   Here we have $(\cos 3t)^2 + (\sin 3t)^2 = 1$ so $x^2 + y^2 = 1$ and we again have motion on a circle. But if we plot points at different times, we see that the particle is moving three times as fast around the circle as before. (See Figure 16.2 and Table 16.2.)

Example 2 is the same as Example 1 except that t is replaced by $3t$. Replacing t by some function of t is called a *change in parameter*. Generally speaking, if we make a change in parameter, the particle follows the same curve but it traces out the curve at a different speed. In Section 16.4 we discuss the speed along a parameterized curve in more detail.

Example 3   Describe the motion of the particle whose x and y coordinates at time t are

$$x = \cos(e^{-t^2}),$$
$$y = \sin(e^{-t^2}).$$

Solution   As in the previous two examples, we still have $x^2 + y^2 = 1$ so the motion of the particle lies on the unit circle. As time t goes from $-\infty$ (way back in the past) to 0 (the present) to ∞ (way off in the future), e^{-t^2} goes from 0 to 1 back to 0. Thus, the angle of which we are taking the cosine and sine goes from 0 to 1 and back to 0. Thus, we trace out the part of the circle, with the angle between 0 and 1 radian. The particle does not actually reach the point $(1, 0)$ at any finite time. (See Figure 16.3 and Table 16.3.)

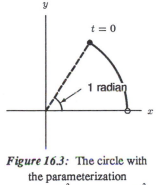

Figure 16.3: The circle with the parameterization $x = \cos(e^{-t^2})$, $y = \sin(e^{-t^2})$

Table 16.3: Table: Points on circle with $x = \cos(e^{-t^2})$, $y = \sin(e^{-t^2})$

t	x	y
-100	∼ 1	∼ 0
0	∼ 0.54	∼ 0.84
100	∼ 1	∼ 0

Parametric Representation of Graphs of Functions

The graph of any function $y = f(x)$ can be given parametrically simply by letting the parameter t be x:

$$x = t,$$
$$y = f(t).$$

Example 4   Give parametric equations for the motion of a particle that traces out the curve $y = x^3 - x$.

Solution   Let $x = t$, $y = t^3 - t$. At any time t, the particle's position (x, y) satisfies $y = t^3 - t = x^3 - x$. Since $x = t$, as time increases the x-coordinate moves from left to right, thus, the particle traces out the curve $y = x^3 - x$ from left to right.

Parametric Equations of Lines

Suppose a particle is moving along a straight line at a constant speed. One way to parameterize the line is to think of it as the graph of the function $y = mx + b$ and let $x = t$. Suppose we know that the particle starts at the point (x_0, y_0) when $t = 0$ and that the x-coordinate is increasing at a rate of a and the y-coordinate is increasing at a rate of b. Then the x-coordinate will be given by $x = x_0 + at$. Similarly the y-coordinate will be given by $y = y_0 + bt$. The slope of the line is the ratio of the y-change in a given time to the x-change in that time, namely b/a. The parametric equations

$$x = x_0 + at, \quad y = y_0 + bt$$

represent motion in the direction of the vector $a\vec{i} + b\vec{j}$. (See Figure 16.4). Thus, we can rewrite the equations in the following form:

The parametric equation of a line in the direction of the vector $a\vec{i} + b\vec{j}$ passing through the point (x_0, y_0) at time $t = 0$ is:

$$(x, y) = (x_0, y_0) + t(a\vec{i} + b\vec{j}).$$

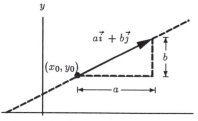

Figure 16.4: Line through the point (x_0, y_0) and in the direction of $a\vec{i} + b\vec{j}$

Example 5   Find parametric equations for motion along the line $y = 3x + 7$ such that the x-coordinate decreases by 2 units for each unit of time.

Solution We should have $x = x_0 - 2t$. Since no initial position at time $t = 0$ is given, we can use any point on the line $y = 3x + 7$ as (x_0, y_0). We choose the y-intercept $(0, 7)$. Then $x = 0 - 2t$ and $y = 7 + bt$. Since the slope of the line is 3 and the x-coordinate increases by 2 units for each unit of time, we know that the y-coordinate increases by -6 units for each unit of time. Therefore, $b = -6$. Our equations are $x = -2t$, $y = 7 - 6t$.

Using Position Vectors to Write Parameterized Curves as Vector Functions

Recall that a point with coordinates (x, y) can be represented by a position vector $\vec{r}$ with its tail at the origin and its tip at the point (x, y) as shown in Figure 16.5. Since $\vec{r} = x\vec{i} + y\vec{j}$, we can write a pair of parametric equations $x = f(t)$, $y = g(t)$ as a single vector equation $\vec{r} = \vec{F}(t)$. For example, the circular motion $x = \cos t$, $y = \sin t$ can be written as $\vec{r} = \cos t\vec{i} + \sin t\vec{j}$. The equation of a line can then be written as

$$\vec{r} = \vec{r}_0 + t(a\vec{i} + b\vec{j})$$

where $\vec{r}_0 = x_0\vec{i} + y_0\vec{j}$.

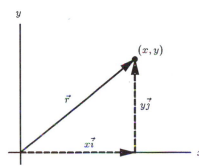

Figure 16.5: Position vector $\vec{r}$ for the point (x, y)

Explicit, Implicit, and Parametric Representations

In this section we have introduced the natural way to describe motion in the plane, giving the x and y coordinates of a particle as functions of the time parameter t. At the same time, we have discovered a different way of looking at a curve in the plane. Up to this point, if we wanted to give an algebraic formulation for, say, a circle of radius 1 centered at the origin, we could write down the equation $x^2 + y^2 = 1$ or we could solve for y to get the upper and lower halves of the circle, $y = \sqrt{1 - x^2}$ and $y = -\sqrt{1 - x^2}$. Now we have a third way: introduce a parameter t (which can be thought of as time) and give the parametric equations $x = \cos t$, $y = \sin t$, $0 \le t \le 2\pi$. These three ways of representing the circle are called (respectively) implicit, explicit, and parametric. That is,

---

- An **implicit** representation of a curve in the xy-plane is given by a single equation in x and y, $f(x, y) = 0$.
- An **explicit** representation of a curve in the xy-plane is given by a single equation expressing y in terms of x or x in terms of y, $y = g(x)$ or $x = h(y)$.
- A **parametric** representation of a curve in the xy-plane is given by a pair of equations expressing x and y in terms of a third variable, usually denoted t.

Example 6   Give implicit, explicit, and parametric representations of the line passing through the points $(3, 0)$ and $(0, 5)$.

Solution   An implicit representation is $x/3 + y/5 - 1 = 0$ (check that the x-intercept is 3 and y-intercept is 5). An explicit representation is $y = 5 - (5/3)x$. A parametric representation is given by $x = 3t$, $y = -5t + 5$.

Example 7   Give implicit and explicit representations of the curve having the parametric representation

$$x = 3 + 5\sin t, \ y = 1 + 2\cos t.$$

Solution   We need to eliminate the parameter t. Solving for $\sin t$ and $\cos t$, we get $(x - 3)/5 = \sin t$, $(y - 1)/2 = \cos t$. Since $\sin^2 t + \cos^2 t = 1$, we have

$$\left(\frac{x - 3}{5}\right)^2 + \left(\frac{y - 1}{2}\right)^2 = 1,$$

which is an implicit representation for an ellipse centered at the point $(3, 1)$. To get an explicit representation, we solve for one variable in terms of the other

$$\left(\frac{y - 1}{2}\right)^2 = 1 - \left(\frac{x - 3}{5}\right)^2$$

$$\frac{y - 1}{2} = \pm\sqrt{1 - \left(\frac{x - 3}{5}\right)^2}$$

$$y = 1 \pm 2\sqrt{1 - \frac{(x - 3)^2}{25}}.$$

We do not get one explicit representation for the whole ellipse; rather, we get one for the upper half (the positive square root) and one for the lower half (the negative square root).

Explicit and parametric equations are easier to plot than implicit equations. For example, to sketch a curve given explicitly as $y = f(x)$, we simply pick some x values, compute the corresponding y values by substituting the x values into $f(x)$, and plot points. Similarly for a curve given parametrically, we pick some values for t, compute the corresponding x and y values, and plot points. For an implicit representation, however, we can try values for x, but then we must solve the implicit equation for y. There may be many values for y for each value of x; there may be no y values for some values of x. Moreover, it may be impossible to solve the equation for y algebraically. For each x value we may have to compute the corresponding y-values numerically. This is so complicated that computers usually sketch implicitly defined curves by indirect methods.

Complicated Curves

Using parametric equations, we can plot curves that we do not usually see when graphing equations in x and y.

Example 8   Sketch the curve traced out by the particle whose motion is given by

$$x = \cos 3t,$$
$$y = \sin 5t.$$

Solution   The x-coordinate oscillates back and forth between 1 and -1, completing 3 full oscillations every 2π seconds (assuming t is time in seconds). The y-coordinate oscillates up and down between 1 and -1, completing 5 full oscillations every 2π seconds. Since both the x and y coordinates return to their original values every 2π seconds, the curve is retraced every 2π seconds. The result is a pattern called a Lissajous figure . It is difficult to sketch the graph by hand, but a graphing calculator set on "parametric" graphing mode can draw the figure quickly. You might wish to experiment with other Lissajous figures $x = \cos at$, $y = \sin bt$, for different values of a and b (see Problems 32–35).

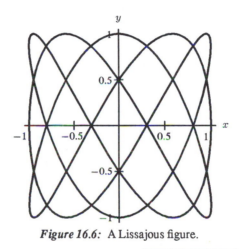

Figure 16.6: A Lissajous figure.

Differential Equations and Examples Other Than Motion in Space

In all the examples we have given so far, we are thinking of the xy-plane as 2-dimensional space with both x and y representing space coordinates. There are many applications where the x-coordinate measures one quantity and the y-coordinate measures another. For example, if a particle is moving along the x-axis, in such a way that $x = x(t)$, we are often interested both in where the particle is at time t and in what the velocity is at time t. To draw a picture that captures the position and velocity of the particle simultaneously, we use the parametric equations

$$x = x(t),$$
$$v = x'(t).$$

Now the horizontal x-axis represents position but the vertical v-axis represents velocity. This viewpoint is frequently used to understand solutions to second-order differential equations arising in physics. The resulting parametric curve is called the *phase portrait* of the solution.

Example 9   Suppose $x(t)$ is the solution of the harmonic oscillator equation

$$\frac{d^2x}{dt^2} + 4x = 0, \text{ where } x(0) = 1, \; v(0) = x'(0) = 0.$$

For example, x might represent the position of a mass attached to a spring on a frictionless surface, where $x = 0$ indicates the spring is neither stretched nor compressed. Sketch the phase portrait.

Solution   You can check that a solution to the differential equation with these initial conditions is $x = \cos 2t$. Thus, writing $v = x'$, we want to sketch the equations

$$x = \cos 2t,$$
$$v = -2\sin 2t.$$

Observe that $x^2 + \left(v/2\right)^2 = \cos^2 2t + (-\sin 2t)^2 = 1$ so the resulting curve is an ellipse

$$x^2 + \left(v^2/4\right) = 1,$$

which intersects the x-axis at $x = 1$ and $x = -1$ and intersects the v-axis at $v = 2$ and $v = -2$. As we trace around the curve starting at $x = 1$, $v = 0$, the position of the mass (the x-coordinate) varies from 1 to 0 to -1 back to 0. At the same time, the velocity of the mass (the v-coordinate) begins at 0 but becomes negative (since $v = dx/dt$ and x is decreasing). The velocity reaches its most negative value of -2 precisely when the position is $x = 0$, the equilibrium position of the spring. Then, as position of the mass continues to move back to $x = -1$, the velocity v becomes less negative (as the spring compresses) until the mass comes to rest when $x = -1$, the point at which the spring is most compressed. The velocity becomes positive as the mass moves back to the right; the largest positive value for v occurs when $x = 0$. Finally, the velocity decreases and reaches 0 again when $x = 1$.

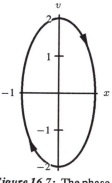

Figure 16.7: The phase portrait of a solution to $\frac{d^2x}{dt^2} + 4x = 0$.

The systems of differential equations used to model interacting populations are another source of parametrically defined curves whose coordinates do not represent position in space.

Example 10 The Lanchester model for a battle between two armies proposes the system of differential equations

$$\frac{dx}{dt} = -ay,$$

$$\frac{dy}{dt} = -bx,$$

where x and y represent the number of soldiers, in thousands, in each army at time t measured in weeks, say. Assume $a = b = 1$. Show that

$$x = e^t + 9e^{-t},$$

$$y = -e^t + 9e^{-t},$$

is a solution of the system of differential equation satisfying the initial condition $x(0) = 10$, $y(0) = 8$. Sketch the solution curve. When is the battle over and what is the final result?

Solution To show that these expressions for x and y are solutions to the differential equation, we check that

$$\frac{dx}{dt} = \frac{d}{dt}(e^t + 9e^{-t}) = e^t - 9e^{-t} = -y,$$

$$\frac{dy}{dt} = \frac{d}{dt}(-e^t + 9e^{-t}) = -e^t - 9e^{-t} = -x.$$

To check that this solution satisfies the initial conclusions, we observe that

$$x(0) = 1 + 9 = 10, \ y(0) = -1 + 9 = 8.$$

To sketch the curve, we use a graphing calculator or computer. Alternatively, we can observe that $x + y = 18e^{-t}$ and $x - y = 2e^t$ so $(x + y)(x - y) = (18e^{-t})(2e^t) = 36$. Thus, the underlying curve is the hyperbola $x^2 - y^2 = 36$. Since the battle ends when one army has no soldiers left, we look for the point where one coordinate is 0. The curve intersects the x-axis at $x = 6$, $y = 0$, so the final result is a victory for the x-army, with six thousand of its soldiers surviving. The battle ends at the time when $y = 0$, that is when $-e^t + 9e^{-t} = 0$. Solving for t gives $e^t = 9e^{-t}$ so $e^{2t} = 9$ and $t = \frac{1}{2}\ln 9 = 1.09$ weeks.

If we were not given the solution, we could still have obtained the underlying curve (but not how it is traced out with time) by using the chain rule

$$\frac{dy}{dx} = \frac{dy/dt}{dx/dt} = \frac{x}{y}$$

and separating variables

$$\int y \, dy = \int x \, dx$$

$$y^2/2 = x^2/2 + C.$$

The initial conditions $x = 10$, $y = 8$ give $64/2 = 100/2 + C$ so $C = -36/2$. Therefore, $y^2/2 = x^2/2 - 36/2$ so $x^2 - y^2 = 36$, as before.

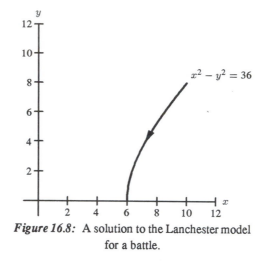

Figure 16.8: A solution to the Lanchester model for a battle.

Problems for Section 16.1

Write a parameterization for each of the curves in the xy-plane in Problems 1–10.

1. A circle of radius 3 centered at the origin and traced out clockwise.
2. A line through the point $(1, 3)$ and parallel to the vector $-2\vec{i} - 4\vec{j}$.
3. A vertical line through the point $(-2, -3)$.
4. A circle of radius 5 centered at the point $(2, 1)$ and traced out counterclockwise.
5. A horizontal line through the point $(0, 5)$.
6. A circle of radius 2 centered at the origin traced out clockwise starting at the point $(-2, 0)$ when $t = 0$.
7. A circle of radius 2 centered at the origin starting at the point $(0, 2)$ when $t = 0$.
8. A circle of radius 4 centered at the point $(4, 4)$, starting on the x-axis, when $t = 0$.
9. An ellipse centered at the origin and crossing the x-axis at ± 5 and the y-axis at ± 7.
10. An ellipse centered at the origin, crossing the x-axis at ± 3 and the y-axis at ± 7. The parameterization should start at the point $(-3, 0)$ and trace out the ellipse counterclockwise.
11. True or false? The equations $x = \cos(t^3)$, $y = \cos(t^3)$ parameterize a line segment.
12. If t is allowed to take on all real values, the parametric equations

$$(x, y) = (2 + 3t, 4 + 7t)$$

describe a line in the plane.

   (a) What part of the line is obtained by restricting t to nonnegative numbers?
   (b) What part of the line is obtained if t is restricted to $-1 \le t \le 0$?
   (c) How should t be restricted to give the part of the line to the left of the y-axis?

1. $x(t) = 3\sin t$ 3. $x(t) = -2$ 5. $x(t) = t$ 7. $x(t) = 2\sin t$ 9. $x(t) = 5\cos t$ 16.1

   $y(t) = 3\cos t$ $y(t) = t$ $y(t) = 5$ $y(t) = 2\cos t$ $y(t) = 7\sin t$ 1-15 odd, 19-27 odd

11. false   13a. $x = 2 + t \Rightarrow t = x - 2$ $x = 1 - 2t \Rightarrow 2t = 1 - x \Rightarrow t = \frac{1-x}{2}$ 31, 35-37 odd

   $\quad x - 1 = -2t$

   $y = 4 + 3t \Rightarrow y = 4 + 3(x - 2)$ $y = 1 - 6t \Rightarrow y = 1 - 6\left(\frac{1-x}{2}\right)$

   $y = 4 + 3x - 6$ $y = 1 - \left(\frac{6 - 6x}{2}\right)$

   $y = 3x - 2$ $y = 1 - 3 + 3x$

   b. slope: 3 ; y-int.: -2 $y = 3x - 2$

15a. $a = b = 0$; $k = 5$ 19. parabola ; $y = x^2 - 4$ 21. The graph moves clockwise when $t > 0$

   b. $a = 0$; $b = 5$; $k = 5$ 23.   25.   27.

   c. $a = 10$; $b = -10$; $k = \sqrt{200}$

(?)

31.   35.   37.

13. (a) Explain why the two pairs of equations

$$(x, y) = (2 + t, 4 + 3t),$$

$$(x, y) = (1 - 2t, 1 - 6t)$$

parameterize the same line.

(b) What are the slope and y intercept of this line?

14. Suppose $a, b, c, d, m, n, p, q > 0$ and consider the two pairs of parametric equations below: Match the equations with two of the lines l_1, l_2, l_3, l_4 in Figure 16.9.

(I) $\begin{cases} x = a + ct, \\ y = -b + dt \end{cases},$

(II) $\begin{cases} x = m + pt, \\ y = n - qt \end{cases}.$

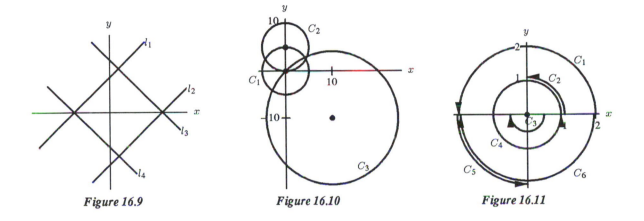

Figure 16.9   Figure 16.10   Figure 16.11

15. What can you say about the values of a, b and k if the equations

$$x = a + k \cos t, \quad y = b + k \sin t$$

trace out each of the circles in Figure 16.10 (a) C_1?   (b) C_2?   (c) C_3?

16. Consider the parametric equations below for $0 \le t \le \pi$.

   I. $\vec{r} = \cos(2t)\vec{i} + \sin(2t)\vec{j}$ II. $\vec{r} = 2\cos t\,\vec{i} + 2\sin t\,\vec{j}$
   III. $\vec{r} = \cos(t/2)\vec{i} + \sin(t/2)\vec{j}$ IV. $\vec{r} = 2\cos t\,\vec{i} - 2\sin t\,\vec{j}$

   (a) Match the equations above with four of the curves C_1, C_2, C_3, C_4, C_5 and C_6 in Figure 16.11. (Each curve is part of a circle.)

   (b) Give parametric equations for the curves which have not been matched, again assuming $0 \le t \le \pi$.

What curves do the parametric equations in Problems 17–19 trace out? Find the Cartesian equation for each curve.

17. $x = 2 + \cos t, y = 2 - \sin t.$ 18. $x = 2 + \cos t, y = 2 - \cos t.$ 19. $x = 2 + \cos t, y = \cos^2 t.$

In Problems 20–22, the curve being traced out is a circle. Describe in words how the circle is traced out, including when and where the particle is moving clockwise and when and where the particle is moving counterclockwise.

20. $x = \cos(t^3 - t)$, $y = \sin(t^3 - t)$ 21. $x = \cos(\ln t)$, $y = \sin(\ln t)$

22. $x = \cos(\cos t)$, $y = \sin(\cos t)$

On a graphing calculator or a computer, plot the curves given by the equations in Problems 23–27.

23. $x = (\cos 3t)(\cos t)$, $y = (\cos 3t)(\sin t)$ 24. $x = t^2 - t$, $y = t^3 + t^2$

25. $x = 3\cos t + \cos 3t$, $y = 3\sin t + \sin 3t$ 26. $x = 3e^t + 5e^{2t}$, $y = 2e^t - 7e^{2t}$

27. $x = t + 2\sin t$, $y = 2\cos t$

28. Suppose a particle is moving along the x-axis, say, $x = t^3 - t$ and that we want to "see" the motion. If all we do is plot where the particle is at various points along the x-axis we will not see much; we will just trace out the x-axis, as in Figure 16.12. To see the full motion of the particle, we introduce a y-coordinate and let it slowly increase, giving Figure 16.13. Try the following on a graphing calculator or computer. Let $y = t$. Now plot the pair of parametric equations $x = t^3 - t$, $y = t$ for, say, $-2 \le t \le 3$. What happens?

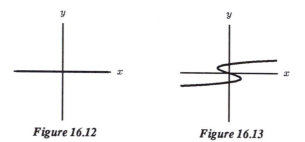

Figure 16.12   *Figure 16.13*

For Problems 29–31, graph the parametric equations by the method of Problem 28.

29. $x = \cos t$, $-10 \le t \le 10$ 30. $x = t^4 - 2t^2 + 3t - 7$, $-3 \le t \le 2$

31. $x = t\ln t$, $0.01 \le t \le 10$

Graph the Lissajous figures in Problems 32–35 using a computer or graphing calculator.

32. $x = \cos 2t$, $y = \sin 5t$ 33. $x = \cos 3t$, $y = \sin 7t$

34. $x = \cos 2t$, $y = \sin 4t$ 35. $x = \cos 2t$, $y = \sin \sqrt{3}t$

36. (a) Check that $x = e^{-t}\cos 7t$ is a solution of the second-order differential equation for a damped harmonic oscillator (that is, a spring with friction):

$$\frac{d^2x}{dt^2} + 2\frac{dx}{dt} + 50x = 0$$

(b) With a graphing calculator or a computer, sketch the phase portrait of the solution $x = e^{-t} \cos t$ for the differential equation given in part (a). That is, in the xv-plane sketch the curve given by the parametric equations:

$$x = e^{-t} \cos 7t$$
$$v = \frac{dx}{dt} = -e^{-t} \cos 7t + e^{-t}(-7 \sin 7t).$$

(c) Explain in words what the phase portrait shows about the position x and velocity v of a mass attached to a spring with friction satisfying the differential equation in part (a).

State whether the equations in Problems 37–39 represent a curve parametrically, implicitly, or explicitly. Give the two other types of representations for the same curve.

37. $xy = 1$ for $x > 0$ 38. $x^2 - 2x + y^2 = 0$ for $y < 0$ 39. $x = e^t, y = e^{2t}$ all t

16.2 PARAMETRIC EQUATIONS IN THREE DIMENSIONS

To describe a curve in 3-dimensional space parametrically, we need an extra equation giving z in terms of t.

Example 1   Sketch the curve given parametrically by

$$x = \cos t, \ y = \sin t, \ z = t.$$

Solution   To picture the curve, look at how the equations for $x, y,$ and z control the particle's motion front to back, right to left, and up and down, respectively. In this case, the particle's x and y coordinates oscillate back and forth while the z-coordinates steadily increases. (See Figure 16.14). This spiral-shaped curve which looks like a slinky is called a *helix*.

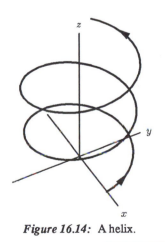

Figure 16.14: A helix.

Changing Parameterization

As in two dimensions, if the parameter t in a parametric representation of a 3-dimensional curve is replaced by some other function of t, the underlying curve remains the same but it is traced out at a different speed.

Example 2   Describe how the helix in Example 1 is traced out for each of the following parameterizations.

(a)   Replace t by $3t$, so that $x = \cos 3t$, $y = \sin 3t$, $z = 3t$.

(b)   Replace t by t^2, so that $x = \cos(t^2)$, $y = \sin(t^2)$, $z = t^2$.

(c)   Replace t by e^t, so that $x = \cos(e^t)$, $y = \sin(e^t)$, $z = e^t$.

Solution   (a)   The particle travels 3 times as far in one unit of the new time, t, as the particle in Example 1. Thus, the helix is traced out 3 times as fast.

(b)   The quantity t^2 goes from ∞ to 0 then back to ∞, as t goes from $-\infty$ to ∞. When $t \to \pm\infty$ the particle is far above the xy-plane. Thus, the particle comes down the helix, stops at the point $(1, 0, 0)$ when $t = 0$, then reverses direction and goes back up the helix. Also, since a unit change in t causes larger and larger changes in t^2 as t approaches ∞ or $-\infty$, the particle is traveling faster when it is further above the xy-plane.

(c)   As t goes from $-\infty$ to ∞, e^t starts from near 0, slowly increases to 1 at time $t = 0$ and then grows quite rapidly for $t > 0$. Thus, the particle begins way back in negative time near the point $(1, 0, 0)$, slowly climbs the helix reaching $(\cos 1,\ \sin 1,\ 1)$ at time $t = 0$ (about 1/6 of the first turn of the helix above the xy-plane), then climbs more and more rapidly for $t > 0$. By the time $t = 10$, one unit change in time t changes the z-coordinate from $e^{10} \cong 22{,}000$ to $e^{11} \cong 60{,}000$. In that same interval, the particle will have circled around the z-axis $38{,}000/(2\pi) \cong 6{,}000$ times!

The Equations of Lines in Three Dimensions

If the line is traced out at a constant speed, then each unit change in time t produces the same change in the x-coordinate. Thus, x is a linear function of time, so $x = x_0 + at$. Similarly, $y = y_0 + bt$ and $z = z_0 + ct$. These are the parametric equations of a line in the direction of the vector $a\vec{i} + b\vec{j} + c\vec{k}$. Thus, we can rewrite these equations in the following form:

The **parametric equation of a line** through the point (x_0, y_0, z_0) in the direction of the vector $a\vec{i} + b\vec{j} + c\vec{k}$ is:

$$(x, y, z) = (x_0, y_0, z_0) + t(a\vec{i} + b\vec{j} + c\vec{k})$$

Example 3   Find parametric equations for the following lines:

(a)   The line passing through the points $(2, -1, 3)$ and $(-1, 5, 4)$.

(b)   The line of intersection of the planes $z = 4 + 2x + 5y$ and $z = 3 + x + 3y$.

Solution   (a)   Let us choose a parameterization such that at time $t = 0$ the particle is at the point $(2, -1, 3)$ and at time $t = 1$ the particle is at the point $(-1, 5, 4)$. Then the x-change per unit of time is $(-1) - 2 = -3$, the y-change per unit time is $5 - (-1) = 6$, and the z-change per unit time is $4 - 3 = 1$. Thus, the parametric equation is

$$(x, y, z) = (2, -1, 3) + t(-3\vec{i} + 6\vec{j} + \vec{k}).$$

(b)   We can find this equation in two ways. First we could find two points on the line of intersection and then proceed as we did in part (a). To find two points just substitute two different values for z and solve for x and y for each value of z. Alternatively, assuming the line isn't horizontal (which it turns out not to be), we could take z to be the parameter t, so $z = t$. To find x and y as functions of t we solve the two equations for x and y in terms of t. We have

$$t = 4 + 2x + 5y$$
$$t = 3 + x + 3y.$$

Eliminating x we get

$$t = 3 + a + 3(-2+t)$$
$$t = 3 + a - 6 + 3t$$
$$t = -3 + a + 3t \implies x = 3 - 2t$$

$$-t = -2 - y \quad \text{and} \quad y = -2 + t.$$

Substituting $-2 + t$ for y in the second equation and solving for x, we get

$$x = 3 - 2t.$$

Our equations are therefore

$$x = 3 - 2t, \ y = -2 + t, \ z = t.$$

or

$$(x, y, z) = (3, -2, 0) + t(-2\vec{i} + \vec{j} + \vec{k}).$$

The Intersection of a Line and a Plane: An Application to Computer Graphics

An important application of the parametric representation of a line in three dimensions is to find where the line between two points intersects a given plane. For example, the pictures of curves and surfaces in three-dimensional space in this book are drawn by computer. To do this, the computer first calculates the xyz-coordinates of some points on the curve or surface. For each such point, it then computes the line of sight from that point to the eye of some fixed imaginary viewer and determines where that line intersects a fixed imaginary window (the plane of the computer screen) lying between the point and the viewer's eye. The two-dimensional screen coordinates of that point of intersection are computed so the point can be plotted on the screen. The following problem derives some of the equations the computer uses.

Example 4   Find formulas for the coordinates of the point of intersection of the plane $Ax + By + Cz = D$ with the line from the point (a, b, c) to a viewer at the point (A, B, C).

Solution   Suppose we choose a parameterization in which the particle is at the point (a, b, c) when $t = 0$ and at the point (A, B, C) when $t = 1$. Then the x-coordinate changes by $(A - a)$ in unit time, the

y-coordinate changes by $(B - b)$, and the z-coordinate by $(C - c)$. Thus, parametric equations for the line of sight are

$$x = a + t(A - a)$$
$$y = b + t(B - b)$$
$$z = c + t(C - c).$$

We now substitute these expressions into the equation $Ax + By + Cz = D$ and solve for t:

$$A(a + t(A - a)) + B(b + t(B - b)) + C(c + t(C - c)) = D$$
$$Aa + Bb + Cc + t(A^2 + B^2 + C^2 - Aa - Bb - Cc) = D.$$

Writing $E = Aa + Bb + Cc$ and $F = A^2 + B^2 + C^2$, we get

$$t = \frac{D - E}{F - E}.$$

Then we have:

$$x = a + \frac{D - E}{F - E}(A - a)$$
$$y = b + \frac{D - E}{F - E}(B - b)$$
$$z = c + \frac{D - E}{F - E}(C - c)$$

The problem of converting the xyz-coordinates of the point of intersection into two-dimensional screen coordinates is considered in Problem 20.

More Complicated Curves

As in two dimensions, we can represent very complicated curves parametrically. For example, you might like to look at 3-dimensional Lissajous curves such as $x = \cos 3t$, $y = \sin 5t$, $z = \cos 4t$ on a computer.

Example 5   Describe the curve given by the parametric equations

$$x = (4 + \sin 3t) \cos 2t,$$
$$y = (4 + \sin 3t) \sin 2t,$$
$$z = \cos 3t.$$

Solution   First look at the x and y equations only. If we ignore the $\sin 3t$ term we get $x = 4 \cos 2t$, $y = 4 \sin 2t$, which is a circle of radius 4. If we now include the $\sin t$ term we have a "circle" whose "radius" of $(4 + \sin 3t)$ oscillates once between 3 and 5 as the circle goes twice around the origin. Meanwhile, the z-coordinate oscillates three times between -1 and 1. The result is a curve that goes two times around the z-axis, meanwhile going through three full twists. Figure 16.15 shows the curve in xyz-space and the shadow it casts in the xy-plane.

1. $x(t) = 2\cos t$

5. $(x, y, z) = (1, 1, 1) + t(2, -3, 5)$

7.

16.2

1, 5, 7, 11, 13, 15

$y(t) = 0$

$z(t) = 2\sin t$

11.

13. $x(t) = 1 + 3t$

$y(t) = 2 - 3t$

$z(t) = 3 + t$

(?)

15. $x - y + z = 3 \Rightarrow t = 3 - x + y$

$2x + y - z = 5 \Rightarrow t = -5 + 2x + y$

$t = -2 + x \Rightarrow x = 2 + t$

$t = 3 - x + y$

$t = 3 - (2 + t) + y$

$t = 3 - 2 - t + y \Rightarrow y = -1 + 2t$

$z = t$

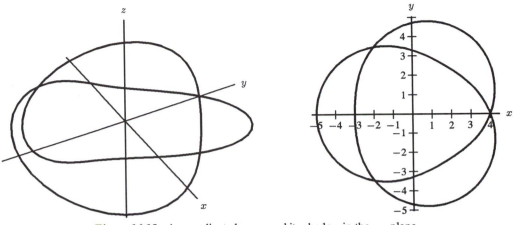

Figure 16.15: A complicated curve and its shadow in the xy-plane

The curve in the above example is called a knot, because it can be obtained by taking a piece of string, tying a knot in the string, and then splicing the free ends together.

Problems for Section 16.2

Write a parameterization for each of the curves in Problems 1–5.

1. A circle of radius 2 in the xz-plane, centered at the origin.
2. A circle of radius 3 centered at the point $(0, 0, 2)$ parallel to the xy-plane.
3. The line through the points $(2, -1, 4)$ and $(1, 2, 5)$.
4. The line perpendicular to the xz-plane through the point $(1, 3, 2)$.
5. The line perpendicular to the plane $2x - 3y + 5z = 4$ and through the point $(1, 1, 1)$.
6. The equation $\vec{r} = 10\vec{k} + t(\vec{i} + 2\vec{j} + 3\vec{k})$ parameterizes a line.
   (a) Suppose we restrict $t < 0$. What do we get?
   (b) Suppose we restrict $0 \le t \le 1$. What do we get?
7. Imagine a light shining on the helix of Example 1 from far down each of the axes. Sketch the shadow cast by the helix on each of the coordinate planes: xy, xz, and yz.

For Problems 8–11, use a computer to draw the given curve.

8. $x = \cos 3t, y = \sin 5t, z = \cos(2t - \pi/6)$ 9. $x = t \cos t, y = t \sin t, z = t$

10. $x = t^2 + t, y = t^2 - t, z = t^2$
11. $x = (4 + \sin 5t) \cos 3t, y = (4 + \sin 5t) \sin 3t, z = \cos 3t.$

For Problems 12–16, give parametric equations for the given line.

12. The line through the points $(2, 3, -1)$ and $(5, 2, 0)$.
13. The line pointing in the direction of the vector $3\vec{i} - 3\vec{j} + \vec{k}$ and through the point $(1, 2, 3)$.
14. The line parallel to the z-axis passing through the point $(1, 0, 0)$.

15. The line of intersection of the planes $x - y + z = 3$ and $2x + y - z = 5$.

16. The line perpendicular to the surface $z = x^2 + y^2$ at the point $(1, 2, 5)$.

17. Do the lines in Problems 12 and 13 intersect?

18. Find the equation of the line passing through the points $(1, 2, 3)$, $(3, 5, 7)$ and calculate the shortest distance to the origin.

19. Is the point $(-3, -4, 2)$ visible from the point $(4, 5, 0)$ if there is an opaque ball of radius 1 centered at the origin?

20. In Example 4, the xyz-coordinates are computed for the point at which the line of sight from a viewer at (A, B, C) to a point (a, b, c) meets a viewing plane (the screen) $Ax + By + Cz = D$. The computer actually needs to compute screen coordinates of this point, not the xyz-space coordinates. To compute screen coordinates, we need an origin for the screen and two coordinate axes, that is, two vectors $\vec{u}$ and $\vec{v}$ at right angles to each other beginning at the screen origin and lying in the plane of the screen. We choose for the screen origin the point Q where the line of sight from the viewer (A, B, C) to the xyz-origin $(0, 0, 0)$ intersects the viewing plane $Ax + By + Cz = D$. We choose $\vec{u}$ to be a unit vector parallel to the xy-plane pointing to the viewer's right and $\vec{v}$ to be a unit vector at right angles to $\vec{u}$ and pointing up (its z-component is positive). The uv-screen coordinates are found by taking the dot product with $\vec{u}$ and $\vec{v}$ of the vector from the screen origin Q to the point P of intersection computed in Example 4 on page 289.

(a) Find the xyz-coordinates of the screen origin Q in terms of A, B, C, D.
(b) Find the vector $\vec{u}$ in terms of A, B, C.
(c) Find the vector $\vec{v}$ in terms of A, B, C.
(d) Find the coordinates of the point P.
(e) Find the uv-screen coordinates of the point computed in Example 4 . That is, find $\vec{u} \cdot (\vec{P} - \vec{Q})$ and $\vec{v} \cdot (\vec{P} - \vec{Q})$. [Hint: Use the fact that $\vec{u} \cdot (A\vec{i} + B\vec{j} + C\vec{k}) = 0$ and $\vec{v} \cdot (A\vec{i} + B\vec{j} + C\vec{k}) = 0$.]

16.3 PARAMETRIC REPRESENTATION OF SURFACES

We have already seen how a curve can be represented in several ways—explicitly, implicitly, and parametrically. In this section we consider how to represent a surface parametrically. We have already seen how a surface can be represented explicitly (for example, $z = x^2 + y^2$) or implicitly (for example, $x^2 + y^2 + z^2 = 1$).

You probably think of a curve as a one-dimensional object, and this is reflected in the fact that the curve can be described in terms of one parameter (usually t). A surface is a two-dimensional object, so we expect to describe points on a surface in terms of *two* parameters. We can think of this as assigning "map coordinates" to points on the surface, like longitude and latitude on the globe. A point on the surface is a point (x, y, z) in 3-space, but to make a two-dimensional map of the surface we want x, y and z each to be functions of two independent parameters. We want to express the coordinates of points on the surface in the form

$$x = f_1(p, q) \quad y = f_2(p, q) \quad z = f_3(p, q),$$

for appropriate functions f_1, f_2, f_3. The p and q are called the *parameters*; they are the "map coordinates". This way of describing points on a surface is called a *parametric representation* of the surface; we say that we have a *parameterized surface*.

Parameterizing a Surface of the Form $z = f(x, y)$

Suppose we have a surface which is the graph of a function of the form $z = f(x, y)$. Is this also a parametric surface? Yes, and in the easiest way possible. We write the surface parametrically as the three equations

$$x = p, \quad y = q, \quad z = f(p, q).$$

In practice we usually regard the two independent variables x and y as parameters rather than to introduce new variables p and q. Thus, we write $x = x, y = y, z = f(x, y)$, even though the first two equations don't really add anything.

Example 1   Give a parametric description of the lower hemisphere of the sphere $x^2 + y^2 + z^2 = 1$.

Solution   We say that the surface is the graph of the function $z = -\sqrt{1 - x^2 - y^2}$ over the region $x^2 + y^2 \leq 1$ in the plane. Then parametric equations are $x = p, y = q, z = -\sqrt{1 - p^2 - q^2}$, where the parameters p and q vary inside the unit circle.

Parameterizing a Plane

Though a plane can also be given as the graph of a function, there are good reasons to parameterize it in another way. Geometrically, a plane is determined by a point P_0, and two non-parallel vectors $\vec{v}_1$ and $\vec{v}_2$ lying in the plane. Starting at P_0 we can get to any other point P on the plane by moving some distance in the direction $\vec{v}_1$ and another distance in the direction $\vec{v}_2$. Thus, the vector $\overrightarrow{P_0P}$ will be some scalar multiple of $\vec{v}_1$ plus another scalar multiple of $\vec{v}_2$. See Figure 16.16. These multiples are the parameters that locate P. They give map coordinates for the plane.

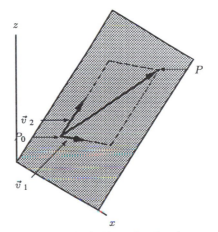

Figure 16.16: Plane showing the parameterization depending on the point P_0 and the vectors $\vec{v}_1$ and $\vec{v}_2$.

If $P_0 = (x_0, y_0, z_0)$, the coordinates of any point (x, y, z) on the plane can be expressed in the following form:

Parametric equation of a plane through the point (x_0, y_0, z_0) and containing the vectors $\vec{v}_1$ and $\vec{v}_2$:

$$(x, y, z) = (x_0, y_0, z_0) + p\vec{v}_1 + q\vec{v}_2$$

If $\vec{v}_1 = a_1\vec{i} + a_2\vec{j} + a_3\vec{k}$, and $\vec{v}_2 = b_1\vec{i} + b_2\vec{j} + b_3\vec{k}$ then the equations of the plane can be written in the form:

$$x = x_0 + pa_1 + qb_1$$
$$y = y_0 + pa_2 + qb_2$$
$$z = z_0 + pa_3 + qb_3$$

Example 2   Planetown is a city on a plane. The center of town is the point $C_0 = (2, -1, 3)$ and all the city blocks are parallelograms formed by the vectors $\vec{v}_1 = 2\vec{i} + 3\vec{j} - \vec{k}$ and $\vec{v}_2 = \vec{i} - 4\vec{j} + 5\vec{k}$. "East" is in the direction of $\vec{v}_1$ and "north" is in the direction of $\vec{v}_2$.

  (a) Starting at the center of town you walk 3 blocks east, 5 blocks north, 1 block west and 2 blocks south. What are the parameters of the point where you end up? What are the x, y and z coordinates of that point? See Figure 16.17.

  (b) Write the parametric equations for Planetown.

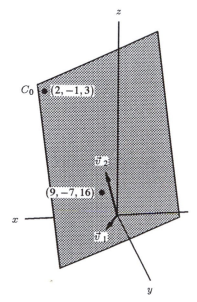

Figure 16.17: Planetown showing the center C_0 and the vectors $\vec{v}_1$ and $\vec{v}_2$.

Solution   Since east is in the direction of $\vec{v}_1$, west is in the direction of $-\vec{v}_1$; likewise south is in the direction of $-\vec{v}_2$. The parameters, which we will call p and q, represent the number of blocks a point is east and north of C_0, respectively.

(a)   According to the directions, your final displacement from the center of town C_0 is $3\vec{v}_1 + 5\vec{v}_2 - \vec{v}_1 - 2\vec{v}_2 = 2\vec{v}_1 + 3\vec{v}_2$. In other words, you are 2 blocks east and 3 blocks north of where you started. The parameters of the point where you end up are $(2, 3)$. These are your map coordinates relative to C_0, the center of town. The parameters for C_0 are $(0, 0)$.

To get the x, y, z coordinates of where you end up, you start with the point $C_0 = (2, -1, 3)$ and add to it the vector $2\vec{v}_1 + 3\vec{v}_2$:

$$
\begin{aligned}
(x, y, z) &= (2, -1, 3) + 2\vec{v}_1 + 3\vec{v}_2 \\
&= 2(2\vec{i} + 3\vec{j} - \vec{k}) + 3(\vec{i} - 4\vec{j} + 5\vec{k}) \\
&= (2, -1, 3) + (7\vec{i} - 6\vec{j} + 13\vec{k}) \\
&= (9, -7, 16).
\end{aligned}
$$

Thus, you end up at the point $(9, -7, 16)$.

(b)   The parametric equations for Planetown are

$$
\begin{aligned}
x &= 2 + 2p + q, \\
y &= -1 + 3p - 4q, \\
z &= 3 - p + 5q.
\end{aligned}
$$

Notice that x, y, and z are each *linear* functions of the parameters p and q.

Parameterizing Spheres

We have already seen that we can parameterize part of a sphere by regarding it as the graph of a function, for example $z = \sqrt{1 - x^2 - y^2}$. However if we think of the sphere as a globe we can parameterize it by longitude and latitude, and these two numbers can serve as parameters.

We determine the latitude of a point by measuring an angle ϕ (in radians) down from the north pole. The north pole has latitude $\phi = 0$, the equator has latitude $\phi = \pi/2$ and the south pole has latitude $\phi = \pi$. We determine longitude θ by measuring around the equator from a fixed starting point. Thus, θ goes from 0 to 2π and we can locate any point on the sphere by specifying ϕ and θ. (We have defined the latitude ϕ slightly differently than is done on globes or maps. There the equator is 0° latitude and the two poles are at 90° latitude.) See Figure 16.18.

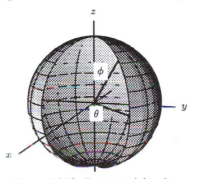

Figure 16.18: Parameterizing the sphere by latitude, ϕ, and longitude, θ.

Example 3   You are at a point on a sphere with latitude $3\pi/4$. Are you in the northern or southern hemisphere? If your latitude decreases have you moved closer to or farther from the equator?

Solution   The equator has latitude $\pi/2$. Since your latitude is greater than this you are in the southern hemisphere. If your latitude decreases you are moving closer to the equator.

Example 4   On a sphere, you are standing at a point with longitude and latitude θ_0 and ϕ_0, respectively. Your *antipodal* point is the point on the other side of the sphere on a line through you and the center. What are the longitude and latitude of your antipodal point?

Solution   See Figure 16.19. Your antipodal point is on the opposite longitude, which is $\pi + \theta_0$. The latitude of your antipodal point is the reflection of your latitude in the equator. This is $\pi - \phi_0$. Notice that if you are on the equator, then so is your antipodal point.

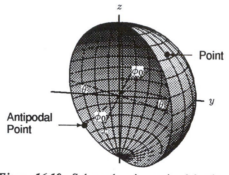

Figure 16.19: Sphere showing antipodal points (at opposite ends of a diameter).

What Are the Parametric Equations of the Sphere in Terms of ϕ and θ?

Suppose we examine the sphere centered at the origin with radius 1. Then the x, y, and z coordinates for a point on the sphere in terms of ϕ and θ are given by

$$x = \sin\phi\cos\theta$$
$$y = \sin\phi\sin\theta$$
$$z = \cos\phi.$$

As a check, verify that $x^2 + y^2 + z^2 = \sin^2\phi(\cos^2\theta + \sin^2\theta) + \cos^2\phi = \sin^2\phi + \cos^2\phi = 1$. Notice that the z-coordinate depends only on the parameter ϕ. Geometrically, this means that all points on the same latitude have the same z-coordinate.

We can write these equations in vector form. The tip of the radial vector $\vec{r} = x\vec{i} + y\vec{j} + z\vec{k}$ lies on the sphere if we use the equations above and write

$$\vec{r}(\theta, \phi) = \sin\phi\cos\theta\,\vec{i} + \sin\phi\sin\theta\,\vec{j} + \cos\phi\,\vec{k}.$$

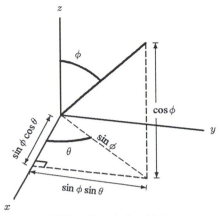

Figure 16.20: The relationship between
x, y, z and ϕ, θ of a sphere of radius 1.

Example 5   Find parametric equations for the sphere:
  (a)   Centered at the origin and having radius 2.

  (b)   Centered at the point $(2, -1, 3)$ and having radius 2.

Solution   In each case we describe a point on the sphere by its longitude and latitude, ϕ and θ.

  (a)   We must scale the distance from the origin by 2. Thus, we have

$$x = 2 \sin \phi \cos \theta$$
$$y = 2 \sin \phi \sin \theta$$
$$z = 2 \cos \phi.$$

  (b)   We shift the center of the sphere from the origin to the point $(2, -1, 3)$. We take the displacement vector from the origin to the point $(2, -1, 3)$ and add the radial vector corresponding to the parameterization we found in part (a). As we see in Figure 16.21, this leads to

$$x = 2 + 2 \sin \phi \cos \theta$$
$$y = -1 + 2 \sin \phi \sin \theta$$
$$z = 3 + 2 \cos \phi.$$

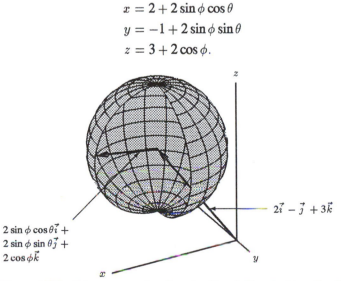

$2 \sin \phi \cos \theta \vec{i} +$
$2 \sin \phi \sin \theta \vec{j} +$
$2 \cos \phi \vec{k}$

$2\vec{i} - \vec{j} + 3\vec{k}$

Figure 16.21: Sphere centered at the point $(2, -1, 3)$ and with radius 2

You should notice that the map maker's notions of longitude and latitude aren't always well-defined. For example, points with $\theta = 0$ also have $\theta = 2\pi$. To avoid this we restrict θ to $0 \leq \theta < 2\pi$. Also, the north pole, at $\phi = 0$, and the south pole, at $\phi = \pi$, don't have a well-defined longitude, so we restrict ϕ to $0 < \phi < \pi$. Each point on the globe other than the north and south pole is then described uniquely.

Parameterizing Surfaces of Revolution

Many surfaces that come up in applications have an axis of rotational symmetry and circular cross-sections perpendicular to that axis. These surfaces are usually referred to as *surfaces of revolution*.

Example 6   Find a parameterization of a circular cylinder of radius r whose axis is along the z-axis, from $z = 0$ to a height $z = h$. See Figure 16.22.

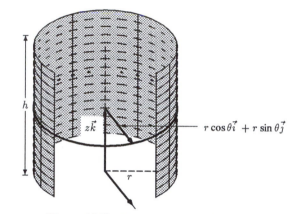

Figure 16.22: Parameterization of a cylinder

Solution   The cross sections of the cylinder perpendicular to the z-axis are circles which are vertical translates of the circle $x^2 + y^2 = r^2$, which is given parametrically by $x = r\cos\theta$, $y = r\sin\theta$. The vector $r\cos\theta\vec{i} + r\sin\theta\vec{j}$ traces out the circle, at any height. We get to a point on the surface by adding that vector to the vector $z\vec{k}$. Hence, the parameters are θ, with $0 \leq \theta \leq 2\pi$, and z, with $0 \leq z \leq h$. The parametric equations for the cylinder are

$$x\vec{i} + y\vec{j} + z\vec{k} = r\cos\theta\vec{i} + r\sin\theta\vec{j} + z\vec{k}$$

which can be written as

$$x = r\cos\theta,$$
$$y = r\sin\theta,$$
$$z = z.$$

A modification of this idea enables us to find parametric equations for a cone.

Example 7   Find a parameterization of the cone whose base is the the circle $x^2 + y^2 = r^2$ in the xy-plane and whose vertex is at a height h above the xy-plane. See Figure 16.23.

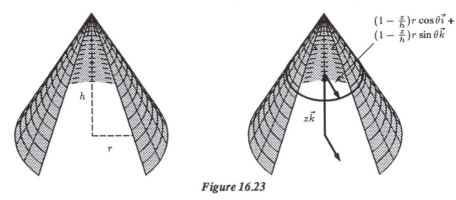

Figure 16.23

Solution The radius of the circular cross section decreases linearly as we move up the axis. At a height z the radius of the circle is $(1 - z/h)r$ and so this circle is traced out by the vector $(1 - z/h)r \cos \theta \vec{i} + (1 - z/h)r \sin \theta \vec{j}$ as θ goes from 0 to 2π. We get to a point on the cone by adding this to the vector $z\vec{k}$ going up the axis. This gives

$$x\vec{i} + y\vec{j} + z\vec{k} = r\left(1 - \frac{z}{h}\right)\cos\theta\vec{i} + r\left(1 - \frac{z}{h}\right)\sin\theta\vec{j} + z\vec{k},$$

which can be written as

$$x = \left(1 - \frac{z}{h}\right)r\cos\theta,$$
$$y = \left(1 - \frac{z}{h}\right)r\sin\theta,$$
$$z = z.$$

Other examples of surfaces of revolution are obtained by revolving the graph of a function of one variable about an axis. Imagine a curve given by the graph of $z = f(x)$ with $f(x) \geq 0$ on an interval $a \leq x \leq b$. This curve is rotated about the x-axis to obtain the surface in Figure 16.24.

The cross-section at a point x is a circle of radius $f(x)$ parallel to the yz-plane. The circle is traced out by the vector $f(x)\cos\theta\vec{j} + f(x)\sin\theta\vec{k}$. We get to a point on the surface by adding this to the vector $x\vec{i}$. This leads to the parametric equations

$$x = x,$$
$$y = f(x)\cos\theta,$$
$$z = f(x)\sin\theta,$$

where $0 \leq \theta \leq 2\pi$ and $a \leq x \leq b$.

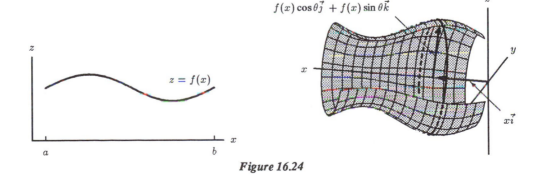

Figure 16.24

Parameter Curves

When a surface is described parametrically there are two families of curves that are naturally associated with the parameterization. Each curve is obtained by setting one parameter equal to a constant.

Example 8 Consider Planetown of Example 2 on page 294. Main Street is a one-way street through the center of town running west to east. State Street is a one-way street through the center of town running south to north. First Street is a one-way street parallel to Main Street, one block north of Main running in the opposite direction to Main. California Avenue is a one-way street parallel to State, one block west and running in the same direction as State. See Figure 16.25.

   (a) Find parametric equations for Main Street and for State Street.

   (b) Find parametric equations for First Street and for California Ave.

   (c) Write the equations of these four streets in terms of the parameters.

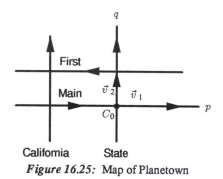

Figure 16.25: Map of Planetown

Solution (a) Main street goes through the center of town $C_0 = (2, 1, 3)$ and is in the direction of the vector $\vec{v_1} = 2\vec{i} + 3\vec{j} - \vec{k}$. Therefore, writing the parameter as p, the parametric equations for Main Street are

$$x = 2 + 2p, \quad y = 1 + 3p, \quad z = 3 - p.$$

Likewise, State Street goes through C_0 and is in the direction $\vec{v_2} = \vec{i} - 4\vec{j} + 5\vec{k}$. Writing the other parameter as q, the parametric equations for State Street are

$$x = 2 + q, \quad y = 1 - 4q, \quad z = 3 + 5q.$$

   (b) Follow State Street one block north from the center of town. This puts us at the point $(3, -3, 8)$. First Street goes through this point in the direction $-\vec{v}_1 = -2\vec{i} - 3\vec{j} + \vec{k}$ (opposite to Main Street). The parametric equations for First Street are therefore

$$x = 3 - 2p, \quad y = -3 - 3p, \quad z = 8 + p.$$

To get to California Avenue, we walk one block west from the center of town to the point $(0, -2, 4)$. California Avenue runs through this point in the direction $\vec{v_2} = \vec{i} - 4\vec{j} + 5\vec{k}$, so its parametric equations are

$$x = q, \quad y = -2 - 4q, \quad z = 4 + 5q.$$

(c)   Any point in Planetown is given parametrically by the equation

$$(x, y, z) = (2, 1, 3) + p\vec{v_1} + q\vec{v_2},$$

which can be written as

$$x = 2 + 2p + q, \quad y = 1 + 3p - 4q, \quad z = 3 - p + 5q.$$

Main street consists of points obtained by taking $q = 0$ and any value of p. Thus, Main Street corresponds to the equation $q = 0$ and parameterized by p. Similarly, State Street is described by the equation $p = 0$ and parameterized by q. First Street is described by the equation $q = 1$ and is parameterized by $-p$. California Ave. is described by the equation $p = -1$ and parameterized by q.

Example 9   Suppose that the earth is represented by the sphere $x^2 + y^2 + x^2 = 1$.
(a)   New York City and Rome are on the same latitude, roughly $21°$ from the north pole. What is the equation of the circle of latitude that passes through both cities? What is the parametric equation of the part of the latitude that goes just from Rome to New York?

(b)   Istanbul, Turkey, and Johannesburg, South Africa, are on approximately the same longitude $30°$ E. What is the equation of the longitude that runs between the two cities?

Solution   (a)   In radians, the circle of latitude passing through New York City and Rome is $\phi = 21\pi/180 = 7\pi/60$. We get parametric equations for the entire circle as a curve on the sphere in terms of the single parameter θ by setting $\phi = 7\pi/60$ in the parametric equations for the sphere and letting θ vary from 0 to 2π. In Cartesian coordinates, we can write the curve as

$$(x, y, z) = (\sin\frac{7\pi}{60}\cos\theta, \sin\frac{7\pi}{60}\sin\theta, \cos\frac{7\pi}{60}).$$

Since Rome is on longitude $12°$E (that is, $\theta = -12\pi/180$) and New York is $74°$W (that is, $\theta = 74\pi/180$) the equation of the part of the latitude between Rome and New York corresponds to

$$\phi = \frac{7\pi}{60} \quad \text{and} \quad \frac{-12\pi}{180} \leq \theta \leq \frac{74\pi}{180}.$$

(b)   Since $30°$E corresponds to $\theta = -30\pi/180 = -\pi/6$, the longitude through Istanbul and Johannesburg is $\theta = -\pi/6$. In Cartesian coordinates the curve is

$$(x, y, z) = (\sin\phi \cos(-\frac{\pi}{6}), \sin\phi \sin(-\frac{\pi}{6}), \cos\phi) = \left(\frac{\sqrt{3}}{2}\sin\phi, -\frac{1}{\sqrt{2}}\sin\phi, \cos\phi \right).$$

The streets in Planetown and the curves on the sphere in the two examples are *parameter curves*. One parameter is held fixed and the other is free to vary. Anytime we work with a parametric representation of a surface, say

$$x = f_1(p, q), \quad y = f_2(p, q), \quad z = f_3(p, q)$$

we have corresponding parameter curves. In fact, through each point $(f_1(p_0, q_0), f_2(p_0, q_0), f_3(p_0, q_0))$ on the surface there are two curves, a "p-curve" given by setting $q = q_0$:

$$(x, y, z) = (f_1(p, q_0), f_2(p, q_0), f_3(p, q_0)),$$

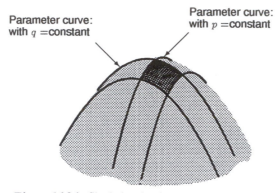

Parameter curve: with q =constant

Parameter curve: with p =constant

Figure 16.26: Shaded region is parameter rectangle

and a "q-curve" given by setting $p = p_0$:

$$(x, y, z) = (f_1(p_0, q), f_2(p_0, q), f_3(p_0, q)).$$

We have already seen parameter curves before. The traces obtained by setting $x = a$ and $y = b$ on a surface $z = f(x, y)$ are examples of parameter curves. So are the grid lines on a computer sketch of a surface. The small rectangular regions surrounded by nearby pairs of parameter curves are called *parameter rectangles*. See Figure 16.26.

Problems for Section 16.3

1. Describe in words the curve $\phi = \pi/4$ on the surface of the globe.
2. Describe in words the curve $\theta = \pi/4$ on the surface of the globe.
3. You are at a point on the earth with longitude 80° west of Greenwich, England, and latitude 40° north of the equator.

   (a) Are you in the northern or southern hemisphere?
   (b) If your latitude decreases have you moved nearer to or farther from the equator?
   (c) If your latitude decreases, have you moved nearer or farther from the north pole?
   (d) If your longitude increases, have you moved nearer to or farther from Greenwich?

4. A city is described parametrically by the equation

$$(x, y, z) = (x_0, y_0, z_0) + p\vec{v_1} + q\vec{v_2}$$

   where $\vec{v}_1 = 2\vec{i} - 3\vec{j} - \vec{k}$ and $\vec{v}_2 = \vec{i} + 4\vec{j} + 5\vec{k}$. A city block is a parallelogram determined by $\vec{v}_1$ and $\vec{v}_2$. East is in the direction of $\vec{v}_1$ and north is in the direction of $\vec{v}_2$. Starting at the point (x_0, y_0, z_0) you walk 5 blocks east, 4 blocks north, 1 block west and 2 blocks south. What are the parameters of the point where you wind up? What are your x, y and z coordinates at that point?

5. Find a parameterization for the plane through $(1, 3, 4)$ and orthogonal to $\vec{n} = 2\vec{i} + \vec{j} - \vec{k}$.
6. Find parametric equations for the sphere centered at the origin and having radius 2.
7. Find parametric equations for the sphere centered at the point $(2, -1, 3)$ and with radius 5.

1. a horizontal circle in the northern hemisphere

16.3

3a. northern   5. a plane through $(1,3,4)$ and orthogonal to $\vec{v} = 2\hat{i} + \hat{j} - \hat{k}$

1-15 odd

b. nearer   is $2(x-1) + (y-3) - (z-4) = 0 \Rightarrow 2x + y - z = 0$

c. farther   7. $x = 2 + 5\sin\phi\cos\theta$ 9. $x = (1 - \frac{z}{h})r\cos\theta$

d. farther $y = -1 + 5\sin\phi\sin\theta$ $y = (1 - \frac{z}{h})r\sin\theta$

$z = -3 + 5\cos\phi$ $z = z$

(?)
11. $\phi = \pi$; $\theta = \frac{\pi}{2}$ 13a. When z increases, the x-coordinate value will increase

so it will spread out

b. When b increases, the y-coordinate will increase,

(?)
15a.   so it will widen

c. if c increases, the x-coordinate will increase

so it will compress

It can be flipped over by making c negative

8. Find parametric equations for the sphere $(x - a)^2 + (y - b)^2 + (z - c)^2 = d^2$.

9. Find parametric equations for the cone $x^2 + y^2 = z^2$.

10. Adapt the parameterization for the sphere to find a parameterization for the ellipsoid

$$\frac{x^2}{a^2} + \frac{y^2}{b^2} + \frac{z^2}{c^2} = 1.$$

11. Suppose you are standing at a point on the equator. If you go halfway around the equator and halfway up toward the north pole along a longitude, what will be your new latitude and longitude in terms of your old latitude and longitude?

12. Find a parameterization for the following surfaces.
    (a) The spherical cap $x^2 + y^2 + z^2 = 1$, with $1/2 \leq z \leq 1$
    (b) The spherical cap $x^2 + y^2 + z^2 = 1$, with $-1 \leq x \leq -1/3$
    (c) The intersection of the spherical caps in part (a) and part (b)

13. Figure 16.27 is a picture of the parametric surface

$$x = a(p + q), \quad y = b(p - q), \quad z = 4cq^2$$

for $a = 1$, $b = 1$ and $c = 1$. What happens if you increase a? Increase b? Increase c? How could we flip the surface upside down?

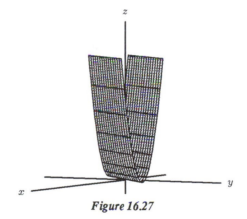

Figure 16.27

14. You are standing at the point $(-5, 0, 5)$, your friend Jane is standing at the point $(0, -5, -5)$ and her friend Jo is standing at the point $(10, 5, 0)$. Find parametric equations for the plane you are all standing on so that your parameters are $(0, 0)$, Jane's are $(1, 0)$ and Jo's are $(0, 1)$. Find a parameterization so Jane's and Jo's parameters are switched. Find a parameterization so that your parameters and Jane's are switched.

15. There is a famous way to project the plane onto the sphere (or the sphere onto the plane) called *stereographic projection*. We work with the sphere $X^2 + Y^2 + Z^2 = 1$. Draw a line from a point (x, y) in the plane to the north pole $(0, 0, 1)$. This line intersects the sphere in a point (X, Y, Z). This gives a parameterization of the sphere by points in the plane.
    (a) Which point corresponds to the south pole?
    (b) Which points correspond to the equator?
    (c) Do we get all the points of the sphere by this parameterization?
    (d) Which points correspond to the upper hemisphere?
    (e) Which points correspond to the lower hemisphere?

16.4   MOTION IN SPACE: VELOCITY VECTORS

We can represent the velocity of a moving particle by a vector with the following properties:

> **The velocity vector of a moving object has:**
>
> - Magnitude equal to the speed of the object
> - Direction equal to the direction of motion; the velocity vector is tangent to the object's path.

Example 1   Find the velocity vector of an object moving with the circular motion specified by

$$x = \cos 3t, \ y = \sin 3t.$$

Solution   The particle moves at a constant speed around a circle of radius 1, going around the circle three times in 2π seconds. Three revolutions around a circle of radius one is a distance of 6π, so the particle's speed is $6\pi/2\pi = 3$. Hence, the magnitude of the velocity vector is 3. The direction of motion is tangent to the circle, and hence perpendicular to the radius at that point. Figure 16.28 shows the vector at two different points on the circle.

Figure 16.28:   The velocity vector for counterclockwise circular motion and constant speed.

Example 2   An object moving with constant velocity in three-space passes the point $(1, 1, 1)$, and then passes the point $(2, -1, 3)$ five seconds later. What is its velocity vector?

Solution   The displacement vector from $(1, 1, 1)$ to $(2, -1, 3)$ in five seconds is $\vec{d} \doteq (2, -1, 3) - (1, 1, 1) = \vec{i} - 2\vec{j} + 2\vec{k}$. The velocity vector has the same direction as $\vec{d}$, and is given by

$$\vec{v} = \frac{\vec{d}}{5} = 0.2\vec{i} - 0.4\vec{j} + 0.4\vec{k}.$$

In general:

> The velocity vector of an object moving with constant velocity is
>
> $$\vec{v} = \frac{\vec{d}}{t} = \frac{1}{t}\vec{d},$$
>
> where $\vec{d}$ is its displacement vector in time t.

What if the Velocity is Not Constant?

If the velocity is not constant, we find the velocity as in one-variable calculus: by taking a limit. Suppose a particle is moving along a curve in the xy-plane and we want to determine the velocity vector at a certain instant t. We let a small amount of time Δt go by and see how the position of the particle changes. If the position vector of the particle is $\vec{r}(t)$ at time t, then the displacement vector between times t and $t + \Delta t$ is $\Delta \vec{r} = \vec{r}(t + \Delta t) - \vec{r}(t)$. See Figure 16.29. Thus, the velocity vector is given approximately by

$$\vec{v}(t) \approx \frac{\Delta \vec{r}}{\Delta t} = \frac{\vec{r}(t + \Delta t) - \vec{r}(t)}{\Delta t}.$$

Taking the limit as Δt goes to zero, we have the following result:

The velocity vector of a moving object at time t is

$$\vec{v}(t) = \lim_{\Delta t \to 0} \frac{\vec{r}(t + \Delta t) - \vec{r}(t)}{\Delta t},$$

where $\vec{r}(t)$ is the position vector of the object at time t.

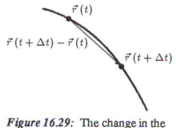

Figure 16.29: The change in the position vector for a particle moving on a curve.

The Components of the Velocity Vector

If we represent the motion of a particle in two-space parametrically by $x(t)$ and $y(t)$, then we can write the components of its position vector: $\vec{r}(t) = x(t)\vec{i} + y(t)\vec{j}$. Knowing the components of the position vector enables us to compute the components of the velocity vector:

$$\begin{aligned}
\vec{v}(t) &= \lim_{\Delta t \to 0} \frac{\vec{r}(t + \Delta t) - \vec{r}(t)}{\Delta t} \\
&= \lim_{\Delta t \to 0} \frac{(x(t + \Delta t)\vec{i} + y(t + \Delta t)\vec{j}) - (x(t)\vec{i} + y(t)\vec{j})}{\Delta t} \\
&= \lim_{\Delta t \to 0} \left(\frac{x(t + \Delta t) - x(t)}{\Delta t}\vec{i} + \frac{y(t + \Delta t) - y(t)}{\Delta t}\vec{j} \right) \\
&= x'(t)\vec{i} + y'(t)\vec{j}.
\end{aligned}$$

We have the following result:

The Components of the Velocity Vector in Two-Space

The velocity vector, $\vec{v}(t)$, at time t of a particle moving in two-space with motion described parametrically by $x(t), y(t)$ is given by

$$\vec{v} = x'(t)\vec{i} + y'(t)\vec{j} = \frac{dx}{dt}\vec{i} + \frac{dy}{dt}\vec{j}.$$

Example 3 Find the components of the velocity vector for the circular motion

$$x = \cos 3t, \ y = \sin 3t,$$

and show that they agree with the description of this vector in Example 1.

Solution We have:

$$\vec{v} = \frac{dx}{dt}\vec{i} + \frac{dy}{dt}\vec{j} = -3\sin 3t\vec{i} + 3\cos 3t\vec{j}.$$

We observe that the magnitude of $\vec{v}$ is $\|\vec{v}\| = \sqrt{9\sin^2 3t + 9\cos^2 3t} = 3\sqrt{\sin^2 3t + \cos^2 3t} = 3$, as we expected. To see that the direction is correct, we should show that the vector $\vec{v}$ at any time t is perpendicular to the vector pointing from the origin to the position of the particle at time t. The radius vector from the origin to the particle is $\vec{r} = \cos 3t\vec{i} + \sin 3t\vec{j}$ (see Figure 16.30).

To test perpendicularity, we see that the dot product of $\vec{v}$ and $\vec{r}$ is zero:

$$\begin{aligned}
\vec{v} \cdot \vec{r} &= (-3\sin 3t\vec{i} + 3\cos 3t\vec{j}) \cdot (\cos 3t\vec{i} + \sin 3t\vec{j}) \\
&= -3\sin 3t\cos 3t + 3\cos 3t\sin 3t \\
&= 0.
\end{aligned}$$

Thus, the velocity vector is perpendicular to $\vec{r}$ and hence tangent to the circle, as expected.

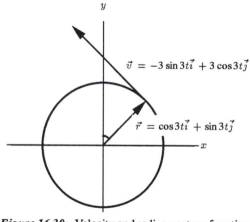

Figure 16.30: Velocity and radius vector of motion around a circle

Example 4 Consider the motion of the particle given by the parametric equations

$$x = t^3 - 3t, \ y = t^2 - 2t.$$

(a) Does the particle ever come to a stop? If so, when?

(b) Is the particle ever moving straight up or down? If so, where?

(c) Is the particle ever moving straight horizontally right or left? Is so, where?

(d) Give parametric equations for the tangent line to the curve at time $t = -2$.

Solution (a) In order for the particle to be stopped, its velocity vector $\vec{v} = (dx/dt)\vec{i} + (dy/dt)\vec{j}$ must be zero. That is, we must have simultaneously $dx/dt = 0$ and $dy/dt = 0$ (that is, no motion in the x-direction and no motion in the y-direction). Thus, we need to find a value of t which satisfies both of these equations:

$$\frac{dx}{dt} = 3t^2 - 3 = 3(t-1)(t+1) = 0,$$

$$\frac{dy}{dt} = 2t - 2 = 2(t-1) = 0.$$

The value $t = 1$ is the only one that works. The particle at time $t = 1$ is at the position $(t^3 - 3t, \ t^2 - 2t)\Big|_{t=1} = (-2, -1)$.

(b) In order for the particle to be traveling straight up or down the x-component of the velocity vector must be 0. Thus, we solve $dx/dt = 3t^2 - 3 = 0$ and get $t = \pm 1$. However, at $t = 1$ the particle has no vertical motion, as we saw in part (a). Thus, the particle is moving straight up or down only when $t = -1$. The position at that time is $(t^3 - 3t, \ t^2 - 2t)\Big|_{t=-1} = (2, 3)$.

(c) For straight horizontal motion we need $dy/dt = 0$. That happens when $dy/dt = 2t - 2 = 0$, $t = 1$. But we already saw in part (a) that $dx/dt = 0$ also at $t = 1$ so the particle is not moving at all when $t = 1$. Thus, there is no point where the motion is horizontal.

(d) The velocity vector is tangent to the curve. At $t = -2$, $\vec{v} = (3t^2 - 3)\vec{i} + (2t - 2)\vec{j}\Big|_{t=-2} = 9\vec{i} - 6\vec{j}$. Thus, the tangent line should have parametric equations where the x value changes by 9 for each unit change in time, and the y value changes by -6 for each unit change in time. Also, the tangent line must pass through the point where the particle is at time $t = -2$, namely the point $(t^3 - 3t, \ t^2 - 2t)\Big|_{t=-2} = (-2, 8)$. Therefore, parametric equations for the tangent line are

$$x = -2 + 9(t + 2)$$
$$y = 8 - 6(t + 2).$$

(Other parameterizations of the line are possible.) A sketch of the curve with its tangent line is given in Figure 16.31.

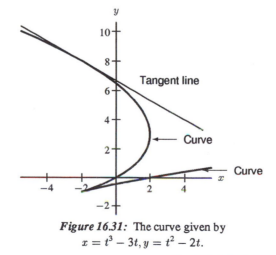

Figure 16.31: The curve given by $x = t^3 - 3t, y = t^2 - 2t$.

Local Linearity

We can see the meaning of the velocity vector by looking at a table of values of x and y coordinates in terms of a parameter t.

Example 5 Give a table of values near $t = 1$ for the circular motion with position vector

$$\vec{r}(t) = \cos t \vec{i} + \sin t \vec{j}$$

and interpret the table in terms of the velocity vector at time $t = 1$.

Solution Table 16.4 shows values near $t = 1$ with t changing by increments of 0.01.

TABLE 16.4
Values for the position vector
$\vec{r} = \cos t \vec{i} + \sin t \vec{j}$ *near* $t = 1$

t	$\vec{r}$
0.98	$0.5570\vec{i} + 0.8305\vec{j}$
0.99	$0.5487\vec{i} + 0.8360\vec{j}$
1.00	$0.5403\vec{i} + 0.8415\vec{j}$
1.01	$0.5319\vec{i} + 0.8468\vec{j}$
1.02	$0.5234\vec{i} + 0.8521\vec{j}$

As we go down the table, the x-values are decreasing by about 0.0084 and the y-values are increasing by about 0.0054. Thus, a change in time of $\Delta t = 0.01$ produces the change in position vector $\Delta x \vec{i} + \Delta y \vec{j} = -0.0084\vec{i} + 0.0054\vec{j}$. Thus, the velocity vector is approximately

$$\vec{v} \approx \frac{1}{\Delta t}\left(\Delta x \vec{i} + \Delta y \vec{j}\right) = -0.84\vec{i} + 0.54\vec{j}.$$

Note that this velocity vector is perpendicular to the radius vector from the origin $(0, 0)$ to the position $(0.54, 0.84)$ at time $t = 1$. Finally, the x and y values in the table are almost indistinguishable from those of linear motion given by $x = 0.5403 - 0.84t$, $y = 0.8415 + 0.54t$, which are the parametric equations for the tangent line through the point $(0.5403, 0.8415)$.

Speed and the Length of a Curve

The speed, or how fast a particle is moving, is given by the magnitude of the velocity vector,

$$\text{Speed} = \|\vec{v}\| = \sqrt{\left(\frac{dx}{dt}\right)^2 + \left(\frac{dy}{dt}\right)^2}.$$

Just as we can find the total distance traveled by a particle moving along the x-axis by integrating the absolute value of its velocity, we can find the distance traveled along a curve by integrating the magnitude of its velocity vector. Thus,

$$\text{Distance traveled} = \int_a^b \|\vec{v}(t)\|\, dt$$

If the particle doesn't stop and back up as it moves along the curve, the distance it travels will be the same as the length of a curve. Thus, we have that:

If the curve C is given parametrically for $a \leq t \leq b$ and if $\vec{v}$ is never $\vec{0}$ (i.e. particle does not stop) then

$$\text{Length of } C = \int_a^b \|\vec{v}\| \, dt = \int_a^b \sqrt{\left(\frac{dx}{dt}\right)^2 + \left(\frac{dy}{dt}\right)^2} \, dt.$$

The integral giving the length of a curve is usually computed numerically.

Example 6 Find the circumference of the ellipse given by the parametric equations

$$x = 2\cos t, \ y = \sin t, \ 0 \leq t \leq 2\pi.$$

Solution The length of this curve is given by the integral

$$\int_0^{2\pi} \sqrt{\left(\frac{dx}{dt}\right)^2 + \left(\frac{dy}{dt}\right)^2} \, dt = \int_0^{2\pi} \sqrt{(-2\sin t)^2 + (\cos t)^2} \, dt$$

$$= \int_0^{2\pi} \sqrt{4\sin^2 t + \cos^2 t} \, dt.$$

Simpson's rule with $n = 20$ gives a value of 9.688449 and with $n = 40$ gives a value of 9.688448. Thus, the length of the curve is approximately 9.68845. Since the given ellipse is inscribed in a circle of radius 2 and circumscribes a circle of radius 1, we would expect the length of the ellipse to be between $2\pi(2) \cong 12.56$ and $2\pi(1) \cong 6.28$ so the value of 9.68845 is reasonable.

Velocity in Three Dimensions

Everything we have said about velocity vectors in two dimensions generalizes to three dimensions. Given parametric equations in terms of time t for the x, y, and z coordinates of a moving particle, the velocity vector is given by:

$$\vec{v} = \frac{dx}{dt}\vec{i} + \frac{dy}{dt}\vec{j} + \frac{dz}{dt}\vec{k}.$$

The velocity vector is tangent to the curve of motion: that is, if we zoom in on a small piece of the curve, the curve looks like a straight line lying over the velocity vector. The length of the velocity vector gives us the speed at which the particle is moving:

$$\text{Speed} = \|\vec{v}\| = \sqrt{\left(\frac{dx}{dt}\right)^2 + \left(\frac{dy}{dt}\right)^2 + \left(\frac{dz}{dt}\right)^2}.$$

If we integrate the speed with respect to time we have the distance traveled by the particle between $t = a$ and $t = b$:

$$\text{Distance traveled} = \int_a^b \sqrt{\left(\frac{dx}{dt}\right)^2 + \left(\frac{dy}{dt}\right)^2 + \left(\frac{dz}{dt}\right)^2} \, dt.$$

If the particle never retraces its steps, the distance it travels equals the length of the curve.

Example 7   Consider the motion of the particle tracing out the curve:

$$x = (4 + \sin t)(\cos 3t),$$
$$y = (4 + \sin t)(\sin 3t),$$
$$z = \cos t.$$

(a)   Find the speed of the particle at the times when the particle's motion is parallel to the xy-plane.

(b)   Find the length of the curve from $t = 0$ to $t = 2\pi$.

Solution   We find the velocity vector and then the speed at any time t.

$$\frac{dx}{dt} = \cos t \cos 3t + (4 + \sin t)(-3 \sin 3t),$$
$$\frac{dy}{dt} = \cos t \sin 3t + (4 + \sin t)(3 \cos 3t),$$
$$\frac{dz}{dt} = -\sin t.$$

The velocity vector is

$$\vec{v}(t) = \frac{dx}{dt}\vec{i} + \frac{dy}{dt}\vec{j} + \frac{dz}{dt}\vec{k}.$$

In this example, it turns out that to find an expression for the speed, it is helpful first to find the sum of the squares of dx/dt and dy/dt. Notice that both dx/dt and dy/dt are the sum of two terms and that the product of these two terms are the same in both cases, but with opposite signs. Thus, when we find the sum of the squares the cross terms cancel. We get:

$$\left(\frac{dx}{dt}\right)^2 + \left(\frac{dy}{dt}\right)^2 = \cos^2 t \, \cos^2 3t + (4 + \sin t)^2 \, 9 \sin^2 3t$$
$$+ \cos^2 t \, \sin^2 3t + (4 + \sin t)^2 \, 9 \cos^2 3t$$
$$= \cos^2 t \, (\cos^2 3t + \sin^2 3t) + 9 \, (4 + \sin t)^2 \, (\sin^2 3t + \cos^2 3t)$$
$$= \cos^2 t + 9 \, (4 + \sin t)^2.$$

Adding in the square of $\frac{dz}{dt} = (-\sin t)$ we get:

$$\left(\frac{dx}{dt}\right)^2 + \left(\frac{dy}{dt}\right)^2 + \left(\frac{dz}{dt}\right)^2 = \cos^2 t + 9 \, (4 + \sin t)^2 + \sin^2 t$$
$$= 1 + 9 \, (4 + \sin t)^2.$$

Thus, the speed at time t is

$$\text{Speed} = \|\vec{v}\| = \sqrt{1 + 9 \, (4 + \sin t)^2}.$$

We now answer parts (a) and (b).

(a)   The motion is parallel to the xy-plane when $dz/dt = 0$. Thus, we want $\sin t = 0$. The speed is $\sqrt{1 + 9(4 + 0)^2} = \sqrt{145}$ units per second.

1. $x = t^2 \Rightarrow x'(t) = 2t$

$y = t^3 \Rightarrow y'(t) = 3t^2$

$\vec{v} = 2t\,\hat{\imath} + 3t^2\,\hat{\jmath}$

$\|\vec{v}\| = \sqrt{(2t)^2 + (3t^2)^2}$

$\quad = \sqrt{4 + 9t^2} \cdot |t|$

3. $x = \cos 2t \Rightarrow x'(t) = -2\sin 2t$

$y = \sin t \Rightarrow y'(t) = \cos t$

$\vec{v} = (-2\sin 2t)\,\hat{\imath} + \cos t\,\hat{\jmath}$

$\|\vec{v}\| = \sqrt{4\sin^2(2t) + \cos^2(t)}$

5. $x = t \Rightarrow x'(t) = 1$

$y = t^2 \Rightarrow y'(t) = 2t$

$z = t^3 \Rightarrow z'(t) = 3t^2$

$\vec{v} = \hat{\imath} + 2t\,\hat{\jmath} + 3t^2\,\hat{k}$

$\|\vec{v}\| = \sqrt{1 + 4t^2 + 9t^4}$

7. $\vec{v} = 2t\,\hat{\imath} + 3t^2\,\hat{\jmath} \Rightarrow t = 2$

$\quad = 2(2)\,\hat{\imath} + 3(2)^2\,\hat{\jmath}$

$\quad = 4\,\hat{\imath} + 12\,\hat{\jmath}$

$(t^2, t^3)\big|_{t=2} = (4, 8)$

$x = 4 + 4(t-2)$

$y = 8 + 12(t-2)$

9. $x = 3 + 5t$

$y = 1 + 4t$

$z = 3 - t$

$D = \int_1^2 \sqrt{(13-8)^2 + (9-5)^2 + (1-2)^2}$

$\quad = \sqrt{42} = $ length of line from

$(13, 9, 1)$ to $(8, 5, 2)$

11. $x = \cos 3t$

$y = \sin 5t$

$0 \le t \le 2\pi$

(b) Since the x, y, z coordinates repeat their values with period 2π, we need only integrate the speed from $t = 0$ to $t = 2\pi$.

$$\text{Length} = \int_0^{2\pi} \sqrt{1 + 9(4 + \sin t)^2}\, dt.$$

We apply Simpson's rule. With $n = 20$ and $n = 40$ we get 75.6680606397 and 75.668060640, so we can safely say the length of the curve is about 75.668. Since the curve is basically going around the z-axis 3 times at a distance that averages around 4, we would expect the length to be around $3(2\pi \cdot 4) = 24(3.1416) = 75.398$ so our numerical answer makes sense.

Problems for Section 16.4

For Problems 1–6, find the velocity vector $\vec{v}(t)$ for the motion given by the parametric equations. In each case, also give the speed $\|\vec{v}(t)\|$ and any times when the particle comes to a stop.

1. $x = t^2, y = t^3$

2. $x = \cos(t^2), y = \sin(t^2)$

3. $x = \cos 2t, y = \sin t$

4. $x = \cos 3t, y = \sin 5t$

5. $x = t, y = t^2, z = t^3$

6. $x = t^2 - 2t, y = t^3 - 3t, z = 3t^4 - 4t^3$

7. Find the parametric equations for the tangent line at $t = 2$ for Problem 1.

8. Find the parametric equations for the tangent line at $t = 2$ for Problem 5.

Find the length of the curves in Problems 9–11. In each case, explain why your answer is reasonable.

9. $x = 3 + 5t, y = 1 + 4t, z = 3 - t$ for $1 \le t \le 2$.

10. $x = \cos(e^t), y = \sin(e^t)$ for $0 \le t \le 1$.

11. $x = \cos 3t, y = \sin 5t$ for $0 \le t \le 2\pi$.

12. Table 16.5 gives x and y values of the position of a particle at time t. Assuming the path is continuous, estimate the following quantities:

    (a) The velocity vector and speed at time $t = 2$.
    (b) Any times when the particle is moving straight up or down.
    (c) Any times when the particle has come to a stop.

TABLE 16.5

t	0	0.5	1.0	1.5	2.0	2.5	3.0	3.5	4.0
x	1	4	6	7	6	3	2	3	5
y	3	2	3	5	8	10	11	10	9

13. When a dangerous iceberg is spotted in the North Atlantic, it is important to be able to predict where the iceberg is likely to be a day or a week later. To do this, one needs to know the ocean currents. These currents give us the velocity vector for the motion of the iceberg, and

we want to determine the path of the iceberg from its velocity vector. Ocean currents can be measured at a number of different locations. Most maps of the ocean indicate the direction of the strongest currents, such as the Gulf Stream, with arrows.

For each of the following currents, sketch a picture of the current. Sketch the path of an iceberg in this current. Determine the location of an iceberg at time $t = 7$ if it is at the point $(1, 3)$ at time $t = 0$.

(a) The current everywhere is $\vec{i}$.

(b) The current at (x, y) is $2x\vec{i} + y\vec{j}$.

(c) The current at (x, y) is $-y\vec{i} + x\vec{j}$.

14. The currents studied in Problem 13 are examples of vector fields. Chapter 17 considers vector fields in greater detail. Explain how the Lanchester model for combat discussed in Section 16.1 is related to the iceberg problem.

15. Suppose $\vec{r} = \cos t\,\vec{i} + \sin t\,\vec{j} + 2t\,\vec{k}$ represents the position of a particle on a spiral, where z is the height of the particle above the ground.

(a) Is the particle ever moving downwards? When?

(b) When does the particle reach a point 10 units above the ground?

(c) What is the velocity of the particle when it is 10 units above the ground?

(d) Suppose the particle leaves the spiral and moves along the tangent line to the spiral at this point. What is the equation of the tangent line?

16. Mr. Skywalker is traveling along the curve given by:

$$\vec{F}(t) = -2e^{3t}\vec{i} + 5\cos t\vec{j} - 3\sin(2t)\vec{k}.$$

If the power thrusters are turned off, his ship flies off on a tangent line to $\vec{F}(t)$. He is almost out of power when he notices that a station on Xardon is open at the point with coordinates $(1.5, 5, 3.5)$. Quickly calculating his position, he turns off the thrusters at $t = 0$. Does he make it to the Xardon station? Explain.

17. Suppose you bicycle along a flat road with a safety light attached to one foot. Suppose that your bike moves at a speed of 25 km/hr and that your foot moves in a circle of radius 20 cm centered 30 cm above the ground, making one revolution per second.

(a) Find parametric equations which describe the path traced out by the light.

(b) Sketch a graph of the light's path.

(c) How fast (in revolutions/sec) would your foot have to be rotating if an observer standing at the side of the road sees the light moving backwards?

18. Suppose $F(x, y) = 1/(x^2 + y^2)$ gives the temperature at the point (x, y) in the plane. Suppose a ladybug moves along a parabola according to the parametric equations

$$x = t, \ y = t^2.$$

Find the rate of change in the temperature of the ladybug at time t.

19. This problem generalizes the result of Problem 18. Suppose $F(x, y)$ gives the temperature at any point (x, y) in the plane and that a ladybug moves in the plane with position vector at time t given by $\vec{r}(t) = x(t)\vec{i} + y(t)\vec{j}$ and velocity vector $\vec{v}(t)$. Use the chain rule to show that

Rate of change in the temperature of the bug at time $t\ = \nabla F(x(t), y(t)) \cdot \vec{v}(t).$

20. Emily is standing on the outer edge of a merry-go-round, 30 feet from the center. The merry-go-round completes one full revolution every 15 seconds. As Emily passes over point P on the ground, she drops a ball from 9 feet above the ground.

    (a) How fast is Emily going?

    (b) How far from P does the ball hit the ground? (The acceleration of gravity is 32 ft/sec^2; you will need to compute how long it takes for the ball to fall 9 feet).

    (c) How far from Emily does the ball hit the ground?

21. Problem 7 on page 324 studies the motion of a hypothetical moon which revolves around an earth which in turn revolves around a sun. Suppose the moon revolves around the earth 12 times in the time it takes for the earth to revolve once about the sun. In this problem we will investigate whether the combination of the moon's motion around the earth and the earth's motion around the sun could bring the moon to a stop at some instant. (See Figure 16.32).

    (a) Suppose the radius of the moon's orbit around the earth is 1 unit and the radius of the earth's orbit around the sun is R units. Explain why the motion of the moon relative to the sun can be described by the parametric equations:

    $$x = R\cos t + \cos 12t,$$
    $$y = R\sin t + \sin 12t.$$

    (b) Find a value for R and t such that the moon stops relative to the sun at time t.

    (c) On a graphing calculator, plot the path of the moon for the value of R you obtained in part (b). Experiment with other values for R.

22. A lighthouse L is located on an island in the middle of a lake as shown in Figure 16.33. Consider the motion of the point where the light beam from L hits the shore of the lake as the light rotates around.

    (a) Suppose the beam rotates counterclockwise about L at a constant angular velocity. At which point A, B, C, D, or E do you think the speed of the point is greatest and at which point do you think it is smallest?

    (b) Now suppose the beam rotates counterclockwise around L so that the beam sweeps out equal areas of the lake in equal times. At which point, A, B, C, D or E, do you think the light is moving fastest and slowest?

    (c) What happens if you place the lighthouse at different points in the lake. Can the speed of the point on the shore ever be infinite for part (a)?

    (d) Suppose now that the lake is rectangular instead. What happens to the velocity vector at the corners? For part (b) show that the speed is constant along each side (possibly a different constant on each side).

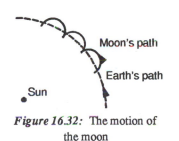

Figure 16.32: The motion of the moon

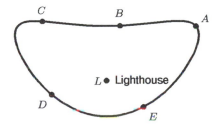

Figure 16.33: The lighthouse in the lake

16.5   THE ACCELERATION VECTOR IN TWO AND THREE DIMENSIONS

Just as the velocity of a particle moving in the xy-plane is a vector quantity, so is the rate of change of the velocity of the particle, namely its acceleration.

Limit Definition of the Acceleration Vector

Figure 16.34 shows a particle moving along a curve at some time t with velocity vector $\vec{v}(t)$ and then at some time a little later $t + \Delta t$.

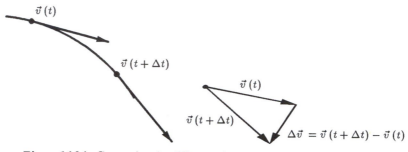

Figure 16.34: Computing the difference between two velocity vectors

Then the vector $\Delta\vec{v} = \vec{v}(t+\Delta t) - \vec{v}(t)$ points approximately in the direction of the acceleration:

$$\text{Acceleration} \approx \frac{\Delta\vec{v}}{\Delta t} = \frac{\vec{v}(t + \Delta t) - \vec{v}(t)}{\Delta t}.$$

In the limit as $\Delta t \to 0$, we have the instantaneous acceleration at time t:

The acceleration vector of a moving object at time t, denoted $\vec{a}(t)$, is

$$\vec{a}(t) = \lim_{\Delta t \to 0} \frac{\vec{v}(t + \Delta t) - \vec{v}(t)}{\Delta t}$$

where $\vec{v}(t)$ is the velocity at time t.

Components of the Acceleration Vector

If the motion is described parametrically by $x(t), y(t)$, we can express the acceleration in components. The velocity vector $\vec{v}(t)$ has components $\vec{v}(t) = x'(t)\vec{i} + y'(t)\vec{j}$. From the definition of the acceleration vector, we have

$$\begin{aligned}
\vec{a}(t) &= \lim_{\Delta t \to 0} \frac{\vec{v}(t + \Delta t) - \vec{v}(t)}{\Delta t} \\
&= \lim_{\Delta t \to 0} \left[\frac{(x'(t + \Delta t)\vec{i} + y'(t + \Delta t)\vec{j}) - (x'(t)\vec{i} + y'(t)\vec{j})}{\Delta t} \right] \\
&= \lim_{\Delta t \to 0} \left[\frac{(x'(t + \Delta t) - x'(t))\vec{i} + (y'(t + \Delta t) - y'(t))\vec{j}}{\Delta t} \right] \\
&= \lim_{\Delta t \to 0} \left[\frac{x'(t + \Delta t) - x'(t)}{\Delta t}\vec{i} + \frac{y'(t + \Delta t) - y'(t)}{\Delta t}\vec{j} \right] \\
&= x''(t)\vec{i} + y''(t)\vec{j}.
\end{aligned}$$

Components of the Acceleration Vector in Two-Space

The acceleration vector, $\vec{a}\,(t)$, at time t of a particle with motion described parametrically by $x(t)$, $y(t)$ is given by

$$\vec{a}\,(t) = x''(t)\vec{i} + y''(t)\vec{j} = \frac{d^2x}{dt^2}\vec{i} + \frac{d^2y}{dt^2}\vec{j}\,.$$

Example 1   Give parametric equations for motion in a circle of radius 7 at a constant speed of 5. Compute the acceleration vector $\vec{a} = x''(t)\vec{i} + y''(t)\vec{j}$.

Solution   Since the circle has radius 7, we use parametric equations of the form

$$x = 7\cos kt, \qquad y = 7\sin kt$$

where the value of k will be chosen to make the speed 5. Since

$$\frac{dx}{dt} = -7k\sin kt, \qquad \frac{dy}{dt} = 7k\cos kt,$$

we have (assuming $k > 0$)

$$\text{Speed} = \sqrt{\left(\frac{dx}{dt}\right)^2 + \left(\frac{dy}{dt}\right)^2} = \sqrt{(-7k\sin t)^2 + (7k\cos t)^2} = 7k.$$

So we take $k = 5/7$. Then the acceleration vector is

$$\begin{aligned}
\vec{a} &= \frac{d^2x}{dt^2}\vec{i} + \frac{d^2y}{dt^2}\vec{j} \\
&= -(7\cdot\frac{5}{7}\cdot\frac{5}{7}\cos\frac{5}{7}t)\vec{i} - (7\cdot\frac{5}{7}\cdot\frac{5}{7}\sin\frac{5}{7}t)\vec{j} \\
&= -\frac{5^2}{7}\cos\frac{5}{7}t\vec{i} - \frac{5^2}{7}\sin\frac{5}{7}t\vec{j}\,.
\end{aligned}$$

Observe that the acceleration vector is $\left(\frac{5}{7}\right)^2$ times the negative of the position vector $x\vec{i} + y\vec{j} = 7\cos\frac{5}{7}t\vec{i} + 7\sin\frac{5}{7}t\vec{j}$ and thus points toward the origin. The length of the acceleration vector is $\frac{5^2}{7}$.

In general, the following result is always true:

A particle traveling around a circle of radius r with constant speed $\|\vec{v}\|$ has its acceleration vector pointing to the center of the circle, with magnitude $\|\vec{a}\| = \dfrac{\|\vec{v}\|^2}{r}$.

In uniform circular motion, the acceleration vector reflects the fact that the velocity vector does not change in magnitude, only in direction. We now look at linear motion where the velocity vector always has the same direction but the magnitude changes. We expect that the acceleration vector will point in the same direction as the velocity vector if the speed is increasing and in the opposite direction as the velocity vector if the speed is decreasing. The following example verifies these expectations.

Example 2   Consider the motion given by the parametric equations

$$x = 2 + 4(t^3 + t), \; y = 6 + 3(t^3 + t).$$

Show that this is linear motion in the direction of the vector $4\vec{i} + 3\vec{j}$ and relate the acceleration vector to the velocity vector.

Solution   We observe that the given equations are obtained by making a parameter change of $t^3 + t = s$ in the parametric equations

$$x = 2 + 4s, \; y = 6 + 3s.$$

Thus, we have motion in a straight line through the point $(2, 6)$ and in the direction of the vector $4\vec{i} + 3\vec{j}$. The velocity vector is

$$
\begin{aligned}
\vec{v} &= 4(3t^2 + 1)\vec{i} + 3(3t^2 + 1)\vec{j} \\
&= (3t^2 + 1)(4\vec{i} + 3\vec{j}).
\end{aligned}
$$

Since $(3t^2 + 1)$ is a scalar which does not change the direction of $\vec{v}$, the velocity vector always points in the direction of the vector $4\vec{i} + 3\vec{j}$. In addition,

$$\text{Speed} = \|\vec{v}\| = (3t^2 + 1)\sqrt{4^2 + 3^2} = 5(3t^2 + 1).$$

Notice that the speed increases for $t > 0$. The acceleration vector is

$$
\begin{aligned}
\vec{a} &= 4(6t)\vec{i} + 3(6t)\vec{j} \\
&= 6t(4\vec{i} + 3\vec{j}).
\end{aligned}
$$

The acceleration vector for $t > 0$ points in the same direction as $4\vec{i} + 3\vec{j}$, which is the same direction as $\vec{v}$. For $t < 0$, the acceleration vector $6t(4\vec{i} + 3\vec{j})$ points in the opposite direction of the velocity vector. Finally, the magnitude of the acceleration vector is

$$\|\vec{a}\| = |6t|\sqrt{4^2 + 3^2} = |6t|5 = |30t|,$$

which is the absolute value of the rate of change of the speed $5(3t^2 + 1)$.

Acceleration in Three Dimensions

The acceleration vector in three dimensions is defined and computed exactly the same way as in two dimensions:

$$\vec{a}\,(t) = \frac{d^2x}{dt^2}\vec{i} + \frac{d^2y}{dt^2}\vec{j} + \frac{d^2z}{dt^2}\vec{k}$$

Example 3   Find the velocity and acceleration vectors for the helical motion:

$$x = 3\cos{(t^2)}, \; y = 3\sin{(t^2)}, \; z = t^2.$$

Solution   We have

$$\vec{v}(t) = -6t \sin(t^2)\vec{i} + 6t \cos(t^2)\vec{j} + 2t\vec{k},$$
$$\vec{a}(t) = (-6 \sin(t^2) - 12t^2 \cos(t^2))\vec{i} + (6 \cos(t^2) - 12t^2 \sin(t^2))\vec{j} + 2\vec{k}.$$

Problems for Section 16.5

For Problems 1–5, find the velocity and acceleration vectors for the parameterized motions.

1. $x = 3\cos t, y = 4\sin t$

2. $x = t, y = t^3 - t$

3. $x = 2 + 3t, y = 4 + t, z = 1 - t$

4. $x = 2 + 3t^2, y = 4 + t^2, z = 1 - t^2$

5. $x = t, y = t^2, z = t^3$

6. The moon is circling around the earth which is itself circling around the sun. If we assume that both motions are circular, we can write the position of the moon relative to the sun parametrically as

$$\vec{r} = x\vec{i} + y\vec{j} = (93\cos(2\pi t) + 0.24\cos(24\pi t))\vec{i} + (93\sin(2\pi t) + 0.24\sin(24\pi t))\vec{j}$$

where x and y are in millions of miles.

(a) What are the units of t? What is the period of each of the two motions?
(b) How far is it from the earth to the sun and from the earth to the moon?
(c) What are the longest and shortest distances between the moon and the sun?
(d) What is the velocity, speed, and acceleration of the moon at $t = 1/3$?

7. The moon takes about 27 days to circle once around the earth at a distance of about 240,000 miles from the center of the earth. Assuming the orbit of the moon around the earth is a circle and the speed of the moon is constant, compute the acceleration of the moon in ft/sec^2. Newton used this computation to verify his law of gravitation, which says the acceleration caused by the gravitational pull of the earth should be proportional to the inverse of the square of the distance from the center of the earth. Compare the acceleration of the moon with the acceleration of gravity at the surface of earth, namely 32 ft/sec^2 (The surface of the earth is about 4,000 miles from the center of the earth). Do you have rough agreement with Newton's law?

8. Determine the position vector $\vec{F}(t)$ for a rocket which is launched from the origin at time $t = 0$ seconds, reaches its highest point of $(1000, 3000, 10000)$ ft, and is only subject to the force of gravity after the launch.

16.6   NOTES ON CURVATURE

We have seen that there are two factors that influence the acceleration vector of a moving particle: change in speed and change in direction of the velocity vector. If a particle is moving in a straight line, all of its acceleration is explained by change of speed. If a particle is moving at a constant speed, as in uniform motion in a circle, all of the acceleration is explained by change in the direction of the velocity vector. Is there a way to analyze the acceleration of a particle that is following a curved path and is changing its speed? We are especially interested in this because it gives us a chance to measure the curvature of the path traveled by the particle, independently of the speed at which the particle traverses the path. This curvature tells us something about the geometry of the curve along which the particle is moving.

The basic idea is to write the velocity vector at time t as the product of the speed $s(t)$ of the particle at time t (a scalar) and a vector of unit length $\vec{T}(t)$ tangent to the curve, pointing in the direction of the velocity vector:

$$\vec{v}(t) = s(t)\vec{T}(t), \text{ where } s(t) \text{ is the speed and } \vec{T}(t) \text{ is the unit tangent.}$$

If we differentiate this equation with respect to t we will get the acceleration vector. We use an extension of the product rule for derivatives to get:

$$\vec{a} = \frac{d\vec{v}}{dt} = \frac{d}{dt}(s\vec{T}) = \frac{ds}{dt}\vec{T} + s\frac{d\vec{T}}{dt}.$$

The two vectors on the right hand side of this equation, $\vec{T}$ and $d\vec{T}/dt$, are perpendicular to each other. To see why this is true, remember that $\vec{T}$ is a unit vector. Therefore, if we think of $\vec{T}$ as a vector based at the origin, its tip traces out a piece of the unit circle (see Figure 16.35). Thus, the derivative of $\vec{T}$ is tangential to the circle, and so is perpendicular to $\vec{T}$. (Another way of seeing this is indicated in Problem 2 on page 320.)

Thus, we have a decomposition of the the acceleration vector into a tangential component of acceleration, $(ds/dt)\vec{T}$, and a normal component, $s(d\vec{T}/dt)$, as follows:

$$\text{Acceleration} = \vec{a} = \frac{ds}{dt}\vec{T} + s\left(\frac{d\vec{T}}{dt}\right) = \text{Tangential component} + \text{Normal component}$$

For motion in a straight line, $\vec{T}$ is a unit vector always pointing in the same direction, so $\vec{T}$ is constant and $d\vec{T}/dt = 0$. That is, all the acceleration is explained by the tangential component. For motion at constant speed, $ds/dt = 0$ and all the acceleration is explained by the normal component $s(d\vec{T}/dt)$.

Figure 16.35: The unit tangent vector at two different times, explaining why $d\vec{T}/dt$ is perpendicular to $\vec{T}$

A case of particular interest is uniform motion in a circle of radius r. We have already seen that the magnitude of the acceleration is the square of the speed over the radius, s^2/r. Suppose we have motion at constant speed but not in a circle. Then the magnitude of the normal component should still tell us how much we are curving, that is how tight a corner or circle we are turning around. If r measures our "turning radius" then we should have

$$\frac{s^2}{r} = \left\| s\frac{d\vec{T}}{dt} \right\| = s\left\| \frac{d\vec{T}}{dt} \right\|.$$

Solving for r gives

$$r = \frac{s^2}{s\left\| d\vec{T}/dt \right\|} = \frac{s}{\left\| d\vec{T}/dt \right\|}.$$

We therefore define the *radius of curvature*, denoted by p, to be:

$$\boxed{\text{Radius of curvature} = p = \frac{s}{\left\| d\vec{T}/dt \right\|}.}$$

Notice that a small radius means a tight turn, namely a lot of curvature. Thus, we define the *curvature* κ to be the reciprocal of the radius of curvature:

$$\boxed{\text{Curvature} = \kappa = \frac{1}{p} = \frac{1}{s}\left\| \frac{d\vec{T}}{dt} \right\|.}$$

There is another way to make sense of this definition of curvature. As a particle travels along a curve the rate at which it changes direction, namely the magnitude of the rate of change of the unit tangent, $\left\| d\vec{T}/dt \right\|$, should tell us the curvature, — except for one problem. If the particle moves slowly, the rate of change of direction will be small even if the path is very curvy; similarly if the particle moves fast, the rate of change of direction will be large. Thus, we need to compensate for the speed, s. That is why the curvature is $\left\| d\vec{T}/dt \right\|/s$, not simply $\left\| d\vec{T}/dt \right\|$.

Example 1 For $t > 0$, find the tangential and normal components of the acceleration and the curvature for the helical motion of Example 3 on page 316 given by the parametric equations:

$$x = 3\cos(t^2), \; y = 3\sin(t^2), \; z = t^2.$$

Solution We have already computed

$$\vec{v}(t) = -6t\sin(t^2)\vec{i} + 6t\cos(t^2)\vec{j} + 2t\vec{k}.$$

Earlier, we differentiated again to get the acceleration. This time we first factor out $2t$ in $\vec{v}(t)$, giving

$$\vec{v}(t) = 2t(-3\sin(t^2)\vec{i} + 3\cos(t^2)\vec{j} + \vec{k}).$$

Thus, we have

$$s(t) = \|\vec{v}(t)\| = 2t\sqrt{9\sin^2(t^2) + 9\cos^2(t^2) + 1^2}$$
$$= (2\sqrt{10})t.$$

Therefore, $\vec{T}(t) = \vec{v}(t)/s(t) = \frac{1}{\sqrt{10}}(-3\sin(t^2)\vec{i} + 3\cos(t^2)\vec{j} + \vec{k})$. Also $ds/dt = 2\sqrt{10}$ and

$$\frac{d\vec{T}}{dt} = \frac{1}{\sqrt{10}}(-6t\cos(t^2)\vec{i} - 6t\sin(t^2)\vec{j}).$$

Thus,

$$\vec{a}(t) = \text{Tangential component} + \text{Normal component}$$

$$= \frac{ds}{dt}\vec{T} + s\left(\frac{d\vec{T}}{dt}\right)$$

$$= 2\sqrt{10}\cdot\frac{(-3\sin(t^2)\vec{i} + 3\cos(t^2)\vec{j} + \vec{k})}{\sqrt{10}} + (2\sqrt{10})t\cdot\frac{(-6t\cos(t^2)\vec{i} - 6t\sin(t^2)\vec{j})}{\sqrt{10}}.$$

You should verify that this agrees with the expression for $\vec{a}(t)$ obtained in Example 3. You should also verify that the two component vectors for $\vec{a}(t)$ are perpendicular, that is, their dot product is 0. Finally, the curvature is

$$\kappa = \frac{1}{s}\left\|\frac{d\vec{T}}{dt}\right\| = \frac{1}{(2\sqrt{10})t}\cdot\frac{1}{\sqrt{10}}\sqrt{(-6t)^2\cos^2(t^2) + (-6t)^2\sin^2(t^2)}$$

$$= \frac{6t}{2\cdot 10t} = \frac{3}{10}.$$

Therefore, the radius of curvature is

$$p = \frac{10}{3}.$$

This makes sense, since the x and y coordinates of the helix describe a circle of radius 3 and the motion up the z-axis "uncoils" the circular motion to make it a little straighter. Thus, the curvature is a little smaller and the radius of curvature a little larger than that for a circle of radius 3.

Problems for Section 16.6

1.   Compute the normal and tangential components of the acceleration for Problem 1 on page 317.

2.   Show that $d\vec{T}/dt$ is normal to $\vec{T}$ by differentiating the equation $\vec{T}\cdot\vec{T} = 1$ [Hint: $\vec{T}$ is a unit vector.]

3.   You might think that the curvature for the graph of a function $y = f(x)$ should be given simply by $f''(x)$, since the second derivative measures the concavity. It is not quite as simple as that. For example, the parabola $y = x^2$ has concavity $y'' = 2$ everywhere but has much greater curvature at $x = 0$ than at $x = 100$. The reason is that the curvature depends not only on $f''(x)$, but also on $f'(x)$. Parameterize the curve $y = f(t)$ by

$$x = t, \; y = f(t).$$

Derive the formula for the curvature

$$K = \frac{|f''(x)|}{(1 + f'(x)^2)^{3/2}}.$$

16.7   NOTES ON TANGENT PLANES AND VECTOR AREAS

In Chapter 18 we will use parameterizations of surfaces to enable us to calculate integrals on a surface. To do this, we will need to be able to measure the size of a parameter rectangle. In this section we will compute the size of these rectangles, as well as find the equation of a tangent plane to the surface.

Finding the Tangent Plane to a Parameterized Surface

We can use parameter curves to find the equation of a tangent plane at a point on the surface. The idea is to calculate tangent vectors to the two parameter curves through the point. Since both vectors lie in the tangent plane, we can use them to find the equation of the plane.

Example 1   Find the equation of the tangent plane to the unit sphere at the point P where $\phi = \pi/4, \theta = \pi/3$.

Solution   We define $\vec{v}_\phi$ to be the vector tangent to the curve parameterized by ϕ and obtained by setting $\theta = \pi/3$ in the parametric equations of the sphere. See Figure 16.36. Then the $\theta = \pi/3$ curve is parameterized by ϕ and given by

$$\vec{r} = x\vec{i} + y\vec{j} + z\vec{k} = \sin\phi\cos\frac{\pi}{3}\vec{i} + \sin\phi\sin\frac{\pi}{3}\vec{j} + \cos\phi\vec{k} = \frac{1}{2}\sin\phi\vec{i} + \frac{\sqrt{3}}{2}\sin\phi\vec{j} + \cos\phi\vec{k}$$

Thus,

$$\vec{v}_\phi = \frac{\partial\vec{r}}{\partial\phi} = \frac{1}{2}\cos\phi\vec{i} + \frac{\sqrt{3}}{2}\cos\phi\vec{j} - \sin\phi\vec{k}$$

At the point P, where $\phi = \pi/4$, we have

$$\vec{v}_\phi(P) = \frac{1}{2}\cos\frac{\pi}{4}\vec{i} + \frac{\sqrt{3}}{2}\cos\frac{\pi}{4}\vec{j} - \sin\frac{\pi}{4}\vec{k} = \frac{1}{2\sqrt{2}}\vec{i} + \frac{\sqrt{3}}{2\sqrt{2}}\vec{j} - \frac{1}{\sqrt{2}}\vec{k}\ .$$

The curve through P parameterized by θ is given by

$$\vec{r} = x\vec{i} + y\vec{j} + z\vec{k} = \sin\frac{\pi}{4}\cos\theta\vec{i} + \sin\frac{\pi}{4}\sin\theta\vec{j} + \cos\frac{\pi}{4}\vec{k} = \frac{1}{\sqrt{2}}(\cos\theta\vec{i} + \sin\theta\vec{j} + \vec{k})$$

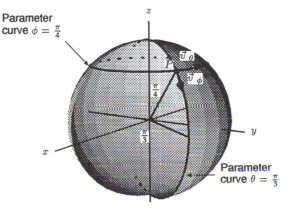

Figure 16.36: Sphere with tangent vectors $\vec{v}_\phi$ and $\vec{v}_\theta$ at the point P

A similar calculation shows that $\vec{v}_\theta$, the tangent vector to this curve is

$$\vec{v}_\theta = \frac{\partial \vec{r}}{\partial \theta} = \frac{1}{\sqrt{2}}(-\sin\theta\vec{i} + \cos\theta\vec{j}).$$

At the point P, where $\theta = \pi/3$, we have

$$\vec{v}_\theta(P) = \frac{1}{\sqrt{2}}(-\sin\frac{\pi}{3}\vec{i} + \cos\frac{\pi}{3}\vec{j}) = -\frac{\sqrt{3}}{2\sqrt{2}}\vec{i} + \frac{1}{2\sqrt{2}}\vec{j}.$$

Since $\vec{v}_\phi$ and $\vec{v}_\theta$ are in the tangent plane, the normal to the tangent plane is given by the cross product

$$\vec{n} = \vec{v}_\phi \times \vec{v}_\theta = \left(\frac{1}{2\sqrt{2}}\right)^2 \begin{vmatrix} \vec{i} & \vec{j} & \vec{k} \\ 1 & \sqrt{3} & -2 \\ -\sqrt{3} & 1 & 0 \end{vmatrix} = \frac{1}{8}(-2\vec{i} + 2\sqrt{3}\vec{j} + 4\vec{k}).$$

Since the coordinates of point P are

$$P = \left(\sin\frac{\pi}{4}\cos\frac{\pi}{3}, \sin\frac{\pi}{4}\sin\frac{\pi}{3}, \cos\frac{\pi}{4}\right) = \left(\frac{1}{2\sqrt{2}}, \frac{\sqrt{3}}{2\sqrt{2}}, \frac{1}{\sqrt{2}}\right),$$

the equation of the tangent planes is

$$\frac{1}{8}(2\vec{i} + 2\sqrt{3}\vec{j} + 4\vec{k}) \cdot (x\vec{i} + y\vec{j} + z\vec{k}) = \frac{1}{8}(2\vec{i} + 2\sqrt{3}\vec{j} + 4\vec{k}) \cdot (\frac{1}{2\sqrt{2}}\vec{i} + \frac{\sqrt{3}}{2\sqrt{2}}\vec{j} + \frac{1}{\sqrt{2}}\vec{k})$$

$$2x + 2\sqrt{3}y + 4z = \frac{1}{\sqrt{2}} + \frac{3}{\sqrt{2}} + 2\sqrt{2}$$

$$2x + 2\sqrt{3}y + 4z = 4\sqrt{2}.$$

Calculating a Vector Area on a Parameterized Surface

We can also use tangent vectors to calculate the vector area of a patch in the surface. Suppose we have a surface given by an equation of the form $z = f(x, y)$. Imagine a parameter rectangle with

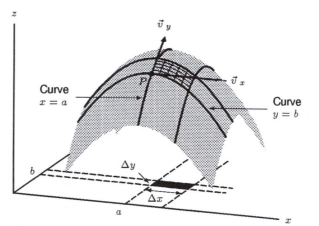

Figure 16.37: Surface showing parameter rectangle and tangent vectors $\vec{v}_x$ and $\vec{v}_y$ at P.

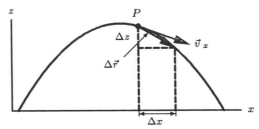

Figure 16.38: Cross-section of Surface showing
vector $\Delta\vec{r}$ along the edge of the parameter rectangle

one corner at the point P on the surface where $x = a, y = b$. See Figure 16.37. Then since a point
on the surface has position vector $\vec{r} = x\vec{i} + y\vec{j} + z\vec{k} = x\vec{i} + y\vec{j} + f(x,y)\vec{k}$, the curve $y = b$ has
a tangent vector at P:

$$\vec{v}_x(P) = \frac{\partial\vec{r}}{\partial x} = \vec{i} + f_x(a,b)\vec{k}.$$

Similarly, the curve $x = a$ has tangent vector at P

$$\vec{v}_y(P) = \frac{\partial\vec{r}}{\partial y} = \vec{j} + f_y(a,b)\vec{k}.$$

We will now consider the vector $\Delta\vec{r}$ along the $y = b$ side of the parameter rectangle. Figure 16.38
shows that $\Delta\vec{r}$ has components Δx in the x-direction and Δz in the z-direction. By local linearity,

$$\Delta z \approx f_x(a,b)\Delta x$$

so

$$\Delta\vec{r} \approx \Delta x\vec{i} + f_x(a,b)\Delta x\vec{k} = \vec{v}_x\Delta x$$

Similarly, the vector along the other side of the parameter rectangle is approximately

$$\Delta y\vec{j} + f_y(a,b)\Delta y\vec{k} = \vec{v}_y\Delta y.$$

Thus, the vector area of the parameter rectangle $\Delta\vec{A}(P)$ is approximated by the cross-product:

$$\begin{aligned}
\Delta\vec{A}(P) &\approx (\vec{v}_x\Delta x) \times (\vec{v}_y\Delta y)\\
&= \begin{vmatrix} \vec{i} & \vec{j} & \vec{k} \\ 1 & 0 & f_x(a,b) \\ 0 & 1 & f_y(a,b) \end{vmatrix}\Delta x\Delta y\\
&= (-f_x(a,b)\vec{i} - f_y(a,b)\vec{j} + \vec{k})\Delta x\Delta y.
\end{aligned}$$

We will often write this relationship in its infinitesimal form:

Vector area of a parameter rectangle at the point (a,b) **on the surface** $z = f(x,y)$:

$$d\vec{A}(a,b) = (-f_x(a,b)\vec{i} - f_y(a,b)\vec{j} + \vec{k})dx\,dy$$

Problems for Section 16.7 ——————————————————

1. Find the vector area element for Planetown.
2. Find the vector area element of the sphere

$$x = 2 + 2\cos\phi\cos\theta,$$
$$y = -1 + 2\cos\phi\sin\theta,$$
$$z = 3 + 2\sin\phi.$$

   centered at the point $(2, -1, 3)$ and having radius 2.

3. Find the equation of the tangent plane to the parabolic bowl $z = x^2 + y^2$ at the point where $x = a$, $y = b$.

4. Find the equation of the tangent plane to the torus

$$x = 5\cos\theta_2 + 2\cos\theta_1\cos\theta_2$$
$$y = 5\sin\theta_2 + 2\cos\theta_1\sin\theta_2$$
$$z = 2\sin\theta_1$$

   at the point $P = (\frac{5}{\sqrt{2}} + 1, \frac{5}{\sqrt{2}} + 2, \frac{2}{\sqrt{2}})$ corresponding to the parameter values $\theta_1 = \pi/4$ and $\theta_2 = \pi/4$.

REVIEW PROBLEMS FOR CHAPTER SIXTEEN

Write parametric equations for the curves in Problems 1–3.

1. A circle of radius 1 in the xy-plane centered at the origin, traversed counterclockwise when viewed from above.

2. A circle of radius 2 parallel to the xy-plane, centered at the point $(0, 0, 1)$, traversed counterclockwise when viewed from below.

3. A circle of radius 3 parallel to the xz-plane, and centered at the point $(0, 5, 0)$, traversed counterclockwise when viewed from $(0, 10, 0)$.

4. On a graphing calculator or a computer, plot $x = 2t/(t^2 + 1)$, $y = (t^2 - 1)/(t^2 + 1)$ first for $-50 \le t \le 50$ then for $-5 \le t \le 5$. Explain what you see. Is the curve really a circle?

5. Let $f(x, y) = \dfrac{x^2 - y^2}{x^2 + y^2}$.

   (a) In which direction should you move from the point $(1, 1)$ to obtain the maximum rate of increase of f?

   (b) Find a direction in which the directional derivative at the point $(1, 1)$ is equal to zero.

   (c) Suppose you move along the curve $x = e^{2t}$, $y = 2t^3 + 6t + 1$. What is df/dt at $t = 0$?

6. Plot the Lissajous figure given by $x = \cos 2t$, $y = \sin t$ using a graphing calculator or computer. Explain why it looks like part of a parabola. [Hint: Use a double angle identity from trigonometry.]

7. Suppose that a planet P in the xy-plane moves around the star S in a circle of radius 10 units, completing one revolution in 2π units of time. Suppose in addition a moon M moves around planet P in a circle of radius 3 units, completing one revolution in $2\pi/8$ units of time. The

star S is fixed at the origin $x = 0, y = 0$, and at time $t = 0$ the planet P is at the point $(10, 0)$ and the moon M is at the point $(13, 0)$.

(a) Find parametric equations for the x and y coordinates of the planet at time t.

(b) Find parametric equations for the x and y coordinates of the moon at time t. [Hint: For the moon's position at time t, take a vector from the sun to the planet at time t and add a vector from the planet to the moon].

(c) Plot the path of the planet using a graphing calculator or computer.

(d) Experiment with different radii for the moon's circle around the planet and different speeds of revolution for the moon around the planet.

8. A particle travels along a line in such a way that its position at time t is given by the equation

$$\vec{r}(t) = (2 + 5t)\vec{i} + (3 + t)\vec{j} + 2t\vec{k} .$$

(a) Where is the particle when $t = 0$?

(b) At what time does the particle reach the point $(12, 5, 4)$?

(c) Does the particle ever reach the point $(12, 4, 4)$? Why or why not?

9. An ant, starting at the origin, moves at 2 units/sec along the x-axis to the point $(1, 0)$. The ant then moves counterclockwise around the circumference of the circle of radius 1, centered at the origin, to the point $(0, 1)$ at a speed of $3\pi/2$ units/sec. It then returns to the origin at a speed of 2 units/sec.

(a) Express the ant's coordinates as a function of time, t, in secs.

(b) Express the reverse path as a function of time.

10. A basketball player shoots the ball from 6 feet above the ground towards a basket that is 10 feet above the ground and 15 feet away horizontally.

(a) Suppose she shoots the ball at an angle of A degrees above the horizontal with an initial velocity of V. Give the x and y-coordinates of the position of the basketball at time t. Assume the x-coordinate of the basket is 0 and that the x-coordinate of the shooter is -15. [Hint: There is an acceleration of -32 ft/sec^2 in the y-direction; there is no acceleration in the x-direction. Ignore air resistance.]

(b) Using the parametric equations you obtained above, experiment with different values for V and A, plotting the path of the ball on a graphing calculator or computer to see how close the ball comes to the basket. (The tick marks on the y-axis can be used to locate the basket.)

(c) Find the angle A that minimizes the velocity needed for the ball to reach the basket (This is a lengthy computation. First find an equation in V and A that holds if the path of the ball to passes through the point 15 feet from the shooter and 10 feet above the ground. Then minimize V.)

11. A cheerleader has a 0.4 m long baton with a light on one end. She throws the baton in such a way that its center moves along a parabola, and the baton rotates clockwise around the center with a constant angular velocity. The baton is thrown from a position where its center is 1.5 m above the ground; its initial velocity is 8 m/sec horizontally and 10 m/sec vertically, and its angular velocity is 2 revolutions per second. Also, the baton is horizontal when released. Find parametric equations describing the following motions:

(a) The center of the baton relative to the ground.

(b) The end of the baton relative to its center.

(c) The path traced out by the end of the baton relative to the ground.

(d) Sketch a graph of motion of the end of the baton.

CHAPTER SEVENTEEN

VECTOR FIELDS AND LINE INTEGRALS

Some physical quantities are best represented by vectors; others by scalars. If $f(x)$ is a scalar quantity we say f is a scalar-valued function. So far we have studied scalar-valued functions. Now we will study vector-valued functions, which are also called vector fields. Particularly important are gradient fields, which are obtained by taking the gradient of a scalar function. In this chapter we will define a line integral and see how to use one to recover functions from their gradient fields, using a theorem analogous to the one-variable Fundamental Theorem of Calculus.

17.1 VECTOR FIELDS

Introduction to Vector Fields

A *vector field* is a function that assigns a vector to each point in the plane or in space. One example of a vector field is the gradient of a function $f(x, y)$: at each point (x, y) the vector grad $f(x, y)$ points in the direction of maximum increase of f. Gradients are not the only kinds of vector fields; in this section we shall look at examples of vector fields representing velocities and forces.

Velocity Vector Fields

Vector fields arise in fluid dynamics, the study of how liquids and gases flow. The vector field in this situation is a *velocity vector field*: each vector shows the velocity of the fluid at that point. Figure 17.1 shows an example of water flowing through a pipe. (The vector field here is really three-dimensional; the figure shows a two-dimensional cross section.) You can see that the water flows slowly where the pipe is wide and fast where it is narrow.

Figure 17.1 shows how we graph a vector field: we select a few points and draw the vector at each point with its tail at the point. Of course there are an infinite number of vectors which make up the field. We sketch just enough to show the *qualitative* behavior of the field.

As another example, consider the flow of water in a full bathtub with its drain plug removed. Near the drain, the velocity of the water is large, and it decreases as you get further from the drain. The familiar spiraling behavior is evident in the picture in Figure 17.2.

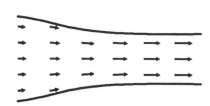

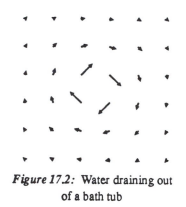

Figure 17.1: Water flowing through a pipe

Figure 17.2: Water draining out of a bath tub

Force Fields

Another physical quantity represented by a vector is force. When we experience a force, sometimes it results from direct contact with the object that supplies the force (for example, a hug). Many forces, however, (for example, electric and gravitational forces) can be felt at all points in space. For example, the earth exerts a gravitational pull on other masses: the direction of the attractive force is toward the center of the earth and the magnitude decreases with increasing distance from the earth. Such forces can be modeled by vector fields.

The gravitational force felt by a mass of one kilogram at different points in space as shown in Figure 17.3. This is a sketch of the vector field in three-space. You can see that the vectors all point towards the earth (which is left out of the diagram for clarity) and that the vectors further from the earth are smaller in magnitude. Since the perspective in the drawing distorts the lengths and directions of the arrows, most of our sketches of vector fields will be in two-space.

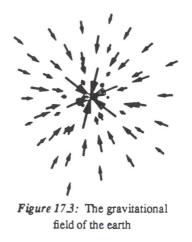

Figure 17.3: The gravitational
field of the earth

Definition of a Vector Field

Now that you have seen some examples of vector fields we give a more formal definition.

> A vector field in 2-space is a function $\vec{F}(x, y)$ whose value at a point (x, y) is a 2-vector.
> Similarly, a vector field in 3-space is a function $\vec{F}(x, y, z)$ whose values are 3-vectors; in
> general, a vector field in n-space is a function $\vec{F}(x_1, \ldots, x_n)$ whose values are n-vectors.

Notice that the function, $\vec{F}$, itself has an arrow over it to indicate that its value is a vector, not
a scalar. Because it is easier to write, we shall often represent the point (x, y) or (x, y, z) by its
position vector $\vec{r}$. The position vector $\vec{r}$ of a point P has its tail at the origin and its tip at P. Then,
using the position vector a vector field can also be written $\vec{F}(\vec{r})$.

Vector Fields Given By Formulas

Since a vector field is a function that assigns a vector to each point, we can often represent a vector
field using a formula.

Example 1   Draw a picture of the vector field in 2-space given by $\vec{F}(x, y) = x\vec{j}$.

Solution   The vector $x\vec{j}$ points in the y-direction; it points up when x is positive and down when x is negative.
Also, the larger $|x|$ is, the longer the vector. The vectors in the field are constant along vertical lines
since the vector field does not depend on y. See Figure 17.4.

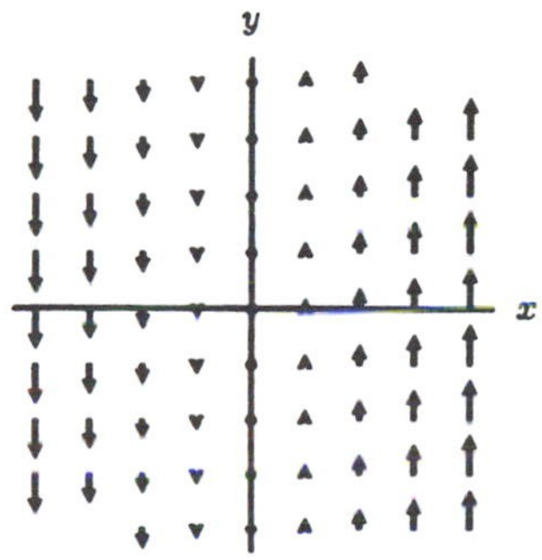

Figure 17.4: The vector field
$\vec{F}(x, y) = x\vec{j}$

Example 2   Draw a picture of the vector field in 2-space given by $\vec{F}(x, y) = -y\vec{i} + x\vec{j}$.

Solution   We evaluate the vector field at a small number of points as shown in Table 17.1. We then plot each of these eight vectors $\vec{F}(x, y)$ with their tails placed at the point (x, y); see Figure 17.5.

TABLE 17.1

(x, y)	(1,0)	(-1,0)	(0,1)	(0,-1)	(1,1)	(1,-1)	(-1,1)	(-1,-1)
$\vec{F}(x, y)$	$\vec{j}$	$-\vec{j}$	$-\vec{i}$	$\vec{i}$	$-\vec{i} + \vec{j}$	$\vec{i} + \vec{j}$	$-\vec{i} - \vec{j}$	$\vec{i} - \vec{j}$

We get a more global view of this vector field by looking carefully at its definition. The magnitude of the vector at (x, y) is $\|\vec{F}(x, y)\| = \|-y\vec{i} + x\vec{j}\| = \sqrt{x^2 + y^2}$, which is the distance of the point (x, y) from the origin. Therefore, this vector field has the property that all points at the same distance from the origin (that is, on circles centered at the origin) are assigned vectors of equal magnitude. As we look at larger circles, the magnitude increases. Since $(x\vec{i} + y\vec{j}) \cdot (-y\vec{i} + x\vec{j}) = 0$ for all (x, y), each vector is perpendicular to the vector joining the origin to the point. This means that the vectors in this field are all tangent to circles centered at the origin and get longer as we go out. For a computer-generated plot of this vector field, which has been scaled so that the vectors aren't so large that they obscure each other, see Figure 17.6.

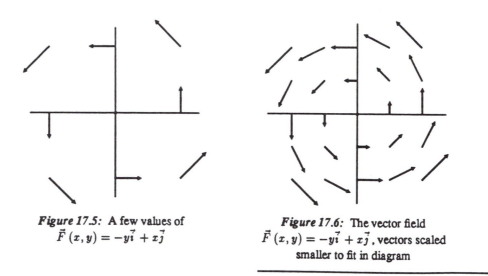

Figure 17.5: A few values of
$\vec{F}(x, y) = -y\vec{i} + x\vec{j}$

Figure 17.6: The vector field
$\vec{F}(x, y) = -y\vec{i} + x\vec{j}$, vectors scaled
smaller to fit in diagram

Gradient Vector Fields

The gradient of a scalar function f is a function that assigns a vector to each point, and is therefore a vector field. It is called the *gradient field* of f. Many vector fields in physics arise in this way.

Example 3   Sketch the gradient field of the functions in Figures 17.7–17.9.

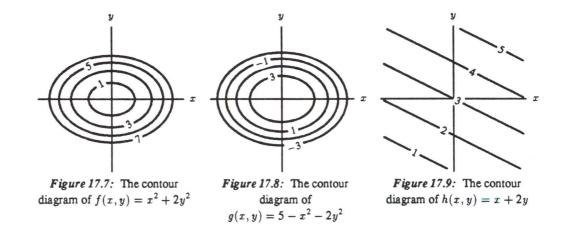

Figure 17.7: The contour diagram of $f(x, y) = x^2 + 2y^2$

Figure 17.8: The contour diagram of $g(x, y) = 5 - x^2 - 2y^2$

Figure 17.9: The contour diagram of $h(x, y) = x + 2y$

Solution   We know that the gradient vector of a function, f, at a point is the vector that is perpendicular to the contours in the direction of increasing f, and whose magnitude is the rate of change in that direction. The rate of change is large when the contours are close together and small when they are far apart. Using these ideas, we get the sketches in Figures 17.10–17.12. Notice that in Figure 17.10 the vectors all point outward, away from the local minimum of f. This is because the gradient vector points in the direction of increasing f. For the same reason the vectors of grad g in Figure 17.11 all point toward the local maximum of g. Finally, since h is a linear function, its gradient is constant, so grad h in Figure 17.12 is a constant vector field.

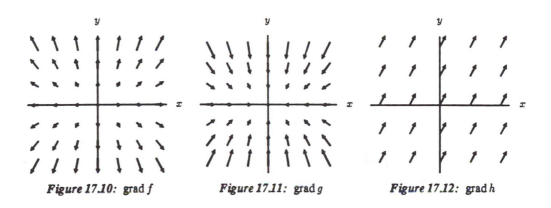

Figure 17.10: grad f

Figure 17.11: grad g

Figure 17.12: grad h

Example 4   Draw a picture of the vector field in 3-space given by $\vec{F}(\vec{r}) = \vec{r}$, where $\vec{r} = x\vec{i} + y\vec{j} + z\vec{k}$.

Solution   The $\vec{r}$ inside the function is the position vector of a point in 3-space. The value of the function is the same position vector, but now it is to be regarded as a vector with its tail at the point. The result is shown in Figure 17.13. Note that the lengths of the vectors have been scaled down so as to fit into the diagram.

This vector field can also be written without using the position vector $\vec{r}$ as $\vec{F}(x, y, z) = x\vec{i} + y\vec{j} + z\vec{k}$. You can see that the position vector notation is more concise.

Figure 17.13: The vector field $\vec{F}(\vec{r}) = \vec{r}$

Example 5   Newton's Law of Gravitation states that the gravitational force between two objects is proportional to their masses and inversely proportional to the square of the distance between them. The direction of the force is along the line connecting the two objects (see Figure 17.14). Find a formula for the vector field $\vec{F}(\vec{r})$ that represents the force exerted by a mass M located at the origin on an object of mass m located at a point with position vector $\vec{r}$.

Solution   If the mass m is located at $\vec{r}$, then Newton's law says that the magnitude of force is

$$\frac{GMm}{\|\vec{r}\|^2}$$

where G is a constant of proportionality. A unit vector in the direction of the force is $-\vec{r} / \|\vec{r}\|$, where the negative sign indicates that the direction of force is towards the origin (gravity is attractive). By taking the product of the magnitude of the forces and a unit vector in the direction of the force we obtain an expression for the force vector field:

$$\vec{F}(\vec{r}) = \frac{GMm}{\|\vec{r}\|^2}\left(-\frac{\vec{r}}{\|\vec{r}\|}\right) = \frac{-GMm\vec{r}}{\|\vec{r}\|^3}.$$

We have already seen a picture of this vector field in Figure 17.3.

Figure 17.14: Force exerted on
mass m by mass M

Problems for Section 17.1

1. Each vector field in Figures (I)–(IV) represents the force on a particle at different points in space as a result of another particle. Match up the vector fields with the descriptions below.
   (a) A repulsive force, decreasing as distance increases, such as between electric charges of the same sign.
   (b) A repulsive force, increasing with distance.
   (c) An attractive force, decreasing as distance increases, such as gravity.
   (d) An attractive force, increasing with distance.

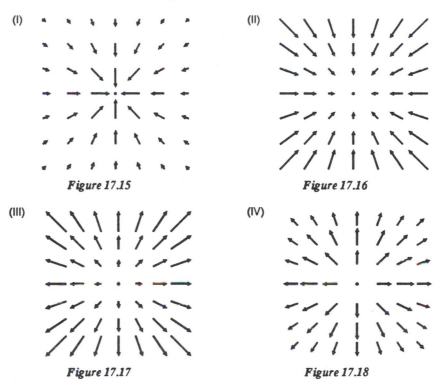

(I)

Figure 17.15

(II)

Figure 17.16

(III)

Figure 17.17

(IV)

Figure 17.18

Sketch graphs of the vector fields in Problems 2–7.

2. $\vec{F}(x, y) = 2\vec{i} + 3\vec{j}$

3. $\vec{F}(x, y) = y\vec{i}$

4. $\vec{F}(x, y) = 2x\vec{i} + x\vec{j}$

5. $\vec{F}(\vec{r}) = 2\vec{r}$

6. $\vec{F}(\vec{r}) = \dfrac{\vec{r}}{\|\vec{r}\|}$

7. $\vec{F}(x, y) = (x+y)\vec{i} + (x-y)\vec{j}$

For Problems 8–13, find formulas for the vector fields. (There are many possible answers.)

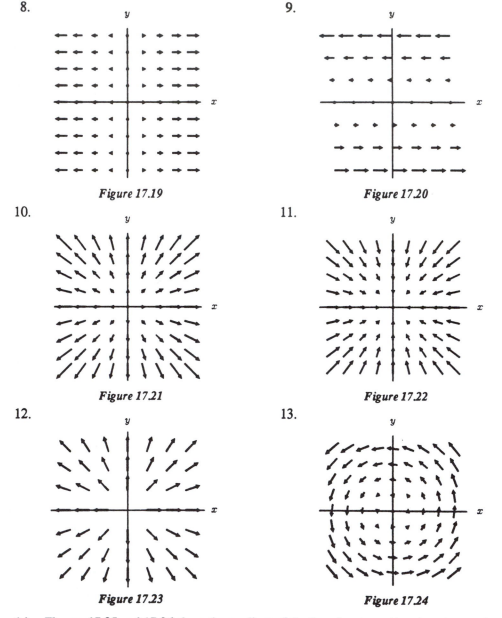

8.

Figure 17.19

9.

Figure 17.20

10.

Figure 17.21

11.

Figure 17.22

12.

Figure 17.23

13.

Figure 17.24

14. Figures 17.25 and 17.26 show the gradient of the functions $z = f(x, y)$ and $z = g(x, y)$.

(a) For each function, draw a rough sketch of the level curves, showing possible z values.

(b) The xz-plane cuts each of the surfaces $z = f(x, y)$ and $z = g(x, y)$ in a curve. Sketch each of these curves, making clear how they are similar and how they are different from one another.

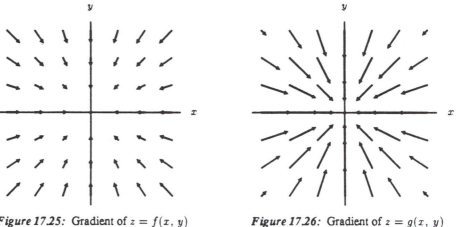

Figure 17.25: Gradient of $z = f(x, y)$ **Figure 17.26:** Gradient of $z = g(x, y)$

In Problems 15–17, use a computer to print out vector fields with the given properties. Show on your printout the formula used to generate it. (There are many possible answers.)

15. All vectors are parallel to the x-axis, and all those on a vertical line have the same magnitude.

16. All vectors point towards the origin and have constant length.

17. All vectors of unit length are perpendicular to the position vector at that point.

18. Imagine a wide, steadily flowing river in the middle of which there is a fountain that spews out water horizontally in all directions.

   (a) Suppose that the river flows in the $\vec{i}$-direction in the xy-plane and that the fountain is at the origin. Explain why the expression

$$\vec{V} = A\vec{i} + K(x^2 + y^2)^{-1/2}(x\vec{i} + y\vec{j}), \quad A > 0, K > 0$$

   is a natural candidate for the velocity vector field of the combined flow of the river and the fountain.

   (b) What is the significance of the constants A and K?

   (c) Using a computer, sketch the vector field $\vec{v}$ for $K = 1$ and $A = 1$ and $A = 2$.

17.2   FLOW OF A VECTOR FIELD

Suppose that we have the velocity vector field of a moving fluid. If we drop a small object into the fluid flow, what path will it take? In this section, we shall see how to answer this question by associating a system of differential equations with the vector field. The family of solution curves will be called the *flow* of the field.

Suppose that $\vec{v}(x, y) = v_1\vec{i} + v_2\vec{j}$ is the velocity field of water on the surface of a creek. Then the value of $\vec{v}(x, y)$ is the velocity of a seed floating in the water as it passes the point (x, y). Now imagine the path a seed would travel in this velocity field: this is called a *flowline*[1]. The collection of all flowlines in the creek is the *flow* of the field. Since the seed is moving with velocity $\vec{v}$, if its path is parameterized $x(t), y(t)$, with the parameter t being time, then

$$x'(t) = v_1 \quad \text{and} \quad y'(t) = v_2.$$

---
[1] This path is also referred to as an *integral curve* or a *streamline*.

Thus, we make the following definition

A **flowline** of a vector field $\vec{F}(x, y) = F_1\vec{i} + F_2\vec{j}$ is a curve C whose parameterization $(x(t), y(t))$ satisfies

$$x'(t) = F_1(x(t), y(t)) \quad \text{and} \quad y'(t) = F_2(x(t), y(t)).$$

The **flow** of a vector field is the family of all of its flowlines.

This definition allows us to treat any vector field as if it were a velocity field and to think of how a particle would travel if it moved with velocity given by the field. It turns out to be profitable to study the flow of fields (for example, electric and magnetic) that are not velocity fields.

The problem of finding flowlines is equivalent to solving a system of differential equations. In general, we know that this can be difficult, if not impossible, to do exactly. However, we can use numerical methods, such as Euler's method, to approximate the solutions and compute flowlines as accurately as we wish.

Example 1   Given the constant velocity field $\vec{v} = 3\vec{i} + 4\vec{j}$ ft/sec, find the flowline passing through the point $(1, 2)$.

Solution   We are looking for the path of a particle placed in the field at the point $(1, 2)$. Considering the component of the flow in the x-direction we have $x'(t) = 3$ or $x(t) = 3t + x_0$. Similarly, $y(t) = 4t + y_0$, where x_0 and y_0 are constants of integration. Therefore, the flowline is given by

$$(x(t), y(t)) = (x_0 + 3t, y_0 + 4t).$$

Since the path starts at the point $(1, 2)$, we have $x_0 = 1$ and $y_0 = 2$ and so

$$(x(t), y(t)) = (1 + 3t, 2 + 4t).$$

Equivalently, we can write the position at time t as

$$x(t) = 1 + 3t \quad \text{and} \quad y(t) = 2 + 4t.$$

Eliminating t between these expressions gives

$$\frac{x - 1}{3} = \frac{y - 2}{4} \quad \text{or} \quad y = \frac{4}{3}x + \frac{2}{3}.$$

This straight line is the path which a particle would follow if placed at the point $(1, 2)$.

Example 2   Determine the flow corresponding to the vector field $\vec{v} = -y\vec{i} + x\vec{j}$.

Solution   Figure 17.27 suggests that the flow consists of concentric counterclockwise circles, centered at the origin. If we write down the system of differential equations for this field we get

$$x'(t) = -y \quad \text{and} \quad y'(t) = x$$

It is not hard to verify that $(x(t), y(t)) = (a \cos t, a \sin t)$ is a parameterized family that satisfies the system. These curves *are* the counterclockwise circles of radius a, centered at the origin.

The next example shows a numerical approximation to a flowline.

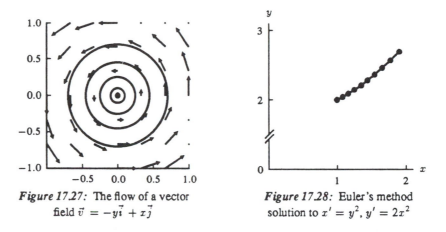

Figure 17.27: The flow of a vector field $\vec{v} = -y\vec{i} + x\vec{j}$

Figure 17.28: Euler's method solution to $x' = y^2,\ y' = 2x^2$

Example 3   Use Euler's method to determine the flowline through $(1, 2)$ for the vector field $\vec{F} = y^2\vec{i} + 2x^2\vec{j}$.

Solution   The flow is determined by the differential equations

$$x'(t) = y^2$$
$$y'(t) = 2x^2.$$

We use Euler's method with $\Delta t = 0.02$, giving

$$x_{n+1} = x_n + 0.02 y_n{}^2,$$
$$y_{n+1} = y_n + 0.02 \cdot 2x_n{}^2.$$

Initially, that is when $t = 0$, we have $(x_0, y_0) = (1, 2)$. Then

$$x_1 = x_0 + 0.02 \cdot y_0{}^2 = 1 + 0.02 \cdot 2^2 = 1.08,$$
$$y_1 = y_0 + 0.02 \cdot 2x_0^2 = 2 + 0.02 \cdot 2 \cdot 1^2 = 2.04.$$

Accordingly, we see that after one step $x(0.02) = 1.08$ and $y(0.02) = 2.04$. Similarly, $x(0.04) = x(2\Delta t) \approx 1.16323, y(2\Delta t) \approx 2.08666$ and so on. Further values along the flowline are given in Table 17.2 and plotted in Figure 17.28.

TABLE 17.2   *The computed path for the vector field $\vec{v} = y^2\vec{i} + 2x^2\vec{j}$ starting at the point $(1, 2)$*

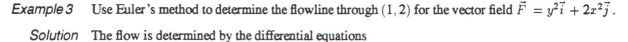

| x | 1 | 1.08 | 1.16323 | 1.25031 | 1.34917 | 1.43907 | 1.54261 | 1.65383 | 1.77421 | 1.90557 |
| y | 2 | 2.04 | 2.08666 | 2.14078 | 2.20331 | 2.27535 | 2.35818 | 2.45337 | 2.56278 | 2.68869 |

As the preceding examples illustrate, we can take any vector field and find its flow by treating the field as a velocity vector field and solving the resulting system of differential equations. The process is reversible: any system of differential equations of the form

$$x'(t) = F_1(x(t), y(t)) \quad \text{and} \quad y'(t) = F_2(x(t), y(t))$$

can be thought of as a velocity field $\vec{v} = F_1\vec{i} + F_2\vec{j}$.

Example 4   Consider a model for two interacting populations of robins and worms. Let $r(t)$ and $w(t)$ represent the robin population and the worm population as functions of time. Their interaction is governed by the system

$$w'(t) = aw - cwr, \quad r'(t) = -br + kwr$$

(where a, b, c, and k are constants). Find a vector field whose flow is the set of solutions to this system.

Solution   If we think of $(r(t), w(t))$ as a point in the phase plane then $r'(t)\vec{i} + w'(t)\vec{j}$ is a velocity vector, and the system gives a vector field

$$\vec{v} = (aw + crw)\vec{i} + (-br + krw)\vec{j}.$$

Notice that the vector field obtained in the last example contains more information than the slope fields sometimes used to study first order differential equations. In particular, the arrowheads of the vector field show the direction of the flow. If we draw slope fields for systems, we do so by eliminating t, so the direction of the trajectories is lost.

The Divergence and Volume Change

We have seen how to think of any vector field in terms of a flow. We will exploit this interpretation to arrive at a measure of compression or expansion of the fluid. This measure is called the *divergence* of a vector field, and has significant application in the theory of fluids, heat flow, and electricity and magnetism. Most instances in which the divergence is applied to a vector field will be in 3-space, but for simplicity we shall explore the concept in two-dimensional vector fields — the results carry over to three dimensions with little change. Assume that $\vec{F}$ is a velocity field and imagine that it describes the flow of a gas, such as air, that can expand and contract as it flows. To see the expansion or contraction, we follow a small area of gas as it moves. See Figure 17.29.

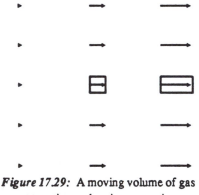

Figure 17.29: A moving volume of gas
at two times, showing expansion

Example 5   Determine whether the gas in the regions depicted in the velocity fields of Figure 17.30 is expanding or contracting.

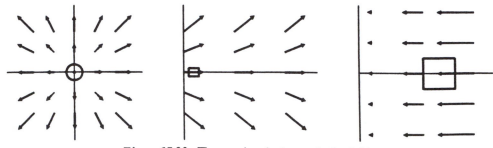

Figure 17.30: Three regions in three velocity fields

Solution   In Figure 17.31 we have sketched the regions the gas will occupy after flowing along with the velocity fields for a short time. In the first two diagrams, the gas occupies a larger volume than originally, so it is expanding. The third diagram shows a shrinking volume, so the gas is contracting.

Figure 17.31: The three regions in Figure 17.30 a short period of time later

Now we will measure the rate a vector field is expanding. Suppose we want to calculate the rate of expansion described by a velocity vector field

$$\vec{v} = a(x,y)\vec{i} + b(x,y)\vec{j}.$$

Since the rate may vary from point to point, we will fix a point $P = (x_0, y_0)$, and compute the rate at P. We focus on the gas at time t_0 in a small box of dimensions $\Delta x \times \Delta y$ with edges parallel to the axes and with one corner at P, as shown in Figure 17.32. Now imagine the small box and the gas inside moving according to the flow of the velocity field. The rectangle will change shape (and area) if the gas is getting compressed or expanding. It is the rate of change of the area of this small box that we shall now estimate. We let $A(t)$ denote the area of the small box at time t and we will compute

$$\frac{dA}{dt} = \lim_{\Delta t \to 0} \frac{A(t_0 + \Delta t) - A(t_0)}{\Delta t}.$$

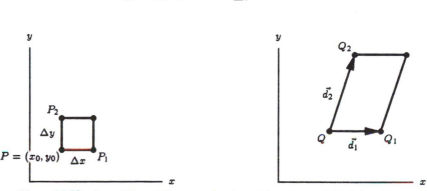

Figure 17.32: A small box at time t_0 and at $t_0 + \Delta t$, a short period of time later

At time t_0, the area is given by

$$A(t_0) = \Delta x \Delta y.$$

By time $t_0 + \Delta t$, the flow will have moved the gas from the box to a new region whose area $A(t_0 + \Delta t)$ we wish to approximate. To find the corners of the new region, we begin at the corners of the original box and move with velocity $\vec{v}$ for time Δt. The point P will flow (approximately) to the point

$$Q = P + \vec{v}(P)\Delta t = (x_0, y_0) + (a(P)\vec{i} + b(P)\vec{j})\Delta t.$$

Similarly, the point $P_1 = (x_0 + \Delta x, y_0)$ will flow (approximately) to

$$Q_1 = P_1 + \vec{v}(P_1)\Delta t.$$

Using the linear approximation

$$\vec{v}(P_1) = a(x_0 + \Delta x, y_0)\vec{i} + b(x_0 + \Delta x, y_0)\vec{j} \approx (a(P) + a_x(P)\Delta x)\vec{i} + (b(P) + b_x(P)\Delta x)\vec{j},$$

we find that

$$Q_1 \approx (x_0 + \Delta x, y_0) + (a(P) + a_x(P)\Delta x)\vec{i} + (b(P) + b_x(P)\Delta x)\vec{j}.$$

The edge of the region from Q to Q_1 can be described (approximately) by the displacement vector

$$\vec{d}_1 = (\Delta x + a_x \Delta x \Delta t)\vec{i} + b_x \Delta x \Delta t \vec{j},$$

where a_x, b_x are evaluated at P. A similar computation shows the displacement vector for the edge QQ_2 is approximated by

$$\vec{d}_2 = a_y \Delta y \Delta t \vec{i} + (\Delta y + b_y \Delta y \Delta t)\vec{j},$$

where a_y, b_y are evaluated at P. See Figure 17.32.

The region occupied by the gas at time $t_0 + \Delta t$, where Δt is a short time, will be approximately a parallelogram with sides $\vec{d}_1$ and $\vec{d}_2$. Now the area of a parallelogram [2] with sides given by the vectors $u_1\vec{i} + u_2\vec{j}$ and $v_1\vec{i} + v_2\vec{j}$ is $u_1 v_2 - u_2 v_1$. So the area of the parallelogram is

$$\begin{aligned} A(t_0 + \Delta t) &\approx (\Delta x + a_x \Delta x \Delta t)(\Delta y + b_y \Delta y \Delta t) - (b_x \Delta x \Delta t)(a_y \Delta y \Delta t) \\ &= \Delta x \Delta y[(1 + a_x \Delta t)(1 + b_y \Delta t) - (b_x \Delta t)(a_y \Delta t)] \\ &= \Delta x \Delta y[1 + a_x \Delta t + b_y \Delta t + (a_x b_y - b_x a_y)(\Delta t)^2] \end{aligned}$$

Finally, we can compute the difference quotient for dA/dt:

$$\begin{aligned} A'(t_0) &= \lim_{\Delta t \to 0} \frac{(A(t_0 + \Delta t) - A(t_0))}{\Delta t} \\ &\approx \lim_{\Delta t \to 0} \frac{\Delta x \Delta y(1 + a_x \Delta t + b_y \Delta t + (a_x b_y - b_x a_y)(\Delta t)^2) - (\Delta x \Delta y)}{\Delta t} \\ &= \lim_{\Delta t \to 0} \Delta x \Delta y(a_x + b_y + (a_x b_y - b_x a_y)\Delta t). \end{aligned}$$

Thus, as $\Delta t \to 0$, we see that the rate of change of the area or the rectangle is given by

$$\frac{dA}{dt} = \Delta x \Delta y(a_x + b_y).$$

To get a result that is independent of the area of the original rectangle at P, we compute the rate of change of area *per unit area*, that is $A'(t_0)/A(t_0)$, getting

$$\frac{A'(t_0)}{A(t_0)} \approx a_x + b_y$$

---

[2] This is a *signed* area, which is what we want.

where the approximation is better and better as Δx and Δy approach 0. In the limit we imagine the flow of an infinitesimally small box. We have then arrived at the definition of the *divergence* of a vector field:

The **divergence** of a velocity field $\vec{v}$ in 2-space at a point P is
- The rate of change of area per unit area of an infinitesimal region flowing with the field at P.
- If the field is $\vec{v} = a(x, y)\vec{i} + b(x, y)\vec{j}$ then the divergence at P can be computed by

$$\text{div } \vec{v} = \frac{\partial a}{\partial x} + \frac{\partial b}{\partial y}.$$

The discussion generalizes to 3-space: there divergence is defined as the rate of change of volume with respect to time of a small volume of stuff per unit volume. If the field is $\vec{v} = a(x, y, z)\vec{i} + b(x, y, z)\vec{j} + c(x, y, z)\vec{k}$, then

$$\text{div } \vec{v} = \frac{\partial a}{\partial x} + \frac{\partial b}{\partial y} + \frac{\partial c}{\partial z}.$$

The divergence of a field is a scalar function that measures the rate of expansion of the flow; its units are 1/(time). If the divergence is positive, the vector field models an expansion. If the divergence is negative, the vector field models a contraction.

Example 6   The velocity field $\vec{v} = x\vec{i}$ is shown in Figure 17.33. Discuss the flow and divergence of this field.

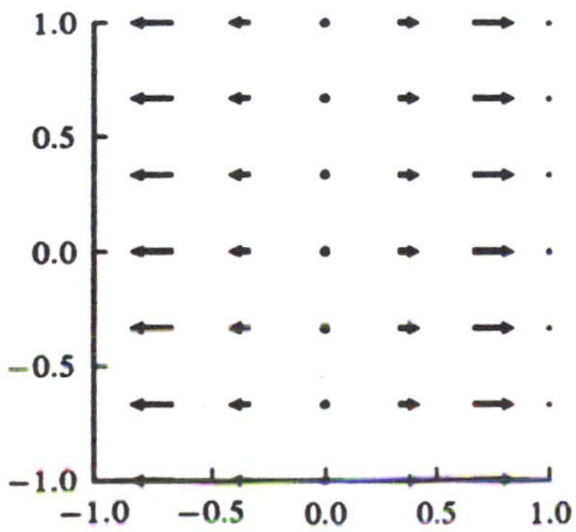

Figure 17.33: The vector field $\vec{v} = x\vec{i}$

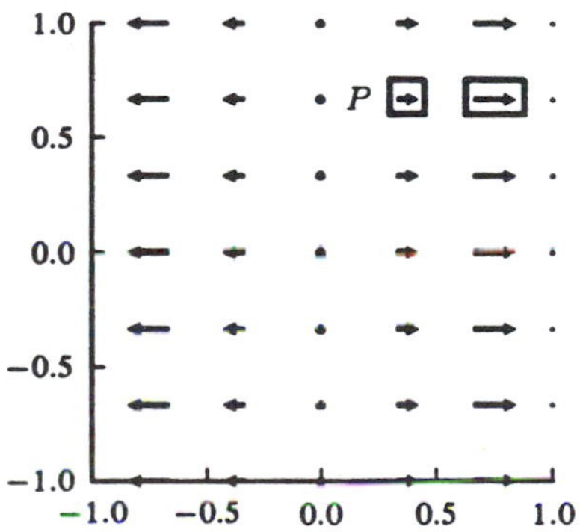

Figure 17.34: Motion of small volume with the vector field $\vec{v} = x\vec{i}$

Solution   The field represents the flow horizontally outward from a source along the y-axis. The flow of this vector field consists of horizontal lines, with speed increasing as distance from the y-axis increases. If we place a small box at a point P in the flow in Figure 17.34, then in time Δt the box will become elongated due to the higher velocity at its right end. Thus, there is a positive rate of change of area (that is, fluid is expanding), and we expect the divergence to be positive. To confirm, compute the divergence exactly as

$$\text{div } \vec{v} = a_x + b_y = \frac{\partial x}{\partial x} + \frac{\partial 0}{\partial y} = 1.$$

This tells us that the rate of expansion is constant throughout the flow.

Example 7   The velocity field $\vec{v} = x\vec{j}$ is shown in Figure 17.35. Discuss the flow and divergence of this field.

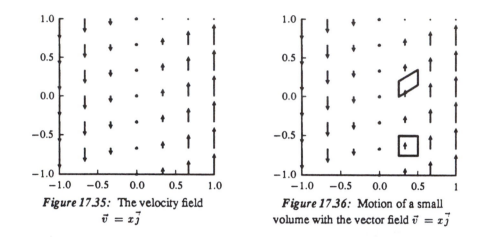

Figure 17.35: The velocity field
$\vec{v} = x\vec{j}$

Figure 17.36: Motion of a small
volume with the vector field $\vec{v} = x\vec{j}$

Solution   The field in Figure 17.35 represents a flow parallel to the y-axis, with no apparent source (this kind of flow is called a *shear*). The flow of this vector field consists of vertical lines, with constant speed along the lines. If we place a small box at a point P in the flow, then in time Δt the box will become a parallelogram due to the higher velocity at its right end and lower velocity at its left end (See Figure 17.36.) However, this parallelogram has the same area as the original box, so it appears there is a zero rate of change of area. If we compute the divergence exactly we see that

$$\text{div } \vec{v} = \frac{\partial 0}{\partial x} + \frac{\partial x}{\partial y} = 0.$$

This tells us that there is no expansion or contraction anywhere in this flow.

Problems for Section 17.2

Sketch the vector field and flow for the vector fields in Problems 1 – 3.

1. $\vec{v} = 3\vec{i}$ 2. $\vec{v} = 2\vec{j}$ 3. $\vec{v} = 3\vec{i} - 2\vec{j}$

For each of the vector fields in Problems 4 – 7, sketch the field and flow. Then find the system of differential equations associated with the field and verify that the flow satisfies the system.

4. $\vec{v}(t) = y\vec{i} + x\vec{j}$; $x(t) = a(e^t + e^{-t})$, $y(t) = a(e^t - e^{-t})$.

5. $\vec{v}(t) = y\vec{i} - x\vec{j}$; $x(t) = a\sin t$, $y(t) = a\cos t$.

6. $\vec{v}(t) = x\vec{i} + y\vec{j}$; $x(t) = ae^t$, $y(t) = be^t$.

7. $\vec{v}(t) = x\vec{i} - y\vec{j}$; $x(t) = ae^t$, $y(t) = be^{-t}$.

8. Use a computer or calculator with Euler's method to determine the parameterized curve through $(1, 2)$ for the vector field $\vec{v} = y^2\vec{i} + 2x^2\vec{j}$ using 10 steps with a time interval $\Delta t = 0.1$.

17.3 LINE INTEGRALS

Finding The Total Change In f From grad f

In one-variable calculus, we saw how the definite integral could be used to reconstruct the total change in a function from its derivative. This is the Fundamental Theorem of Calculus:

$$\int_a^b f'(t)\, dt = f(b) - f(a).$$

In words, the Fundamental Theorem tells us that the definite integral of a rate of change gives the total change.

Can we do the same thing for a function of two (or more) variables? Now, the quantity that describes the rate of change is a vector quantity, namely the gradient. If we know the gradient of a function f, can we compute the total change in f between two points? That is, given grad f and two points P and Q in the plane, can we find $f(Q) - f(P)$? The answer is yes, and we will see how trying to reconstruct the total change leads to a new kind of integral called a *line integral*. We begin with the case where the gradient is constant.

If grad f is Constant

If grad f is constant, then the contours of f are equally spaced with the same slope everywhere, so f is a linear function. (See Figure 17.37). Going along the displacement vector $\Delta\vec{r}$ takes us to

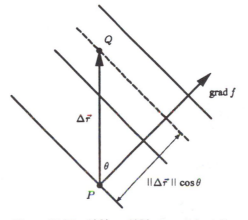

Figure 17.37: $f(Q) - f(P) = \text{grad } f \cdot \Delta\vec{r}$

the same contour as if we had gone a distance of $\|\Delta \vec{r}\| \cos \theta$ perpendicular to the contours. Since the rate of change perpendicular to the contours is $\|\text{grad } f\|$, the change in f is

$$\text{Distance} \times \text{Rate of change} = \|\Delta \vec{r}\| \cos \theta \, \|\text{grad } f\| = \text{grad } f \cdot \Delta \vec{r}.$$

Thus, we have the following result:

If grad f is constant, then Δf, the change in f, is given by

$$\Delta f = f(Q) - f(P) = \text{grad } f \cdot \Delta \vec{r},$$

where $\Delta \vec{r}$ is the displacement vector from P to Q.

If grad f is Not Constant

Suppose now that we want to find the change in f between two points P and Q when grad f is not constant. As we often did in one-variable calculus, we divide the path from P to Q into many small pieces. Then we use grad f to approximate the change of f on each piece and add the changes up to get a Riemann sum. However, there is one complication that was not present in the one-variable case: *there are many different ways* to get from P to Q. See Figure 17.38. So we first choose a path which begins at P and ends at Q. Figure 17.39 shows the smooth path C divided into small pieces. The position vectors of the points that divide the path are $\vec{r}_0, \vec{r}_1, \ldots, \vec{r}_{n-1}, \vec{r}_n$.

How do we estimate the change in f between the points $\vec{r}_i$ and $\vec{r}_{i+1}$? Assuming grad f is approximately constant on this small interval, and letting $\Delta \vec{r}_i$ be the displacement vector from $\vec{r}_i$ to $\vec{r}_{i+1}$, we have

$$\Delta f = f(\vec{r}_{i+1}) - f(\vec{r}_i) \approx \text{grad } f(\vec{r}_i) \cdot \Delta \vec{r}_i.$$

To find an approximation for the total change in f from P to Q, we add up the changes on each subinterval, giving

$$f(Q) - f(P) \approx \sum \text{grad } f(\vec{r}_i) \cdot \Delta \vec{r}_i.$$

Finally we let the size of each piece $\Delta \vec{r}_i$ shrink (and the number of subintervals n get larger and larger), to obtain a better and better approximation to the change in f:

$$f(Q) - f(P) = \lim_{\|\Delta \vec{r}\| \to 0} \sum \text{grad } f(\vec{r}_i) \cdot \Delta \vec{r}_i.$$

Figure 17.38: Different paths from P to Q

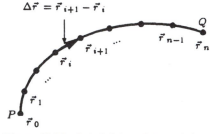

Figure 17.39: Subdivision of the path from P to Q

We define the limit of sums to be the **line integral of grad f** along the path C. We write

$$\int_C \operatorname{grad} f \cdot d\vec{r} = \lim_{\|\Delta\vec{r}\| \to 0} \sum \operatorname{grad} f(\vec{r}_i) \cdot \Delta\vec{r}_i.$$

Compare the definition of the line integral with the definition of the one-variable definite integral. Here, instead of subdividing an interval on the x-axis, we subdivided a path in the plane; instead of multiplying a scalar function $f(x)$ by a scalar increment Δx, we took the dot product of a vector function grad f with a vector increment $\Delta\vec{r}$.

Example 1   Suppose that grad f is shown in Figure 17.40. Explain why $f(P) = f(Q)$:
    (a)   Using a line integral,   (b)   By looking at contours.

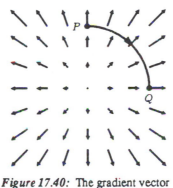

Figure 17.40: The gradient vector field of the function f

Solution   (a)   The change in f from P to Q, namely $f(Q) - f(P)$, is given by the line integral of grad f along any path joining P to Q. If we choose the path shown in Figure 17.40, we see that the direction of grad f along the path is perpendicular to the path. This means that the component of grad f in the direction of the path is 0 at each point on the path. So the sum of all of these small changes will also be 0.

       (b)   Recall that the gradient vector is perpendicular to the contours of f. Since P and Q lie on a path which is everywhere perpendicular to the gradient vector, this path is part of contour. Thus, P and Q lie on the same contour; hence, $f(P) = f(Q)$.

Definition of the Line Integral for General Vector Fields

The definition of the line integral works for *any* vector field, not just a gradient field, although in general there is no longer an interpretation in terms of total change.

In order to define a line integral of a vector field over a curve C, we need the notion of an *oriented* curve:

> A curve is said to be **oriented** if we have chosen a direction of travel along it.

The orientation is shown graphically by an arrowhead on the curve. Suppose that C is an oriented curve. As before, we subdivide C into small pieces, form a sum of dot products and take a limit.[3] We define

> **The Line Integral of $\vec{F}$** along C as
>
> $$\int_C \vec{F} \cdot d\vec{r} = \lim_{\|\Delta \vec{r}_i\| \to 0} \sum \vec{F}(\vec{r}_i) \cdot \Delta \vec{r}_i.$$

Like a definite integral from single-variable calculus, the value of a line integral is a number. The notation used in the line integral is similar to that used for the definite integral. The $d\vec{r}$ reminds us of the vectors $\Delta \vec{r}_i$ in the sum; the integral sign reminds us of the $\sum$ and shows that we have taken a sum of many small dot products.

What Does the Line Integral Tell Us?

Remember that $\vec{u} \cdot \vec{v}$ is 0 when $\vec{u}$ is perpendicular to $\vec{v}$, positive if $\vec{u}$ and $\vec{v}$ point in the same direction (that is, if the angle between them is less than $\pi/2$), and negative if they point in the opposite direction (that is, if the angle between them is more than $\pi/2$). The line integral of $\vec{F}$ adds up the dot products of $\vec{F}$ with small portions of the path; thus, it gives you a positive number if $\vec{F}$ is mostly pointing in the same direction as the path at all points along the path, a negative number if $\vec{F}$ is mostly pointing in the opposite direction, and zero if $\vec{F}$ is perpendicular to the path at all points. In general, the line integral of a vector field $\vec{F}$ along a curve C tells you the extent to which $\vec{F}$ is pointing along the curve C. Notice that the value of a line integral depends on the values of the vector field along the curve, C, as well as on the orientation of C.

Example 2   The vector field $\vec{F}$ and the oriented curves $C_1, \ldots, C_4$ are shown in Figure 17.41. Which of the line integrals $\int_{C_i} \vec{F} \cdot d\vec{r}$ for $i = 1, \ldots, 4$ are positive? Which are negative? Arrange these line integrals in ascending order.

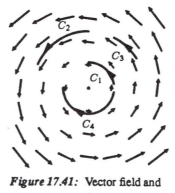

Figure 17.41: Vector field and
paths

---

[3]This limit exists in all reasonable cases, such as when $\vec{F}$ is continuous and defined everywhere and C is made up of a finite number of smooth pieces.

Solution   The vector field $\vec{F}$ and the line segments $\Delta\vec{r}$ are parallel and in the same direction for the curves C_1, C_2, and C_3. So the contributions of each term $\vec{F} \cdot \Delta\vec{r}$ are positive for these curves. Thus, $\int_{C_1} \vec{F} \cdot d\vec{r}$, $\int_{C_2} \vec{F} \cdot d\vec{r}$, and $\int_{C_3} \vec{F} \cdot d\vec{r}$ are each positive. For the curve C_4, the vector field and the line segments are in opposite directions, so each term $\vec{F} \cdot \Delta\vec{r}$ is negative, and therefore the integral $\int_{C_4} \vec{F} \cdot d\vec{r}$ is negative.

Since the magnitude of the vector field is smaller along C_1 than along C_3, and these two curves are the same length, we must have that:

$$\int_{C_1} \vec{F} \cdot d\vec{r} < \int_{C_3} \vec{F} \cdot d\vec{r}.$$

In addition, the magnitude of the vector field is the same along C_2 and C_3, but the curve C_2 is longer than the curve C_3. Thus,

$$\int_{C_3} \vec{F} \cdot d\vec{r} < \int_{C_2} \vec{F} \cdot d\vec{r}.$$

Putting this together, we have

$$\int_{C_4} \vec{F} \cdot d\vec{r} < \int_{C_1} \vec{F} \cdot d\vec{r} < \int_{C_3} \vec{F} \cdot d\vec{r} < \int_{C_2} \vec{F} \cdot d\vec{r}.$$

The Fundamental Theorem of Calculus For Line Integrals

If $\vec{F}$ is the gradient of a function f, then the line integral of $\vec{F}$ gives the total change in f between the beginning and end points of the path. This very important result is the analogue of the Fundamental Theorem in one-variable calculus.

The Fundamental Theorem of Calculus for Line Integrals

Suppose C is an oriented path with P as its starting point and Q as its endpoint. If f is a function whose gradient is continuous in a region containing the path C, then

$$\int_C \text{grad } f \cdot d\vec{r} = f(Q) - f(P).$$

Notice that the value of the line integral $\int_C \text{grad } f \cdot d\vec{r}$ depends on the endpoints of C; it does not depend on where C goes in between.

Example 3   Find the line integral of the constant vector field $\vec{F} = \vec{i} + 2\vec{j}$ along the path from $(1, 1)$ to $(10, 10)$ shown in Figure 17.42, and check your answer using the Fundamental Theorem.

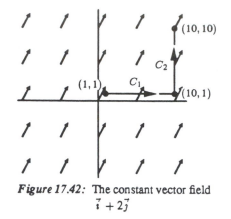

Figure 17.42: The constant vector field
$\vec{i} + 2\vec{j}$

Solution   Let C_1 be the horizontal segment of the path going from $(1, 1)$ to $(10, 1)$. When we break this path into pieces, each piece $\Delta \vec{r}$ is horizontal, so $\Delta \vec{r} = \Delta x \vec{i}$, and $\vec{F} \cdot \Delta \vec{r} = (\vec{i} + 2\vec{j}) \cdot \Delta x \vec{i} = \Delta x$. Hence,

$$\int_{C_1} \vec{F} \cdot d\vec{r} = \int_{x=1}^{x=10} dx = 9.$$

Similarly, along the vertical segment C_2, we have $\Delta \vec{r} = \Delta y \vec{j}$ and $\vec{F} \cdot \Delta \vec{r} = (\vec{i} + 2\vec{j}) \cdot \Delta y \vec{j} = 2\Delta y$, so

$$\int_{C_2} \vec{F} \cdot d\vec{r} = \int_{y=1}^{y=10} 2 \, dy = 18.$$

Thus,

$$\int_C \vec{F} \cdot d\vec{r} = \int_{C_1} \vec{F} \cdot d\vec{r} + \int_{C_2} \vec{F} \cdot d\vec{r} = 9 + 18 = 27.$$

Alternatively, observe that $\vec{F}$ is the gradient of the function $f(x, y) = x + 2y$, and that $f(10, 10) - f(1, 1) = 30 - 3 = 27$. Thus, our explicit calculation of the line integral agrees with the result from the Fundamental Theorem.

Example 4   Is the vector field $\vec{F}$ shown in Figure 17.43 a gradient field? At any point $\vec{F}$ has magnitude equal to the distance from the origin and direction perpendicular to the line joining the point to the origin.

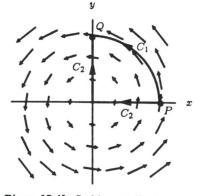

Figure 17.43: Is this a gradient vector
field?

Solution   If $\vec{F}$ = grad f for some function f, then by the Fundamental Theorem, the integral of $\vec{F}$ along any path is the change in f between the endpoints of the path. In particular, if we choose two different paths between the same endpoints, then the integral should be the same on both paths. Let's pick two points, $P = (1, 0)$ and $Q = (0, 1)$, and investigate line integrals over two different paths that connect the points. We will choose C_1 to be a quarter circle of radius 1 and C_2 to be a path formed by parts of the x and y-axes. (See Figure 17.43.) Along C_1, the line integral $\int_{C_1} \vec{F} \cdot d\vec{r} > 0$, since $\vec{F}$ points along the direction of the curve. Along C_2, however, we have $\int_{C_2} \vec{F} \cdot d\vec{r} = 0$, since $\vec{F}$ is perpendicular to C_2 everywhere. So $\vec{F}$ cannot be a gradient field.

Physical Interpretation of the Line Integral: Work

In this section we will use the line integral to define the work done by a variable force. If a constant force $\vec{F}$ pushes an object through a displacement $\vec{d}$ the work done by the force on the object is given by the following expression:

$$\text{Work done} = \vec{F} \cdot \vec{d} = \|\vec{F}\| \|\vec{d}\| \cos \theta.$$

See Figure 17.44. Now suppose we want to find the work done in pushing an object far above the surface of the earth against the force of gravity. Since the force of gravity varies with distance from the earth, we can't just use the formula $\vec{F} \cdot \vec{d}$. Also, what do we do if the object doesn't move along a straight path? We will approximate the path by line segments which are small enough that the force is approximately constant on each one. Then we can approximate the work done on each segment using the formula $\vec{F} \cdot \vec{d}$ and add. Taking the limit as the segments become smaller gives as accurate an approximation to the work as we want.

Let's suppose that the force at a point with position vector $\vec{r}$ is given by the vector $\vec{F}(\vec{r})$ and that the body is moving along the oriented curve C in Figure 17.45. Suppose the curve C is approximated by small segments represented by $\Delta \vec{r}$. Then

$$\begin{array}{cc}\text{Work done by force } \vec{F}(\vec{r}) \\ \text{over small displacement } \Delta\vec{r}\end{array} \approx \vec{F}(\vec{r}) \cdot \Delta\vec{r}.$$

Thus,

$$\begin{array}{cc}\text{Total work done by force} \\ \text{along curve } C\end{array} \approx \sum \vec{F}(\vec{r}) \cdot \Delta\vec{r}.$$

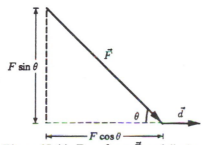
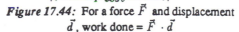

Figure 17.44: For a force $\vec{F}$ and displacement $\vec{d}$, work done = $\vec{F} \cdot \vec{d}$

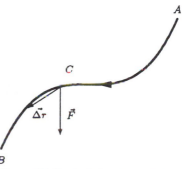
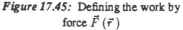

Figure 17.45: Defining the work by force $\vec{F}(\vec{r})$

Since the work is approximated more and more closely as the length of the line segments tends to zero (that is, as $\|\Delta \vec{r}\| \rightarrow 0$), we see that the work done is represented by a line integral:

$$\text{Work done by force } \vec{F}(\vec{r}) \atop \text{along curve } C = \lim_{\|\Delta \vec{r}\| \rightarrow 0} \sum \vec{F}(\vec{r}) \cdot \Delta \vec{r} = \int_C \vec{F} \cdot d\vec{r}.$$

Example 5   On a flat table lies a mass attached to a spring whose other end is fastened to the wall. (See Figure 17.46.) The spring is extended 20 cm beyond its rest position and released. If the axes are as shown in Figure 17.46, when the spring is extended by a distance of x, the force exerted by the spring on the mass is given by

$$\vec{F}(x) = -kx\vec{i}.$$

Suppose the mass moves back to the rest position. How much work is done by the force exerted by the spring?

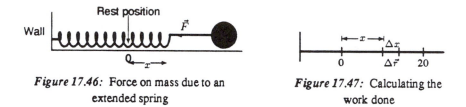

Figure 17.46: Force on mass due to an extended spring   **Figure 17.47:** Calculating the work done

Solution   The path from $x = 20$ to $x = 0$ is divided as shown in Figure 17.47, with a typical segment represented by

$$\Delta \vec{r} = \Delta x \vec{i}.$$

Since we are moving from $x = 20$ to $x = 0$, the quantity Δx will be negative. The work done by the force in moving the mass through this segment is approximated by

$$\text{Work done} \approx \vec{F} \cdot \Delta \vec{r} = (-kx\vec{i}) \cdot (\Delta x \vec{i}) = -kx\, \Delta x.$$

Thus, we have

$$\text{Total work done} \approx \sum -kx\, \Delta x.$$

In the limit, as $\Delta x \rightarrow 0$, this sum becomes an ordinary definite integral. Since the path starts at $x = 20$, this is the lower limit of integration; $x = 0$ is the upper limit. Thus, we get

$$\text{Total work done} = \int_{x=20}^{x=0} -kx\, dx = -\frac{kx^2}{2}\bigg|_{20}^{0} = \frac{k(20)^2}{2} = 200k.$$

The previous example shows that line integrals over paths parallel to a coordinate axis reduce to one-variable integrals. Section 17.4 shows how to convert *any* line integral into a one-variable integral.

Example 6   Suppose a particle has position vector $\vec{r}$. The particle is subject to a force $\vec{F}$ due to gravity. What is the *sign* of the work done by $\vec{F}$ when the particle is moved along the path C_1, a radial line through the center of the earth, starting 8000 km from the center and ending 10,000 km from the center (see Figure 17.48).

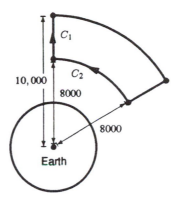

Figure 17.48: The earth

Solution   We divide the path into small radial segments $\Delta\vec{r}$, pointing away from the center of the earth, and parallel to the gravitational force. The vectors $\vec{F}$ and $\Delta\vec{r}$ point in opposite directions, so each term $\vec{F} \cdot \Delta\vec{r}$ is negative. Adding up all of these negative quantities and taking the limit results in a negative value for the total work. The negative sign indicates that we would have to do work *against* gravity to move the particle along the path C_1

Example 7   Find the work done by gravity in moving a particle along C_2, an arc of a circle 8000 km long at a distance of 8000 km from the center of the earth (see Figure 17.48).

Solution   Since C_2 is everywhere perpendicular to the gravitational force, $\vec{F} \cdot \Delta\vec{r} = 0$ for all $\Delta\vec{r}$ along C_2. Thus,

$$\text{Work done} = \int_{C_2} \vec{F} \cdot d\vec{r} = 0,$$

so the work done is zero. This is why satellites can remain in orbit without expending any fuel.

Example 8   Find the sign of the work done by gravity along the curve C_1 in Example 6, but with the opposite orientation.

Solution   Tracing a curve in the opposite direction changes the sign of the line integral because all the segments $\Delta\vec{r}$ change direction, and so every term $\vec{F} \cdot \Delta\vec{r}$ changes sign. Thus, the result will be the negative of the answer found in Example 6. Therefore, the work done by gravity moving a particle along C_1 toward the center of the earth is positive.

Circulation

We are often interested in the line integral of a vector field $\vec{F}$ around an oriented closed curve C.

> If C is a closed curve (that is, one that starts and ends at the same point), the line integral of $\vec{F}$ around C is called the **circulation** of $\vec{F}$ around C.

Circulation[4] is a measure of the net tendency of the vector field to rotate around the curve C.

Example 9   Describe the rotation of the vector fields in Figures 17.49 and Figure 17.50 . Find the sign of the of circulation of the vector fields around the indicated paths.

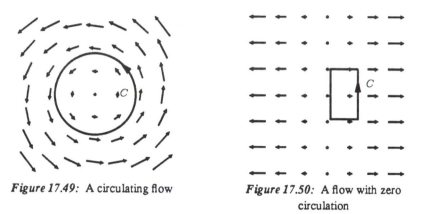

Figure 17.49: A circulating flow   *Figure 17.50:* A flow with zero circulation

Solution   First, consider Figure 17.49. If you think of this as the velocity vector field of water flowing in a pond, you can see that the water is rotating. The line integral around C, measuring the circulation around C, is positive, because the vectors of the field are all pointing in the same direction as the direction of the path. By way of contrast, look at the vector field shown in Figure 17.50. Here the line integral around C is zero because the vertical portions of the path are perpendicular to the field and the contributions from the two horizontal portions cancel out. This means that there is no tendency for the water to circulate around C.

It turns out that the vector field in Figure 17.50 has the property that its circulation around *any* closed path is zero. Water moving according to this vector field has no tendency to rotate around any point and a leaf dropped anywhere in the flow will not spin around. We'll look at such special fields again later when we introduce the notion of the *curl* of a vector field.

Properties of Line Integrals

The definition of the line integral of $\vec{F}$ over an oriented curve C is similar to the definition of the definite integral of a single-variable function. Consequently, the integrals share several properties. The first property says we can take a constant out of a line integral.

---

[4]To emphasize that C is closed, the circulation is sometimes denoted $\oint_C \vec{F} \cdot d\vec{r}$.

If λ is a scalar, then

$$\int_C \lambda \vec{F} \cdot d\vec{r} = \lambda \int_C \vec{F} \cdot d\vec{r}.$$

In words:

> The line integral of a constant times a vector field is the constant times the line integral.

The second property tells us how to deal with sums of vector fields.

If $\vec{F}$ and $\vec{G}$ are two vector fields defined on C, then

$$\int_C (\vec{F} + \vec{G}) \cdot d\vec{r} = \int_C \vec{F} \cdot d\vec{r} + \int_C \vec{G} \cdot d\vec{r}.$$

In words:

> The line integral of a sum is the sum of the line integrals.

The next two properties show the dependence of line integrals on the path of integration. If C_1 and C_2 are oriented curves that meet (so that, C_1 ends where C_2 begins) we can construct a new oriented curve by joining them together. This new curve is called $C_1 + C_2$. See Figure 17.51.

If C_1 and C_2 are two curves, then

$$\int_{C_1+C_2} \vec{F} \cdot d\vec{r} = \int_{C_1} \vec{F} \cdot d\vec{r} + \int_{C_2} \vec{F} \cdot d\vec{r}.$$

In words:

> The line integral of $\vec{F}$ along the curve obtained by joining C_1 and C_2 is the sum of the line integrals along the curves separately.

This property is the analogue for line integrals of the property for definite integrals which says that

$$\int_a^b f(x)\, dx = \int_a^c f(x)\, dx + \int_c^b f(x)\, dx.$$

Finally, if C is an oriented curve, we let $-C$ be the same curve traversed in the opposite direction, that is, with the opposite orientation. See Figure 17.52.

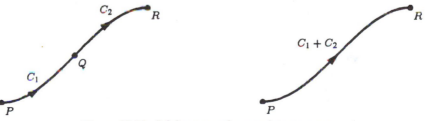

Figure 17.51: Joining two paths to make a new one

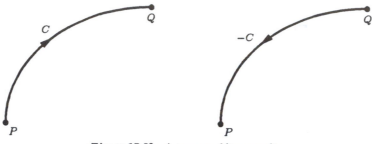

Figure 17.52: A curve and its opposite

If C is an oriented curve, then

$$\int_{-C} \vec{F} \cdot d\vec{r} = -\int_{C} \vec{F} \cdot d\vec{r}.$$

In words:

The line integral of $\vec{F}$ along C with the opposite orientation is the negative of the line integral along C.

This property holds because if we integrate in the opposite direction, the vectors $\Delta \vec{r}_i$ are facing in the opposite direction from before, so the dot products $\vec{F} \cdot \Delta \vec{r}_i$ are the negatives of what they were before.

Problems for Section 17.3

1. Consider the vector field $\vec{F}$ shown in Figure 17.53, together with the paths C_1, C_2, and C_3. Arrange the line integrals $\int_{C_1} \vec{F} \cdot d\vec{r}$, $\int_{C_2} \vec{F} \cdot d\vec{r}$ and $\int_{C_3} \vec{F} \cdot d\vec{r}$ in ascending order.

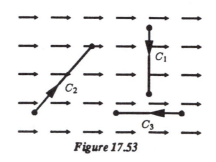

Figure 17.53

2. Let $\vec{F} = x\vec{i} + y\vec{j}$, and let C_1 be the line joining the point $(1, 0)$ to the point $(0, 2)$ and let C_2 be the line joining the point $(0, 2)$ to the point $(-1, 0)$. Is $\int_{C_1} \vec{F} \cdot d\vec{r} = -\int_{C_2} \vec{F} \cdot d\vec{r}$? Explain.

3. What is the approximate value of $\int_{C} \vec{F} \cdot d\vec{r}$ if C is a curve that runs from the point $(2, -6)$ to the point $(4, 4)$ and if $\vec{F} \approx 6\vec{i} - 7\vec{j}$ on C?

4. The properties

$$\int_{C_1+C_2} \vec{F} \cdot d\vec{r} = \int_{C_1} \vec{F} \cdot d\vec{r} + \int_{C_2} \vec{F} \cdot d\vec{r}.$$

and

$$\int_{-C} \vec{F} \cdot d\vec{r} = -\int_{C} \vec{F} \cdot d\vec{r}.$$

are similar to properties of the definite integral of one-variable calculus. What are these properties?

5. Suppose P and Q both lie on the same contour of f. What can you say about the total change in f from P to Q? Explain your answer in terms of $\int_C$ grad $f \cdot d\vec{r}$ where C is a portion of the contour that goes from Q to P.

6. Suppose that grad $f = (2xe^{x^2} \cos y)\vec{i} - (e^{x^2} \sin y)\vec{j}$. Find the change in f between the points $(0,0)$ and $(1, \pi)$ by guessing a formula for f.

7. Draw a curve C and a vector field $\vec{F}$ along C that is not always perpendicular to C, but for which $\int_C \vec{F} \cdot d\vec{r} = 0$.

8. Given the force field $\vec{F}(x, y) = y\vec{i} + x^2\vec{j}$ and the right-angle curve, C, from the points $(0, -1)$ to $(4, -1)$ to $(4, 3)$, shown in Figure 17.54:

    (a) Evaluate $\vec{F}$ at the points $(0, -1)$, $(1, -1)$, $(2, -1)$, $(3, -1)$, $(4, -1)$, $(4, 0)$, $(4, 1)$, $(4, 2)$, $(4, 3)$.
    (b) Make a sketch showing the force field along C.
    (c) Estimate the work done by the indicated force field on an object traversing the curve C.

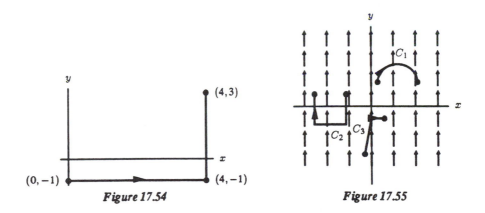

Figure 17.54   Figure 17.55

9. If $\vec{F}$ is the constant force field $\vec{j}$, consider the work done by the field on particles traveling on paths C_1, C_2 and C_3 of Figure 17.55. On which of these paths will zero work be done? Explain.

10. Example 4 on Page 348 showed that the vector field in Figure 17.56 could not be a gradient field by showing that it is not conservative. Here is another way to see the same thing. Suppose that the vector field were the gradient of a function f. Draw a diagram showing what the contours of f would have to look like, and explain why this diagram cannot be the contour diagram of any function.

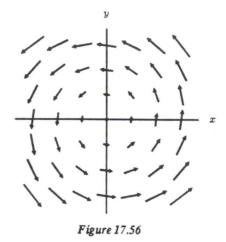

Figure 17.56

11. It is a physical fact that any electric current gives rise to a magnetic field — this is the basis for any electric motor. Ampere's Lawx relates the magnetic field $\vec{B}$ to the current I. It says:

$$\int_C \vec{B} \cdot d\vec{r} = kI$$

where I is the current[5] flowing through a closed curve C, and k is a constant. Figure 17.57 shows a rod carrying current and the magnetic field induced around the rod. If the rod is very long and thin, experiments show that the magnetic field $\vec{B}$ is tangent to every circle (like C in Figure 17.57) perpendicular to the rod and with center on the axis of the rod. The magnitude of $\vec{B}$ is constant along every such circle.

Use Ampere's Law to show that along a circle of radius r, the magnetic field due to a current I has magnitude $kI/2\pi r$. (That is, the strength of the field is inversely proportional to the radial distance from the rod.)

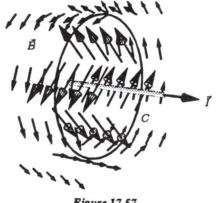

Figure 17.57

For Problems 12–17, use a computer to calculate the line integrals.

---

[5]More precisely, I is the net current through any surface that has C as its boundary.

12. For $\vec{F} = x\vec{i} + y\vec{j}$,

    (a) Find the line integral of $\vec{F}$ around any rectangle, any ellipse, and any polygon. What do you get?

    (b) Find the line integral of $\vec{F}$ along three curves, each of which starts at the origin and ends at $(\frac{1}{2}, \frac{1}{2})$. What do you notice?

For Problems 13–15, answer the same questions as in Problems 12 for the given vector field.

13. $\vec{F} = -y\vec{i} + x\vec{j}$.   14. $\vec{F} = \vec{i} + y\vec{j}$.   15. $\vec{F} = \vec{i} + x\vec{j}$.

16. As a result of your answers to Problems 12–15, you should have noticed that the following statement is true: Whenever the line integral around any closed curve is zero, the line integral along a curve with fixed endpoints has a constant value (that is, the line integral is independent of the path the curve takes between the endpoints). Can you explain why this is so?

17. As a result of your answers to Problems 12–15, you should have noticed that the converse to the statement in Problem 16 is also true: Whenever the line integral depends only on endpoints and not on paths, the circulation is always zero. Can you explain why this is so?

17.4 COMPUTING LINE INTEGRALS USING PARAMETRIZATIONS

So far our discussion of the line integral has centered on a graphical understanding of its definition, and some useful physical interpretations. The goal of this section is to show how we can, in some cases, convert a line integral into a single-variable definite integral. The resulting integral can then be evaluated numerically or analytically.

The key to computing $\int_C \vec{F} \cdot d\vec{r}$ is to choose a parameterization for C, the oriented path over which the line integral is taken.

How Does a Parameterization of C Help In Evaluating $\int_C \vec{F} \cdot d\vec{r}$?

Recall the definition of the line integral: we chop up the curve C into little pieces and compute the sum of the dot products of $\vec{F}$ with $\Delta\vec{r}$ (the displacement vector between two nearby points on the curve):

$$\int_C \vec{F} \cdot d\vec{r} = \lim_{||\Delta\vec{r}_i|| \to 0} \sum \vec{F}(\vec{r}_i) \cdot \Delta\vec{r}_i$$

With a parameterization of C in hand, $\vec{r}(t) = (x(t), y(t))$ for $a \le t \le b$, we can run through the same process: we divide the interval $a \le t \le b$ up into n pieces, each of size $\Delta t = (b-a)/n$. This has the effect of dividing the curve C into n pieces (see Figure 17.58). Now at each point $\vec{r}_i = \vec{r}(t_i)$ we want to compute

$$\vec{F}(\vec{r}_i) \cdot \Delta\vec{r}_i.$$

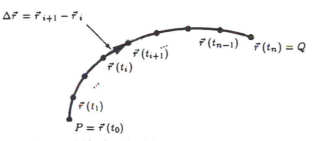

Figure 17.58: Subdivision of a parameterized path

We use the velocity vector to approximate $\Delta \vec{r}_i$. Over a short interval, the velocity vector $\vec{v}(t) = x'(t)\vec{i} + y'(t)\vec{j}$ is approximately constant. Thus, the displacement vector $\Delta \vec{r}_i$ between $\vec{r}(t_i)$ and $\vec{r}(t_{i+1})$ is approximately $\vec{v}(t_i)\Delta t = (x'(t)\vec{i} + y'(t)\vec{j})\Delta t$. Then

$$\int_C \vec{F} \cdot d\vec{r} = \lim_{\|\Delta \vec{r}_i\| \to 0} \sum \vec{F}(\vec{r}_i) \cdot \Delta \vec{r}_i$$

$$\approx \lim_{\Delta t \to 0} \sum \vec{F}(x(t_i), y(t_i)) \cdot (x'(t_i)\vec{i} + y'(t_i)\vec{j})\Delta t.$$

The function, $\vec{F}(x(t_i), y(t_i)) \cdot (x'(t_i)\vec{i} + y'(t_i)\vec{j})$, is a one-variable function of t that is evaluated at each of the points t_i in our subdivision of $a \leq t \leq b$. So this last sum is really a one-variable Riemann sum. Thus, in the limit as $\Delta t \to 0$, we get a definite integral:

$$\lim_{\Delta t \to 0} \sum \vec{F}(x(t_i), y(t_i)) \cdot (x'(t_i)\vec{i} + y'(t_i)\vec{j})\Delta t = \int_a^b \vec{F}(x(t), y(t)) \cdot (x'(t)\vec{i} + y'(t)\vec{j})\,dt.$$

We have shown:

If $(x(t), y(t))$, $a \leq t \leq b$ is any parameterization of C, then

$$\int_C \vec{F} \cdot d\vec{r} = \int_a^b \vec{F}(x(t), y(t)) \cdot (x'(t)\vec{i} + y'(t)\vec{j})\,dt$$

In words:

To compute the line integral of $\vec{F}$ over C, take the definite integral of $\vec{F}$ evaluated on C dotted with the velocity vector of the parameterization of C.

Now let's see some examples of this approach.

Example 1   Compute $\int_C \vec{F} \cdot d\vec{r}$ where $\vec{F} = (x + y)\vec{i} + y\vec{j}$ and C is the quarter unit circle, oriented counterclockwise as shown in Figure 17.59.

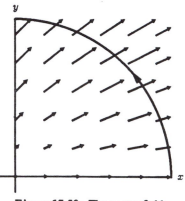

Figure 17.59: The vector field
$\vec{F} = (x + y)\vec{i} + y\vec{j}$ and the quarter
circle C

Solution   Since all of the vectors in $\vec{F}$ along C point in the direction opposite to the orientation of C, we expect our answer to be negative. The first step is to parameterize C by

$$(x(t), y(t)) = (\cos t, \sin t), \quad 0 \le t \le \frac{\pi}{2}.$$

Substituting the parameterization into $\vec{F}$ we get $\vec{F}(x(t), y(t)) = (\cos t + \sin t)\vec{i} + \sin t\vec{j}$. The velocity vector is $\vec{v} = x'(t)\vec{i} + y'(t)\vec{j} = -\sin t\vec{i} + \cos t\vec{j}$. Then

$$\int_C \vec{F} \cdot d\vec{r} = \int_0^{\pi/2} ((\cos t + \sin t)\vec{i} + \sin t\vec{j}) \cdot (-\sin t\vec{i} + \cos t\vec{j})dt$$

$$= \int_0^{\pi/2} (-\cos t \sin t - \sin^2 t + \sin t \cos t)dt$$

$$= \int_0^{\pi/2} -\sin^2 t\, dt$$

$$= \frac{1}{2}\sin t \cos t - \frac{t}{2}\Big|_0^{\pi/2}$$

$$= -\frac{\pi}{4} \approx -0.7854.$$

So the answer is negative, as expected. But why is the line integral small in value? Looking at Figure 17.59 you can see that the vector field is nearly perpendicular to the curve C for much of its length. That means that the contributions made by the vectors $\vec{F}$ in the direction of C are all small.

Example 2   Consider the vector field $\vec{F} = x\vec{i} + y\vec{j}$.

(a)   Suppose C_1 is the line segment joining $(1, 0)$ to $(0, 2)$ and C_2 is a part of a parabola with its vertex at $(0, 2)$, joining the same points in the same order. (See Figure 17.60.) Verify that

$$\int_{C_1} \vec{F} \cdot d\vec{r} = \int_{C_2} \vec{F} \cdot d\vec{r}.$$

(b)   If C is the triangle shown in Figure 17.61, show that $\int_C \vec{F} \cdot d\vec{r} = 0$.

(c)   Find a scalar function f with grad $f = \vec{F}$. Hence, find an easy way to calculate the line integrals in part (a).

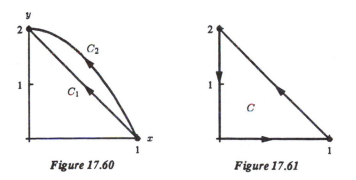

Figure 17.60   *Figure 17.61*

Solution   (a)   The first step is to parameterize C_1 by $(x(t), y(t)) = (1 - t, 2t)$ with $0 \leq t \leq 1$. Then

$$\int_{C_1} \vec{F} \cdot d\vec{r} = \int_0^1 \vec{F}(1 - t, 2t) \cdot (-\vec{i} + 2\vec{j}) \, dt$$

$$= \int_0^1 ((1 - t)\vec{i} + 2t\vec{j}) \cdot (-\vec{i} + 2\vec{j}) \, dt$$

$$= \int_0^1 (t - 1 + 4t) \, dt = \int_0^1 (5t - 1) \, dt$$

$$= \frac{5}{2}t^2 - t \Big|_0^1 = \frac{5}{2} - 1 - (0 - 0) = \frac{3}{2}.$$

To parameterize C_2, we use the fact that it is a parabola with vertex at $(0, 2)$, so its equation is of the form $y = -kx^2 + 2$ for some k. Since the parabola crosses the x-axis at $(1, 0)$, we find that $k = 2$ and $y = -2x^2 + 2$. Therefore, we use the parameterization $(x(t), y(t)) = (t, -2t^2 + 2)$ with $0 \leq t \leq 1$. This traces out C_2 in reverse, since it begins at the point $(0, 2)$ and ends at $(1, 0)$. Thus, if we use this parameterization we must take $t = 1$ as the lower limit of integration and $t = 0$ as the upper:

$$\int_{C_2} \vec{F} \cdot d\vec{r} = \int_1^0 \vec{F}(t, -2(t^2 - 1)) \cdot (\vec{i} - 4t\vec{j}) \, dt$$

$$= \int_1^0 (t\vec{i} - 2(t^2 - 1)\vec{j}) \cdot (\vec{i} - 4t\vec{j}) \, dt$$

$$= \int_1^0 (t + 8t(t^2 - 1)) \, dt = \int_1^0 (8t^3 - 7t) \, dt$$

$$= (2t^4 - \frac{7}{2}t^2) \Big|_1^0 = (0 - 0) - (2 - \frac{7}{2}) = \frac{3}{2}.$$

So the line integrals along C_1 and C_2 are the same.

(b)   We break $\int_C \vec{F} \cdot d\vec{r}$ up into three pieces, one of which we have already computed (namely, the piece connecting $(1, 0)$ to $(0, 2)$, where the line integral has value $3/2$). The piece running from $(0, 2)$ to $(0, 0)$ can be parameterized by $(x(t), y(t)) = (0, 2 - t)$ where $0 \leq t \leq 2$. The piece running from $(0, 0)$ to $(1, 0)$ can be parameterized by $(x(t), y(t)) = (t, 0)$ where $0 \leq t \leq 1$. Then

$$\int_C \vec{F} \cdot d\vec{r} = \frac{3}{2} + \int_0^2 \vec{F}(0, (2 - t)) \cdot (-\vec{j}) \, dt + \int_0^1 \vec{F}(t, 0) \cdot \vec{i} \, dt$$

$$= \frac{3}{2} + \int_0^2 (2 - t)\vec{j} \cdot (-\vec{j}) \, dt + \int_0^1 t\vec{i} \cdot \vec{i} \, dt$$

$$= \frac{3}{2} + \int_0^2 (t - 2) \, dt + \int_0^1 t \, dt$$

$$= \frac{3}{2} + \frac{1}{2}t^2 - 2t \Big|_0^2 + \frac{1}{2}t^2 \Big|_0^1$$

$$= \frac{3}{2} + (2 - 4) + \frac{1}{2} = 0.$$

(c)   One possibility for f is $f(x, y) = x^2/2 + y^2/2$. You can check that grad $f = x\vec{i} + y\vec{j}$. Now we can use the Fundamental Theorem to compute the line integral in part (a). The Fundamental

Theorem says that the line integral of grad f is the total change in the function f between the endpoints. Thus, if $\vec{F} = \operatorname{grad} f$ we have

$$\int_{C_1} \vec{F} \cdot d\vec{r} = \int_{C_2} \vec{F} \cdot d\vec{r} = f(0,2) - f(1,0) = 2 - \frac{1}{2} = \frac{3}{2}.$$

Example 3 Let C be the closed curve consisting of the upper half-circle of radius 1 and the line forming its diameter along the x-axis, oriented counterclockwise. (See Figure 17.62.) Find $\int_C \vec{F} \cdot d\vec{r}$ where $\vec{F}(x,y) = -y\vec{i} + x\vec{j}$.

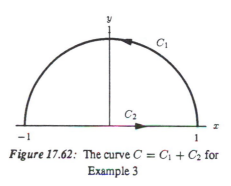

Figure 17.62: The curve $C = C_1 + C_2$ for Example 3

Solution We write $C = C_1 + C_2$ where C_1 is the half-circle and C_2 is the line, and compute $\int_{C_1} \vec{F} \cdot d\vec{r}$ and $\int_{C_2} \vec{F} \cdot d\vec{r}$ separately.

We parameterize C_1 by $(x(t), y(t)) = (\cos t, \sin t)$, $0 \le t \le \pi$. Then

$$\int_{C_1} \vec{F} \cdot d\vec{r} = \int_0^\pi (-\sin t\,\vec{i} + \cos t\,\vec{j}) \cdot (-\sin t\,\vec{i} + \cos t\,\vec{j})\,dt$$

$$= \int_0^\pi (\sin^2 t + \cos^2 t)\,dt = \int_0^\pi 1\,dt = \pi.$$

For C_2, we have $\int_{C_2} \vec{F} \cdot d\vec{r} = 0$, since the vector field $\vec{F}$ has no $\vec{i}$ component along the x-axis (where $y = 0$) and is therefore perpendicular to C_2 at all points.

Finally, we can write

$$\int_C \vec{F} \cdot d\vec{r} = \int_{C_1} \vec{F} \cdot d\vec{r} + \int_{C_2} \vec{F} \cdot d\vec{r} = \pi + 0 = \pi.$$

It is no accident that the result for $\int_{C_1} \vec{F} \cdot d\vec{r}$ is the same as the length of the curve C_1. See Problem 11 and Problem 10 on Page 365.

Example 4 Suppose that grad $f = 2xe^{x^2} \sin y\,\vec{i} + e^{x^2} \cos y\,\vec{j}$. Find the change in f between $(0,0)$ and $(1, \pi/2)$:
(a) By computing a line integral   (b) By computing f.

Solution (a) To find the change in f by computing a line integral, we first choose a path C between the points; the simplest is a line. We parameterize the line by $(x(t), y(t)) = (t, \pi t/2)$, $0 \leq t \leq 1$. Then the Fundamental Theorem of Line Integrals tells us that

$$
\begin{aligned}
f(1, \frac{\pi}{2}) - f(0,0) &= \int_C \operatorname{grad} f \cdot d\vec{r} \\
&= \int_0^1 \operatorname{grad} f\left(t, \frac{\pi t}{2}\right) \cdot \left(\vec{i} + \frac{\pi}{2}\vec{j}\right) dt \\
&= \int_0^1 \left(2te^{t^2} \sin\left(\frac{\pi t}{2}\right)\vec{i} + e^{t^2}\cos\left(\frac{\pi t}{2}\right)\vec{j}\right) \cdot \left(\vec{i} + \frac{\pi}{2}\vec{j}\right) dt \\
&= \int_0^1 \left(2te^{t^2}\sin\left(\frac{\pi t}{2}\right) + \frac{\pi e^{t^2}}{2}\cos\left(\frac{\pi t}{2}\right)\right) dt.
\end{aligned}
$$

This last integral can be broken down into two separate integrals, which are best approximated numerically:

$$
\int_0^1 2te^{t^2}\sin\left(\frac{\pi t}{2}\right) dt \approx 1.481 \quad \text{and} \quad \int_0^1 \frac{\pi e^{t^2}}{2}\cos\left(\frac{\pi t}{2}\right) dt \approx 1.237.
$$

Thus, we have an approximate value of $1.481 + 1.237 = 2.718$. (This looks suspiciously like e.)

(b) The other way to find the change in f between these two points is to first find f. To do this, observe that

$$
2xe^{x^2}\sin y\,\vec{i} + e^{x^2}\cos y\,\vec{j} = \frac{\partial}{\partial x}\left(e^{x^2}\sin y\right)\vec{i} + \frac{\partial}{\partial y}\left(e^{x^2}\sin y\right)\vec{j} = \operatorname{grad}\left(e^{x^2}\sin y\right).
$$

So one possibility for f is $f(x,y) = e^{x^2}\sin y$. Thus,

$$
\text{Change in } f\bigg|_{(0,0)}^{(1,\pi/2)} = e^{x^2}\sin y\bigg|_{(0,0)}^{(1,\pi/2)} = e^1\sin\left(\frac{\pi}{2}\right) - e^0\sin 0 = e.
$$

The exact answer confirms our earlier suspicions.

Notice that if you have a formula for f, it is *much* easier to compute the change in f directly than to compute the line integral. The problem, of course, is that it may be difficult to find the formula for f.

The next example illustrates the computation of a line integral over a path in 3-space.

Example 5 A particle is traveling along the circular helix C given by $\vec{r}(t) = (\cos t, \sin t, 2t)$ and is subject to a force $\vec{F} = x\vec{i} + z\vec{j} - yx\vec{k}$. Find the total work done on the particle by the force for $0 \leq t \leq 3\pi$.

Solution Since the path is given in parametric form, we evaluate the line integral

$$
\int_C \vec{F} \cdot d\vec{r} = \int_0^{3\pi} \vec{F}\left(\vec{r}(t)\right) \cdot \vec{r}'(t)\, dt
$$

$$= \int_0^{3\pi} (\cos t \vec{i} + 2t \vec{j} - \cos t \sin t \vec{k}) \cdot (-\sin t \vec{i} + \cos t \vec{j} + 2 \vec{k}) \, dt$$

$$= \int_0^{3\pi} (-\cos t \sin t + 2t \cos t - 2 \cos t \sin t) \, dt$$

$$= \int_0^{3\pi} (-3 \cos t \sin t + 2t \cos t) \, dt$$

Using integration by parts to see that $\int t \cos t \, dt = t \sin t + \cos t$, we get

$$\int_C \vec{F} \cdot d\vec{r} = -\frac{3}{2} \sin^2 t + 2(t \sin t + \cos t) \Big|_0^{3\pi} = -4.$$

Independence of Parameterization

Since there are many different ways of parameterizing a given oriented curve, you may be wondering what happens to the value of a given line integral if you choose another parameterization. The answer is simple: *The choice of parameterization makes no difference.* Since we initially defined the line integral without reference to any particular parameterization, this is exactly as we would expect.

Example 6 Consider the oriented path which is a straight line L running from $(0, 0)$ to $(1, 1)$. Calculate the line integral of the vector field $\vec{F} = (3x - y)\vec{i} + y\vec{j}$ along L using each of the parameterizations:

(a) $A(t) = (t, t), \quad 0 \le t \le 1,$ (b) $B(t) = (2t, 2t), \quad 0 \le t \le 1/2,$

(c) $C(t) = \left(\frac{t^2-1}{3}, \frac{t^2-1}{3}\right), \quad 1 \le t \le 2,$ (d) $D(t) = (e^t - 1, e^t - 1), \quad 0 \le t \le \ln 2.$

Solution First, make sure that you see why each of these gives a parameterization of the line L: each has both coordinates equal (as do all points on L) and each begins at $(0,0)$ and ends at $(1,1)$. Now let's calculate the line integral of the vector field $\vec{F} = (3x - y)\vec{i} + y\vec{j}$ using each parameterization.

(a) Using $A(t)$, we get

$$\int_L \vec{F} \cdot d\vec{r} = \int_0^1 ((3t - t)\vec{i} + t\vec{j}) \cdot (\vec{i} + \vec{j}) \, dt = \int_0^1 3t \, dt = \frac{3t^2}{2} \Big|_0^1 = \frac{3}{2}.$$

(b) Using $B(t)$ gives

$$\int_L \vec{F} \cdot d\vec{r} = \int_0^{1/2} ((6t - 2t)\vec{i} + 2t\vec{j}) \cdot (2\vec{i} + 2\vec{j}) \, dt = \int_0^{1/2} 12t \, dt = 6t^2 \Big|_0^{1/2} = \frac{3}{2}.$$

Both $A(t)$ and $B(t)$ are parameterizations that traverse L at a constant speed: $A(t)$ has speed 1 and $B(t)$ has speed 2. You can see how the speed factor of 2 cancels when evaluating the definite integral.

(c) Now we use $C(t)$:

$$\int_L \vec{F} \cdot d\vec{r} = \int_1^2 \left(\left(\frac{3(t^2-1)}{3} - \frac{(t^2-1)}{3} \right) \vec{i} + \frac{t^2-1}{3} \vec{j} \right) \cdot \left(\frac{2t}{3} \vec{i} + \frac{2t}{3} \vec{j} \right) dt$$

$$= \int_1^2 \frac{2t}{3}(t^2 - 1)dt = \frac{2}{3} \int_1^2 (t^3 - t)dt$$

$$= \frac{2}{3}\left(\frac{t^4}{4} - \frac{t^2}{2}\right)\Bigg|_1^2 = \frac{3}{2}.$$

(d)   Finally, using $D(t)$, we get

$$\int_L \vec{F} \cdot d\vec{r} = \int_0^{\ln 2} \left((3(e^t - 1) - (e^t - 1))\,\vec{i} + (e^t - 1)\vec{j}\right) \cdot (e^t\vec{i} + e^t\vec{j})dt$$

$$= \int_0^{\ln 2} 3e^t(e^t - 1)dt = 3\int_0^{\ln 2} (e^{2t} - e^t)dt$$

$$= 3\left(\frac{e^{2t}}{2} - e^t\right)\Bigg|_0^{\ln 2} = \frac{3}{2}.$$

The fact that the four computations yield the same result illustrates the fact that the value of a line integral is independent of the parameterization of the path. See Problems 12-14 at the end of this section for another interpretation of this fact.

Problems for Section 17.4

In Problems 1–6, compute the line integral of the given vector field along the given path.

1. $\vec{F}(x, y) = x^2\vec{i} + y^2\vec{j}$ and C is the line from the point $(1, 2)$ to the point $(3, 4)$.

2. $\vec{F}(x, y) = \ln y\,\vec{i} + \ln x\,\vec{j}$ and C is the curve given by $(2t, t^3)$ for $2 \leq t \leq 4$.

3. $\vec{F}(x, y) = e^x\vec{i} + e^y\vec{j}$ and C is the part of the ellipse $x^2 + 4y^2 = 4$ joining the point $(0, 1)$ to the point $(2, 0)$ in the clockwise direction.

4. $\vec{F}(x, y) = xy\vec{i} + (x - y)\vec{j}$ and C is the triangle joining the points $(1, 0)$, $(0, 1)$ and $(-1, 0)$ in the clockwise direction.

5. $\vec{F} = x\vec{i} + 2zy\vec{j} + x\vec{k}$ and C is given by $\vec{r} = t\vec{i} + t^2\vec{j} + t^3\vec{k}$ for $1 \leq t \leq 2$.

6. $\vec{F} = e^y\vec{i} + \ln(x^2 + 1)\vec{j} + \vec{k}$ and C is the circle of radius 2 in the yz-plane centered at the origin and traversed as shown in Figure 17.63.

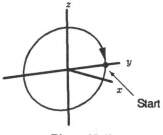

Figure 17.63

7. Find parameterizations for the oriented curves shown in Figure 17.64. C_1 is a semicircle of radius 1, centered at the point $(1, 0)$. C_2 is a portion of a parabola, with vertex at the point $(1, 0)$ and y-intercept -2. C_3 is one arc of the sine curve.

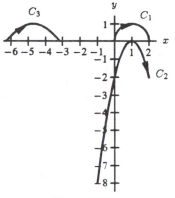

Figure 17.64

8. Let C be the oriented line segment from the point $(0, 0)$ to the point $(4, 12)$, and suppose that $\vec{F} = xy\vec{i} + x\vec{j}$.

    (a) Is $\int_C \vec{F} \cdot d\vec{r}$ greater than, less than, or equal to zero? Give a geometric explanation.
    (b) A simple parameterization of C is $(x(t), y(t)) = (t, 3t)$ for $0 \le t \le 4$. Use this to compute $\int_C \vec{F} \cdot d\vec{r}$.
    (c) Suppose a particle leaves the point $(0, 0)$, moves along the line towards the point $(4, 12)$, stops before reaching it and backs up, stops again and reverses direction, then completes its journey to the endpoint. All travel takes place along the line segment joining the point $(0, 0)$ to the point $(4, 12)$. If we call this path C', explain why $\int_{C'} \vec{F} \cdot d\vec{r} = \int_C \vec{F} \cdot d\vec{r}$.
    (d) A parameterization for a path like C' is given by
    $$(x(t), y(t)) = (\frac{1}{3}(t^3 - 6t^2 + 11t), (t^3 - 6t^2 + 11t)), \ 0 \le t \le 4.$$
    Check that this begins at the point $(0, 0)$ and ends at the point $(4, 12)$. Check also that all points of C' lie on the line segment connecting the point $(0, 0)$ to the point $(4, 12)$. What are the values of t where the particle changes direction?
    (e) Find $\int_{C'} \vec{F} \cdot d\vec{r}$ using the vector field $xy\vec{i} + x\vec{j}$ and the parameterization in part (d). Do you get the same answer as in part (b)?

9. In Example 6 on page 363 the vector field $\vec{F} = (x + y)\vec{i} + x\vec{j}$ was integrated over several parameterizations of the line going from the point $(0, 0)$ to the point $(1, 1)$, always getting $3/2$. In this problem we compute the line integral along several paths with the same endpoints. Show that the line integral of $\vec{F}$ along each of the paths below is $3/2$:

    (a) The path (t, t^2), with $0 \le t \le 1$
    (b) The path (t^2, t), with $0 \le t \le 1$
    (c) The path (t, t^n), with $n > 0$ and $0 \le t \le 1$

10. Consider the vector field $\vec{F} = -y\vec{i} + x\vec{j}$. Let C be the unit circle oriented counterclockwise.

    (a) Show that $\vec{F}$ has a constant magnitude of 1 on the circle C.
    (b) Show that $\vec{F}$ is always tangent to the circle C.
    (c) Show that $\int_C \vec{F} \cdot d\vec{r} = $ Length of C.

11.   Suppose that along a curve C, a vector field $\vec{F}$ has direction always tangent to C in direction of orientation and has constant magnitude $m = \|\vec{F}\|$. Use the definition of the line integral to explain why

$$\int_C \vec{F} \cdot d\vec{r} = m \times (\text{Length of } C).$$

Problems 12–14 concern the line integral in Example 6 on Page 363. In the example, several different parameterizations are used to convert the line integral into a definite integral. In each of the following cases, show that the two definite integrals corresponding to the two given parameterizations are equal by finding a change of variable (that is, a substitution) which converts one integral to the other. Since making a substitution into a one-variable integral does not change its value, the value of the line integral is unaltered. This gives us another way of seeing why changing the parameterization of the curve does not change the value of the line integral.

12. $A(t)$ and $B(t)$ 13. $A(t)$ and $C(t)$ 14. $A(t)$ and $D(t)$

17.5   CONSERVATIVE VECTOR FIELDS

In Section 17.3, we saw how to recover the total change in f between two points P and Q by taking the line integral of its gradient field grad f over any path that joins P to Q. In particular, we saw that the integral of grad f from P to Q does not depend on the path taken from P to Q. We give vector fields with this property a special name.

> A vector field $\vec{F}$ is said to be **conservative** if for any two points P and Q, the line integral has the same value along any path from P to Q.

Being conservative thus means that the line integral of $\vec{F}$ is the same along *any* path with the same endpoints. We say that the line integral is *path-independent*, meaning that the value of the line integral depends on the endpoints of the path, but not on where the path goes in between.

Example 1   The vector field $\vec{F}(x, y) = x\vec{i} + y\vec{j}$ is conservative. Compute the line integrals over the three paths A, B and C shown in Figure 17.65 from $(0, 0)$ to $(1, 1)$ and verify that they are equal. Here A is a line segment, B is the graph of $f(x) = x^2$ and C consists of two line segments meeting at a right angle.

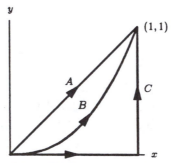

Figure 17.65: A conservative
vector field has same line integral
along these three paths

Solution   We parameterize A by (t, t) where $0 \le t \le 1$. Then

$$\int_A \vec{F} \cdot d\vec{r} = \int_0^1 (t\vec{i} + t\vec{j}) \cdot (\vec{i} + \vec{j}) \, dt$$

$$= \int_0^1 2t \, dt = t^2 \Big|_0^1 = 1.$$

The path B has the parameterization (t, t^2) where $0 \le t \le 1$. Then we have

$$\int_B \vec{F} \cdot d\vec{r} = \int_0^1 (t\vec{i} + t^2\vec{j}) \cdot (\vec{i} + 2t\vec{j}) \, dt$$

$$= \int_0^1 (t + 2t^3) \, dt = \frac{t^2}{2} + \frac{2t^4}{4} \Big|_0^1 = 1.$$

We have to break the path C into two separate parameterizations: $(t, 0)$ where $0 \le t \le 1$ and $(1, t)$ where $0 \le t \le 1$. Then

$$\int_C \vec{F} \cdot d\vec{r} = \int_0^1 (t\vec{i} \cdot \vec{i}) \, dt + \int_0^1 (\vec{i} + t\vec{j}) \cdot \vec{j} \, dt$$

$$= \int_0^1 t \, dt + \int_0^1 t \, dt = \frac{1}{2} + \frac{1}{2} = 1.$$

Example 2   The vector field $\vec{F}(x, y) = x\vec{i} + y\vec{j}$ is conservative. Compute geometrically the line integrals over the three paths A, B and C shown in Figure 17.66 from $(1, 0)$ to $(0, 1)$ and verify that they are equal. Here A is a portion of a circle, B is a line, and C consists of two line segments meeting at a right angle.

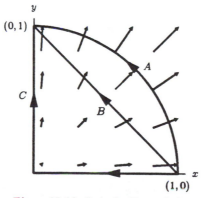

Figure 17.66: Paths for Example 2

Solution   The vector field $\vec{F}$ points radially outward, and so is everywhere perpendicular to A; thus, $\int_A \vec{F} \cdot d\vec{r} = 0$.

Along the first half of B, the terms $\vec{F} \cdot \Delta \vec{r}$ are negative; along the second half the terms $\vec{F} \cdot \Delta \vec{r}$ are positive. By symmetry the positive and negative contributions cancel out, giving a Riemann sum and a line integral of 0. The line integral is also 0 along C, by cancellation.

Here the values of $\vec{F}$ along the x-axis have the same magnitude as those along the y-axis. On the first half of C the path is traversed in the opposite direction to $\vec{F}$; on the second half of C the path is traversed in the same direction as $\vec{F}$. So the two halves cancel.

Of course, we cannot prove that a vector field is conservative by showing that the line integral along a few paths between a few points give the same answer. We will consider the question of how to show that a field is conservative later in this section.

Why do We Care about Conservative Vector Fields?

It turns out that some of the fundamental vector fields of nature are conservative — two important examples are the gravitational field and the electric field of particles at rest. The fact that the gravitational field is conservative means that the work done in moving an object subject to gravity depends only on the starting and ending points, and not on the path taken. For example the work done (computed by the line integral) in carrying a bicycle to a sixth floor apartment is the same whether it is carried up the stairs or hauled up the side of the building and in the window.

When a force field is conservative we can define the *potential energy* of a body to depend only on its position. When the body is moved to another position, the potential energy changes by the work done by the force, which then depends only on the start and end positions. If the work done had not been path independent, the potential energy would depend both on the body's current position *and* on how it got there. It turns out that the potential energy function is related to the force field by means of a gradient, as we see below.

Conservative Fields and Gradient Fields

If $\vec{F} = \operatorname{grad} f$, then the Fundamental Theorem of Line Integrals tells us that if C is a path from P to Q, then

$$\int_C \vec{F} \cdot d\vec{r} = f(Q) - f(P),$$

Notice that the right hand side of this equation does not depend on the path, but only on the endpoints of the path. Thus, $\vec{F}$ is conservative. We have shown the following:

$$\boxed{\text{Every gradient field is conservative.}}$$

In particular, if a field is not conservative, it cannot be a gradient field. What about the converse? That is, given a conservative vector field $\vec{F}$, can we always find a function f such that $\vec{F} = \operatorname{grad} f$? The answer is yes, and we will show how to construct f from $\vec{F}$.

First, notice that there are many different choices for f, since you can always add a constant to f without changing $\operatorname{grad} f$. If we pick a fixed point P arbitrarily, then by adding or subtracting a constant to f we can make sure that $f(P) = 0$. Now suppose we want to know $f(Q)$ for some variable point Q. By the Fundamental Theorem of Line Integrals, we have

$$f(Q) = f(Q) - f(P) = \int_C \operatorname{grad} f \cdot d\vec{r},$$

where C is a path from P to Q. Since grad f is supposed to be $\vec{F}$, we can turn this into a definition of $f(Q)$:

$$f(Q) = \int_C \vec{F} \cdot d\vec{r}.$$

As long as $\vec{F}$ is conservative, this definition makes sense, because the expression on the right does not depend on which path we choose from P to Q. On the other hand, if $\vec{F}$ is not conservative, then different choices might give different values for $f(Q)$, so f would not be a function (a function has to have one and only one value at each point).

Although we will not do so here, it is now necessary to demonstrate here that the gradient of the function f is indeed $\vec{F}$. Thus, by constructing a function f in this manner, we can show the following:

> If $\vec{F}$ is conservative, then $\vec{F}$ = grad f for some f.

The function f is sufficiently important that it is given a special name:

> If a vector field $\vec{F}$ is of the form $\vec{F}$ = grad f for some scalar function f, then f is called a **potential function** for the field $\vec{F}$

Example 3   Show that the field $\vec{F}(x, y) = y \cos x \vec{i} + \sin x \vec{j}$ is conservative.

Solution   If we can find a potential function f, then $\vec{F}$ must be conservative. We want $\partial f / \partial x = y \cos x$, so f must be of the form $y \sin x + g(y)$ where $g(y)$ is a function of y only. Now, $\partial f / \partial y = \sin x + g'(y)$. For this to be the same as the second component of $\vec{F}$, that is, $\sin x$, we must have $g'(y) = 0$, so $g(y) = C$ where C is some constant. Thus,

$$f(x, y) = y \sin x + C$$

is a potential function for $\vec{F}$ and therefore $\vec{F}$ is conservative.

Example 4   Show that the gravitational field

$$\vec{F} = -\frac{GM}{r^3} \vec{r}$$

of an object of mass M is a gradient field, so is therefore conservative, and find a potential function for $\vec{F}$.

Solution   All the force vectors point in towards the origin, and if they are going to be gradient vectors of some function f they must be perpendicular to the level surfaces of f, so the level surfaces of f must be spheres. Also, the magnitude of the vector is GM/r^2, and this is the rate of change of the function f in the inward direction. Since

$$\frac{d}{dr}\left(\frac{1}{r}\right) = \frac{-1}{r^2},$$

we might guess that $\vec{F}$ is the gradient of the function

$$f(x, y, z) = \frac{GM}{r} = \frac{GM}{\sqrt{x^2 + y^2 + z^2}}.$$

We will try this:

$$f_x = \frac{\partial}{\partial x} \frac{GM}{\sqrt{x^2 + y^2 + z^2}} = \frac{-GMx}{(x^2 + y^2 + z^2)^{3/2}},$$

$$f_y = \frac{\partial}{\partial y} \frac{GM}{\sqrt{x^2 + y^2 + z^2}} = \frac{-GMy}{(x^2 + y^2 + z^2)^{3/2}},$$

$$f_z = \frac{\partial}{\partial z} \frac{GM}{\sqrt{x^2 + y^2 + z^2}} = \frac{-GMz}{(x^2 + y^2 + z^2)^{3/2}}.$$

So

$$\text{grad } f = f_x \vec{i} + f_y \vec{j} + f_z \vec{k} = \frac{GM}{(x^2 + y^2 + z^2)^{\frac{3}{2}}}(x\vec{i} + y\vec{j} + z\vec{k})$$

$$= -\frac{GM}{r^3}\vec{r} = \vec{F}.$$

Our computations show that $\vec{F}$ is a gradient field and that $f = GM/r$ is a potential function for $\vec{F}$. The function f is called the *gravitational potential*.

Conservative Fields and Circulation

If $\vec{F}$ is a conservative vector field, what can we say about its circulation around a closed curve C? Suppose P and Q are any two points on the path, then (see Figure 17.67) we can think of C as made up of the path C_1 followed by $-C_2$. Since $\vec{F}$ is conservative

$$\int_{C_1} \vec{F} \cdot d\vec{r} = \int_{C_2} \vec{F} \cdot d\vec{r},$$

so

$$\int_{C_1} \vec{F} \cdot d\vec{r} - \int_{C_2} \vec{F} \cdot d\vec{r} = 0.$$

Thus, the circulation around C is zero because the contributions from C_1 and C_2 cancel, so

$$\int_C \vec{F} \cdot d\vec{r} = \int_{C_1} \vec{F} \cdot d\vec{r} - \int_{C_2} \vec{F} \cdot d\vec{r} = 0.$$

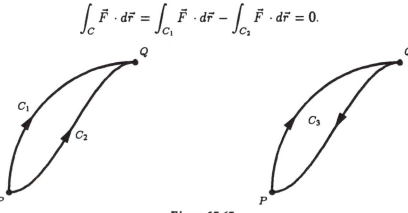

Figure 17.67

A similar argument shows that if the circulation around any closed curve is zero, the line integral must be path-independent and so the vector field is conservative. We define a vector field to be *circulation free* if and only if it has zero circulation around any closed path. Thus, we have the following result:

> A vector field is conservative if and only if it is circulation free.

Summary

We have studied three apparently different sorts of vector fields: conservative vector fields, gradient vector fields, and circulation free vector fields. It turns out that these are all the same thing. It is still useful, however, to have the different terminology, since depending on the context we may want to emphasize one or another property of these types of fields. Here is a summary of the properties and the connections between them:

- **Conservative vector fields** are fields with the property that for any two points P and Q, the integral along a path from P to Q is the same no matter what path you choose.

- **Gradient vector fields** are fields of the form grad f for some scalar function f, which is called the potential function of the vector field.

- **Circulation free vector fields** are vector fields with the property that the integral around any closed path (also called the circulation around the path) is zero.

- **Conservative and circulation free** vector fields are the same thing because any time you have two different paths joining the same pair of points, you can make a closed path by going out along one and back along the other. On the other hand, any time you have a closed path, you can break it up into two different paths between the same pair of points just by picking any two points on the path. If the line integral of a vector field is the same along both paths, then its circulation around the closed path is zero, and vice versa.

- **Gradient vector fields are conservative** by the Fundamental Theorem of Line Integrals, which says that integral of a gradient field is the change in its potential between the beginning and ending points, and so must be zero for a closed path (where the beginning and ending points are the same).

- **Conservative vector fields are gradient fields** because you can use path independence to construct the potential function using a line integral.

Problems for Section 17.5

For Problems 1–4, decide whether or not the given vector fields could be gradient vector fields. Give a justification for your answer.

1. $\vec{F}(x, y) = x\vec{i}$

2. $\vec{F}(x, y, z) = \dfrac{-z}{\sqrt{x^2 + z^2}}\vec{i} + \dfrac{y}{\sqrt{x^2 + z^2}}\vec{j} + \dfrac{x}{\sqrt{x^2 + z^2}}\vec{k}$

3. $\vec{G}(x, y) = (x^2 - y^2)\vec{i} - 2xy\vec{j}$

4. $\vec{F}(\vec{r}) = \vec{r}/r^3$, where $\vec{r} = x\vec{i} + y\vec{j} + z\vec{k}$

5. Consider the vector field $\vec{F}$ graphed in Figure 17.68.

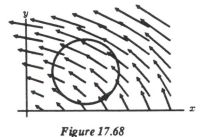

Figure 17.68

(a) Is the line integral $\int_C \vec{F} \cdot d\vec{r}$ positive, negative, or zero?

(b) From your answer to part (a), can you determine whether or not $\vec{F} = \text{grad } f$ for some function f?

(c) Which of the following formulas best fits this vector field?

$$\vec{F_1} = \frac{x}{x^2 + y^2}\vec{i} + \frac{y}{x^2 + y^2}\vec{j}, \quad \vec{F_2} = -y\vec{i} + x\vec{j}, \quad \vec{F_3} = \frac{-y}{(x^2 + y^2)^2}\vec{i} + \frac{x}{(x^2 + y^2)^2}\vec{j}.$$

6. Consider the vector field $\vec{F}(x, y) = x\vec{j}$ shown in Figure 17.69. The field is constant along lines $x = a$.

(a) Find paths C_1, C_2 and C_3 from P to Q such that

$$\int_{C_1} \vec{F} \cdot d\vec{r} = 0, \qquad \int_{C_2} \vec{F} \cdot d\vec{r} > 0, \qquad \text{and} \int_{C_3} \vec{F} \cdot d\vec{r} < 0.$$

(b) Is $\vec{F}$ a gradient field?

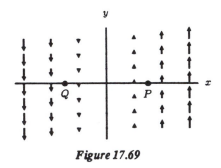

Figure 17.69

For Problems 7–9, decide if the given vector field is the gradient of a function f. If so, find such an f. If not, explain why not.

7. $x\vec{i} + y\vec{j}$

8. $y\vec{i} + y\vec{j}$

9. $(x^2 + y^2)\vec{i} + 2xy\vec{j}$

In Problems 10–14, each of the statements is *false*. Explain why or give a counterexample.

10. If $\int_C \vec{F} \cdot d\vec{r} = 0$ for one particular closed path C, then $\vec{F}$ is conservative.

11. $\int_C \vec{F} \cdot d\vec{r}$ is the total change in $\vec{F}$ along C.

12. If the vector fields $\vec{F}$ and $\vec{G}$ have $\int_C \vec{F} \cdot d\vec{r} = \int_C \vec{G} \cdot d\vec{r}$ for a particular path C, then $\vec{F} = \vec{G}$.

13. If the vector fields $\vec{F}$ and $\vec{G}$ have $\int_C \vec{F} \cdot d\vec{r} = \int_C \vec{G} \cdot d\vec{r}$ for a particular path C, then for each point (x_c, y_c) on the curve C we must have $\vec{F}(x_c, y_c) = \vec{G}(x_c, y_c)$.

14. If the total change of a function f along a curve C is zero, then C must be a contour of f.

15. Let $\vec{F}(x, y)$ be the conservative vector field in Figure 17.70. The vector field $\vec{F}$ associates with each point a unit vector pointing radially outward. The curves $C_1, C_2, \ldots, C_7$ have the directions shown. Consider the line integrals $\displaystyle\int_{C_i} \vec{F} \cdot d\vec{x}, i = 1, \ldots, 7$. Without computing any integrals:

    (a) List all the line integrals which are zero.
    (b) List all the positive line integrals in ascending order.
    (c) List all the negative line integrals.

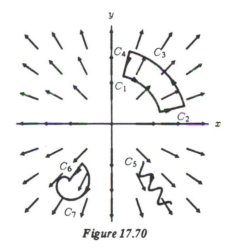

Figure 17.70

16. Repeat Problem 10 for the vector field in Problem 6.

17. Suppose a particle subject to a force $\vec{F}(x, y) = y\vec{i} - x\vec{j}$ moves along a portion of a unit circle, centered at the origin, that begins at (-1,0) and ends at (0,1).

    (a) Find the work done by $\vec{F}$. Explain the sign of your answer.
    (b) Is $\vec{F}$ conservative? Explain.

18. A *central field* is a vector field whose direction is always toward (or away from) a fixed point C (the center) and whose magnitude at a point P is a function only of the distance from P to C. In two dimensions this means that the field has constant magnitude on circles centered at C. The gravitational and electrical fields are both central fields.

    (a) Sketch an example of a central field $\vec{F}$.
    (b) Suppose that the central field $\vec{F}$ is a gradient field, i.e. $\vec{F} = \text{grad } f$. What must be the shape of the contours of f? Sketch in some contours for this case.

(c)   Is every gradient field a central field? Explain.

(d)   In Figure 17.71, two paths are shown between the points Q and P. Assuming that there is a central force $\vec{F}$ with center C, explain why the work done by $\vec{F}$ is the same for either path.

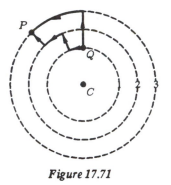

Figure 17.71

(e)   It is in fact true that every central field is a gradient field. Use an argument suggested by Figure 17.71 to explain why any central field must be path independent.

REVIEW PROBLEMS FOR CHAPTER SEVENTEEN

1.   (a)   What is meant by a vector field?

     (b)   Which of the following are vector fields? Why?

           (i) $\vec{r} + \vec{a}$

           (ii) $\vec{r} \cdot \vec{a}$

           (iii) $x^2\vec{i} + y^2\vec{j} + z^2\vec{k}$

           (iv) $x^2 + y^2 + z^2$

           where

$$\vec{r} = x\vec{i} + y\vec{j} + z\vec{k}$$
$$\vec{a} = a_1\vec{i} + a_2\vec{j} + a_3\vec{k}$$

a_1, a_2, a_3 all constant.

Sketch the vector fields in Problems 2–4.

2. $\vec{F} = y\vec{i} - x\vec{j}$

3. $\vec{F} = \left(\dfrac{y}{\sqrt{x^2 + y^2}}\right)\vec{i} - \left(\dfrac{x}{\sqrt{x^2 + y^2}}\right)\vec{j}$

4. $\vec{F} = \left(\dfrac{y}{x^2 + y^2}\right)\vec{i} - \left(\dfrac{x}{x^2 + y^2}\right)\vec{j}$

5.   The following problem concerns the vector field $\vec{F} = \vec{r}/\|\vec{r}\|^3$, where $\vec{r} = x\vec{i} + y\vec{j} + z\vec{k}$. In each case find the given quantity in terms of x, y, z, or t.

     (a) $\|\vec{F}\|$

(b) $\vec{F} \cdot \vec{r}$

(c)   A unit vector parallel to $\vec{F}$ and pointing in the same direction.

(d)   A unit vector parallel to $\vec{F}$ and pointing in the opposite direction.

(e) $\vec{F}$ if $\vec{r} = \cos t\vec{i} + \sin t\vec{j} + \vec{k}$

(f) $\vec{F} \cdot \vec{r}$ if $\vec{r} = \cos t\vec{i} + \sin t\vec{j} + \vec{k}$

6.   Explain why, when $\vec{F} = F_1\vec{i} + F_2\vec{j}$, and C is the path below, with Δx and Δy very small, we have

$$\int_C \vec{F} \cdot d\vec{r} \approx F_1(0,0)\Delta x + F_2(\Delta x, 0)\Delta y - F_1(\Delta x, \Delta y)\Delta x - F_2(0, \Delta y)\Delta y.$$

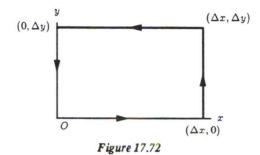

Figure 17.72

For Problems 7–8, consider the vector field $\vec{F}$ graphed in Figure 17.73 and Figure 17.74. Determine if the line integral $\int_C \vec{F} \cdot d\vec{r}$ is positive, negative, or zero along the given paths.

7.   (a) A

     (b) C_1, C_2, C_3, C_4

     (c) C, the closed curve consisting of all the C's together.

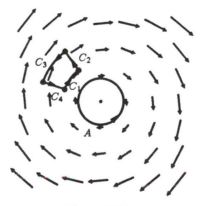

Figure 17.73

8.   (a) A

     (b) C_1, C_2, C_3, C_4

     (c) C, the closed curve consisting of all the C's together.

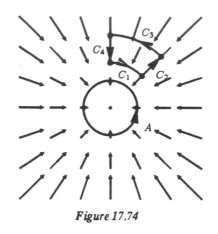

Figure 17.74

For Problems 9–11, compute $\int_C \vec{F} \cdot d\vec{r}$ where $\vec{F}$ and C are given.

9. $\vec{F} = (x^2 - y)\vec{i} + (y^2 + x)\vec{j}$, C is the parabola $y = x^2 + 1$ traversed from $(0, 1)$ to $(1, 2)$.

10. $\vec{F} = (2x - y + 4)\vec{i} + (5y + 3x - 6)\vec{j}$, C is the triangle with vertices $(0, 0), (3, 0), (3, 2)$ traversed counterclockwise.

11. $\vec{F} = (3x - 2y)\vec{i} + (y + 2z)\vec{j} - x^2\vec{k}$, C is the path consisting of the straight line joining the points $(0, 0, 0)$ to $(1, 1, 1)$.

Are the statements in Problems 12–15 true or false? Explain why or give a counterexample.

12. True or false? Explain your answer. $\int_C \vec{F} \cdot d\vec{r}$ is a vector.

13. $\int_C \vec{F} \cdot d\vec{r} = \vec{F}(Q) - \vec{F}(P)$ when P and Q are the endpoints of C.

14. The fact that the line integral of a vector field $\vec{F}$ is zero around the unit circle $x^2 + y^2 = 1$ means that $\vec{F}$ must be a gradient vector field.

15. Suppose C_1 is the unit square joining the points $(0, 0), (1, 0), (1, 1), (0, 1)$ oriented clockwise and C_2 is the same square but traversed twice in the opposite direction. If $\int_{C_1} \vec{F} \cdot d\vec{r} = 3$, then $\int_{C_2} \vec{F} \cdot d\vec{r} = -6$.

CHAPTER EIGHTEEN

CALCULUS OF VECTOR FIELDS

In the previous chapter we saw how to integrate vector fields along curves. So far, we haven't differentiated a vector field. In this chapter we will see that there is more than one way to do this. If we view the vector field as representing the velocity of a fluid flow, then one method of differentiation (the divergence) tells us about the creation or expansion of fluid and the other method (the curl) tells us about the rotation. We will also meet a new sort of integral, the flux integral, which goes over a surface rather than along a curve. Finally, we will put the integration and the differentiation together in the vector analogues of the Fundamental Theorem of Calculus: the Divergence theorem and Stokes' theorem.

18.1 FLUX INTEGRALS

Suppose we want to measure the flow rate of a fluid through a porous surface; for example, imagine water flowing through a fishing net stretched across a stream. (See Figure 18.1.) This flow rate is called the *flux* of the fluid through the surface. It is the volume of fluid that passes through the surface per unit of time. If we know the velocity vector field $\vec{v}\,(x, y, z)$, which gives the velocity of the fluid at any point (x, y, z), we should be able to estimate the flux through any given surface.

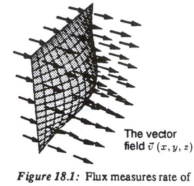

The vector
field $\vec{v}\,(x, y, z)$

Figure 18.1: Flux measures rate of
flow through a surface

Orientation of a Surface

There are usually two directions in which fluid can flow through a surface. See Figure 18.2. To find the flux through the surface we need to decide which direction is the direction of positive flow and which is negative.

A surface is said to be **oriented** if one direction of flow through the surface has been chosen as the positive direction. This choice is called a choice of **orientation**.

At any point on a surface, there are two possible normal directions, pointing in opposite directions to one another. Often the orientation of a surface is indicated by putting on the surface a normal vector that points in the direction of positive flow. In some cases the orientation will be determined by the problem; in others you will have to make your own choice of orientation.

Direction of
positive flow

Direction of
positive flow

Figure 18.2: Two different orientations of the same surface

Vector Area

As we saw in Section 12.4 on page 98, any oriented, flat surface has a *vector area*. The vector area has the same direction as the normal vector that points in the positive direction, and has magnitude equal to the area of the surface.

Flow Through a Surface

The Flux of a Constant Flow Through a Flat Surface

We first consider the situation where the velocity vector field of the fluid is constant and the surface is flat, so the fluid is all flowing in the same direction at the same velocity. Then the flux through the area is the volume of fluid that flows past in one unit of time—the volume of the skewed cylinder in Figure 18.3. It has cross-sectional area $\|\vec{A}\|$ and height $\|\vec{v}\| \cos\theta$, so its volume is $(\|\vec{v}\| \cos\theta) \|\vec{A}\| = \vec{v} \cdot \vec{A}$. Thus,

$$\boxed{\text{Flux through } \vec{A} = (\|\vec{v}\| \cos\theta) \|\vec{A}\| = \vec{v} \cdot \vec{A}.}$$

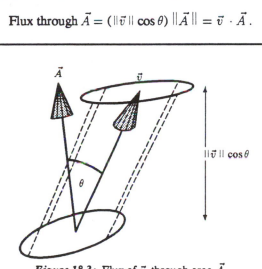

Figure 18.3: Flux of $\vec{v}$ through area $\vec{A}$

Example 1 Water is flowing down a cylindrical pipe 2 cm in radius with a velocity of 3 cm/sec. Find the flux of the vector field through the following regions:

   (a) A 2 cm diameter circular region perpendicular to the pipe

   (b) An ellipse-shaped region whose normal makes an angle of θ with the direction of flow, and which cuts all the way across the pipe. The area of this region is $4\pi / \cos\theta$ cm^2.

Solution The answer to both parts of this question will be the same, because in both cases the flux tells us the rate at which water is flowing down the pipe. Therefore,

$$\text{Flux through circle} = \text{Flux through ellipse}$$
$$= \frac{\text{Rate of flow}}{\text{of water}} \times \frac{\text{Area of}}{\text{circle}}$$
$$= \left(3\,\frac{\text{cm}}{\text{sec}}\right)(\pi 2^2\,\text{cm}^2) = 12\pi\,\text{cm}^3/\text{sec}.$$

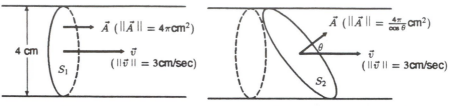

Figure 18.4: Flux through circular region   ***Figure 18.5:*** Flux through ellipse-shaped region

(a) Let's calculate the flux using the formula above to confirm that water is flowing down the pipe at 12π cm^3/sec. Let $\vec{A}$ be the vector area of the circle, and let $\vec{v}$ be the velocity vector of the fluid. Since $\vec{v}$ and $\vec{A}$ are parallel and in the same direction (see Figure 18.4), the flux through the circular region is given by

$$\vec{v} \cdot \vec{A} = \|\vec{v}\|\|\vec{A}\| = 3(\text{Area of circle}) = 3(\pi 2^2) = 12\pi \text{ cm}^3/\text{sec}.$$

(b) For the ellipse-shaped region in Figure 18.5,

$$\vec{v} \cdot \vec{A} = \|\vec{v}\| \cdot \|\vec{A}\| \cos\theta = 3(\text{Area of ellipse})\cos\theta$$

$$= 3\left(\frac{4\pi}{\cos\theta}\right)\cos\theta = 12 \text{ cm}^3/\text{sec}.$$

The Flux of a Variable Flow through a Curved Surface

To calculate the flux of a variable fluid flow through a curved surface, we divide the surface into small patches, or pieces, each of which is approximately flat. If the patches are small enough, we can assume that the flow is approximately constant on each piece. We let $\Delta\vec{A}$ be the vector area of a small piece. See Figure 18.6. Thus,

$$\text{Flux through } \Delta\vec{A} \approx \vec{v} \cdot \Delta\vec{A}$$

and

$$\text{Flux through whole surface} \approx \sum \vec{v} \cdot \Delta\vec{A},$$

where the sum adds up the fluxes through all the small pieces. In the limit as each of the patches becomes smaller and smaller, and $\|\Delta\vec{A}\| \to 0$, the approximation gets better and better, and we get

$$\text{Flux through } S = \lim_{\|\Delta\vec{A}\|\to 0} \sum \vec{v} \cdot \Delta\vec{A}.$$

We can write $\|\Delta\vec{A}\| = \Delta A$, where ΔA represents the area of a patch.

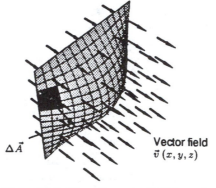

$\Delta\vec{A}$

Vector field
$\vec{v}(x, y, z)$

Figure 18.6: Flux of a vector field through a curved surface divided into patches of area $\Delta\vec{A}$

Definition of the Flux Integral

The limit of sums that we used to compute the flux of a fluid through a surface can be taken for any vector field, whether or not it represents a fluid flow. We define this limit to be an integral called a *flux integral*.

The **flux integral** of the vector field $\vec{v}$ through the surface S is

$$\int_S \vec{v} \cdot d\vec{A} = \lim_{\|\Delta\vec{A}\|\to 0} \sum \vec{v} \cdot \Delta\vec{A}.$$

How Does This Limit Work?

When we take the limit as $\|\Delta\vec{A}\| \to 0$, we mean that the area of every patch tends to zero. In addition, we usually require that the diameter of every patch tend to zero also—thus, ensuring that each patch becomes approximately flat. Thus, provided $\vec{v}$ is continuous, $\vec{v}$ is becoming approximately constant on each patch. In the case that the vector field is the velocity field of a fluid, we have

$$\begin{array}{c} \text{Rate fluid flows through} \\ \text{whole surface} \end{array} = \begin{array}{c} \text{Flux of } \vec{v} \\ \text{through } S \end{array} = \int_S \vec{v} \cdot d\vec{A}$$

In the next example, we give a more realistic model for water flowing down a pipe.

Example 2   Suppose water is flowing down a cylindrical pipe of radius 2 cm, and that the velocity is zero at the edge and increases linearly to 3 cm/sec at the center. Find the flux through the circular cross-section of the pipe.

Solution   Suppose $\vec{i}$ is the unit vector parallel to the direction of flow. Then, at a distance r from the center of the pipe, the velocity is given by

$$\vec{v} = \left(3 - \frac{3}{2}r\right)\vec{i} \text{ cm/sec}.$$

Divide the circular cross-section into concentric rings of width Δr, so that the velocity is approximately constant on each one. The area of a typical ring is $\Delta A \approx 2\pi r \Delta r$. Then since $\vec{v}$ and $\Delta\vec{A}$ are parallel (see Figure 18.7), we have

$$\vec{v} \cdot \Delta\vec{A} = \|\vec{v}\| \cdot \|\Delta\vec{A}\| \approx \left(3 - \frac{3}{2}r\right)\frac{\text{cm}}{\text{sec}} \cdot (2\pi r \Delta r)\,\text{cm}^2.$$

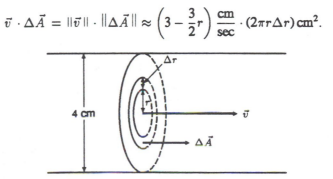

Figure 18.7: Flux through pipe when velocity varies with distance from the center

Thus, the flux through the circular cross-section is given by

$$\lim_{\|\Delta\vec{A}\|\to 0}\sum\vec{v}\cdot\Delta\vec{A} = \lim_{\Delta r\to 0}\sum\left(3-\frac{3}{2}r\right)(2\pi r\Delta r) = \int_{r=0}^{r=2}\left(3-\frac{3}{2}r\right)2\pi r\,dr$$

$$= 6\pi\int_0^2\left(r-\frac{r^2}{2}\right)dr$$

$$= 4\pi\,\mathrm{cm}^3/\mathrm{sec}.$$

In the Example 2, the flow was variable, but the surface was still flat. Later, we will learn how to deal with curved surfaces.

Example 3   The vector field $\vec{v}$ in Figure 18.8 represents a fluid rotating around the z-axis. The magnitude of the vector at a distance of r units from the z-axis is $2r$, in the direction shown. Find the flux of the vector field through the square S of side length 2 in Figure 18.9. The square is oriented by a normal vector in the negative x-direction.

Figure 18.8: The vector field $\vec{v}$

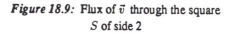

Figure 18.9: Flux of $\vec{v}$ through the square S of side 2

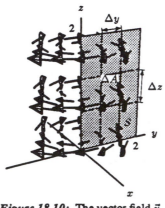

Figure 18.10: The vector field $\vec{v}$

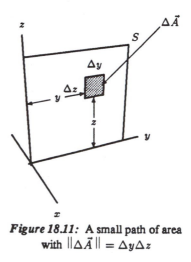

Figure 18.11: A small path of area with $\|\Delta\vec{A}\| = \Delta y\Delta z$

Solution Consider a small vector area $\Delta\vec{A}$ in S, with sides Δy and Δz so that $\|\Delta\vec{A}\| = \Delta y\,\Delta z$. At the point $(0, y, z)$ in S, the vector $\vec{v}$ is perpendicular to the plane, in the opposite direction to $\Delta\vec{A}$ because $\Delta\vec{A}$ points in the direction of the orientation of S. Thus, $\vec{v}\cdot\Delta\vec{A} = -\|\vec{v}\|\,\|\Delta\vec{A}\|$. Since y is the distance from the z-axis, $\vec{v}$ has magnitude $2y$. Hence,

$$\vec{v}\cdot\Delta\vec{A} = -\|\vec{v}\|\,\|\Delta\vec{A}\| = -2y\,\Delta y\,\Delta z.$$

So

$$\int_{S}\vec{v}\cdot d\vec{A} = \lim_{\|\Delta\vec{A}\|\to 0}\sum\vec{v}\cdot\Delta\vec{A} = \lim_{\substack{\Delta y\to 0\\\Delta z\to 0}}\sum -2y\,\Delta y\,\Delta z$$

$$= \int_{0}^{2}\int_{0}^{2}(-2y)\,dy\,dz = -8.$$

The answer is negative, as we would expect, since the vector field is passing through the surface in the direction opposite to the orienting normal vector.

Example 4 Find the flux of the vector field $\vec{F} = (x^2 + y^2)\vec{i} + xy\vec{j}$ through the square region in the xy-plane with corners at $(1, 1, 0)$, $(-1, 1, 0)$, $(1, -1, 0)$, and $(-1, -1, 0)$.

Solution All the vectors in the vector field point horizontally (because their z-component is zero), and the surface is horizontal, so there is no flow through the surface and the flux is zero.

Flux through a Closed Surface

If S is a closed surface (that is, a surface with no boundary), as in Figure 18.12, it is conventional to orient S so that the positive direction of flow is the direction from inside to outside. Thus, in computing a flux integral, the vector areas $\Delta\vec{A}$ all point outwards. Thus,

> For a closed surface S, where $\vec{v}$ is a velocity field,
>
> Flux through $S = \displaystyle\int_{S}\vec{v}\cdot d\vec{A} = $ Net flow out of region enclosed by the surface.

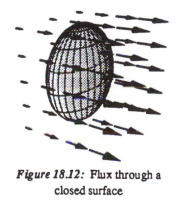

Figure 18.12: Flux through a closed surface

Example 5   Each of the vector fields in Figure 18.13 consists entirely of vectors parallel to the xy-plane, and is constant in the z direction. For each one, decide whether the flux through a closed surface surrounding the origin is positive, negative, or zero. In (a) the surface is a cube with faces parallel to the axes; in (b) and (c) the surface is a sphere. Cross sections of each surface are shown in Figure 18.13.

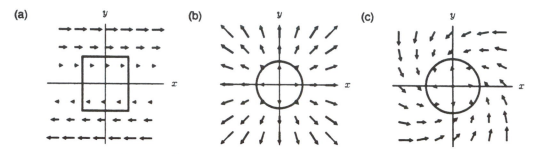

Figure 18.13: Flux of a vector field through a closed surface

Solution   (a)   Since the vector field is parallel to the faces of the cube which are perpendicular to the y- and z-axes, the flux through these faces is zero. The fluxes through the two faces which are perpendicular to the x-axis are equal in magnitude and opposite in sign, so the net flux is zero.

(b)   Using the outward normal, $\vec{v}$ and $\Delta\vec{A}$ are everywhere parallel and in the same direction, so each term $\vec{v} \cdot \Delta\vec{A}$ is positive, and therefore the flux integral $\int_S \vec{v} \cdot d\vec{A}$ is positive.

(c)   In this case $\vec{v}$ and $\Delta\vec{A}$ are not parallel, but since the fluid is flowing inwards as well as swirling, each term $\vec{v} \cdot \Delta\vec{A}$ is negative, and so the flux integral is negative.

Example 6   Find the flux of the vector field $\vec{v} = \vec{r}$ out of the sphere of radius R.

Figure 18.14: Flux of $\vec{v}$ through surface of a
sphere

Solution Since this vector field points radially outward from the origin, it always points in the same direction as the surface area vector, $\Delta \vec{A}$. In addition, $\|\vec{v}\| = R$ on the surface, S, so writing $\|\Delta \vec{A}\| = \Delta A$, we have

$$\vec{v} \cdot \Delta \vec{A} = \|\vec{v}\| \|\Delta \vec{A}\| = R \Delta A.$$

Thus,

$$\int_S \vec{v} \cdot d\vec{A} = \lim_{\|\Delta \vec{A}\| \to 0} \sum \vec{v} \cdot \Delta \vec{A} = \lim_{\Delta A \to 0} \sum R \Delta A = R \lim_{\Delta A \to 0} \sum \Delta A.$$

The last sum adds up all the little areas ΔA, so it approximates the area of the sphere. The approximation gets better as the subdivisions get finer, so that in the limit we have

$$\lim_{\Delta A \to 0} \sum \Delta A = \text{Surface area of sphere}.$$

Thus, the flux is given by

$$\int_S \vec{v} \cdot d\vec{A} = R \lim_{\Delta A \to 0} \sum \Delta A = R(\text{Surface area of sphere}) = R(4\pi R^2) = 4\pi R^3.$$

Problem 24 generalizes this result.

Problems for Section 18.1

Let $\vec{F}(x, y, z) = -z\vec{i} + x\vec{k}$. For each of the surfaces in Problems 1–5, say whether the flux of $\vec{F}$ through the surface is positive, negative, or zero. In each case, the orientation of the surface is indicated by the given normal vector.

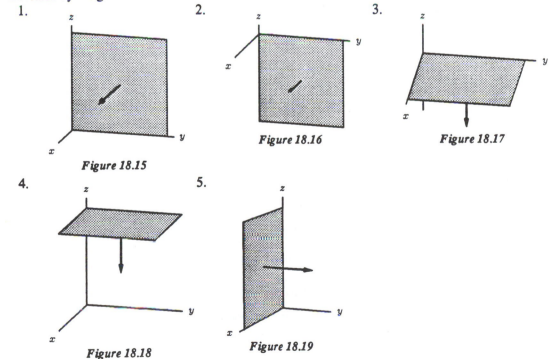

1.

Figure 18.15

2.

Figure 18.16

3.

Figure 18.17

4.

Figure 18.18

5.

Figure 18.19

6. Repeat Problems 1–5 with the vector field $\vec{F}(\vec{r}) = \vec{r}$.

7. Arrange the following flux integrals,

$$\int_{S_i} \vec{F} \cdot d\vec{A},$$

with $i = 1, 2, 3, 4$ in ascending order if $\vec{F} = -\vec{i} - \vec{j} + \vec{k}$ and S_i are the following surfaces:

S_1 is a horizontal square of side 1, facing upward with one corner at $(0, 0, 2)$ and above the first quadrant of the xy-plane.

S_2 is a horizontal square of side 1, facing upward with one corner at $(0, 0, 3)$ and above the third quadrant of the xy-plane.

S_3 is a square of side $\sqrt{2}$ in the xz-plane with one corner at the origin, one edge along the positive x-axis, one along the negative z-axis, and facing in the negative y- direction.

S_4 is a square of side $\sqrt{2}$, one corner at the origin, one edge along the positive y-axis, one corner at $(1, 0, 1)$, and facing upwards.

8. Find the flux of the vector field $\vec{V} = 2\vec{i} + 3\vec{j} + 5\vec{k}$ through each of the rectangular regions in Figures 18.20–18.23, assuming each is oriented in the positive x or the positive z-direction, as appropriate.

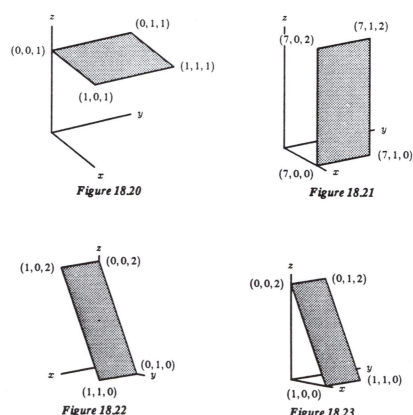

Figure 18.20

Figure 18.21

Figure 18.22

Figure 18.23

9.   Figure 18.24 shows a cross-section of the earth's magnetic field. Say whether the magnetic flux through a horizontal plate, oriented skyward, is positive, negative, or zero if the plate is
(a)   At the north pole.   (b)   At the south pole.

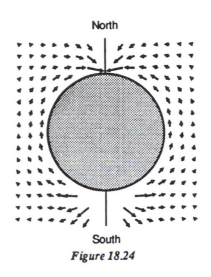

Figure 18.24

10.   What do you think will be the electric flux through the tube-shaped surface that is placed as shown in the electric field in Figure 18.25? Why?

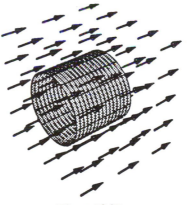

Figure 18.25

For Problems 11–14 find the flux of the constant vector field $\vec{v} = \vec{i} - \vec{j} + 3\vec{k}$ through the given surfaces.

11.   A disk of radius 2 in the xy-plane with upward normal.

12.   A triangular plate of area 4 in the yz-plane with normal in the positive x-direction.

13.   A square plate of area 4 in the yz-plane with normal in the positive x-direction.

14.   The triangular plate with vertices $(1, 0, 0)$, $(0, 1, 0)$, and $(0, 0, 1)$ with normal pointing away from the origin.

For Problems 15-19, compute the flux integral of the given vector field through the given surface.

15. $\vec{F} = 2\vec{i}$ and S is a disk of radius 2 on the plane $x + y + z = 2$, oriented upward.

16. $\vec{F} = -y\vec{i} + x\vec{j}$ and S is the square plate in the yz plane with corners at $(0, 1, 1)$, $(0, -1, 1)$, $(0, 1, -1)$, and $(0, -1, -1)$, oriented in the positive x-direction..

17. $\vec{F} = -y\vec{i} + x\vec{j}$ and S is the disk in the xy plane with radius 2, oriented upwards and centered at the origin.

18. $\vec{F} = \vec{r}$ and S is the disk of radius 2 parallel, to the xy plane oriented upwards and centered at $(0, 0, 2)$.

19. $\vec{F}(x, y, z) = (2 - x)\vec{i}$ and S is the cube whose vertices include the points $(0, 0, 0)$, $(3, 0, 0)$, $(0, 3, 0)$, $(0, 0, 3)$, and oriented outward.

20. Find the flux of $\vec{F}(\vec{r}) = \vec{r}/r^3$ through the sphere of radius R centered at the origin.

21. Find the flux of $\vec{F}(\vec{r}) = \vec{r}/r^2$ through the sphere of radius R centered at the origin.

22. Let S be the cube with side 2, faces parallel to the coordinate planes, and centered at the origin.

    (a) Calculate the total flux of the constant vector field $\vec{v} = -\vec{i} + 2\vec{j} + \vec{k}$ out of S by computing the flux through each face separately.

    (b) Calculate the flux out of S for any constant vector field $\vec{v} = a\vec{i} + b\vec{j} + c\vec{k}$.

    (c) Do your answers in parts (a) and (b) make sense? Explain.

23. Let S be the tetrahedron with vertices at the origin and at $(1, 0, 0)$, $(0, 1, 0)$ and $(0, 0, 1)$.

    (a) Calculate the total flux of the constant vector field $\vec{v} = -\vec{i} + 2\vec{j} + \vec{k}$ out of S by computing the flux through each face separately.

    (b) Calculate the flux out of S in part (a) for any constant vector field $\vec{v}$.

    (c) Do your answers in parts (a) and (b) make sense? Explain.

24. Explain why if $\vec{F}$ has constant magnitude on S and is everywhere normal to S and in the direction of orentation, then

$$\int_S \vec{F} \cdot d\vec{A} = \|\vec{F}\| \text{Area of } S.$$

For Problems 25-26 let $\vec{F}(\vec{r}) = \vec{r}$, and let S be a square plate perpendicular to the z-axis and centered on the z-axis. Sketch as a function of time the flux of $\vec{F}$ through S as S moves in the given manner.

25. S moves from far up the positive z-axis to far down the negative z-axis. Assume S is oriented upward.

26. S rotates about an axis parallel to the x-axis, through the center of S. Assume S is far up the z-axis so that $\vec{r}$ is approximately constant on S as S rotates.

27. Repeat Problems 25-26 with $\vec{F}(\vec{r}) = \vec{r}/r^3$.

28. Consider a body of fluid with an xyz-coordinate system to locate points in it. Let $P(x, y, z)$ be the pressure at the point (x, y, z). Let $\vec{F}(x, y, z) = P(x, y, z)\vec{k}$. Let S be the surface of a body submerged in the fluid. Show that $\int_S \vec{F} \cdot d\vec{A}$ is the buoyant force on the body, that is, the force upwards on the body due to the pressure of the fluid surrounding it. [Hint: $\vec{F} \cdot d\vec{A} = P(x, y, z)\vec{k} \cdot d\vec{A} = (P(x, y, z)\, d\vec{A}) \cdot \vec{k}$.]

29. Suppose a region of space is unevenly heated. Let $T(x, y, z)$ be the temperature at a point (x, y, z). One form of Newton's law of cooling says that grad T is proportional to the vector field $\vec{F}$ that gives the heat flow. (Here $\vec{F}$ points in the direction that heat is flowing and has magnitude equal to the rate of flow.)

    (a) Explain why this form of Newton's Law of Cooling makes sense.
    (b) Suppose $\vec{F} = k$ grad T for some constant k. What is the sign of k?
    (c) Let V be a region of space bounded by the surface S. Explain why the rate of heat loss from V is

    $$k \int_S (\text{grad } T) \cdot d\vec{A}.$$

30. A fluid is flowing along in a cylindrical pipe of radius a running in the $\vec{i}$ direction. The velocity of the fluid at a distance r from the center of the pipe is $\vec{v} = u(1 - r^2/a^2)\vec{i}$.

    (a) What is the significance of the constant u?
    (b) What is the velocity of the fluid at the wall of the pipe?
    (c) Find the flux through a circular cross-section of the pipe.

31. Consider the function $\rho(x, y, z)$ which gives the electrical charge density at all points in space. The vector field $\vec{J}(x, y, z)$ gives the electric current density at any point in space and is defined so that the current through a small area $d\vec{A}$ is given by

    $$\text{Current through small area} \approx \vec{J} \cdot d\vec{A}.$$

    Suppose S is a surface enclosing a volume W.

    (a) What does the integral

    $$\int_W \rho \, dV$$

    represent, in terms of electricity?
    (b) What does the integral

    $$\int_S \vec{J} \cdot d\vec{A}$$

    represent, in terms of electricity?
    (c) Using the fact that an electric current is the rate of change of charge with time, explain why

    $$\int_S \vec{J} \cdot d\vec{A} = -\frac{\partial}{\partial t} \left(\int_W \rho \, dV \right).$$

32. The purpose of this problem is to investigate the behavior of the electric field of an infinitely long, straight, uniformly charged wire. (There is no current running through the wire — all charges are fixed.) Making the wire infinitely long, rather than very long but finite, is simpler because in that case it is reasonable to assume that the electric field is normal to any cylinder that has the wire as an axis, and that the magnitude of the field is constant on any such cylinder. Denote by E_r the magnitude of the electric field due to the wire on a cylinder of radius r (see Figure 18.26).

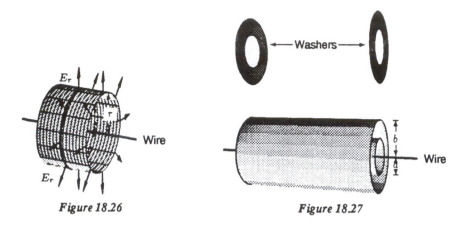

Figure 18.26   *Figure 18.27*

Imagine a closed surface made up of two cylinders, one of radius a and one of larger radius b, both coaxial with the wire, and the two washers that cap the ends (see Figure 18.27).

(a) Explain why the flux of $\vec{E}$ through the washers is 0.

(b) Gauss' Law states that the flux of an electric field through a closed surface S is proportional to the amount of electric charge inside S. Explain why Gauss' Law implies that the flux through the inner cylinder is the same as the flux through the outer cylinder. (Note that the charge on the wire is *not* inside the surface S).

(c) Use part (b) to show that $E_b/E_a = a/b$.

(d) Explain why part (c) shows that the strength of the field due to an infinite charged wire is proportional to $1/r$.

33. Consider an infinite flat sheet of uniform charge. As in the case of the charged wire from Problem 32, symmetry considerations force the electric field to be perpendicular to the sheet, and to have the same magnitude for all points that are at the same distance from the sheet. By considering the flux through a box with sides parallel to the sheet, shown in Figure 18.28, use Gauss' Law (described in Problem 32) to explain why the field due to the charged sheet is the same at all points in space.

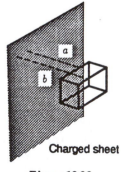

Charged sheet

Figure 18.28

18.2   FLUX INTEGRALS OVER PARAMETERIZED SURFACES

So far we have calculated flux integrals only in a few special cases. In this section we will see how to compute them in more general situations. The simplest case of a flux integral is that of a constant vector field through a flat surface. If $\vec{F}$ is a constant vector field and S is a flat surface with vector area $\vec{A}$, then the flux integral is simply the dot product:

$$\int_S \vec{F} \cdot d\vec{A} = \vec{F} \cdot \vec{A}.$$

In the more general situation where the surface is curved and the vector field is not constant, the flux integral is a limit of sums. These sums give us a way to approximate the flux integral. We divide the surface into small pieces that are each nearly flat and over each of which the vector field is nearly constant. Then the flux through each small piece can be approximated by a dot product, and the flux through the entire surface will be well approximated by the sum of the contributions from the small pieces. This sounds like a Riemann sum, and indeed it is.

But how do we divide a general surface into many small pieces? The key is to first parameterize the surface over which we are integrating, then take the small pieces to be parameter rectangles. This will enable us to express the flux integral as an ordinary double integral. Let us see how this works in a few examples.

When the Surface is of the Form $z = f(x, y)$

Example 1   Compute $\int_S \vec{F} \cdot d\vec{A}$ where $\vec{F}(x, y, z) = -y\vec{i} + z\vec{k}$ and S is the portion of the curved surface $z = f(x, y) = x^2$ above the square region R given by $0 \le x \le 1, 0 \le y \le 1$. Let the orientation of S be given by an upward pointing normal. See Figure 18.29 and Figure 18.30.

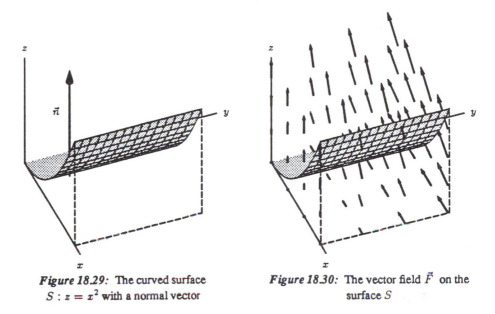

Figure 18.29: The curved surface   *Figure 18.30:* The vector field $\vec{F}$ on the
$S : z = x^2$ with a normal vector   surface S

Solution   Figures 18.29 and 18.30 shows that both $\vec{F}$ and the normal vectors point to the same side of the surface S. This leads us to expect that the flux integral will be a positive number. To compute the

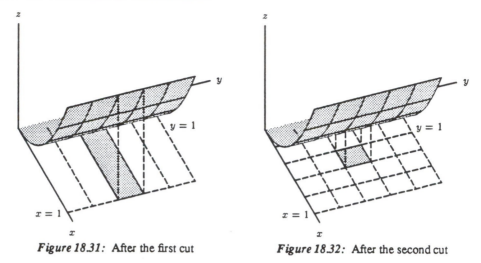

Figure 18.31: After the first cut   **Figure 18.32:** After the second cut

value exactly we divide the surface S into small pieces by cutting it first into thin strips parallel to the x-axis, then cutting across the strips in the y-direction. See Figure 18.31 and Figure 18.32. From Figure 18.32 you can see that the cuts on the surface S correspond to a division of the square $0 \leq x \leq 1, 0 \leq y \leq 1$ into rectangles which we will take to be of dimension $\Delta x \times \Delta y$, with edges parallel to the x and y-axes. The little pieces of the surface S above these rectangles are called parameter rectangles (for the parameters x and y). If Δx and Δy are both small enough, then the parameter rectangles will be nearly flat, and the vector field $\vec{F}$ will be nearly constant on each parameter rectangle, which is what we want.

The surface S is of the form $z = f(x, y)$ (where $f(x, y) = x^2$). In Section 16.7, we saw that the parameter rectangle S_{mn} of S in Figure 18.33, lying above the rectangle with corner (x_m, y_n), is approximately a parallelogram bounded by the vectors

$$\vec{a}_{mn} = (\vec{i} + f_x(x_m, y_n)\vec{k})\Delta x = (\vec{i} + 2x_m\vec{k})\Delta x$$

and

$$\vec{b}_{mn} = (\vec{j} + f_y(x_m, y_n)\vec{k})\Delta y = \vec{j}\,\Delta y.$$

The vector area of this parallelogram is given by the cross product:

$$\Delta \vec{A}_{mn} = \vec{a}_{mn} \times \vec{b}_{mn} = (\vec{i} + 2x_m\vec{k})\Delta x \times \vec{j}\,\Delta y = (-2x_m\vec{i} + \vec{k})\Delta x \Delta y.$$

Notice that $\Delta \vec{A}_{mn}$ has a positive $\vec{k}$-component as we expect for an upward orientation.

Observe next that when both Δx and Δy are sufficiently small, then the vector field $\vec{F}$ will be approximated well at all points of S_{mn} by the constant vector field $\vec{F}(x_m, y_n, f(x_m, y_n))$. It follows that the flux of $\vec{F}$ through the surface S_{mn} will be approximated by

$$\int_{S_{mn}} \vec{F} \cdot d\vec{A} \approx \vec{F}(x_m, y_n, f(x_m, y_n)) \cdot \Delta \vec{A}_{mn}$$

$$= (-y_n\vec{i} + x_m^2\vec{k}) \cdot (-2x_m\vec{i} + \vec{k})\Delta x \Delta y$$

$$= (2x_m y_n + x_m^2)\Delta x \Delta y.$$

Thus, we have the approximation

$$\int_S \vec{F} \cdot d\vec{A} = \sum_{m,n} \int_{S_{mn}} \vec{F} \cdot d\vec{A} \approx \sum_{m,n} (2x_m y_n + x_m^2)\Delta x \Delta y.$$

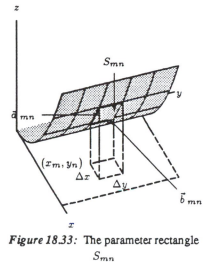

Figure 18.33: The parameter rectangle S_{mn}

Finally, taking the limit as Δx and Δy go to zero, the double sum becomes a double integral which we can evaluate over the region R:

$$\int_S \vec{F} \cdot d\vec{A} = \lim_{\Delta x \to 0, \Delta y \to 0} \sum_{m,n} (2x_m y_n + x_m^2)\Delta x \Delta y$$

$$= \int_R (2xy + x^2)dx\, dy = \frac{5}{6}.$$

In general, we have the following method of computing flux integrals. Suppose that S is part of the surface $z = f(x, y)$, oriented in the positive z-direction. Then on S, in xy-coordinates,

$$d\vec{A} = (\vec{i} + f_x(x,y)\vec{k}) \times (\vec{j} + f_y(x,y)\vec{k})dx\, dy = (-f_x\vec{i} - f_y\vec{j} + \vec{k})dx\, dy.$$

This means that if $\vec{F}(x, y, z)$ is a vector field,

$$\int_S \vec{F} \cdot d\vec{A} = \int_R \vec{F}(x, y, f(x,y)) \cdot (-f_x\vec{i} - f_y\vec{j} + \vec{k})dx\, dy$$

where R is the "shadow" region obtained by projecting the surface S onto the xy-plane.

Example 2   Compute $\int_S \vec{F} \cdot d\vec{A}$ where $\vec{F}(x, y, z) = z\vec{k}$ and S is the rectangular plate with corners $(0, 0, 0)$, $(1, 0, 0)$, $(0, 1, 3)$, and $(1, 1, 3)$, oriented upwards (that is, in the positive z-direction).

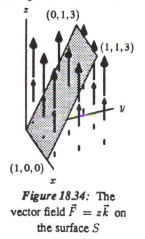

Figure 18.34: The vector field $\vec{F} = z\vec{k}$ on the surface S

Solution   Since S is part of a plane, we can begin by finding the equation for the plane in the form $z = f(x, y)$. Since f is linear, with x-slope equal to 0 and y-slope equal to 3, and $f(0, 0) = 0$, we have

$$z = f(x, y) = 0 + 0x + 3y = 3y.$$

Thus, in xy-coordinates,

$$d\vec{A} = (-f_x\vec{i} - f_y\vec{j} + \vec{k})dx\,dy = (0\vec{i} - 3\vec{j} + \vec{k})dx\,dy = (-3\vec{j} + \vec{k})dx\,dy.$$

The flux integral is therefore

$$\int_S \vec{F} \cdot d\vec{A} = \int_{x=0}^1 \int_{y=0}^1 3y\vec{k} \cdot (-3\vec{j} + \vec{k})dx\,dy$$

$$= \int_{x=0}^1 \int_{y=0}^1 3y\,dx\,dy = 1.5$$

General Parameterized Surfaces

As we saw in Chapter 16, the most convenient parameters for a surface may not be x and y. Fortunately, we can compute flux integrals over a surface given any parameters p and q for the surface. Let us consider the general case of the flux of a vector field $\vec{F}$ through a surface, S, given parametrically by the equations

$$x = x(p, q), \quad y = y(p, q), \quad z = z(p, q)$$

Following the approach of Example 1, we divide the surface S into parameter rectangles as in Figure 18.35, where the rectangles in the parameter space R are of dimension $\Delta p \times \Delta q$. If Δp and Δq are small enough, the parameter rectangles will be nearly flat, and the vector field $\vec{F}$ will be nearly constant on each one, which is what we want. The parameter rectangle S_{mn} shaded in Figure 18.35, comes from a rectangle of dimension $\Delta p \times \Delta q$ in the parameter space with corner (p_m, q_n). As in Example 1 on page 1, the parameter rectangle S_mn is approximately a parallelogram spanned by the vectors

$$\vec{a}_{mn} = \vec{v}_p(p_m, q_n)\Delta p \quad \text{and} \quad \vec{b}_{mn} = \vec{v}_q(p_m, q_n)\Delta q$$

where

$$\vec{v}_p = x_p\vec{i} + y_p\vec{j} + z_p\vec{k} \quad \text{and} \quad \vec{v}_q = x_q\vec{i} + y_q\vec{j} + z_q\vec{k}$$

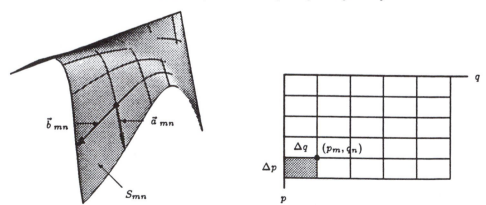

Figure 18.35: The surface S and the parameter space R

are tangents to the parameter curves. The vector area of this parallelogram is

$$\Delta \vec{A}_{mn} = \vec{a}_{mn} \times \vec{b}_{mn} = (\vec{v}_p(p_m, q_n) \times \vec{v}_q(p_m, q_n)) \Delta p \Delta q$$

Summing fluxes over all the parameter rectangles and making a Riemann sum shows that the flux integral $\int_S \vec{F} \cdot d\vec{A}$ can be computed in pq coordinates by making the substitution

$$d\vec{A} = (\vec{v}_p \times \vec{v}_q) dp dq.$$

In other words,

The flux integral over a parameterized surface is given by

$$\int_S \vec{F} \cdot d\vec{A} = \int \int_R \vec{F}(x(p,q), y(p,q), z(p,q)) \cdot (\vec{v}_p \times \vec{v}_q) \, dp \, dq$$

where R is the region in the pq-parameter space and $\vec{v}_p$ and $\vec{v}_q$ are the tangents to the parameter curves. The parameterization $(x(p,q), y(p,q), z(p,q))$ of S must be an oriented parameterization.

Example 3   Compute $\int_S \vec{F} \cdot d\vec{A}$ where $\vec{F}(x, y, z) = y\vec{j}$ and S is the portion of the circular cylinder of radius 2 centered on the z-axis such that $x \geq 0$, $y \geq 0$, and $0 \leq z \leq 3$, and the surface is oriented towards the z-axis.

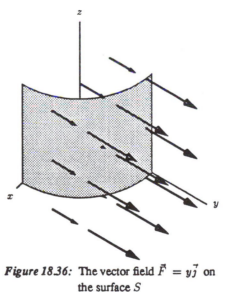

Figure 18.36: The vector field $\vec{F} = y\vec{j}$ on the surface S

Solution   Since the orientation of S is towards the z-axis, Figure 18.36 shows that the flux across S will be negative. To compute the exact value we parameterize the surface S. The shape of S suggests that we parameterize with cylindrical coordinates θ and z. (The coordinate r is unnecessary, because $r = 2$ at all points on the cylinder.) We have the parameterization

$$x(\theta, z) = 2\cos\theta$$
$$y(\theta, z) = 2\sin\theta \qquad \text{for } 0 \leq \theta \leq \pi/2, 0 \leq z \leq 3.$$
$$z(\theta, z) = z$$

The tangents to the parameter curves for this parameterization are

$$\vec{v}_\theta = -2\sin\theta\vec{i} + 2\cos\theta\vec{j} \quad \text{and} \quad \vec{v}_z = \vec{k}$$

and so

$$\vec{v}_\theta \times \vec{v}_z = 2\cos\theta\vec{i} + 2\sin\theta\vec{j}.$$

Since the components of $\vec{v}_\theta \times \vec{v}_z$ in the $\vec{i}$ and $\vec{j}$ directions are positive in the range $0 \le \theta \le \pi/2$, the vector $\vec{v}_\theta \times \vec{v}_z$ points away from the z-axis, and so our parameterization has the wrong orientation. To re-orient it, we reverse the order of θ and z, taking z first. Finally,

$$\begin{aligned}
\int_S \vec{F} \cdot d\vec{A} &= \int_{\theta=0}^{\pi/2} \int_{z=0}^3 2\sin\theta\vec{j} \cdot (\vec{v}_z \times \vec{v}_\theta)\, dz\, d\theta \\
&= \int_{\theta=0}^{\pi/2} \int_{z=0}^3 2\sin\theta\vec{j} \cdot (-2\sin\theta\vec{i} - 2\sin\theta\vec{j})\, dz\, d\theta \\
&= \int_{\theta=0}^{\pi/2} \int_{z=0}^3 -4\sin^2\theta\, dz\, d\theta = -3\pi.
\end{aligned}$$

Note on Surface Area

Although it does not involve a flux integral, we can use parameter rectangles to find surface area. Let us see how to compute the surface area of a parameterized surface $S : x = x(p, q), y = y(p, q), z = z(p, q)$. We expect to do it by some sort of an integral. We can try to set up the integral by thinking of the Riemann sums that will approximate it. To make a Riemann sum for the surface area of S we must first divide S into many tiny pieces. Of course we plan to take the small pieces to be parameter rectangles, as in Figure 18.35. Since the piece S_{mn} in Figure 18.35 is approximated by a parallelogram with vector area

$$\Delta\vec{A}_{mn} \approx (\vec{v}_p(p_m, q_n) \times \vec{v}_q(p_m, q_n))\Delta p\Delta q,$$

we have the approximation

$$\text{Area of } S_{mn} = \|\Delta\vec{A}_{mn}\| \approx \|\vec{v}_p \times \vec{v}_q\|\Delta p\Delta q.$$

Thus,

$$\text{Area of } S = \sum_{m,n} \text{Area of } S_{mn} \approx \sum_{m,n} \|(\vec{v}_p(p_m, q_n) \times \vec{v}_q(p_m, q_n))\|\Delta p\Delta q.$$

In the limit as Δp and Δq go to 0, we have

$$\text{Surface area of } S = \int_R \|\vec{v}_p \times \vec{v}_q\|\, dp\, dq$$

where R is the region in the pq-parameter space and $\vec{v}_p$ and $\vec{v}_q$ are the velocities of the parameter curves. It is suggestive to summarize as follows: The surface area of a parameterized surface S is given by

$$\text{Area of } S = \int_S \|d\vec{A}\|$$

where, in pq coordinates, $\|d\vec{A}\| = \|\vec{v}_p \times \vec{v}_q\|\, dp\, dq$.

Example 4   Compute the surface area of a sphere of radius R.

Solution   We will take the sphere S of radius R centered at the origin, and parameterize it with the spherical coordinates ϕ and θ. The parameterization is

$$S : x = R\cos\theta\sin\phi, y = R\sin\theta\sin\phi, z = R\cos\phi \quad \text{where } 0 \le \theta \le 2\pi, 0 \le \phi \le \pi/2.$$

We compute

$$\vec{v}_\phi \times \vec{v}_\theta = (R\cos\theta\cos\phi\vec{i} + R\sin\theta\cos\phi\vec{j} - R\sin\phi\vec{k}) \times (-R\sin\theta\sin\phi\vec{i} + R\cos\theta\sin\phi\vec{j})$$
$$= R^2(\cos\theta\sin^2\phi\vec{i} + \sin\theta\sin^2\phi\vec{j} + \sin\phi\cos\phi\vec{k})$$

and so

$$\|\vec{v}_\phi \times \vec{v}_\phi\| = R^2\sin\phi.$$

Finally

$$\text{Surface area of } S = \int_{\phi=0}^{\pi} \int_{\theta=0}^{2\pi} R^2\sin\phi\,d\theta\,d\phi = 4\pi R^2.$$

Problems for Section 18.2

1. Consider a sphere of radius 2 parameterized by the usual spherical coordinates θ and ϕ.

    (a) Compute $d\vec{A}$, the vector area of a parameter rectangle, oriented outward.
    (b) Use part (a) to compute the flux integral

    $$\int_S \vec{F} \cdot d\vec{A}$$

    where $\vec{F}(x, y, z) = z\vec{k}$, and S is the upper hemisphere of radius 2 centered at the origin.

2. If $\vec{F} = x\vec{i} + y\vec{j} + z\vec{k}$, evaluate $\int_S \vec{F} \cdot d\vec{A}$, where S is the surface of the sphere $x^2 + y^2 + z^2 = a^2$.

3. Evaluate $\int_S \vec{F} \cdot d\vec{A}$, where $\vec{F} = x^2\vec{i} + y^2\vec{j} + z^2\vec{k}$ and S is the surface of the triangle $\triangle ABC$ as shown in Figure 18.37.

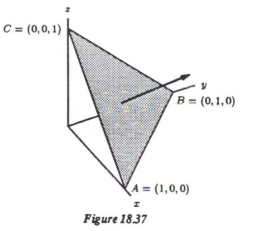

Figure 18.37

4.   Evaluate $\int_S \vec{F} \cdot d\vec{A}$, where $\vec{F} = x^2 y^2 z \vec{k}$ and S is the surface of the cone $\sqrt{x^2 + y^2} = z$ $(0 \leq z \leq R)$. See Figure 18.38.

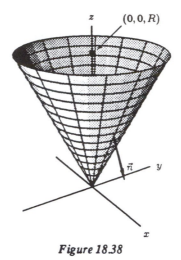

Figure 18.38

5.   If $\vec{F} = x^2 \vec{i} + y^2 \vec{j} + z^2 \vec{k}$ Evaluate $I = \int_S \vec{F} \cdot d\vec{A}$ where S is the surface of the sphere $(x - a)^2 + (y - b)^2 + (z - c)^2 = d^2$.

6.   Evaluate $\int_S \vec{F} \cdot d\vec{A}$ where $\vec{F} = b/ax \vec{i} + a/by \vec{j}$ and S is the curved surface of the elliptic cylinder $x^2/a^2 + y^2/b^2 = 1$, and $|z| \leq c$. Where $\vec{F} = (bx/a)\vec{i} + (ay/b)\vec{j}$.

7.   Evaluate $\int_S \vec{F} \cdot d\vec{A}$ where $\vec{F} = x^3 \vec{i} + y^3 \vec{j} + z^3 \vec{k}$ and S is the surface of the ellipsoid $x^2/a^2 + y^2/b^2 + z^2/c^2 = 1$.

8.   Compute the flux of the radius vector $\vec{r} = x\vec{i} + y\vec{j} + z\vec{k}$ through the lateral surface of the circular cylinder $x^2 + y^2 = 1$ bounded below by the plane $x + y + z = 1$ and from above by the plane $x + y + z = 2$.

9.   Find the area of the ellipse (S) cut on the plane $2x + y + z = 2$ by the circular cylinder $x^2 + y^2 = 2x$.

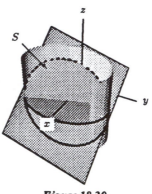

Figure 18.39

18.3   THE DIVERGENCE OF A VECTOR FIELD

Look at the vector fields pictured in Figure 18.40 and 18.41. Imagine that they are velocity vector fields describing the flow of water. Figure 18.40 suggests a spewing out from the origin, which might be the location of a fountain or spring. Figure 18.41 suggests the flow of water into a hole at the origin. You might think of a flow into a drain. In neither case is the quantity of water in the flow conserved. In the first case, we will say that water is created at the origin and that the origin is a *source*. In the second case we will say that water is destroyed at the origin and that the origin is a *sink*.

When a vector field represents the velocity of a steadily moving liquid, the flux integral over a closed surface with outward normal represents the net rate at which liquid is flowing out of the region enclosed by the surface. The net flux through a closed surface is zero unless fluid is being created or destroyed within the enclosed region. Thus, the flux integral tells us whether fluid is being created or destroyed within the enclosed region (or whether there are sources or sinks within the region). We use this idea to measure the rate at which fluid is being created or destroyed at a point. This rate is called the *divergence* of the vector field at the point. It is an extension of the idea of divergence introduced in Section 17.2.

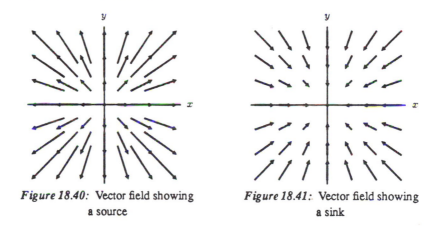

Figure 18.40: Vector field showing a source

Figure 18.41: Vector field showing a sink

Example 1   A liquid has velocity vector field $\vec{v} = \vec{r}$. Find the rate at which fluid is being created at the origin.

Solution   In Example 6 on page 384, we calculated the flux across the sphere of radius R to be $4\pi R^3$. Since fluid is being generated at a rate of $4\pi R^3$ in the enclosed ball of volume $\frac{4}{3}\pi R^3$, we say that fluid is being generated within the sphere at a rate of

$$\frac{\text{Flux}}{\text{Volume}} = \frac{4\pi R^3}{\frac{4}{3}\pi R^3} = 3 \text{ cubic units of fluid per unit time per cubic unit of space.}$$

To get the rate at which liquid is being created *at the origin*, we should consider the limit of this flux to volume ratio as R tends to zero, so that the sphere contracts around the origin. In this case the limit is 3, so we say that liquid is created at the origin at a rate of 3 cubic units of liquid per unit of time per cubic unit of space. We say that

$$\begin{array}{c}\text{Divergence of the} \\ \text{vector field } \vec{v} \text{ at the origin}\end{array} = \begin{array}{c}\text{3 cubic units per unit of time} \\ \text{per cubic units of space.}\end{array}$$

Definition of Divergence

The definition of divergence of a general vector field at a point follows the same idea.

The **divergence of a vector field** $\vec{v}$ at a point P is defined by the limit

$$\text{div}\,\vec{v}\,(P) = \lim_{\text{vol}\to 0} \frac{\int_S \vec{v} \cdot d\vec{A}}{\text{Volume enclosed by } S}.$$

Here S is a closed surface surrounding P that contracts suitably down to P in the limit. We take the outward pointing normals to S.

Notice that a vector field has a divergence at every point, and hence the divergence of a vector field is itself a function. A vector field $\vec{v}$ is said to be *divergence free* or *solenoidal* if $\text{div}\,\vec{v}\,(P) = 0$ at every point P.

What Does the Divergence Tell Us?

If you think of a vector field as a velocity flow of a liquid, then the divergence tells you the outflow per unit time per unit volume at a point. If the vector field represents a flow into a point, the divergence will be negative (or zero) there; if the vector field represents a flow out of a point, the divergence will be positive (or zero) there.

Example 2   Figure 18.42 shows the part of a vector field $\vec{B}$ that lies in the xy-plane. Assume that the vector field is independent of z, so that any horizontal cross section looks the same. What can you say about the divergence of the vector field at the points marked P and Q?

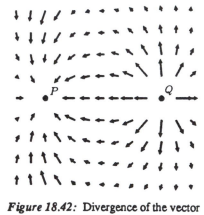

Figure 18.42: Divergence of the vector
field $\vec{B}$

Solution   The flux of $\vec{B}$ through a sphere of radius R around the point marked P will be negative, because all the arrows are pointing into the sphere. The divergence at P is

$$\text{div}\,\vec{B}\,(P) = \lim_{\text{vol}\to 0}\left(\frac{\int_S \vec{B}\cdot d\vec{A}}{\text{Volume of sphere}}\right) = \lim_{R\to 0}\left(\frac{\text{Negative number}}{\frac{4}{3}\pi R^3}\right) \le 0.$$

By a similar argument, the divergence at Q must be positive or zero.

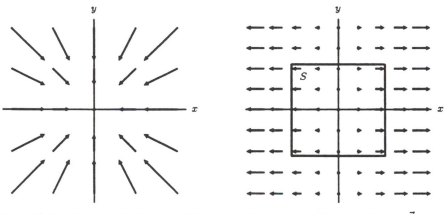

Figure 18.43: The vector field $\vec{v} = -2\vec{r}$

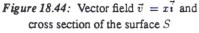

Figure 18.44: Vector field $\vec{v} = x\vec{i}$ and cross section of the surface S

Example 3   Find div $\vec{v}$ at the origin, when $\vec{v}$ is the vector field in 3-space given by
(a) $\vec{v} = -2\vec{r}$ (b) $\vec{v} = x\vec{i}$

Solution   (a)   Figure 18.43 shows a two dimensional cross-section of the vector field $\vec{v} = -2\vec{r}$.

The vector field points radially inwards, so if we take S to be a sphere of radius R centered at the origin, we have,

$$\vec{v} \cdot \Delta\vec{A} = -2R\,\Delta A,$$

where ΔA and dA are the magnitudes of the area s of $\Delta\vec{A}$ and $d\vec{A}$. Therefore,

$$\int_S \vec{v} \cdot d\vec{A} = \int_{\text{Sphere}} -2R\,dA = -2R(\text{Surface area of sphere}) = -2R(4\pi R^2) = -8\pi R^3.$$

Thus, we find that

$$\text{div}\,\vec{v}\,(0,0,0) = \lim_{\text{vol}\to 0}\left(\frac{\int_{\text{sphere}}\vec{v}\cdot d\vec{A}}{\text{Volume of sphere}}\right) = \lim_{R\to 0}\left(\frac{-8\pi R^3}{\frac{4}{3}\pi R^3}\right) = -6.$$

Notice that the divergence is negative. This is what you would expect, since the vector field represents an inward flow at the origin. If the vector field represents the movement of a liquid, then fluid is being destroyed at the origin.

(b)   The vector field $\vec{v} = x\vec{i}$ is parallel to the x axis, as shown in Figure 18.44. Let's take S to be a cube centered at the origin with edges parallel to the axes, of length $2R$. Then the flux through the faces perpendicular to the y and z-axes is zero (because the vector field is parallel to these faces). On the faces perpendicular to the x-axis, the vector field and the outward normal are parallel. On the face at $x = R$, we have

$$\vec{v} \cdot \Delta\vec{A} = R\,\Delta A.$$

On the face at $x = -R$, the dot product is still positive, and

$$\vec{v} \cdot \Delta\vec{A} = R\,\Delta A.$$

Therefore, the flux through the box is given by

$$\int_S \vec{v} \cdot d\vec{A} = \int_{x=-R}\vec{v}\cdot d\vec{A} + \int_{x=R}\vec{v}\cdot d\vec{A}$$

$$= 2\int_{x=R} R\,dA = 2R(\text{Area of one face}) = 2R(2R)^2 = 8R^3.$$

Thus,

$$\operatorname{div} \vec{v}(0,0,0) = \lim_{\text{vol}\to 0}\left(\frac{\int_S \vec{v}\cdot d\vec{A}}{\text{Volume of box}}\right) = \lim_{R\to 0}\left(\frac{8R^3}{(2R)^3}\right) = 1.$$

Since the vector field points outward from the yz-plane, it makes sense that the divergence is positive at the origin.

___

Notice that the divergence will be positive if the *net* outflow from small regions around the point is positive. This could happen even if there is some inflow, as long as there is a greater outflow to counteract it.

___

Example 4   Find the divergence of the vector field $\vec{v} = x\vec{i}$ at the point $(2,2,0)$.

Solution   In Example 3 we computed the divergence at the origin. This time, let's take S to be a cube with edges parallel to the axes, of length $2R$, centered at the point $(2,2,0)$. See Figure 18.45. As before, the flux through the faces perpendicular to the y and z-axes is zero (because the vector field is parallel to these faces). On the face at $x = 2 + R$,

$$\vec{v}\cdot\Delta\vec{A} = (2+R)\,\Delta A.$$

On the face at $x = 2 - R$ with outward normal, the dot product is negative, and

$$\vec{v}\cdot\Delta\vec{A} = -(2-R)\,\Delta A.$$

Therefore, the flux through the box is given by

$$\int_S \vec{v}\cdot d\vec{A} = \int_{x=2-R}\vec{v}\cdot d\vec{A} + \int_{x=2+R}\vec{v}\cdot d\vec{A}$$
$$= ((2+R)-(2-R))(\text{Area of one face}) = 2R(2R)^2 = 8R^3.$$

Then, as before,

$$\operatorname{div}\vec{v}(2,2,0) = \lim_{\text{vol}\to 0}\left(\frac{\int_S \vec{v}\cdot d\vec{A}}{\text{Volume of box}}\right) = \lim_{R\to 0}\left(\frac{8R^3}{(2R)^3}\right) = 1.$$

Note that the vector field does not appear to flow away from the point $(2,2,0)$; however, because the inflow on the left of this point is less than the outflow on the right, the divergence is positive.

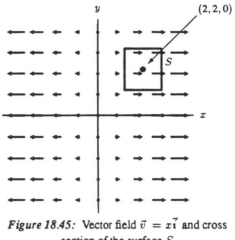

Figure 18.45: Vector field $\vec{v} = x\vec{i}$ and cross
section of the surface S

How Do We Calculate Divergence in Cartesian Coordinates?

Suppose we want to find div $\vec{F}$, where $\vec{F} = F_1\vec{i} + F_2\vec{j} + F_3\vec{k}$. To find div $\vec{F}$ at the point (x_0, y_0, z_0) we take a small cube-shaped surface S with one corner at this point and edges of length Δx, Δy, and Δz parallel to the axes; see Figure 18.46. (It turns out not to matter in this case that the point is on S rather than surrounded by it.)

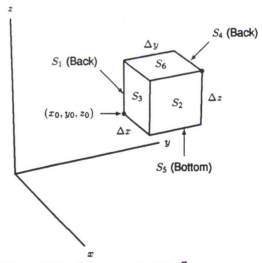

Figure 18.46: Cube used to find div $\vec{F}$ at (x_0, y_0, z_0)

First we'll calculate the flux through the surfaces S_1 and S_2 perpendicular to the x-axis. On S_1 (the back face of the cube shown in Figure 18.46), the outward normal is in the negative x direction, and S_1 has area $\Delta y \Delta z$, so $\Delta \vec{A} = -\Delta y \Delta z \vec{i}$. Therefore, on S_1,

$$\vec{F} \cdot \Delta \vec{A} = (F_1\vec{i} + F_2\vec{j} + F_3\vec{k}) \cdot (-\Delta y \Delta z \vec{i}) = -F_1 \Delta y \Delta z.$$

On S_2 the outward normal points in the positive x direction, and the area of S_2 is again $\Delta y \, \Delta z$, so $\Delta \vec{A} = \Delta y \, \Delta z \vec{i}$. Therefore, on S_2

$$\vec{F} \cdot \Delta \vec{A} = (F_1\vec{i} + F_2\vec{j} + F_3\vec{k}) \cdot (\Delta y \, \Delta z \vec{i}) = F_1 \, \Delta y \, \Delta z.$$

Notice, however, that on S_1 the function F_1 is evaluated at the point (x_0, y, z) whereas on S_2, the function F_1 is evaluated at $(x_0 + \Delta x, y, z)$. But, by local linearity,

$$F_1(x_0 + \Delta x, y, z) \approx F_1(x_0, y, z) + \frac{\partial F_1}{\partial x} \Delta x.$$

Thus, the total contribution to the flux from S_1 and S_2 is

$$-F_1(x_0, y, z) \, \Delta y \, \Delta z + F_1(x_0 + \Delta x, y, z) \, \Delta y \, \Delta z$$

$$\approx -F_1(x_0, y, z) \, \Delta y \, \Delta z + \left(F_1(x_0, y, z) + \frac{\partial F_1}{\partial x} \Delta x \right) \Delta y \, \Delta z$$

$$= \frac{\partial F_1}{\partial x} \Delta x \, \Delta y \, \Delta z.$$

By an analogous argument, the contribution to the flux from S_3 and S_4 (the surfaces perpendicular to the y-axis) is approximately

$$\frac{\partial F_2}{\partial y} \Delta x \, \Delta y \, \Delta z,$$

and the contribution to the flux from S_5 and S_6 is approximately

$$\frac{\partial F_3}{\partial z} \Delta x \, \Delta y \, \Delta z.$$

Thus, adding these contributions we have

$$\text{Total flux through } S \approx \frac{\partial F_1}{\partial x} \Delta x \, \Delta y \, \Delta z + \frac{\partial F_2}{\partial y} \Delta x \, \Delta y \, \Delta z + \frac{\partial F_3}{\partial z} \Delta x \, \Delta y \, \Delta z.$$

Since the volume of the cube is $\Delta x \, \Delta y \, \Delta z$,

$$\operatorname{div} \vec{F} (x_0, y_0, z_0) = \lim_{\text{vol} \to 0} \left(\frac{\text{Flux through } S}{\text{Volume of } S} \right)$$

$$= \lim_{\Delta x \, \Delta y \, \Delta z \to 0} \frac{\left(\frac{\partial F_1}{\partial x} + \frac{\partial F_2}{\partial y} + \frac{\partial F_3}{\partial z} \right) \Delta x \, \Delta y \, \Delta z}{\Delta x \, \Delta y \, \Delta z}$$

$$= \frac{\partial F_1}{\partial x} + \frac{\partial F_2}{\partial y} + \frac{\partial F_3}{\partial z}.$$

Thus, we have the following result:

If $\vec{F} = F_1\vec{i} + F_2\vec{j} + F_3\vec{k}$, then

$$\operatorname{div} \vec{F} = \frac{\partial F_1}{\partial x} + \frac{\partial F_2}{\partial y} + \frac{\partial F_3}{\partial z}.$$

Alternative Notation

Using $\nabla = \frac{\partial}{\partial x}\vec{i} + \frac{\partial}{\partial y}\vec{j} + \frac{\partial}{\partial z}\vec{k}$, we can write

$$\operatorname{div} \vec{F} = \nabla \cdot \vec{F} = (\frac{\partial}{\partial x}\vec{i} + \frac{\partial}{\partial y}\vec{j} + \frac{\partial}{\partial z}\vec{k}) \cdot (F_1\vec{i} + F_2\vec{j} + F_3\vec{k}) = \frac{\partial F_1}{\partial x} + \frac{\partial F_2}{\partial y} + \frac{\partial F_3}{\partial z}.$$

Example 5 Find div $\vec{F}$ for the following vector fields (a) $\vec{F} = -2\vec{r}$ (b) $\vec{F} = x\vec{i}$

Solution In Example 3 we calculated the divergence of these vector fields using the flux definition. Here we will show the same results using Cartesian coordinates.

(a) $\operatorname{div}(-2\vec{r}) = \operatorname{div}(-2x\vec{i} - 2y\vec{j} - 2z\vec{k}) = \frac{\partial}{\partial x}(-2x) + \frac{\partial}{\partial y}(-2y) + \frac{\partial}{\partial z}(-2z) = -6.$

(b) $\operatorname{div}(x\vec{i}) = \operatorname{div}(x\vec{i} + 0\vec{j} + 0\vec{k}) = \frac{\partial}{\partial x}(x) = 1.$

The Divergence and Changes in Volume

We already have one interpretation of the divergence of a vector field as the rate at which a liquid is being created or destroyed at a point. We now consider a second interpretation. For this second interpretation we again imagine that our vector field represents the flow of a gas, such as air, which can expand and contract as it flows. In Section 17.2, we considered a two-dimensional vector field $\vec{v} = v_1\vec{i} + v_2\vec{j}$ and showed that at any point we have

$$\text{Rate of expansion of gas per unit area} = \frac{\partial v_1}{\partial x} + \frac{\partial v_2}{\partial y}.$$

An analogous calculation holds if $\vec{v} = v_1\vec{i} + v_2\vec{j} + v_3\vec{k}$ is a three-dimensional vector field. Then, it can be shown that if V is the volume of gas:

$$\text{Rate of expansion of gas per unit volume} = \frac{1}{V}\frac{dV}{dt} = \frac{\partial v_1}{\partial x} + \frac{\partial v_2}{\partial y} + \frac{\partial v_3}{\partial z} = \operatorname{div}\vec{v}$$

Notice that since div $\vec{v}$ is a function of position, the rate of expansion of gas varies from point to point. The rate of expansion, or divergence, is measured in cubic units (of gas) per unit time per cubic unit (of space).

Incompressible Flow

The flow of a substance is said to be *incompressible* if there is no expansion or contraction. In an incompressible flow, volumes do not change as they move. It follows that a vector field $\vec{v}$ can model an incompressible flow only if div $\vec{v} = 0$.

Comparison of Two Interpretations of the Divergence

If div $\vec{v} \neq 0$, then the flow of $\vec{v}$ produces volume change. This change usually has one of two possible causes. If there are no sources or sinks, so that mass is conserved, then expansion or contraction must produce density changes in the substance as it moves. If, on the other hand, the substance is of constant density everywhere (as for a liquid), then increase in volume indicates the presence of one or more sources that are injecting additional matter into the flow, and similarly decrease in volume indicates the presence of one or more sinks draining matter from the flow.

Divergence in n-Space

So far we have only talked about divergence in 3-space. In 2-space, we can define the divergence to be the outflow per unit time per unit area of a vector field, and in n-space we can define it to be the outflow per unit time per unit volume, although here we mean n-dimensional volume. The formula for the divergence in other dimensions is similar to the one we found for 3-space. In two dimensions the formula is

If $\vec{F} = F_1\vec{i} + F_2\vec{j}$, then

$$\text{div } \vec{F} = \frac{\partial F_1}{\partial x} + \frac{\partial F_2}{\partial y}.$$

and in n-dimensions it is

If $\vec{F} = F_1\vec{i}_1 + F_2\vec{i}_2 + \ldots F_n\vec{i}_n$, then

$$\text{div } \vec{F} = \frac{\partial F_1}{\partial x_1} + \frac{\partial F_2}{\partial x_2} + \cdots + \frac{\partial F_n}{\partial x_n}.$$

Problems for Section 18.3

1. Which of the two vector fields in Figure 18.47 has the greater divergence at the origin? Assume the scales are the same on each.

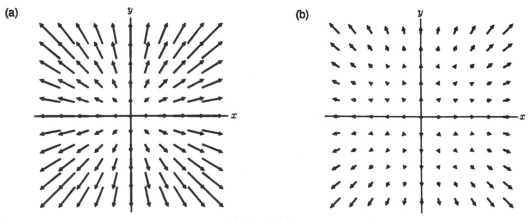

Figure 18.47

2. For each of the following vector fields, say whether the divergence is positive, zero, or negative at the indicated point.

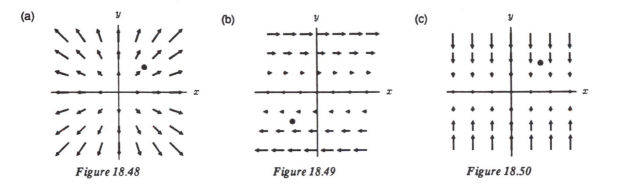

(a)   *Figure 18.48*

(b)   *Figure 18.49*

(c)   *Figure 18.50*

3. Draw two vector fields that have positive divergence everywhere.
4. Draw two vector fields that have negative divergence everywhere.
5. Draw two vector fields that have zero divergence everywhere.

In Problems 6–10, find the divergence of the given vector field.

6. $\vec{F}(x, y) = (x^2 - y^2)\vec{i} + 2xy\vec{j}$

7. $\vec{F}(\vec{r}) = \dfrac{\vec{r}}{r}$ (in 3-space)

8. $\vec{F}(x, y) = -x\vec{i} + y\vec{j}$

9. $\vec{F}(x, y) = -y\vec{i} + x\vec{j}$

10. $\vec{F}(x, y, z) = (-x+y)\vec{i} + (y+z)\vec{j} + (-z+x)\vec{k}$

11. Let $\vec{F}(\vec{r}) = \dfrac{\vec{r}}{r^3}$ (in 3-space).

    (a) Calculate div $\vec{F}$.
    (b) Sketch $\vec{F}$. Does it appear to be diverging? Does this agree with your answer to part (a)?

12. Let $\vec{F}(x, y, z) = z\vec{k}$.

    (a) Calculate div $\vec{F}$.
    (b) Sketch $\vec{F}$. Does it appear to be diverging? Does this agree with your answer to part (a)?.

13. (a) Find the flux of the vector field $\vec{F} = x\vec{i}$ through a box of side c in the first octant with one corner at the origin and sides along the axes.
    (b) Use your answer to part (a) to find div $\vec{F}$ at the origin.

14. (a) Find the flux of $\vec{F} = 2\vec{i} + y\vec{j} + 3\vec{k}$ through a box of side c in the first octant, with one corner at the origin and sides along the axes.
    (b) Use your answer to part (a) to find div $\vec{F}$ at the origin.

15. (a) Find the flux of the vector field $\vec{F} = x\vec{i} + y\vec{j}$ through a box of side c in the first octant with one corner at the origin, sides parallel to the axes.
    (b) Use your answer to part (a) to find div $\vec{F}$ at the origin.

16. (a) Find the flux of the vector field $\vec{F} = x\vec{i} + y\vec{j}$ through the surface of the cylinder of radius c and height c, centered on the z-axis with base in the xy-plane.

(b) Use your answer to part (a) to find div $\vec{F}$ at the origin.

(c) Compute div $\vec{F}$ at the origin using partial derivatives.

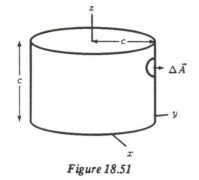

Figure 18.51

Problems 17–18 involve electric fields. Electric charge produces a vector field $\vec{E}$, called the electric field, which represents the force on a unit positive charge placed at the point. Two positive or two negative charges are observed to repel one another, whereas two charges of opposite sign attract one another. In addition it can be shown that the divergence of $\vec{E}$ is the density $\rho(x, y, z)$ of the electric charge (that is, the charge per unit area or charge per unit volume).

17. Suppose a certain distribution of electric charge produces the electric field shown in Figure 18.52. Where are the charges that produced this electric field concentrated? Which concentrations are positive and which are negative?

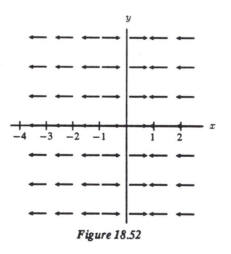

Figure 18.52

18. The electric field produced by a charge located at the origin is given by

$$\vec{E}(\vec{r}) = \frac{\vec{r}}{r^3}.$$

(a) Calculate div $\vec{E}$.

(b) What can you say about div $\vec{E}$ at the point $(0, 0, 0)$?

(c) Explain what your answers mean in terms of charge density.

19. The divergence of a magnetic vector field $\vec{B}$ must be zero everywhere. Which of the following vector fields cannot be a magnetic vector field?

    (a) $\vec{B}(x, y, z) = -y\vec{i} + x\vec{j} + (x+y)\vec{k}$

    (b) $\vec{B}(x, y, z) = -z\vec{i} + y\vec{j} + x\vec{k}$

    (c) $\vec{B}(x, y, z) = (x^2 - y^2 - x)\vec{i} + (y - 2xy)\vec{j}$

20. Show that if $\vec{a}$ is a constant vector and $f(x, y, z)$ is a scalar valued function, and if $\vec{F} = f\vec{a}$, then div $\vec{F} = (\text{grad } f) \cdot \vec{a}$.

21. If $f(x, y, z)$ and $g(x, y, z)$ are scalar valued functions with continuous second derivatives, show that
    $$\text{div}(\text{grad } f \times \text{grad } g) = 0.$$

22. In Problem 29 on page 389 it was shown that the rate of heat loss from a volume V in an unevenly heated region of space was $k \int_S (\text{grad } T) \cdot d\vec{A}$, where k is a constant, S is the surface bounding V, and $T(x, y, z)$ is the temperature at the point (x, y, z) in space. By taking the limit as V contracts to a point, show that $\partial T/\partial t = A \text{ div grad } T$ at that point, where A is a constant with respect to x, y, z, but may depend on t.

23. A vector field in the plane is a *point source* at the origin if its direction is away from the origin at every point, its magnitude depends only on the distance from the origin, and its divergence is zero away from the origin.

    (a) Explain why a source must be of the form $\vec{v} = \left[f(x^2 + y^2) \right] (x\vec{i} + y\vec{j})$ for some positive function f.

    (b) Show that $\vec{v} = K(x^2 + y^2)^{-1}(x\vec{i} + y\vec{j})$ is a point source at the origin (that is, div $\vec{v} = 0$) if $K > 0$.

    (c) Determine the magnitude $\|\vec{v}\|$ of the source in part (b) as a function of the distance from its center.

    (d) Sketch the vector field $\vec{v} = (x^2 + y^2)^{-1}(x\vec{i} + y\vec{j})$.

    (e) Show that $\phi = \frac{K}{2} \log(x^2 + y^2)$ is a potential function for the source in part (b).

24. A vector field in the plane is a *point sink* at the origin if its direction is toward the origin at every point, its magnitude depends only on the distance from the origin, and its divergence is zero away from the origin.

    (a) Explain why a sink must be of the form $\vec{v} = \left[f(x^2 + y^2) \right] (x\vec{i} + y\vec{j})$ for some negative function f.

    (b) Show that $\vec{v} = K(x^2 + y^2)^{-1}(x\vec{i} + y\vec{j})$ is a point sink at the origin (that is, div $\vec{v} = 0$) if $K < 0$.

    (c) Determine the magnitude $\|\vec{v}\|$ of the sink in part (a) as a function of the distance from its center.

    (d) Sketch the vector field $\vec{v} = -(x^2 + y^2)^{-1}(x\vec{i} + y\vec{j})$.

    (e) Show that $\phi = \frac{K}{2} \log(x^2 + y^2)$ is a potential function for the sink in part (b).

25. A vector field is a *point source* at the origin in 3-space if its direction is away from the origin at every point, its magnitude depends only on the distance from the origin, and its divergence is zero away from the origin. (Such a vector field might be used to model the photon flow out of a star or the neutrino flow out of a supernova.)

    (a) Show that $\vec{v} = K(x^2 + y^2 + z^2)^{-3/2}(x\vec{i} + y\vec{j} + z\vec{k})$ is a point source at the origin if $K > 0$.

(b) Determine the magnitude $\|\vec{v}\|$ of the source in (a) as a function of the distance from its center.

(c) Compute the flux of $\vec{v}$ through a sphere of radius r centered at the origin.

(d) Compute the flux of $\vec{v}$ through a closed surface that does not contain the origin.

18.4 THE DIVERGENCE THEOREM

If $\vec{v}$ is the velocity field of an incompressible fluid flow, the flux integral of $\vec{v}$ through a closed surface S tells you the net outflow of fluid from the region enclosed by the surface. On the other hand, the divergence of $\vec{v}$ tells you the outflow per unit volume at a point. Thus, we should be able to compute the flux from the divergence.

Example 1 Calculate the flux of the vector field $\vec{F}(\vec{r}) = \vec{r}$ through the sphere of radius R, using the fact that $\operatorname{div} \vec{F} = 3$.

Solution In Example 6 on page 384 we computed the flux integral directly:

$$\text{Flux through sphere} = \int_{\text{Sphere}} \vec{r} \cdot d\vec{A} = 4\pi R^3.$$

Now we will compute it another way. We calculate the divergence

$$\operatorname{div} \vec{F} = \operatorname{div}(x\vec{i} + y\vec{j} + z\vec{k}) = \frac{\partial x}{\partial x} + \frac{\partial y}{\partial y} + \frac{\partial z}{\partial z} = 1 + 1 + 1 = 3.$$

Thus, fluid is being created at the rate of 3 units of fluid per unit volume at every point. Hence, the total production of fluid in the ball of radius R enclosed by the sphere is

$$3(\text{Volume of ball}) = 3(\frac{4}{3}\pi R^3) = 4\pi R^3.$$

This is the same answer we found before.

What if the Divergence is Not Constant?

In the previous example it was easy to obtain the total flux from the divergence because the divergence was constant. If the divergence is not constant, we can still compute the flux out of a volume W by means of a volume integral over W. Break W up into many small pieces, as shown in Figure 18.53, and for each piece multiply the divergence $\operatorname{div} \vec{F}(x, y, z)$ at a point (x, y, z) in the piece by the volume ΔV of the piece. This gives an approximation

$$\text{Flux out of small piece} \approx \operatorname{div} \vec{F}(x, y, z)\, \Delta V.$$

What happens if we add together the fluxes for two adjacent pieces? The adjacent wall has the flux through it counted twice, once out of the cube on one side and once out of the cube on the other. These two contributions cancel. (See Figure 18.54.) Thus, if we add up the fluxes for all the pieces, we get the flux out of the entire volume. So

$$\text{Flux out of the entire volume} = \sum \text{Flux out of small pieces} \approx \sum \operatorname{div} \vec{F}\, \Delta V.$$

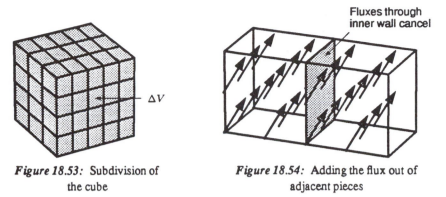

Figure 18.53: Subdivision of the cube

Figure 18.54: Adding the flux out of adjacent pieces

We have approximated the flux by a Riemann sum. As the subdivision gets finer, the sum approaches an integral, so we can say

$$\text{Flux out of entire volume} = \int_W \text{div } \vec{F} \ dV.$$

We now have two ways of calculating the flux of a vector field $\vec{F}$ through a closed surface S: we can calculate the flux integral, or we can calculate the integral of div $\vec{F}$ over the volume W enclosed by S. Thus we have the following result:

The Divergence Theorem

If $\vec{F}$ is a vector field, and if S is a surface bounding a volume W, then

$$\int_S \vec{F} \cdot d\vec{A} = \int_W \text{div } \vec{F} \ dV,$$

assuming S is oriented outward and div $\vec{F}$ is defined at every point of W.

The Divergence Theorem is like the Fundamental Theorem of Calculus. The Fundamental Theorem of Calculus says that the integral of the rate of change of a function of one variable gives us the total change. The Divergence Theorem says that the integral over a volume of the outflow per unit volume of a vector field gives the total flux out of the volume.

The Divergence Theorem often provides an easy way of computing a flux integral for a closed surface.

Example 2   Calculate the flux of the vector field

$$\vec{F}(x, y, z) = (x^2 + y^2)\vec{i} + (y^2 + z^2)\vec{j} + (x^2 + z^2)\vec{k}$$

through the cube in Figure 18.55.

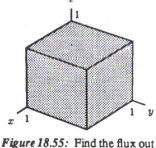

Figure 18.55: Find the flux out of this cube

Solution   The divergence of $\vec{F}$ is div $\vec{F} = 2x + 2y + 2z$. Since div $\vec{F}$ is positive everywhere in the first quadrant, the flux through S will be positive. By the Divergence Theorem,

$$\int_S \vec{F} \cdot d\vec{A} = \int_0^1 \int_0^1 \int_0^1 2(x + y + z)\, dx\, dy\, dz = \int_0^1 \int_0^1 x^2 + 2x(y + z)\Big|_0^1 dy\, dz$$

$$= \int_0^1 \int_0^1 1 + 2(y + z)\, dy\, dz = \int_0^1 y + y^2 + 2yz\Big|_0^1 dz$$

$$= \int_0^1 (2 + 2z)\, dz = 2z + z^2\Big|_0^1 = 3.$$

Example 3   Use the Divergence Theorem to calculate the flux of the vector field $\vec{F}(x, y, z) = -z\vec{i} + x\vec{k}$ through the sphere of radius R centered at the origin.

Solution   The divergence of the field is

$$\text{div } \vec{F} = \frac{\partial(-z)}{\partial x} + \frac{\partial(0)}{\partial y} + \frac{\partial x}{\partial z} = 0.$$

Hence,

$$\int_{\text{Sphere}} \vec{F} \cdot d\vec{A}\, dV = \int_{\text{Ball}} \text{div } \vec{F}\, dV = \int_{\text{Ball}} 0\, dV = 0,$$

and so the flux through the sphere is zero. This makes sense, because, as Figure 18.56 shows, the vector field is flowing around the y-axis and is always tangent to the surface.

Figure 18.56: Find the flux out of the
sphere of radius R

It should be noted that the Divergence Theorem applies to any volume W and its boundary S, even in cases where the boundary consists of two or more surfaces.

Harmonic Functions

A function ϕ of three variables x, y, and z is said to be *harmonic* in a region if $\text{div}(\text{grad}\phi) = 0$ at every point in the region. (This equation is also written $\nabla^2\phi = 0$, because $\text{div}(\text{grad}\phi) = \nabla \cdot (\nabla\phi)$.) In the Problem 12 you will show that

$$\nabla^2\phi(x, y, z) = \frac{\partial^2\phi}{\partial x^2} + \frac{\partial^2\phi}{\partial y^2} + \frac{\partial^2\phi}{\partial z^2}$$

Harmonic functions are of great importance in pure and applied mathematics. For example, the temperature in a region of space that is not undergoing temperature change is harmonic, as is the electric potential in a charge-free region of space. Several of the basic properties of harmonic functions can be deduced from the Divergence Theorem.

Example 4   Show that a nonconstant harmonic function ϕ cannot have a local maximum.

Solution   Suppose that a nonconstant function ϕ has a local maximum at a point P. Then the vector field $\text{grad}\phi$ will point approximately towards P at all points near P, because it points in the direction of increasing ϕ. Taking a small sphere S centered at P, oriented outwards, we will therefore have

$$\int_S \text{grad}\,\phi \cdot d\vec{A} < 0$$

On the other hand, if ϕ is harmonic, then by the divergence theorem

$$\int_S \text{grad}\,\phi \cdot d\vec{A} = \int_W \text{div}(\text{grad}\,\phi)dV = 0$$

where W is the ball enclosed by S. Clearly if ϕ has a local maximum, then it cannot be harmonic.

The most famous fact about harmonic functions is their mean value property, discovered by Gauss. If ϕ is a harmonic function in the region enclosed by a sphere, then the value of ϕ at the center of the sphere equals the average value of ϕ on the sphere. For example, the temperature at a point in space equals the average value of the temperature on any sphere centered at the point.

Problems for Section 18.4

For Problems 1–3, compute the flux integral $\int_S \vec{F} \cdot d\vec{A}$ in two ways, if possible, directly and using the Divergence Theorem. In all cases, S is oriented outwards.

1. $\vec{F}(\vec{r}) = \vec{r}$, where S is the cube enclosing the volume $0 \le x \le 2, 0 \le y \le 2$, and $0 \le z \le 2$.

2. $\vec{F}(x, y, z) = y\vec{j}$, where S a vertical cylinder of height 2, with its base a circle of radius 1 on the xy-plane, centered at the origin. S is to include the disks that close it off at top and bottom.

3. $\vec{F}(x, y, z) = -z\vec{i} + x\vec{k}$, where S is a square pyramid with the base on the xy-plane of side length 1 and height 3.

4. Suppose V_1 and V_2 are the two cube-shaped volumes in the first quadrant shown in Figure 18.57. Both have sides of length 1 parallel to the axes; V_1 has one corner at the origin, while V_2 has the corresponding corner at the point $(1, 0, 0)$. Suppose S_1 and S_2 are the six-sided surfaces of V_1 and V_2, respectively. Suppose V is the box-shaped volume consisting of V_1 and V_2 together, and having outside surface S.

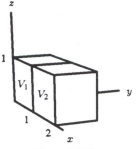

Figure 18.57

Are the following true or false? Give reasons.

(a) If $\vec{F}$ is a constant vector field, $\int_S \vec{F} \cdot d\vec{A} = 0$

(b) If S_1, S_2 and S are all oriented outward and $\vec{F}$ is any vector field:

$$\int_S \vec{F} \cdot d\vec{A} = \int_{S_1} \vec{F} \cdot d\vec{A} + \int_{S_2} \vec{F} \cdot d\vec{A}.$$

5. Calculate $\int_{S_2} \vec{F} \cdot d\vec{A}$ where $\vec{F} = x^2 \vec{i} + 2y^2 \vec{j} + 3z^2 \vec{k}$ and S_2 is as in Problem 4.
   Do this directly and using the Divergence Theorem.

6. Use the Divergence Theorem to evaluate the flux integral $\int_S (x^2 \vec{i} + (y - 2xy)\vec{j} + 10z\vec{k}) \cdot d\vec{A}$, where S is the sphere of radius 5 centered at the origin, oriented outward.

7. Suppose that a vector field $\vec{F}$ satisfies $\text{div}\vec{F} = 0$ everywhere. Show that $\int_S \vec{F} \cdot d\vec{A} = 0$ for every closed surface S.

8. The gravitational field, $\vec{F}$, of a planet of mass m at the origin is given by

$$\vec{F} = -G \frac{m\vec{r}}{r^3}.$$

Use the Divergence Theorem to show that the flux of the gravitational field through the sphere of radius R is independent of R. [Hint: Consider a volume bounded by two concentric spheres.]

9. A basic property of the electric field $\vec{E}$ is that its divergence is zero at points where there is no charge. Suppose that the only charge is along the z-axis, and that the electric field $\vec{E}$ points radially out from the z-axis and its magnitude depends only on the distance r from the z-axis. Use the Divergence Theorem to show that the magnitude of the field is proportional to $1/r$.

10. Some people have suggested that there may be "magnetic charges" which give rise to magnetic fields in the same way that electric charges give rise to electric fields. Suppose the vector field $\vec{B}(x, y, z)$ represents the magnetic field at any point in space. Then one of Maxwell's equations tells us that

$$\text{div}\,\vec{B} = 0.$$

Explain why this equation shows that there can be no such magnetic charges.

11. If a surface S is submerged in an incompressible fluid, a force $\vec{F}$ is exerted on one side of the surface by the pressure in the fluid. The component of force in the direction of a unit vector $\vec{u}$ is given by the following

$$\vec{F} \cdot \vec{u} = -\int_S \rho g z \vec{u} \cdot d\vec{A}$$

where ρ is the density of the fluid (Mass/Volume), g is the acceleration due to gravity, and the surface is oriented away from the side on which the force is exerted. Suppose the z-axis is vertical, with the positive direction upward and the surface level at $z = 0$.

In this problem we will consider a totally submerged closed surface enclosing a volume V. We are interested in the force of the liquid on the external surface, so S is oriented inward.

   (a) Use the Divergence Theorem to show that the force in the $\vec{i}$ and $\vec{j}$ directions is zero.
   (b) Use the Divergence Theorem to show that the force in the $\vec{k}$ direction is $\rho g V$, the weight of the volume of fluid with the same volume as S. This is *Archimedes' Principle*.

12. Show that $\nabla^2 \phi(x, y, z) = \partial^2\phi/\partial x^2 + \partial^2\phi/\partial y^2 + \partial^2\phi/\partial z^2$

13. Show that linear functions are harmonic.

14. What is the condition on the constant coefficients such that $ax^2 + by^2 + cz^2 + dxy + exz + fyz$ is harmonic?

15. Use the Divergence Theorem to show that if ϕ is harmonic in a region W, then $\int_S \nabla\phi \cdot d\vec{A} = 0$ for every closed surface S in W such that the volume enclosed by S lies completely within W.

16. Show that a harmonic function can achieve a maximum value in a closed region only on the boundary of the region.

17. Show that a nonconstant harmonic function can not have a local minimum, and that it can achieve a minimum value in a closed region only on the boundary.

18. Let $\phi = 1/(x^2 + y^2 + z^2)^{1/2}$.

   (a) Show that ϕ is harmonic everywhere except the origin.
   (b) Give a geometric explanation for the fact that ϕ has no local maximum or minimum. [Hint: $\phi = 1/r$.]
   (c) Compute $\int_S \nabla\phi \cdot d\vec{A}$ where S is the sphere of radius 1 centered at the origin. Does your answer contradict the assertion of Problem 15?

19. Show that if ϕ is a harmonic function, then $\text{div}(\phi\text{grad}\phi) = \|\text{grad}\phi\|^2$.

20. Suppose that ϕ is a harmonic function in the region enclosed by a closed surface S, and suppose that such that $\phi = 0$ at all points of S. Show that $\phi = 0$ at all points of the region enclosed by S. [Hint: Apply the Divergence Theorem to $\int_S \phi\text{grad}\phi \cdot d\vec{A}$ and use Problem 19.]

21. Suppose that ϕ_1 and ϕ_2 are harmonic functions in the region enclosed by a closed surface S, and suppose that $\phi_1 = \phi_2$ at all points of S. Show that $\phi_1 = \phi_2$ at all points of the region enclosed by S. [Hint: Use Problem 19.]

22. Show that if u and v are harmonic functions in a region W, then

$$\int_S u\text{grad}v \cdot d\vec{A} = \int_S v\text{grad}u \cdot d\vec{A}$$

for every closed surface S in W such that the volume enclosed by S lies completely within W. [Hint: Use the Divergence Theorem.]

Problems 23 and 24 outline two different proofs of the same result.

23. Let ϕ be a harmonic function, and let $M(R)$ equal the average value of ϕ on the sphere of radius R centered at the origin. This exercise outlines a proof that $M(R) = \phi(0)$ for all R.

   (a) Explain why $M(R) = 1/(4\pi) \int_S \phi(Rx, Ry, Rz)\|d\vec{A}\|$ where S is the sphere of radius 1 centered at the origin.

   (b) Assuming that you can differentiate with respect to R under the integral sign in the formula for $M(R)$ given in part (a), show that $dM/dR = 1/(4\pi) \int_S \nabla\phi \cdot d\vec{A}$.

   (c) Applying the Divergence Theorem to the integral in part (b), show that $dM/dR = 0$, and hence conclude that the average value of ϕ on spheres centered at the origin is the same for all such spheres.

   (d) By considering smaller and smaller values of R and thus contracting the sphere of radius R to the origin, show that $M(R) = \phi(0)$ for all R.

24. Let ϕ be a harmonic function, and let $M(R)$ equal the average value of ϕ on the sphere of radius R centered at the origin. This exercise outlines a proof that $M(R) = \phi(0)$ for all R.

   (a) Explain why $M(R) = -\frac{1}{4\pi} \int_{S(R)} \phi\nabla(1/r) \cdot d\vec{A}$ where $S(R)$ is the sphere of radius R centered at the origin, oriented outwards.

   (b) Show that $\int_S (1/r)\nabla\phi \cdot d\vec{A} = 0$ where S is the surface consisting of two spheres $S(a)$ and $S(b)$ centered at the origin of radii a and b, with $a < b$, where $S(a)$ is oriented towards the origin, and $S(b)$ is oriented away from the origin.

   (c) Show that $M(a) = M(b)$, so that the average value of ϕ on spheres centered at the origin is the same for all such spheres. [Hint: Use Problems 22 and 24(a) and (b).]

   (d) By considering smaller and smaller values of R and thus contracting the sphere of radius R to the origin, show that $M(R) = \phi(0)$ for all R.

18.5   THE CURL OF A VECTOR FIELD

We know that some vector fields, called conservative fields, have the property that their circulation around any closed curve is zero, while other vector fields do not have this property. What is the geometric difference between these two types of vector fields?

The vector fields in Figure 18.58 and Figure 18.59 represent the two types of behavior. Figure 18.58 shows a conservative vector field, with zero circulation around any closed curve; the vector field in Figure 18.59 is not conservative, and has nonzero circulation around many closed curves. One difference between these two vector fields is clear just by looking at them quickly: the non-conservative vector field in Figure 18.59 suggests a rotating or swirling motion, while the conservative one in Figure 18.58 does not. In fact, the flow lines of the non-conservative field do swirl, and those of the other field do not. In this section, we will develop a way of measuring the rotational properties of a vector field.

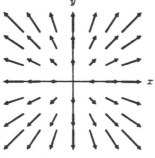

Figure 18.58: A conservative vector field

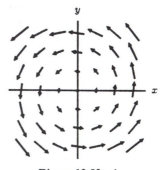

Figure 18.59: A non-conservative vector field

Example 1   Describe the rotation of the following three vector fields.

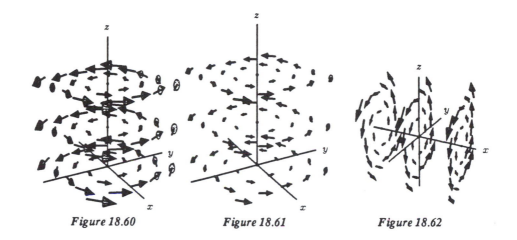

| *Figure 18.60* | *Figure 18.61* | *Figure 18.62* |

Solution   The vector field in Figure 18.60 is rotating or swirling about the z-axis. So is the one in Figure 18.61, but it appears to be rotating at a smaller velocity. The one in Figure 18.62 is rotating about the x-axis, and seems to be rotating at about the same velocity as the one in Figure 18.61.

Example 1 illustrates an important point about the rotation of a vector field. It has both magnitude (the strength or speed of the rotation) and direction (the axis about which the rotation occurs), and so it can be represented by a vector. We call this vector the *curl* of the vector field, but we are not yet in a position to define it precisely.

The Direction of a Rotation

We want to define the *direction* of a rotation. We will say that the direction of a simple rotation is given by a vector parallel to the axis of rotation that is specified by the right-hand rule. As shown in Figure 18.63, the direction of the rotation is indicated by the thumb of the right hand when the fingers curl around the axis of rotation in the direction of motion. The vector fields in Figure 18.60 and Figure 18.61 rotate in the $\vec{k}$ direction, and that in Figure 18.62 rotates in the $\vec{i}$ direction.

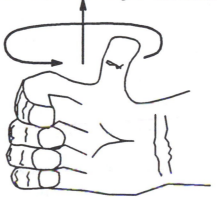

Figure 18.63: A rotation and its
vector direction

Example 2   What is the direction of the rotation of the earth?

Solution   The direction is parallel to the axis. The right-hand rule tells us that the direction is given by a vector pointing from the center of the earth in the direction of the North Pole.

Actually, we should be discussing the direction of rotation of a vector field near a point, for it may well vary from point to point. Think of a windstorm creating eddies, circulating in different directions and at different speeds at different places. Figure 18.64 shows a vector field rotating in the $\vec{k}$-direction near the origin but rotating in the $\vec{j}$-direction near the point P.

At points where a vector field is not zero, the tendency of the flow lines to swirl may be masked by drifting. One way to reveal the rotational tendency is to subtract out the drift.

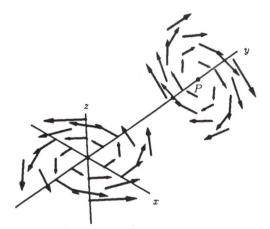

Figure 18.64: A vector field that rotates in the $\vec{k}$-direction at 0 and in the $\vec{j}$-direction at P

Example 3   Describe the motion of a leaf dropped into a stream flowing with the given velocity vector field at the given point:

1. Velocity field $\vec{v} = -y\vec{i} + x\vec{j}$; point $P = (1, 1)$.
2. Velocity field $\vec{v} = y\vec{i}$; point $Q = (0, 1)$.

Solution (a) The leaf will follow the flow of the vector field, circling the origin, but it will rotate counterclockwise (in the $\vec{k}$-direction) as it does so. See Figure 18.65. At the point P, the velocity is $\vec{v}(P) = -\vec{i} + \vec{j}$. To see the rotation around P, we subtract the velocity at P. We get the vector field $\vec{v} - (-\vec{i} + \vec{j}) = (-y + 1)\vec{i} + (x - 1)\vec{j}$, which is sketched in Figure 18.66. This vector field, with the drift at the point P removed, shows the rotation at P clearly.

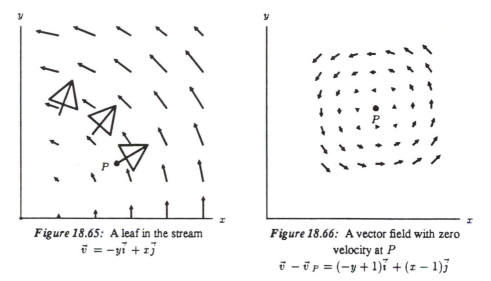

Figure 18.65: A leaf in the stream
$\vec{v} = -y\vec{i} + x\vec{j}$

Figure 18.66: A vector field with zero
velocity at P
$\vec{v} - \vec{v}_P = (-y+1)\vec{i} + (x-1)\vec{j}$

(b) The leaf will follow the flow of the vector field, moving in the y-direction, but it will rotate clockwise (in the $-\vec{k}$ -direction) as it does so. See Figure 18.67. The velocity at Q is $\vec{v}(Q) = \vec{i}$. Subtracting out this velocity, we get the sketch of Figure 18.68 of $\vec{v} - \vec{i} = (y-1)\vec{i}$, which clearly shows the tendency to rotate clockwise at Q.

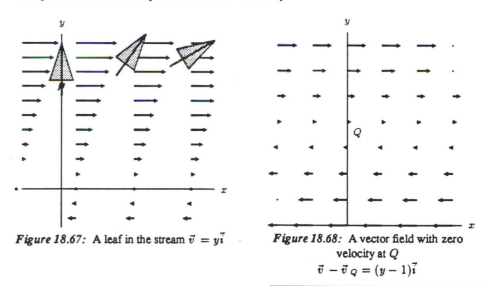

Figure 18.67: A leaf in the stream $\vec{v} = y\vec{i}$

Figure 18.68: A vector field with zero
velocity at Q
$\vec{v} - \vec{v}_Q = (y-1)\vec{i}$

How We Measure the Magnitude of a Rotation

Let's start with a vector field in 3 dimensions, each of whose vectors lies parallel to some fixed plane. For example, let's consider the vector field which looks like Figure 18.60 on page 417. Suppose the vector field at a distance r from the z-axis has magnitude $2r$ in the direction shown. We will calculate the strength of rotation about the z-axis. If C is a circle of radius R in the xy-plane and centered at the origin, then since $\vec{F}$ is tangent to C everywhere, we have

$$\int_C \vec{F} \cdot d\vec{r} = \|\vec{F}\| \cdot (\text{Circumference of circle } C) = 2R(2\pi R) = 4\pi R^2.$$

To see how much the vector field is rotating near a point, we consider the ratio of circulation per unit area:

$$\frac{\text{Circulation around } C}{\text{Area inside } C} = \frac{\int_C \vec{F} \cdot d\vec{r}}{\pi R^2} = \frac{2R \cdot 2\pi R}{\pi R^2} = 4.$$

This ratio of circulation per unit area gives us a measure of how rapidly the vector field is rotating. If the vector field had magnitude $3r$ in the same direction, the ratio would be

$$\frac{\text{Circulation around } C}{\text{Area inside } C} = \frac{\int_C \vec{F} \cdot d\vec{r}}{\pi R^2} = \frac{3R \cdot 2\pi R}{\pi R^2} = 6.$$

If the original vector field had magnitude $5r$, the ratio would be

$$\frac{\text{Circulation around } C}{\text{Area inside } C} = \frac{\int_C \vec{F} \cdot d\vec{r}}{\pi R^2} = \frac{5R \cdot 2\pi R}{\pi R^2} = 10.$$

Of course, in each case we are using a very special vector field whose magnitude is a constant multiple of the distance r. For a more general vector field, we look at the limit of this ratio as the area enclosed by the curve shrinks:

$$\lim_{\text{Area} \to 0} \left(\frac{\text{Circulation around } C}{\text{Area inside } C} \right) = \lim_{\text{Area} \to 0} \left(\frac{\int_C \vec{F} \cdot d\vec{r}}{\text{Area inside } C} \right).$$

This limit gives a measure of how fast the vector field is rotating about the z-axis at the origin, by measuring its circulation around a curve in the xy-plane, as the curve shrinks to the origin.

Now we will define the strength of the rotation of an arbitrary vector field $\vec{F}$ at an arbitrary point P about an arbitrary axis, by taking the limit using closed curves C in a plane perpendicular to the axis. Suppose that $\vec{n}$ is the unit vector pointing in the direction about which we want to measure the rotation (see Figure 18.69). Then we make the following definition:

The **directional circulation** of $\vec{F}$ at the point P about the unit vector $\vec{n}$ is defined by the limit

$$\text{circ}_{\vec{n}} \vec{F}(P) = \lim_{\text{area} \to 0} \left(\frac{\text{Circulation around } C}{\text{Area inside } C} \right) = \lim_{\text{area} \to 0} \left(\frac{\int_C \vec{F} \cdot d\vec{r}}{\text{Area inside } C} \right).$$

Here C is a closed curve about P in the plane perpendicular to $\vec{n}$, traversed in the direction determined by the right-hand rule.

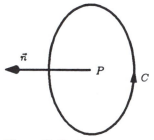

Figure 18.69: Direction of C relates to direction of $\vec{n}$ by the right-hand rule

Example 4 Calculate the directional circulation of the vector field in Figure 18.60 at the origin about the positive x-axis.

Solution We expect the answer to be zero, because, although the vector field is rotating about the z-axis, it does not appear to have any rotation about the x-axis. To calculate the directional circulation, we calculate line integrals around circles C in the yz-plane, centered at the origin.

$$\text{Directional circulation} = \lim_{\text{area} \to 0} \left(\frac{\int_C \vec{F} \cdot d\vec{r}}{\text{Area inside } C} \right).$$

The dot product in the line integral is zero, because at any point in the yz-plane the vector field is perpendicular to the plane, whereas the vector $d\vec{r}$ is in the plane. Hence, the line integral is zero, and so the directional circulation is zero.

The previous example illustrates that we can calculate the directional circulation of a vector field about *any* axis.

Example 5 Calculate the directional circulation of the following vector fields in every direction at every point.
(a) $\vec{F} = 3\vec{i} + 4\vec{j} + 5\vec{k}$ (b) $\vec{G} = ax\vec{i} + by\vec{j} + cz\vec{k}$ (c) $\vec{H} = -y\vec{i} + x\vec{j}$

Solution (a) The vector field $\vec{F}$ is constant. Its sketch in Figure 18.70 suggests a steady drift, but no rotation, so we suspect that the directional circulation is 0 in all directions. To see that this guess is correct, observe that $\vec{F} = \text{grad}(3x + 4y + 5z)$. Since $\vec{F}$ is a gradient field, $\int_C \vec{F} \cdot d\vec{r} = 0$ for *every* closed curve C. It follows from the definition that $\text{circ}_{\vec{n}} \vec{F}(P) = 0$ for every point P and every unit vector $\vec{n}$.

(b) The vector field $\vec{G}$ is also a gradient field,

$$\vec{G} = \text{grad} \left(\frac{a}{2}x^2 + \frac{b}{2}y^2 + \frac{c}{2}z^2 \right).$$

As in part (a), $\text{circ}_{\vec{n}} \vec{G}(P) = 0$ for every point P and every unit vector $\vec{n}$.

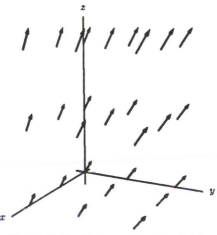

Figure 18.70: The constant vector field
$\vec{F} = 3\vec{i} + 4\vec{j} + 5\vec{k}$

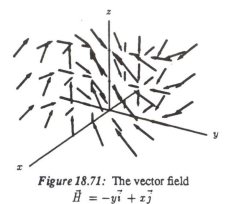

Figure 18.71: The vector field
$\vec{H} = -y\vec{i} + x\vec{j}$

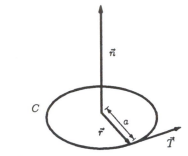

Figure 18.72: A circle of radius a about the origin in the plane normal to $\vec{n}$

(c)   We expect the directional circulation of $\vec{H}$ to be nonzero, for its sketch in Figure 18.71 suggests rotation.

In our computation we will use the expression

$$\vec{H} = -y\vec{i} + x\vec{j} = \vec{k} \times \vec{r}, \text{ where } \vec{r} = x\vec{i} + y\vec{j} + z\vec{k}.$$

Let us begin by calculating the directional circulation of $\vec{H}$ at the origin in the direction of a unit vector $\vec{n}$. We first compute $\int_C \vec{H} \cdot d\vec{r}$ where C is the circle of radius a centered at the origin in the plane perpendicular to $\vec{n}$. Figure 18.72 shows the circle C and a point on the circle, indicated by the tip of the position vector $\vec{r}$, which is perpendicular to $\vec{n}$. The unit tangent vector $\vec{T}$ to the circle at the point $\vec{r}$ is perpendicular to both $\vec{n}$ and $\vec{r}$.

To compute the line integral, we sum dot products of the form $\vec{H} \cdot \Delta\vec{r}$. Since $\Delta\vec{r}$ is parallel to the tangent $\vec{T}$ we will first evaluate the dot product $\vec{H} \cdot \vec{T}$. Since $\vec{H} = \vec{k} \times \vec{r}$, using the identity for the scalar triple product, we have

$$\vec{H} \cdot \vec{T} = (\vec{k} \times \vec{r}) \cdot \vec{T} = \vec{k} \cdot (\vec{r} \times \vec{T}).$$

From Figure 18.72, we see that $\vec{r} \times \vec{T}$ is parallel to $\vec{n}$, and checking magnitudes we find that $\vec{r} \times \vec{T} = a\vec{n}$. Thus,

$$\vec{H} \cdot \vec{T} = \vec{k} \cdot (a\vec{n}) = a\vec{k} \cdot \vec{n}.$$

Notice that the value of $\vec{H} \cdot \vec{T}$ does not depend on the point $\vec{r}$ on C. Hence, factoring the constant $a\vec{k} \cdot \vec{n}$ out of the integral, we get

$$\int_C \vec{H} \cdot d\vec{r} = \int \vec{H} \cdot \vec{T} \, dr = a\vec{k} \cdot \vec{n} \int dr = a\vec{k} \cdot \vec{n} \, (\text{Length of } C)$$

$$= a\vec{k} \cdot \vec{n} \, (2\pi a) = 2\pi a^2 \vec{k} \cdot \vec{n}$$

and so

$$\text{circ}_{\vec{n}} \vec{H} (0) = \lim_{a \to 0} \frac{\int_C \vec{H} \cdot d\vec{r}}{\pi a^2} = \lim_{a \to 0} \frac{2\pi a^2 \vec{k} \cdot \vec{n}}{\pi a^2} = 2\vec{k} \cdot \vec{n}.$$

The case in which P is not the origin is treated in Problem 22 at the end of this section.

Definition of the Curl

A vector field has a directional circulation at every point in every direction. This is analagous to the fact that a function of two or more variables has a directional derivative at every point in every direction. The computation of directional derivatives was simplified by introduction of the gradient vector, from which all the directional derivatives could be obtained by taking a dot product with a unit vector $\vec{u}$:

$$f_{\vec{u}}(P) = (\text{grad } f(P)) \cdot \vec{u}$$

It turns out that a similar simplification exists for directional circulation. We introduce curl $\vec{F}$, a vector with the following properties:

The **curl** of a vector field $\vec{F}$ at a point P is the unique vector curl $\vec{F}(P)$ such that

$$\text{circ}_{\vec{n}} \vec{F}(P) = (\text{curl } \vec{F}(P)) \cdot \vec{n}$$

for every unit vector $\vec{n}$.

Although we haven't proved it, we will assume that curl $\vec{F}$ exists and is unique if $\vec{F}$ is sufficiently well-behaved—for example, if the components of $\vec{F}$ have continuous partial derivatives.

Observe that a vector field has a curl at every point, and hence the curl of a vector field is itself a vector field. A vector field $\vec{F}$ is said to be *curl-free* or *irrotational* if curl $\vec{F}(P) = \vec{0}$ at every point P.

Example 6   Compute the curl of the following vector fields.

    (a) $\vec{F} = 3\vec{i} + 4\vec{j} + 5\vec{k}$ (b) $\vec{G} = ax\vec{i} + by\vec{j} + cz\vec{k}$ (c) $\vec{H} = -y\vec{i} + x\vec{j}$

Solution   (a)   By the preceding example, $\text{circ}_{\vec{n}} \vec{F}(P) = 0 = \vec{0} \cdot \vec{n}$ for every unit vector $\vec{n}$. It follows that curl $\vec{F} = \vec{0}$, the constant zero vector field. Thus, $\vec{F}$ is irrotational.

          (b)   As in part (a), curl $\vec{G} = \vec{0}$, so $\vec{G}$ is irrotational.

          (c)   Since, by the preceding example, $\text{circ}_{\vec{n}} \vec{H}(P) = 2\vec{k} \cdot \vec{n}$ for every unit vector $\vec{n}$ and every point P, we conclude that curl $\vec{H} = 2\vec{k}$.

What Does the Curl Tell Us?

The most important information you can get from the curl is whether or not the original vector field generates a rotation. Suppose $\vec{F}$ represents a velocity vector field of a fluid. Then if curl $\vec{F}$ is nonzero, an object sitting in the fluid will rotate.

Here is a way to visualize curl $\vec{F}$. Suppose that curl $\vec{F}(P) = \vec{v}$. Then for a unit vector $\vec{n}$,

$$\text{circ}_{\vec{n}} \vec{F}(P) = \vec{v} \cdot \vec{n} = \|\vec{v}\| \|\vec{n}\| \cos\theta = \|\vec{v}\| \cos\theta$$

where θ is the angle between $\vec{v}$ and $\vec{n}$. The directional circulation will have its maximum value when $\cos\theta = 1$, that is when $\vec{n}$ is in the direction of $\vec{v}$, the curl; the maximum value of the directional circulation is $\|\vec{v}\|$, the magnitude of the curl. Thus, the curl of a vector field at a point can be described geometrically as follows:

- The direction of the curl vector is the direction in which the directional circulation at the point is the greatest.
- The magnitude of the curl vector equals the directional circulation in that direction.

Notice the analogy with the gradient vector. We defined grad f at a point to be the vector pointing in the direction of maximal rate of increase of f, and whose magnitude is the rate of change of f in that direction.

Example 7   Which of the vector fields in Figure 18.73 appear to have nonzero curl? What is the sign of the z-component of the curl?

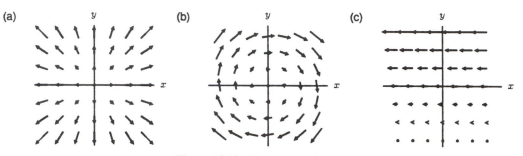

Figure 18.73: Three vector fields

Solution   (a)   This vector field shows no rotation, and the circulation around any closed curve looks like it will be zero, so we suspect a zero curl here.

(b)   This vector field definitely looks like it is swirling, so we expect a nonzero curl here. By the right-hand rule, the direction of the curl will be along the negative z-axis, so the z-component of the curl will be negative.

(c)   At first glance, you might expect this field to have zero curl, as all the vectors are parallel to the x-axis. However, if you find the circulation around the curve C in Figure 18.74, the sides contribute nothing (they are perpendicular to the vector field), the bottom contributes a negative quantity (the curve is in the opposite direction to the vector field), and the top contributes a larger positive quantity (the curve is in the same direction as the vector field and the magnitude of the vector field is larger at the top than at the bottom). Thus, the circulation around C is positive and hence the curl must be nonzero and have a positive z-component.

Another way to see that the curl is nonzero in this case is to imagine the vector field representing the velocity of moving water. A boat sitting in the water will tend to rotate, as the water will be moving faster on one side than the other.

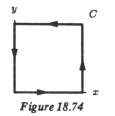

Figure 18.74

How to Calculate the Curl in Cartesian Coordinates

Suppose we have a vector field $\vec{F}$ in 3-space and want to calculate the components of the curl in Cartesian coordinates at some point $\vec{r}_0 = x_0\vec{i} + y_0\vec{j} + z_0\vec{k}$. The formula

$$\text{curl}\,\vec{F} = ((\text{curl}\,\vec{F})\cdot\vec{i}\,)\vec{i} + ((\text{curl}\,\vec{F})\cdot\vec{j}\,)\vec{j} + ((\text{curl}\,\vec{F})\cdot\vec{k}\,)\vec{k}$$

shows that we must compute the three dot products $(\text{curl}\,\vec{F})\cdot\vec{i}$, and $(\text{curl}\,\vec{F})\cdot\vec{j}$ and $(\text{curl}\,\vec{F})\cdot\vec{k}$, which are three directional circulations of $\vec{F}$. First we will calculate the z-component of the curl,

$$(\text{curl}\,\vec{F})\cdot\vec{k} = \text{circ}_{\vec{k}}\,\vec{F} = \lim_{\rightarrow 0}\left(\frac{\int_C \vec{F}\cdot d\vec{r}}{\text{Area inside } C}\right).$$

In this definition, C is a closed curve around the point with position vector $\vec{r}_0$, and in the plane perpendicular to the vector $\vec{k}$. We take C to be a rectangular curve parallel to the xy-plane, with one corner at the point (x_0, y_0, z_0) and edges of length Δx and Δy parallel to the axes. See Figure 18.75. (The fact that the point $\vec{r}_0$ is actually on C rather than enclosed by it turns out not to matter.)

We need to calculate the circulation $\int_C \vec{F}\cdot d\vec{r}$, where $\vec{F} = F_1\vec{i} + F_2\vec{j} + F_3\vec{k}$. Break C into C_1, C_2, C_3, C_4 (see Figure 18.76), and start by calculating the line integrals along C_1 and C_3, where $\Delta\vec{r}$ is parallel to the x-axis, so

$$\Delta\vec{r} = \Delta x\vec{i}.$$

Thus, on C_1 and C_3,

$$\vec{F}\cdot\Delta\vec{r} = (F_1\vec{i} + F_2\vec{j} + F_3\vec{k})\cdot\Delta x\vec{i} = F_1\Delta x.$$

However, the function F_1 is evaluated at (x, y_0, z_0) on C_1 and at $(x, y_0 + \Delta y, z_0)$ on C_3. (In each case, $x_0 \le x \le x_0 + \Delta x$.) By local linearity

$$F_1(x, y_0 + \Delta y, z_0) \approx F_1(x, y_0, z_0) + \frac{\partial F_1}{\partial y}\,\Delta y,$$

so

$$\int_{C_1} \vec{F}\cdot d\vec{r} + \int_{C_3} \vec{F}\cdot d\vec{r} = \int_{C_1} F_1(x, y_0, z_0)\,dx + \int_{C_3} F_1(x, y_0 + \Delta y, z_0)\,dx$$

$$\approx \int_{x=x_0}^{x_0+\Delta x} F_1(x, y_0, z_0)\,dx + \int_{x_0+\Delta x}^{x=x_0}\left(F_1(x, y_0, z_0) + \frac{\partial F_1}{\partial y}\,\Delta y\right)\,dx$$

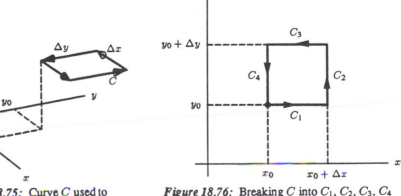

Figure 18.75: Curve C used to calculate $(\text{curl}\,\vec{F})\cdot\vec{k}$

Figure 18.76: Breaking C into C_1, C_2, C_3, C_4

$$= \int_{x=x_0}^{x_0+\Delta x} F_1(x, y_0, z_0)\, dx - \int_{x=x_0}^{x_0+\Delta x} \left(F_1(x, y_0, z_0) + \frac{\partial F_1}{\partial y}\, \Delta y \right) dx$$

$$= - \int_{x=x_0}^{x_0+\Delta x} \left(\frac{\partial F_1}{\partial y}\, \Delta y \right) dx.$$

Thus, we have to integrate the function $\left(\dfrac{\partial F_1}{\partial y} \right) \Delta y$ over the interval $[x_0, x_0 + \Delta x]$. Over this interval, the function $\left(\dfrac{\partial F_1}{\partial y} \right) \Delta y$ is approximately constant, so

$$- \int_{x=x_0}^{x_0+\Delta x} \left(\frac{\partial F_1}{\partial y}\, \Delta y \right) dx \approx - \left(\frac{\partial F_1}{\partial y}\, \Delta y \right) \int_{x=x_0}^{x_0+\Delta x} dx = - \frac{\partial F_1}{\partial y}\, \Delta y\, \Delta x.$$

Similarly, on C_2 and C_4, we have $\Delta \vec{r} = \Delta y \vec{j}$ and so

$$\vec{F} \cdot \Delta \vec{r} = (F_1 \vec{i} + F_2 \vec{j} + F_3 \vec{k}) \cdot \Delta y \vec{j} = F_2\, \Delta y.$$

Now, on C_2 we evaluate F_2 at $(x_0 + \Delta x, y, z_0)$, while on C_4 we evaluate F_2 at (x_0, y, z_0), and

$$F_2(x_0 + \Delta x, y, z_0) \approx F_2(x_0, y, z_0) + \frac{\partial F_2}{\partial x}\, \Delta x.$$

Thus,

$$\int_{C_2} \vec{F} \cdot d\vec{r} + \int_{C_4} \vec{F} \cdot d\vec{r} = \int_{y=y_0}^{y_0+\Delta y} F_2(x_0 + \Delta x, y, z_0)\, dy + \int_{y_0+\Delta y}^{y=y_0} F_2(x_0, y, z_0)\, dy$$

$$\approx \int_{y=y_0}^{y_0+\Delta y} \left(F_2(x_0, y, z_0) + \frac{\partial F_2}{\partial x}\, \Delta x \right) dy - \int_{y=y_0}^{y_0+\Delta y} F_2(x_0, y, z_0)\, dy$$

$$= \int_{y=y_0}^{y_0+\Delta y} \left(\frac{\partial F_2}{\partial x}\, \Delta x \right) dy$$

$$\approx \left(\frac{\partial F_2}{\partial x}\, \Delta x \right) \int_{y=y_0}^{y_0+\Delta y} dy$$

$$= \frac{\partial F_2}{\partial x}\, \Delta x\, \Delta y.$$

Combining these results we get that the circulation is

$$\int_C \vec{F} \cdot d\vec{r} = \int_{C_1} \vec{F} \cdot d\vec{r} + \int_{C_2} \vec{F} \cdot d\vec{r} + \int_{C_3} \vec{F} \cdot d\vec{r} + \int_{C_4} \vec{F} \cdot d\vec{r}$$

$$\approx \frac{\partial F_2}{\partial x}\, \Delta x\, \Delta y - \frac{\partial F_1}{\partial y}\, \Delta y\, \Delta x.$$

Now the area inside C is $\Delta x\, \Delta y$, so the third component of the curl is given by

$$(\operatorname{curl} \vec{F}) \cdot \vec{k} = \lim_{\text{Area} \to 0} \left(\frac{\int_C \vec{F} \cdot d\vec{r}}{\text{Area enclosed by } C} \right)$$

$$= \lim_{\Delta x\, \Delta y \to 0} \frac{\frac{\partial F_2}{\partial x}\, \Delta x\, \Delta y - \frac{\partial F_1}{\partial y}\, \Delta y\, \Delta x}{\Delta x\, \Delta y}$$

$$= \frac{\partial F_2}{\partial x} - \frac{\partial F_1}{\partial y}.$$

The other two components of the curl can be computed in a similar manner, giving the following formula:

If $\vec{F} = F_1\vec{i} + F_2\vec{j} + F_3\vec{k}$, then

$$\text{curl}\, \vec{F} = \left(\frac{\partial F_3}{\partial y} - \frac{\partial F_2}{\partial z}\right)\vec{i} + \left(\frac{\partial F_1}{\partial z} - \frac{\partial F_3}{\partial x}\right)\vec{j} + \left(\frac{\partial F_2}{\partial x} - \frac{\partial F_1}{\partial y}\right)\vec{k}$$

Shorthand Notation for Curl

Using $\nabla = \dfrac{\partial}{\partial x}\vec{i} + \dfrac{\partial}{\partial y}\vec{j} + \dfrac{\partial}{\partial z}\vec{k}$, we can write

$$\text{curl}\, \vec{F} = \nabla \times \vec{F} = \begin{vmatrix} \vec{i} & \vec{j} & \vec{k} \\ \frac{\partial}{\partial x} & \frac{\partial}{\partial y} & \frac{\partial}{\partial z} \\ F_1 & F_2 & F_3 \end{vmatrix}.$$

Example 8   The following formulas are for the three vector fields in Figure 18.73, Example 7.
  (a) $\vec{F} = \vec{r} = x\vec{i} + y\vec{j} + z\vec{k}$ (b) $\vec{F} = y\vec{i} - x\vec{j}$ (c) $\vec{F} = -(y+1)\vec{i}$
  Compute curl $\vec{F}$ for each field and compare your answer with the qualitative answer given in Example 7.

Solution   (a) $\text{curl}\, \vec{F} = \begin{vmatrix} \vec{i} & \vec{j} & \vec{k} \\ \frac{\partial}{\partial x} & \frac{\partial}{\partial y} & \frac{\partial}{\partial z} \\ x & y & z \end{vmatrix} = \left(\frac{\partial z}{\partial y} - \frac{\partial y}{\partial z}\right)\vec{i} + \left(\frac{\partial x}{\partial z} - \frac{\partial z}{\partial x}\right)\vec{j} + \left(\frac{\partial y}{\partial x} - \frac{\partial x}{\partial y}\right)\vec{k} = \vec{0}$, as expected.

  (b) $\text{curl}\, \vec{F} = \begin{vmatrix} \vec{i} & \vec{j} & \vec{k} \\ \frac{\partial}{\partial x} & \frac{\partial}{\partial y} & \frac{\partial}{\partial z} \\ y & -x & 0 \end{vmatrix} = -2\vec{k}$. Notice that curl $\vec{F}$ is parallel to the z-axis — as you would expect from the fact that the z-axis is the axis of rotation — and the component in the z-direction is negative, as predicted on page 424.

  (c) $\text{curl}\, \vec{F} = \begin{vmatrix} \vec{i} & \vec{j} & \vec{k} \\ \frac{\partial}{\partial x} & \frac{\partial}{\partial y} & \frac{\partial}{\partial z} \\ -(y+1) & 0 & 0 \end{vmatrix} = -\frac{\partial}{\partial y}\left(-(y+1)\right)\vec{k} = \vec{k}$.

  On page 424 we argued that curl $\vec{F}$ would be nonzero with a positive z-component, as has turned out to be the case.

Example 9   A flywheel is rotating with angular velocity $\vec{\omega}$, and the velocity of a point P is given by $\vec{v} = \vec{\omega} \times \vec{r}$. See Figure 18.77. Find curl $\vec{v}$.

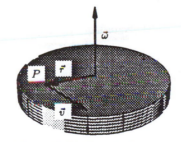

Figure 18.77: Rotating flywheel

Solution   If $\vec{\omega} = \omega_1 \vec{i} + \omega_2 \vec{j} + \omega_3 \vec{k}$, we have

$$\vec{v} = \vec{\omega} \times \vec{r} = \begin{vmatrix} \vec{i} & \vec{j} & \vec{k} \\ \omega_1 & \omega_2 & \omega_3 \\ x & y & z \end{vmatrix} = (\omega_2 z - \omega_3 y)\vec{i} + (\omega_3 x - \omega_1 z)\vec{j} + (\omega_1 y - \omega_2 x)\vec{k}.$$

Thus,

$$
\begin{aligned}
\text{curl}\,\vec{v} &= \begin{vmatrix} \vec{i} & \vec{j} & \vec{k} \\ \frac{\partial}{\partial x} & \frac{\partial}{\partial y} & \frac{\partial}{\partial z} \\ \omega_2 z - \omega_3 y & \omega_3 x - \omega_1 z & \omega_1 y - \omega_2 x \end{vmatrix} \\
&= \left(\frac{\partial}{\partial y}(\omega_1 y - \omega_2 x) - \frac{\partial}{\partial z}(\omega_3 x - \omega_1 z) \right)\vec{i} + \left(\frac{\partial}{\partial z}(\omega_2 z - \omega_3 y) - \frac{\partial}{\partial x}(\omega_1 y - \omega_2 x) \right)\vec{j} \\
&\quad + \left(\frac{\partial}{\partial x}(\omega_3 x - \omega_1 z) - \frac{\partial}{\partial y}(\omega_2 z - \omega_3 y) \right)\vec{k} \\
&= 2\omega_1\vec{i} + 2\omega_2\vec{j} + 2\omega_3\vec{k} = 2\vec{\omega}.
\end{aligned}
$$

Thus, as we would expect, curl $\vec{v}$ is parallel to the axis of rotation of the flywheel (namely, the direction of $\vec{\omega}$) and the magnitude of curl $\vec{v}$ is larger the faster the flywheel is rotating (that is, the larger the magnitude of $\vec{\omega}$).

Problems for Section 18.5

1. Decide whether each of the following vector fields has a nonzero curl. In each case, a cross-section of the vector field is shown; assume that the cross-section is the same in all other planes parallel to the given cross-section.

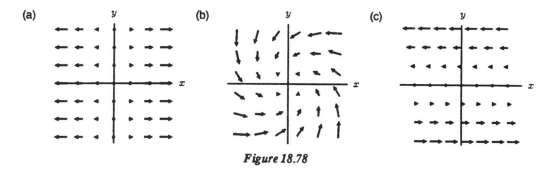

Figure 18.78

Compute the curl of the vector fields in Problems 2–8.

2. $\vec{F}(x, y, z) = (x^2 - y^2)\vec{i} + 2xy\vec{j}$

3. $\vec{F}(\vec{r}) = \vec{r}/r$

4. $\vec{F}(x, y, z) = (-x + y)\vec{i} + (y + z)\vec{j} + (-z + x)\vec{k}$

5. $\vec{F} = x^2\vec{i} + y^3\vec{j} + z^4\vec{k}$.

6. $\vec{F} = 2yz\vec{i} + 3xz\vec{j} + 7xy\vec{k}$.

7. $\vec{F} = e^x\vec{i} + \cos y\vec{j} + e^{z^2}\vec{k}$.

8. $\vec{F} = (x + yz)\vec{i} + (y^2 + xzy)\vec{j} + (zx^3y^2 + x^7y^6)\vec{k}$.

9. Using your answers to Problems 5–7, make a conjecture about the value of curl $\vec{F}$ when the vector field $\vec{F}$ is a certain form. (What form?) Show why your conjecture is true.

10. A large fire becomes a fire-storm when the nearby air acquires a circulatory motion. (This motion has the effect of bringing more air to the fire, causing it to burn faster.) Records show that a fire-storm developed during the Chicago Fire of 1871 and during the Second World War bombing of Hamburg, Germany, but there was no fire-storm during the Great Fire of London in 1666. Explain how a fire-storm could be identified using the curl of a vector field.

11. Sketch the curves you would use to calculate the x-and y- components of curl $\vec{F}$ in Cartesian coordinates. (Note: Figure 18.76 on page 425 contains the curve used to find the z-component.)

12. Derive the expressions for the x- and y-components of curl $\vec{F}$, by a method similar to that used in this section.

13. Assume f is a scalar function with continuous second partial derivatives. Use the Fundamental Theorem of Calculus for Line Integrals to show that $\int_C \operatorname{grad} f \cdot d\vec{r} = 0$ for any closed path C. Deduce that curl grad $f = \vec{0}$.

14. If $\vec{F}$ is any vector field whose components have continuous second partial derivatives, show that div curl $\vec{F} = \vec{0}$.

15. For any constant vector field, $\vec{c}$, and any vector field, $\vec{F}$, show that $\operatorname{div}(\vec{F} \times \vec{c}) = \vec{c} \cdot \operatorname{curl} \vec{F}$.

16. Show that curl $(\vec{F} + \vec{C}) = \operatorname{curl}(\vec{F})$ for a constant vector field $\vec{C}$.

17. Show that curl $(\phi\vec{F}) = \phi(\operatorname{curl} \vec{F}) + (\operatorname{grad} \phi) \times \vec{F}$ where ϕ is a function and $\vec{F}$ is a vector field.

18. A vortex that rotates at constant angular velocity ω about the z-axis has velocity vector field $\vec{v} = \omega(-y\vec{i} + x\vec{j})$.

    (a) Sketch the vector field if $\omega = 1$.
    (b) Sketch the vector field if $\omega = -1$.
    (c) Determine the speed $\|\vec{v}\|$ of the vortex as a function of the distance from its center.
    (d) Compute div $\vec{v}$.
    (e) Compute curl $\vec{v}$.
    (f) Compute the circulation of $\vec{v}$ counterclockwise about the circle of radius R in the xy-plane, centered at the origin.

Problems 19–21 concern the vector fields in Figure 18.79.

19. Three of the vector fields have zero curl. Which are they? How do you know?

20. Three of the vector fields have zero divergence. Which are they? How do you know?

21. Four of the line integrals $\int_{C_i} \vec{F} \cdot d\vec{r}$ are zero. Which are they? How do you know?

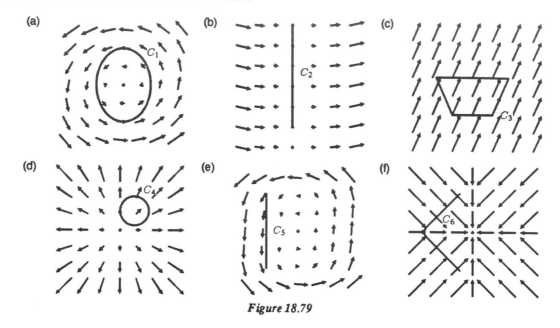

Figure 18.79

22. If $\vec{H} = -y\vec{i} + x\vec{j}$, we showed in Example 3 on page 422 that if P is at the origin, then $\text{circ}_{\vec{n}}\, \vec{H}\,(P) = 2\vec{k}\cdot\vec{n}$ for any $\vec{n}$. Use this to show that this result is true at any point P.

23. Show that if ϕ is a harmonic function, then $\text{grad}\,\phi$ is both curl free and divergence free.

24. It is a theorem of Helmholtz that every vector field $\vec{F}$ equals the sum of a curl free vector field and a divergence free vector field. Show how to do this, assuming that there is a function ϕ such that $\nabla^2\phi = \text{div}\vec{F}$.

25. Express $(3x + 2y)\vec{i} + (4x + 9y)\vec{j}$ as the sum of a curl-free vector field and a divergence-free vector field.

26. Find a vector field $\vec{F}$ such that $\text{curl}\,\vec{F} = 2\vec{i} - 3\vec{j} + 4\vec{k}$. [Hint: Try $\vec{F} = \vec{v}\times\vec{r}$ for a suitable vector $\vec{v}$.]

27. Figure 18.80 gives a sketch of a velocity vector field $\vec{F} = y\vec{i} + x\vec{j}$ in the xy-plane.
    (a) What is the direction of rotation of a thin twig placed at the origin along the x-axis?
    (b) What is the direction of rotation of a thin twig placed at the origin along the y-axis?
    (c) Compute curl $\vec{F}$.

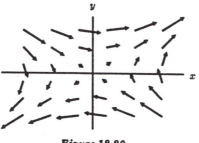

Figure 18.80

28. A radial force field is a vector field of the form $\vec{F} = f(r)\vec{r}$ where f is any function of $r = \|\vec{r}\|$. Show that any radial force field is irrotational.

18.6 STOKES' THEOREM

In Section 18.4, we saw that the flux of a vector field out of a closed surface equals the integral of the divergence of the vector field over the volume enclosed by the surface. This is because the flux integral can be interpreted as a total outflow from the volume, and the divergence at a point is the outflow per unit volume at that point.

In a similar way, we will now find the circulation of a vector field $\vec{F}$ around a closed curve C by computing the flux integral of the curl of the vector field over an oriented surface S bounded by the curve. This works because the normal component of the curl of $\vec{F}$ at a point of the surface equals the circulation of $\vec{F}$ per unit area around small loops in the surface about the point.

Choose an orientation of the curve C, and then choose the orientation of S that is determined from the orientation of C by the right-hand rule. Break the surface S up into many small pieces, as in Figure 18.81. We make certain that the pieces are small enough so that they are each approximately flat. Then we compute the circulation of the vector field $\vec{F}$ around the boundary of each piece. If $\vec{n}$ is an orienting unit vector that is normal to a piece of surface with area ΔA, then the vector $\vec{n}\,\Delta A$ can be thought of as approximately the vector area $\Delta\vec{A}$ of the piece. If the piece is small enough, then

$$\begin{matrix}\text{Circulation of }\vec{F}\text{ around} \\ \text{boundary of the piece}\end{matrix} \approx (\text{circ}_{\vec{n}}\,\vec{F}\,)\Delta A = ((\text{curl}\,\vec{F}\,)\cdot\vec{n}\,)\Delta A = (\text{curl}\vec{F}\,)\cdot\Delta\vec{A}\,.$$

The next step is to add up the circulations around all the small pieces. When we do this, the line integral along the common edge of a pair of adjacent pieces gets counted twice, once each way. See Figure 18.82. Hence, these integrals cancel. All that is left is the line integrals around the outside edges. As we continue to add up all the circulations around all the small pieces, the line integrals along inside edges cancel, and we are left with the circulation around the boundary of the entire surface. Thus,

$$\begin{matrix}\text{Circulation around} \\ \text{boundary of }S\end{matrix} = \sum \begin{matrix}\text{Circulation around} \\ \text{boundary of pieces}\end{matrix} \approx \sum \text{curl}\,\vec{F}\,\cdot\Delta\vec{A}\,.$$

Taking the limit as $\Delta\vec{A}\to 0$, we get

$$\text{Circulation around boundary of }S = \int_S \text{curl}\,\vec{F}\,\cdot d\vec{A}\,.$$

Figure 18.81: The circulations of the pieces add up to the circulation of the whole

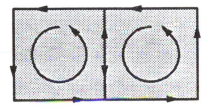

Figure 18.82: Two adjacent pieces of the surface

Since the circulation is also the line integral of $\vec{F}$ around C, the boundary of S, we have following result:

Stokes' Theorem

If $\vec{F}$ is a vector field, and S is an oriented surface with boundary C

$$\text{Circulation around boundary of } S = \int_C \vec{F} \cdot d\vec{r} = \int_S \text{curl } \vec{F} \cdot d\vec{A}.$$

The orientation of S must be determined from the orientation of C by the right-hand rule. curl $\vec{F}$ must be defined at every point of S.

Example 1   Suppose that at a distance r from the z-axis the vector field $\vec{F}$ in Figure 18.83 has magnitude $2r$ in the direction shown. Use Stokes' Theorem to find the circulation of $\vec{F}$ around the circle C of radius R in the xy-plane and centered at the origin. We take the orientation of C to be counterclockwise as viewed from above the xy-plane.

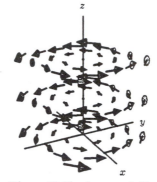

Figure 18.83: A vector field

Solution   First we will compute the line integral of $\vec{F}$ around C. Since $\vec{F}$ is tangent to the circle and $\|\vec{F}\| = 2R$ on C, we have

$$\int_C \vec{F} \cdot d\vec{r} = \|\vec{F}\| \times \text{Length of} C = 2R(2\pi R) = 4\pi R^2.$$

Now we will compute the flux integral of the curl. A formula for $\vec{F}$ is

$$\vec{F}(x, y, z) = -2y\vec{i} + 2x\vec{j}.$$

Computing the curl gives

$$\text{curl } \vec{F}(x, y, z) = 4\vec{k}.$$

Let S be the region in the xy-plane enclosed by the circle. The orientation of S that is determined by that of C from the right hand rule is upward. The curl of $\vec{F}$ is in the same direction as the positive normal vector to the surface, so by Stokes' Theorem,

$$\text{Circulation around } C = \int_S \text{curl } \vec{F} \cdot d\vec{A} = \int_S 4\vec{k} \cdot d\vec{A} = 4(\text{Area of } S) = 4\pi R^2.$$

This is the same answer we obtained before.

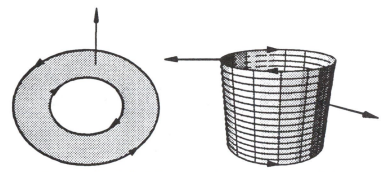

Figure 18.84: Stokes' Theorem applies to a ring and a cylinder

It should be noted that Stokes' Theorem applies to any oriented surface S and its boundary C, even in cases where the boundary may consist of two or more curves. For example,

$$\int_C \vec{F} \cdot d\vec{r} = \int_S \operatorname{curl} \vec{F} \cdot d\vec{A}$$

where S is a ring or a cylinder and C consists of two curves as in Figure 18.84. In such a situation, it is best to orient the surface S first, and then each of the boundary curves, because the right hand relationship must apply to the orientation of the surface and each of the boundary curves.

Curl Fields

Recall that if a vector field $\vec{F}$ is a gradient field, that is $\vec{F} = \operatorname{grad} f$, then $\vec{F}$ has some particularly nice properties. (The line integral $\int_C \vec{F} \cdot d\vec{r}$ depends only on the endpoints of the curve C, and if C is a closed curve then $\int_C \vec{F} \cdot d\vec{r} = 0$.) We will now define *curl fields*, and consider their properties.

> $\vec{F}$ is a **curl field** if $\vec{F} = \operatorname{curl} \vec{G}$ for some vector field $\vec{G}$.

Example 2   Suppose that a vector field $\vec{F}$ equals the curl of another vector field $\vec{G}$.
  (a) Suppose that S_1 and S_2 are two surfaces with the same boundary curve C, both oriented by the right-hand rule with respect to the same orientation of C. Show that the flux integrals of $\vec{F}$ over the two surfaces are equal:

$$\int_{S_1} \vec{F} \cdot d\vec{A} = \int_{S_2} \vec{F} \cdot d\vec{A}.$$

Thus, the flux integral of a curl field over a surface depends only on the boundary of the surface. See Figure 18.85.

  (b) If S is a closed surface, then $\int_S \vec{F} \cdot d\vec{A} = 0$. Thus, the flux integral of a curl field over a closed surface equals zero.

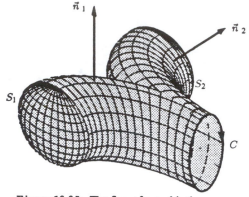

Figure 18.85: The flux of a curl is the same through these two surfaces S_1 and S_2 which have the same boundary, C

Solution   (a)   By Stokes' Theorem,

$$\int_{S_1} \vec{F} \cdot d\vec{A} = \int_{S_1} \operatorname{curl} \vec{G} \cdot d\vec{A} = \int_C \vec{G} \cdot d\vec{r}$$

and

$$\int_{S_2} \vec{F} \cdot d\vec{A} = \int_{S_2} \operatorname{curl} \vec{G} \cdot d\vec{A} = \int_C \vec{G} \cdot d\vec{r}.$$

Thus,

$$\int_{S_1} \vec{F} \cdot d\vec{A} = \int_{S_2} \vec{F} \cdot d\vec{A}$$

(b)   Draw an oriented closed curve C on the surface S, thus dividing S into two oriented surfaces S_1 and S_2 as shown in Figure 18.86.
By part (a) of this Example,

$$\int_{S_1} \vec{F} \cdot d\vec{A} = \int_{S_2} \vec{F} \cdot d\vec{A}.$$

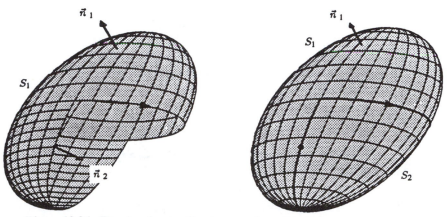

Figure 18.86: The closed curve C and two surfaces S_1 and S_2 which make up S

On the other hand,

$$\int_S \vec{F} \cdot d\vec{A} = \int_{S_1} \vec{F} \cdot d\vec{A} - \int_{S_2} \vec{F} \cdot d\vec{A} = 0$$

where we have subtracted the integral over S_2 because we always choose the outward orientation for a flux integral over a closed surface S, and that is opposite to the orientation of the surface S_2 induced by the boundary C.

Example 3   Evaluate $\int_S \vec{F} \cdot d\vec{A}$ where $\vec{F} = (8yz - z)\vec{j} + (3 - 4z^2)\vec{k}$, and S is the hemisphere of radius 5 shown in Figure 18.87, oriented upwards. Use the fact that $\vec{F} = \text{curl}(4yz^2\vec{i} + 3x\vec{j} + xz\vec{k})$ to simplify the computation.

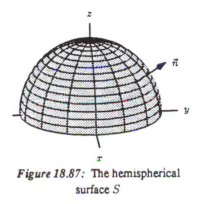

Figure 18.87: The hemispherical
surface S

Solution   Because $\vec{F}$ is a curl field, its flux through S will be the same as its flux through any surface with the same boundary as S. We are free to replace S by another surface with the same boundary as S if it seems convenient. Let us in fact choose the circular disc S_1 of radius 5 in the xy-plane. On the surface S_1, $z = 0$, and so the vector field reduces to $\vec{F} = 3\vec{k}$. Since the vector $3\vec{k}$ is normal to S_1 and in the direction of the orientation of S_1, we have

$$\int_S \vec{F} \cdot d\vec{A} = \int_{S_1} \vec{F} \cdot d\vec{A} = \int_{S_1} 3\vec{k} \cdot d\vec{A} = \|3\vec{k}\|(\text{area of } S_1) = 75\pi.$$

Recall that a function f such that $\vec{F} = \text{grad} f$ is called a potential function for the vector field $\vec{F}$. By analogy, if a vector field $\vec{F}$ is a curl field, then any vector field $\vec{G}$ such that $\vec{F} = \text{curl} \vec{G}$ is called a *vector potential* for $\vec{F}$. In Example 3, $4yz^2\vec{i} + 3x\vec{j} + xz\vec{k}$ was a vector potential for $\vec{F} = (8yz - z)\vec{j} + (3 - 4z^2)\vec{k}$. Only curl fields have a vector potential, and the vector potential of a given curl field is not unique.

How can we tell whether a given vector field is a curl field? It is shown in the problems that $\text{div} \vec{F} = 0$ for every curl field $\vec{F}$, so that if $\text{div} \vec{F} \neq 0$, then the vector field $\vec{F}$ is definitely not a curl field and does not have a vector potential. Conversely, it turns out (though we will not prove it), that if $\text{div} \vec{F} = 0$ at every point of a nice region, such as a ball or a box, then the vector field $\vec{F}$ is a curl field in that region.

Example 4   Which of the following vector fields is a curl field?

(a) $\vec{F} = xy\vec{i} - 2yz\vec{j} + (z^2 - yz)\vec{k}$ (b) $\vec{G} = xy\vec{i} + z^2\vec{j} - 3xz\vec{k}$.

Solution   (a) Since $\text{div}\vec{F} = 0$, we know that $\vec{F}$ is a curl field. In fact, $\vec{F} = \text{curl}\,\vec{H}$, where $\vec{H} = -yz^2\vec{i} - xyz\vec{j}$, so $\vec{H}$ is a vector potential for $\vec{F}$.

(b) Since $\text{div}\vec{G} = y - 3x \neq 0$, the vector field $\vec{G}$ is not a curl field.

Curl Free Vector Fields

We have seen in Section 18.4, Problem 13 that curl $\vec{F} = \vec{0}$ for every gradient vector field $\vec{F}$. What about the converse? What can be said about vector fields $\vec{F}$ such that curl $\vec{F} = \vec{0}$?

Suppose that curl $\vec{F} = \vec{0}$ and let us consider the line integral $\int_C \vec{F} \cdot d\vec{A}$ for a closed curve C in the region where $\vec{F}$ is defined. If C is the boundary curve of an orientable surface S that lies wholly in the domain of curl $\vec{F}$, then Stokes' Theorem asserts that

$$\int_C \vec{F} \cdot d\vec{r} = \int_S \text{curl}\,\vec{F} \cdot d\vec{A} = \int_S \vec{0} \cdot d\vec{A} = 0.$$

If $\int_C \vec{F} \cdot d\vec{r} = 0$ for every closed curve C, then we know that $\vec{F}$ is a gradient field. Thus, we have

> ### The Curl Criterion for a Gradient Vector Field
>
> If curl $\vec{F}$ exists and equals zero wherever $\vec{F}$ is defined, and if every closed curve in the domain of $\vec{F}$ bounds an orientable surface that also lies in the domain of $\vec{F}$, then $\vec{F}$ is a gradient field.

It can be shown that every closed curve is the boundary of some orientable surface. Thus, we get the following result from the curl criterion for a gradient field:

> If a vector field $\vec{F}$ is defined everywhere and its curl is zero everywhere, then $\vec{F}$ is a gradient field.

Example 5   Which of the following vector fields is a gradient field?

(a) $\vec{F} = yz\vec{i} + (xz + z^2)\vec{j} + (xy + 2yz)\vec{k}$ (b) $\vec{G} = -y\vec{i} + x\vec{j}$.

Solution   (a) Since curl $\vec{F} = \vec{0}$, we know that $\vec{F}$ is a gradient field. In fact, $\vec{F} = \text{grad}\,f$, where $f(x, y, z) = xyz + yz^2$, so f is a potential function for $\vec{F}$.

(b) Since curl $\vec{G} = 2\vec{k} \neq \vec{0}$, the vector field $\vec{G}$ is not a gradient field.

Three Operators and Three Integral Theorems

In three-variable calculus we have studied three operators computed in terms of derivatives: grad, curl, and div. Here is a summary of their differences:

- The gradient of a scalar function is a vector field.

- The curl of a vector field is vector field.
- The divergence of a vector field is a scalar function.

There is a fundamental integral theorem for each of these three operators. In all three cases an integral over a region is expressed in terms of the boundary of the region.

Fundamental theorem of line integrals

$$\int_C \text{grad} f \cdot d\vec{r} = f(Q) - f(P)$$

---

expresses the line integral of the gradient of a function f over a (1-dimensional) curve C in terms of the values of f at the boundary (endpoints P and Q) of the curve.

Stokes' Theorem

$$\int_S \text{curl} \, \vec{F} \cdot d\vec{A} = \int_C \vec{F} \cdot d\vec{r}$$

---

expresses the flux integral of the curl of a vector field $\vec{F}$ over a (2-dimensional) surface S in terms of the line integral of $\vec{F}$ over the boundary curve C of the surface.

Divergence Theorem

$$\int_W \text{div} \vec{F} \, dV = \int_S \vec{F} \cdot d\vec{A}$$

---

expresses the integral of the divergence of a vector field $\vec{F}$ over a (3-dimensional) volume W in terms of the flux integral of $\vec{F}$ over the boundary surface S of the volume.

Problems for Section 18.6

In Problems 1-3 compute the given line integral using Stokes' Theorem.

1. $\int_C (-z\vec{i} + y\vec{j} + x\vec{k}) \cdot d\vec{r}$, where C is a circle of radius 2 around the y-axis with orientation indicated in Figure 18.88.

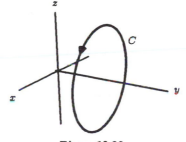

Figure 18.88

2. $\int_C \vec{F} \cdot d\vec{r}$, with $\vec{F} = \vec{r}/r^3$ and C is the path consisting of straight line segments from $(1, 0, 1)$ to $(1, 0, 0)$ to $(0, 0, 1)$ back to $(1, 0, 1)$.

3. Use Stokes' Theorem to evaluate $\int_C \vec{F} \cdot d\vec{r}$ where $\vec{F} = (2x - y)\vec{i} + (x + 4y)\vec{j}$ and C is as given.

   (a) C is the circle of radius 10 in the xy-plane centered at the origin, oriented clockwise as viewed from way up the positive z-axis.

   (b) C is the circle of radius 10 in the yz-plane centered at the origin, oriented clockwise as viewed from way out the positive x-axis.

4. Can you use Stokes' Theorem to compute the line integral $\int_C (2x\vec{i} + 2y\vec{j} + 2z\vec{k}) \cdot d\vec{r}$ where C is the straight line from the point $(1, 2, 3)$ to the point $(4, 5, 6)$? Why or why not?

5. For this exercise, $\vec{F} = -z\vec{j} + y\vec{k}$, C is the circle of radius R in the yz-plane oriented clockwise as viewed from the positive x-axis, and S is the disk in the yz-plane enclosed by C, oriented in the positive x-direction. See Figure 18.89.

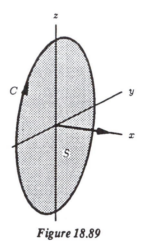

Figure 18.89

   (a)   Evaluate directly $\int_C \vec{F} \cdot d\vec{r}$.

   (b)   Evaluate directly $\int_S \operatorname{curl} \vec{F} \cdot d\vec{A}$.

   (c)   The answers in parts (a) and (b) are not equal. Explain why this does not contradict Stokes' Theorem.

6. Use Stokes' Theorem to show that $\int_S \operatorname{curl} \vec{F} \cdot d\vec{A} = 0$ for any surface S which is the complete boundary surface of a volume V. Deduce that $\operatorname{div} \operatorname{curl} \vec{F} = 0$.

7. A basic property of the magnetic field $\vec{B}$ is that $\operatorname{curl} \vec{B} = 0$ in a region where there are no currents flowing. Consider the magnetic field around a long thin wire carrying a constant current. You are given that the magnitude of the magnetic field depends only on the distance from the wire, and that its direction is always tangent to the circle around the wire traversed in a direction related to the direction of the current by the right hand rule. Use Stokes' Theorem to deduce that the magnitude of the magnetic field is proportional to the reciprocal of the distance from the wire. [Hint: Consider an annulus around the wire. Its boundary has two pieces: the inner circle and the outer circle.]

8. You might guess that the speed of a naturally occurring vortex (tornado, waterspout, whirlpool) is a decreasing function of the distance from its center, so the constant angular velocity model of Problem 18 on page 429 would be inappropriate. A *free vortex* circulating about the z-axis has vector field $\vec{v} = K(x^2 + y^2)^{-1}(-y\vec{i} + x\vec{j})$ where K is a constant.

    (a) Sketch the vector field if $K = 1$.
    (b) Sketch the vector field if $K = -1$.
    (c) Determine the speed $\|\vec{v}\|$ of the vortex as a function of the distance from its center.
    (d) Compute div $\vec{v}$.
    (e) Compute curl $\vec{v}$, thus showing that $\vec{v}$ is irrotational.
    (f) Compute the circulation of $\vec{v}$ counterclockwise about the circle of radius R at the origin.
    (g) The computations in parts (e) and (f) show that $\vec{v}$ is an irrotational vector field with nonzero circulation. Explain why this does not contradict Stokes' Theorem.

9. Is there a vector field $\vec{G}$ such that $\text{curl}\vec{G} = y\vec{i} + x\vec{j}$? How do you know?

10. Determine whether vector potentials for $\vec{F}$ and $\vec{G}$ exist, and if so, find one.

    (a) $\vec{F} = 2x\vec{i} + (3y - z^2)\vec{j} + (x - 5z)\vec{k}$
    (b) $\vec{F} = x^2\vec{i} + y^2\vec{j} + z^2\vec{k}$

11. Use Stokes' Theorem to show that if $u(x, y)$ and $v(x, y)$ are two functions of x and y and C is a closed curve in the xy-plane oriented counterclockwise, then

$$\int_C (u\vec{i} + v\vec{j}) \cdot d\vec{r} = \int_R \left(\frac{\partial v}{\partial x} - \frac{\partial u}{\partial y} \right) dx\,dy$$

where R is the region in the xy-plane enclosed by C. This is known as Green's Theorem.

12. Does Stokes' Theorem apply to the curve C and surface S shown in Figure 18.90. [Hint: Try to use the right-hand rule to orient the surface.]

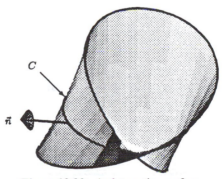

Figure 18.90: An interesting surface

13. Suppose that C is a closed curve in the xy-plane, oriented counterclockwise as viewed from above. Show that $(1/2)\int_C(-y\vec{i} + x\vec{j}) \cdot d\vec{r}$ equals the area of the region R in the xy-plane enclosed by C.

14. Imagine the following vector fields

$$\vec{F} = F_1(x, y)\vec{i} + F_2(x, y)\vec{j} \quad \text{and} \quad \vec{G} = G_1(x, y)\vec{i} + G_2(x, y)\vec{j}$$

sketched in Figures 18.91 and 18.92.

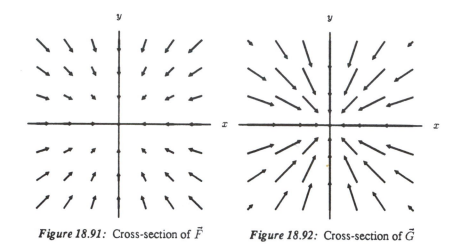

Figure 18.91: Cross-section of $\vec{F}$ *Figure 18.92:* Cross-section of $\vec{G}$

(a) What can you say about div $\vec{F}$ and div $\vec{G}$ at the origin?

(b) What can you say about curl $\vec{F}$ and curl $\vec{G}$ at the origin?

(c) Can you draw a closed surface around the origin such that $\vec{F}$ has a non-zero flux through it?

(d) Repeat part (c) for $\vec{G}$.

(e) Can you draw a closed curve around the origin such that $\vec{F}$ has a non-zero circulation around it?

(f) Repeat part(e) for $\vec{G}$.

15. A vector field $\vec{F}$ is defined everywhere except on the z-axis, and curl $\vec{F} = \vec{0}$ everywhere where $\vec{F}$ is defined. What can you say about $\int_C \vec{F} \cdot d\vec{r}$ if C is a circle of radius 1 in the xy-plane, and if the center of C is at (a) The origin,   (b) The point $(2, 0)$?

16. Compute the line integral $\int_C ((yz^2 - y)\vec{i} + (xz^2 + x)\vec{j} + 2xyz\vec{k}) \cdot d\vec{r}$ where C is the circle of radius 3 in the xy-plane, centered at the origin, oriented counterclockwise as viewed from the positive z-axis. Do it two ways:

(a) Directly and   (b) Using Stokes' Theorem.

REVIEW PROBLEMS FOR CHAPTER EIGHTEEN

1. (a) What is meant by a vector field?

(b) Which of the following are vector fields? Why?

(a) $\vec{r} + \vec{a}$ (b) $\vec{r} \cdot \vec{a}$ (c) $x^2\vec{i} + y^2\vec{j} + z^2\vec{k}$ (d) $x^2 + y^2 + z^2$

where $\vec{r} = x\vec{i} + y\vec{j} + z\vec{k}$ and $\vec{a} = a_1\vec{i} + a_2\vec{j} + a_3\vec{k}$, and where a_1, a_2, a_3 are constant.

2. Evaluate $\int_S \vec{F} \cdot d\vec{A}$, where S is the 2×2 square plate in the yz-plane centered at the origin, oriented in the positive x-direction, and $\vec{F} = (5 + xy)\vec{i} + z\vec{j} + yz\vec{k}$.

3. Can you evaluate the flux integral in Problem 2 on page 441 by application of the Divergence Theorem? Why or why not?

4. Find the flux of the vector field $\vec{F} = x\vec{i} + y\vec{j}$ through the surface of a closed cylinder of radius 2 and height 3 centered on the z-axis with its base in the xy-plane.

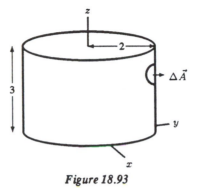

Figure 18.93

5. Find the flux of the vector field $\vec{F} = -y\vec{i} + x\vec{j} + z\vec{k}$ through the surface of a closed cylinder of radius 1 centered on the z-axis with base in the plane $z = -1$ and top in the plane $z = 1$.

6. (a) Find the flux of the vector field $\vec{F} = 2x\vec{i} - 3y\vec{j} + 5z\vec{k}$ through a box of side w with four of its eight corners at the points $(a, b, c), (a + w, b, c), (a, b + w, c), (a, b, c + w)$. See Figure 18.94.

    (b) Use your answer to part (a) to find div $\vec{F}$ at the point (a, b, c).

    (c) Find div $\vec{F}$ using partial derivatives.

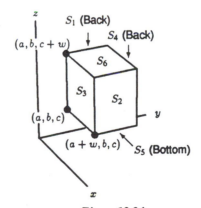

Figure 18.94

7. Suppose $\vec{F} = (3x + 2)\vec{i} + 4x\vec{j} + (5x + 1)\vec{k}$. Use the method of Problem 6 to find div $\vec{F}$ at the point (a, b, c) by two different methods.

8. Compute the flux integral $\int_S (x^3\vec{i} + 2y\vec{j} + 3\vec{k}) \cdot d\vec{A}$, where S is the $2 \times 2 \times 2$ cubical surface centered at the origin, oriented outward. Do this in two ways:

    (a) Directly

    (b) By means of the Divergence Theorem

9. Show that if $g(x, y, z)$ is a scalar valued function and $\vec{F}(x, y, z)$ is a vector field, then

$$\text{div}(g\vec{F}) = (\text{grad } g) \cdot \vec{F} + g \, \text{div} \, \vec{F}.$$

Are the statements in Problems 10–17 true or false? Explain your answer.

10. div $\vec{F}$ is a scalar whose value can vary from point to point.

11. True or false? Explain your answer. If $\int_S \vec{F} \cdot d\vec{A} = 12$ and S is a flat disc of area 4π, then div $\vec{F} = 3/\pi$.

12. True or false? Explain your answer. curl $\vec{F}$ is a vector field.

13. True or false? Explain your answer. If $\vec{F}$ is as shown in Figure 18.95, curl $\vec{F} \cdot \vec{j} > 0$.

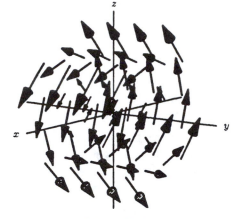

Figure 18.95

14. $\text{div}(\vec{F} + \vec{G}) = \text{div} \, \vec{F} + \text{div} \, \vec{G}$, where $\vec{F} = F_1\vec{i} + F_2\vec{j} + F_3\vec{k}$ and $\vec{G} = G_1\vec{i} + G_2\vec{j} + G_3\vec{k}$ are vector fields in 3-space.

15. $\text{grad}(fg) = (\text{grad } f) \cdot (\text{grad } g)$

16. $\text{grad}(\vec{F} \cdot \vec{G}) = \vec{F}(\text{div} \, \vec{G}) + (\text{div} \, \vec{F})\vec{G}$, where $\vec{F} = F_1\vec{i} + F_2\vec{j} + F_3\vec{k}$ and $\vec{G} = G_1\vec{i} + G_2\vec{j} + G_3\vec{k}$ are vector fields in 3-space.

17. $\text{curl}(f\vec{G}) = (\text{grad } f) \times \vec{G} + f(\text{curl } \vec{G})$, where f is a function of three variables and $\vec{G} = G_1\vec{i} + G_2\vec{j} + G_3\vec{k}$ is a vector field in 3-space.

18. If V is a volume surrounded by a closed surface S, show that

$$\frac{1}{3} \int_S \vec{r} \cdot d\vec{A} = V.$$

19. Use the result of Problem 18 to compute the volume of a sphere given that the surface area of a sphere of radius R is $4\pi R^2$.

20. Use the result of Problem 18 to compute the volume of a cone of base radius b and height h.[Hint: Stand the cone with its point downward and its axis along the positive z-axis.]

21. Due to roadwork ahead, the traffic on a highway slows linearly from 55 miles/hour to 15 miles/hour over a 2000 foot stretch of road, then crawls along at 15 miles/hour for 5000 feet, then speeds back up linearly to 55 miles/hour in the next 1000 feet, after which it moves steadily at 55 miles/hour.

    (a) Sketch a velocity vector field for the traffic flow.
    (b) Write a formula for the velocity vector field $\vec{v}$ (miles/hour) as a function of the distance x feet from the initial point of slowdown. (Take the direction of motion to be $\vec{i}$ and consider the various units separately.)
    (c) Compute div $\vec{v}$ at $x = 1000, 5000, 7500, 10000$. Be sure to include the proper units.

22. The velocity field $\vec{v}$ in Problem 21 on page 443 does not give a complete description of the traffic flow, for it takes no account of the spacing between vehicles. Let ρ be the density (cars/mile) of highway, where we assume that ρ depends only on x.

    (a) Using your highway experience, arrange in ascending order: $\rho(0), \rho(1000), \rho(5000)$.
    (b) What are the units and interpretation of the vector field $\rho\vec{v}$?
    (c) Why would you expect $\rho\vec{v}$ to be constant? What does this mean for div$(\rho\vec{v})$?
    (d) Determine $\rho(x)$ if $\rho(0) = 75$ cars/mile and $\rho\vec{v}$ is constant.
    (e) If the highway has two lanes, find the approximate number of feet between cars at $x = 0, 1000$, and 5000.

23. (a) An idealized model of a river flowing around a rock considers a circular rock of radius 1 in the river, which flows across the xy-plane in the positive x-direction. A simple model begins with the potential function $\phi = x + (x/(x^2 + y^2))$. Compute the velocity vector field, $\vec{v} = \text{grad}\,\phi$.
    (b) Show that div $\vec{v} = 0$.
    (c) Show that the flow of $\vec{v}$ is tangent to the circle $x^2 + y^2 = 1$. This means that no water crosses the circle. The water on the outside must therefore all flow around the circle.
    (d) Use a computer to sketch the vector field $\vec{v}$ in the region outside the unit circle.

24. The relations between the electric field, $\vec{E}$, the magnetic field, $\vec{B}$, the charge density, ρ, and the current density, $\vec{J}$, at a point in space are described by the equations

$$\text{div}\,\vec{E} = 4\pi\rho,$$

$$\text{curl}\,\vec{B} - \frac{1}{c}\frac{\partial\vec{E}}{\partial t} = \frac{4\pi}{c}\vec{J},$$

where c is a constant (the speed of light).

    (a) Using the results of Problem 14 on page 14, show that

$$\frac{\partial\rho}{\partial t} + \text{div}\,\vec{J} = 0.$$

    (b) What does the equation in part (a) say about charge and current density? Explain in intuitive terms why this is reasonable.
    (c) Why do you think the equation in part (a) is called the charge conservation equation?

25. (a) Evaluate

$$\vec{F} = \text{grad}\,\phi + \vec{v} \times \vec{r}$$

where

$$\phi(x, y, z) = \frac{1}{2}(a_1x^2 + b_2y^2 + c_3z^2 + (a_2 + b_1)xy + (a_3 + c_1)xz + (b_3 + c_2)yz)$$

and

$$\vec{v} = \frac{1}{2}((c_2 - b_3)\vec{i} + (a_3 - c_1)\vec{j} + (b_1 - a_2)\vec{k}$$

and

$$\vec{r} = x\vec{i} + y\vec{j} + z\vec{k}$$

Explain why the solution to part (a) shows that every linear vector field can be written in the form grad $\phi + \vec{v} \times \vec{c}$.

(b) Using Problem 13 on page 13, compute $\text{circ}_{\vec{n}}(\vec{F})(P)$ for every unit vector $\vec{n}$.

(c) Using part (b) and the definition of curl, compute the vector field curl $\vec{F}$.

APPENDICES

Appendix A   REVIEW OF POLAR COORDINATES

Polar coordinates are another way of describing points in an xy-plane. The x- and y-coordinates can be thought of as instructions on how to get to the point. To get to the point $(1, 2)$ go 1 unit horizontally and 2 units vertically. Polar coordinates can be thought of the same way. There is an r-coordinate, which tells you how far to go along the ray extending to the point from the origin, and there is the θ-coordinate, which is an angle, and it tells you the angle the ray makes with the positive x-axis. (See Figure A.1.)

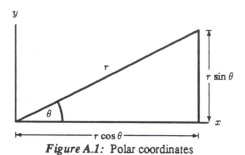

Figure A.1: Polar coordinates

Example 1   Give the polar coordinates of the points $(1, 0)$, $(0, 1)$, $(-1, 0)$, and $(1, 1)$.

Solution   To get to the point $(1, 0)$, you go 1 unit along the horizontal axis, and there you are. So its r-coordinate is 1 and its θ-coordinate is 0, since you go along the x-axis.

The point $(0, 1)$ is also one unit from the origin, so you start out the same way as before, by going 1 unit out along the ray. Then you have to go around the circle of radius 1 through an arc of $\pi/2$ to get to the point $(0, 1)$. So $r = 1$ and $\theta = \pi/2$.

The point $(-1, 0)$ also has r-coordinate equal to 1, but this time you have to go halfway around the circle to get there, so its θ-coordinate is π.

The point $(1, 1)$ is a distance of $\sqrt{2}$ from the origin, and the ray from the origin to it makes an angle of $\pi/4$ with the horizontal ray. Thus, it has r-coordinate equal to $\sqrt{2}$ and θ-coordinate equal to $\pi/4$. See Figure A.2.

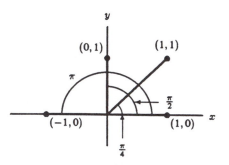

Figure A.2: Four points showing Cartesian
and polar coordinates

Conversion Between Polar and Cartesian Coordinates

Suppose a point has Cartesian coordinates (x, y). Look at Figure A.1. The distance r of the point from the origin is the length of the hypotenuse, which is $\sqrt{x^2 + y^2}$ by Pythagoras' theorem. The angle θ with the positive half of the x-axis satisfies $\tan \theta = y/x$.

On the other hand, if you are given r and θ, then from trigonometry you can see that $x = r \cos \theta$ and $y = r \sin \theta$.

Relation Between Polar and Cartesian Coordinates

$$x = r \cos \theta \qquad r = \sqrt{x^2 + y^2}$$

$$y = r \sin \theta \qquad \tan \theta = y/x.$$

Example 2   Give the Cartesian coordinates of the points with polar coordinates $(2, 3\pi/2)$ and $(2, 1)$.

Solution   Both points have r-coordinate equal to 2, so they are 2 units from the origin. The first one has θ-coordinate equal to $3\pi/2$, which is three quarters of a full revolution, so it is three quarters of the way around the circle of radius 2, at $(0, -2)$.

The second point has θ-coordinate equal to 1. From the formulas above, we see that

$$x = 2 \cos 1 = 1.0806 \quad \text{and} \quad y = 2 \sin 1 = 1.6830.$$

Problems for Appendix A

For Problems 1–7, give Cartesian coordinates for the points with the following polar coordinates (r, θ). The angles are measured in radians.

1. $(1, 0)$ 2. $(0, 1)$ 3. $(2, \pi)$ 4. $(\sqrt{2}, 5\pi/4)$

5. $(5, -\pi/6)$ 6. $(3, \pi/2)$ 7. $(1, 1)$

For Problems 8–15, give polar coordinates for the points with the following Cartesian coordinates. Choose $0 \leq \theta < 2\pi$.

8. $(1, 0)$ 9. $(0, 2)$ 10. $(1, 1)$ 11. $(-1, 1)$

12. $(-3, -3)$ 13. $(0.2, -0.2)$ 14. $(3, 4)$ 15. $(-3, 1)$

16. Every point in the plane can be represented by some pair of polar coordinates, but are the polar coordinates (r, θ) uniquely determined by the Cartesian coordinates (x, y)? In other words, for each pair of Cartesian coordinates, is there one and only one pair of polar coordinates for that point? Why or why not?

Appendix B   DETERMINANTS

We have shown that the cross product $\vec{a} \times \vec{b}$ can be computed using the expression:

$$\vec{a} \times \vec{b} = (a_2 b_3 - a_3 b_2)\vec{i} + (a_3 b_1 - a_1 b_3)\vec{j} + (a_1 b_2 - a_2 b_1)\vec{k}.$$

A useful way of remembering this formula is to write it as a determinant

$$\vec{a} \times \vec{b} = \begin{vmatrix} \vec{i} & \vec{j} & \vec{k} \\ a_1 & a_2 & a_3 \\ b_1 & b_2 & b_3 \end{vmatrix}.$$

The determinant is expanded in the following way: To find the $\vec{i}$-component, cross out the row and the column containing $\vec{i}$, leaving a square of numbers. Now multiply diagonally opposite entries and subtract, as follows:

$$\begin{vmatrix} a_2 & a_3 \\ b_2 & b_3 \end{vmatrix} = a_2 b_3 - a_3 b_2.$$

For the $\vec{j}$ component, leave out the top row and the middle column and put a negative sign in front:

$$-\begin{vmatrix} a_1 & a_3 \\ b_1 & b_3 \end{vmatrix} = -(a_1 b_3 - a_3 b_1) = a_3 b_1 - a_1 b_3.$$

For the $\vec{k}$ component, leave out the top row and the right-hand column:

$$\begin{vmatrix} a_1 & a_2 \\ b_1 & b_2 \end{vmatrix} = a_1 b_2 - a_2 b_1.$$

The signs for the $\vec{i}, \vec{j}$, and $\vec{k}$ components alternate $+, -, +$.

Appendix C   REVIEW OF ONE-VARIABLE INTEGRATION

In single-variable calculus you learned about the definite integral

$$\int_a^b f(x)\, dx$$

of a function $y = f(x)$ on the interval $a \leq x \leq b$. To prepare the way for multivariable integration, we will briefly review the definition and interpretations of the one-variable integral.

Definition of the One-Variable Integral

The one-variable integral

$$\int_a^b f(x)\, dx$$

is defined to be a limit of *Riemann sums*, which may be constructed as follows. We divide the interval $a \leq x \leq b$ into n equal subdivisions, and we call the width of an individual subdivision Δx. Thus,

$$\Delta x = \frac{b - a}{n}.$$

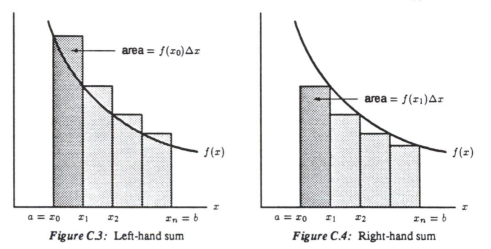

Figure C.3: Left-hand sum   ***Figure C.4:*** Right-hand sum

We will let x_0, x_1, x_2, $\ldots$, x_n be endpoints of the subdivisions, as in Figures C.3 and C.4. We construct two special Riemann sums:

$$\text{Left-hand sum} = f(x_0)\Delta x + f(x_1)\Delta x + \cdots + f(x_{n-1})\Delta x$$

and

$$\text{Right-hand sum} = f(x_1)\Delta x + f(x_2)\Delta x + \cdots + f(x_n)\Delta x.$$

To define the definite integral, we take the limit of these sums as n goes to infinity.

The **definite integral** of f from a to b, written

$$\int_a^b f(x)\,dx,$$

is the limit of the left-hand or right-hand sums with n subdivisions as n gets arbitrarily large. In other words,

$$\int_a^b f(x)\,dx = \lim_{n\to\infty}\,(\text{left-hand sum}) = \lim_{n\to\infty}\left(\sum_{i=0}^{n-1} f(x_i)\Delta x\right)$$

and

$$\int_a^b f(x)\,dx = \lim_{n\to\infty}\,(\text{right-hand sum}) = \lim_{n\to\infty}\left(\sum_{i=1}^{n} f(x_i)\Delta x\right).$$

Each of these sums is called a *Riemann sum*, f is called the *integrand*, and a and b are called the *limits of integration*.

There are other sorts of Riemann sums in addition to left- and right-hand sums; for one thing, you can evaluate the function at any point in each subinterval, not just at the left- or right-hand endpoint. Graphically, this means that the graph of the function can intersect the top of each rectangle at any point, not just at the endpoints. Figure C.5 shows why you might want this to happen: if you want to get a lower estimate for the integral, you need to have all your rectangles sitting below the curve. A Riemann sum where the rectangles all sit just below the graph is called a *lower sum*. Figure C.6 shows an *upper sum* for the same integral.

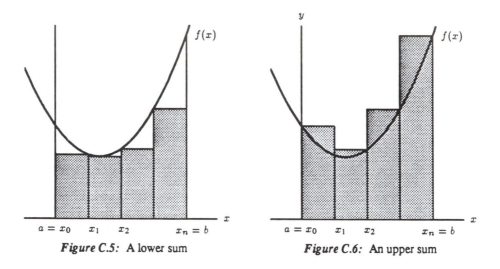

Figure C.5: A lower sum   *Figure C.6:* An upper sum

Interpretations of the Definite Integral

As an Area

If $f(x)$ is positive we can interpret each term $f(x_0)\Delta x$, $f(x_1)\Delta x$, ... in a left- or right-hand Riemann sum as the area of a rectangle, as we saw in the previous section. As the width Δx of the rectangles approaches zero, the rectangles fit the curve of the graph more exactly, and the sum of their areas gets closer and closer to the area under the curve, shaded in Figure C.7. Thus, we conclude that:

If $f(x) \geq 0$ and $a < b$:

$$\begin{array}{c}\text{Area under graph of } f \\ \text{between } a \text{ and } b\end{array} = \int_a^b f(x)\,dx.$$

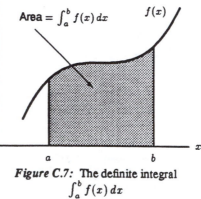

Figure C.7: The definite integral
$$\int_a^b f(x)\,dx$$

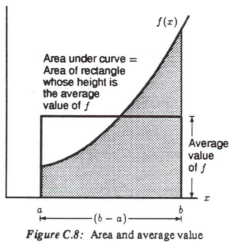

Figure C.8: Area and average value

As an Average Value

The definite integral can be used to compute the average value of a function:

$$\text{Average value of } f \text{ from } a \text{ to } b \;=\; \frac{1}{b-a}\int_a^b f(x)\,dx$$

If we interpret the integral as the area under the graph of f, then we can think of the average value of f as the height of the rectangle with the same area that is on the same base. (See Figure C.8.)

When $f(x)$ Represents a Density

If $f(x)$ represents a density, say a population density or the density of a substance, then we can calculate the total mass or total population using a definite integral. To find the total quantity, we divide the region into small pieces and find the population or mass of each one by multiplying the density by the size of that piece.

If $f(x)$ represents the density of a substance along the interval $a \le x \le b$, the

$$\text{Total mass} = \int_a^b f(x)\,dx.$$

As the Total Change

The integral of the rate of change of any quantity will give the total change in that quantity; for example, the integral of velocity gives the total change in position (or distance moved).

If $r(t)$ is the rate of change of Q with respect to t, then

$$Q(b) - Q(a) = \text{Total Change in } Q \text{ between } t = a \text{ and } t = b \;=\; \int_a^b r(t)\,dt.$$

The Fundamental Theorem of Calculus

Since

$$F'(t) = \text{rate of change of } F(t) \text{ with respect to } t,$$

and since the definite integral of a rate of change of some quantity is the total change in that quantity, we find that

$$\int_a^b F'(t)\, dt = \int_a^b \left[\text{rate of change of } F(t)\right]\, dt$$
$$= \text{Total change in } F(t) \text{ between } a \text{ and } b$$
$$= F(b) - F(a).$$

This is

The Fundamental Theorem of Calculus

If $f = F'$, then

$$\int_a^b f(t)\, dt = F(b) - F(a).$$

The Fundamental Theorem is the basic tool for evaluating definite integrals algebraically; it enables us to avoid using Riemann sums whenever we can find an antiderivative for the integrand.

Problems for Appendix C

In Exercises 1—20, find an antiderivative for each of the functions given.

1. $\displaystyle\int \left(x^2 + 2x + \frac{1}{x}\right) dx.$

2. $\displaystyle\int \frac{t+1}{t^2}\, dt.$

3. $\displaystyle\int \frac{(t+2)^2}{t^3}\, dt.$

4. $\displaystyle\int \sin t\, dt.$

5. $\displaystyle\int \cos 2t\, dt.$

6. $\displaystyle\int \frac{x}{x^2+1}\, dx.$

7. $\displaystyle\int \tan\theta\, d\theta.$

8. $\displaystyle\int e^{5z}\, dz.$

9. $\displaystyle\int te^{t^2+1}\, dt.$

10. $\displaystyle\int \frac{dz}{1+z^2}.$

11. $\displaystyle\int \frac{dz}{1+4z^2}.$

12. $\displaystyle\int \sin^2\theta\cos\theta\, d\theta.$

13. $\displaystyle\int \sin 5\theta \cos^3 5\theta\, d\theta.$

14. $\displaystyle\int \sin^3 z \cos^3 z\, dz.$

15. $\displaystyle\int \frac{(\ln x)^2}{x}\, dx.$

16. $\displaystyle\int \cos\theta\sqrt{1+\sin\theta}\, d\theta.$

17. $\displaystyle\int xe^x\, dx.$

18. $\displaystyle\int t^3 e^t\, dt.$

19. $\displaystyle\int x\ln x\, dx.$

20. $\displaystyle\int \frac{1}{\cos^2\theta}\, d\theta.$

In Exercises 21—25 find the definite integral by two methods (Fundamental Theorem and Numerically).

21. $\int_{1}^{3} x(x^2 + 1)^{70}\, dx$.

22. $\int_{0}^{1} \dfrac{dx}{x^2 + 1}$.

23. $\int_{0}^{10} z e^{-z}\, dz$.

24. $\int_{-\pi/3}^{\pi/4} \sin^3 \theta \cos \theta\, d\theta$.

25. $\int_{1}^{4} \dfrac{e^{\sqrt{x}}}{\sqrt{x}}\, dx$.

26. Water is leaking out of a tank at a rate of $R(t)$ gallons/hour, where t is measured in hours.

    (a) Write a definite integral that expresses the total amount of water that leaks out in the first two hours.

    (b) Figure C.9 is a graph of $R(t)$. On a sketch, shade in the region whose area represents the total amount of water that leaks out in the first two hours.

    (c) Give an upper and lower estimate of the total amount of water that leaks out in the first two hours.

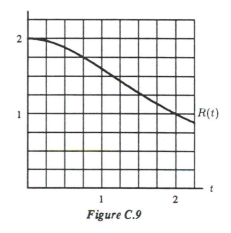

Figure C.9

27. The rate at which the world's oil is being consumed is continuously increasing. Suppose the rate (in billions of barrels per year) is given by the function $r = f(t)$, where t is measured in years and $t = 0$ is the start of 1990.

    (a) Write a definite integral which represents the total quantity of oil used between the start of 1990 and the start of 1995.

    (b) Suppose $r = 32e^{0.05t}$. Using a left-hand sum with five subdivisions, find an approximate value for the total quantity of oil used between the start of 1990 and the start of 1995.

    (c) Interpret each of the five terms in the sum from part (b) in terms of oil consumption.

28. Figure C.10 shows the graph of the derivative $g'(x)$ of a function $g(x)$. It is given that $g(0) = 50$. Sketch the graph of $g(x)$, showing all critical points and inflection points of g and giving their coordinates.

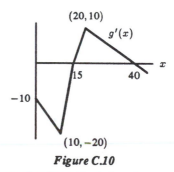

Figure C.10

29. The graph of dy/dt against t is in Figure C.11. Suppose the three shaded regions each have area 2. Given that $y = 0$ when $t = 0$, draw the graph of y against t, indicating all special features the graph might have (known heights, maxima and minima, inflection points, etc.). Pay particular attention to the relationship between the graphs. Mark $t_1, t_2, \ldots, t_5$ on the t

axis.[1]

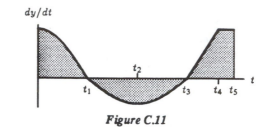

Figure C.11

30. The Quabbin Reservoir in the western part of Massachusetts provides most of Boston's water. The graph in Figure C.12 represents the flow of water in and out of the Quabbin Reservoir throughout 1993.

   (a) Sketch a possible graph for the quantity of water in the reservoir, as a function of time.

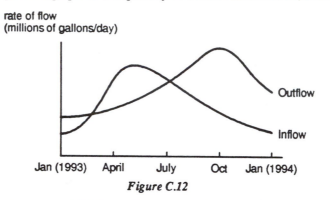

Figure C.12

   (b) When, in the course of 1993, was the quantity of water in the reservoir largest? Smallest? Mark and label these points on the graph you drew in part (a).
   (c) When was the quantity of water decreasing most rapidly? Again, mark and label this time on both graphs.
   (d) By July 1994 the quantity of water in the reservoir was about the same as in January 1993. Draw plausible graphs for the flow into and the flow out of the reservoir for the first half of 1994. Explain your graph.

31. A rod has length 2 meters. At a distance x meters from its left end, the density of the rod is given by

$$\rho(x) = 2 + 6x \text{ g/m}.$$

   (a) Write a Riemann sum approximating the total mass of the rod.
   (b) Find the exact mass by converting the sum into an integral.

32. The density of cars (in cars per mile) down a 20-mile stretch of the Pennsylvania Turnpike can be approximated by

$$\rho(x) = 300 \left(2 + \sin\left(4\sqrt{x + 0.15}\right)\right),$$

   where x is the distance in miles from the Breezewood toll plaza.

   (a) Sketch a graph of this function for $0 \le x \le 20$.
   (b) Write a sum that approximates the total number of cars on this 20-mile stretch.
   (c) Find the total number of cars on the 20-mile stretch.

---

[1]From *Calculus: The Analysis of Functions*, by Peter D. Taylor (Toronto: Wall & Emerson, Inc., 1992)

33. Circle City, a typical metropolis, is very densely populated near its center, and its population gradually thins out toward the city limits. In fact, its population density is $10,000(3 - r)$ people/square mile at distance r miles from the center.

    (a) Assuming that the population density at the city limits is zero, find the radius of the city.
    (b) What is the total population of the city?

34. The density of oil in a circular oil slick on the surface of the ocean at a distance r meters from the center of the slick is given by $\rho(r) = 50/(1 + r)$ kg/m^2.

    (a) If the slick extends from $r = 0$ to $r = 10,000$ m, find a Riemann sum approximating the total mass of oil in the slick.
    (b) Find the exact value of the mass of oil in the slick by turning your sum into an integral and evaluating it.
    (c) Within what distance r is half the oil of the slick contained?

35. An exponential model for the density of the earth's atmosphere says that if the temperature of the atmosphere were constant, then the density of the atmosphere as a function of height, h (in meters), above the surface of the earth would be given by

$$\rho(h) = 1.28e^{-0.000124h} \text{ kg/m}^3.$$

    (a) Write (but do not evaluate) a sum that approximates the mass of the portion of the atmosphere from $h = 0$ to $h = 100$ m (i.e., the first 100 meters above sea level). Assume the radius of the earth is 6370 km.
    (b) Find the exact answer by turning your sum in part (a) into an integral. Evaluate the integral.

36. Water is flowing in a cylindrical pipe of radius 1 inch. Because water is viscous and sticks to the pipe, the rate of flow varies with the distance from the center. The speed of the water at a distance r inches from the center is $10(1 - r^2)$ inches per second. What is the rate (in cubic inches per second) at which water is flowing through the pipe?

37. The reflector behind a car headlight is made in the shape of the parabola, $x = \frac{4}{9}y^2$, with a circular cross-section, as shown in Figure C.13.

    (a) Find a Riemann sum approximating the volume contained by this headlight.
    (b) Find the volume exactly.

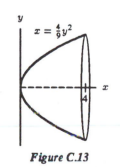

Figure C.13

38. Rotate the bell-shaped curve $y = e^{-x^2/2}$ shown in Figure C.14 around the y-axis, forming a hill-shaped solid of revolution. By slicing horizontally, find the volume of this hill.

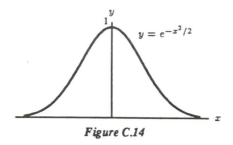

Figure C.14

39. The circumference of the trunk of a certain tree at different heights above the ground is given in the following table.

Height (feet)	0	20	40	60	80	100	120
Circumference (feet)	26	22	19	14	6	3	1

Assume all horizontal cross-sections of the trunk are circles. Estimate the volume of the tree trunk using the trapezoid rule.

40. Most states expect to run out of space for their garbage soon. In New York, solid garbage is packed into pyramid-shaped dumps with square bases. (The largest such dump is on Staten Island.) A small community has a dump with a base length of 100 yards. One yard vertically above the base, the length of the side parallel to the base is 99 yards; the dump can be built up to a vertical height of 20 yards. (The top of the pyramid is never reached.) If 65 cubic yards of garbage arrive at the dump every day, how long will it be before the dump is full?

Appendix D   CHANGE OF VARIABLES

You can often make more sense of a function if you simplify it with a substitution. For example, consider the function in Figure 11.47 on page 32 that gives corn production as a function of climate.

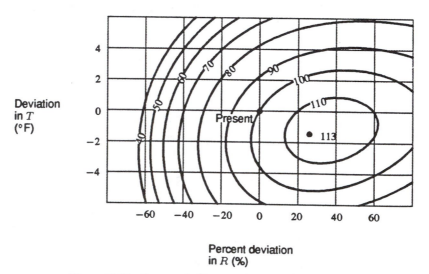

Figure D.15: Corn production as a function of climate change

The main concern of the people who produced this graph was the effect on corn production of a *change* in the climate, so they presented the function in a slightly different way. Figure D.15 shows the new way.

Notice that the point marked "Present" is at the point with coordinates $(15, 76)$ on the old graph, and is at the origin of the new graph, i.e., it is the point with coordinates $(0, 0)$. In the new graph, everything is expressed in terms of deviations from the present values, which is the most sensible way of expressing things when you are trying to understand the problem of the effect of climatic change. Thus, for example, the point on the vertical axis that used to be marked 78 is now marked 2, since it is 2 more than the current value. In addition, on the horizontal axis the rainfall deviations are expressed as as percentages. Thus, since a rainfall of 12 inches would represent a 20% decrease from the current rainfall of 15 inches, the point that used to be marked 12 on this axis is now marked -20.

This is an example of a change of coordinates. The underlying dependence of corn production on rainfall and temperature has not changed, but the coordinates we use to express the rainfall and temperature have changed. In terms of the contour map, this means that the contours stay the same, but the numbers marked on the axes change.

Shifts and Stretches

In the example on corn production above, we made two sorts of changes in our coordinates: first we set things up so that the current temperature and rainfall became the origin of our new coordinate system, and then we scaled the rainfall so that it was expressed as a percentage rather than a number of inches. Let's consider these two sorts of changes separately.

How Do You Change the Origin?

> If you are working with coordinates x and y, and you want the point (a, b) to become the origin of your coordinate system, you define new coordinates
>
> $$u = x - a, \quad v = y - b.$$
>
> Then the point that used to have x-coordinate a and y-coordinate b now has u- and v-coordinates both equal to zero, i.e., it is at the origin of your new coordinate system.

Example 1   Choose better coordinates for the equation

$$z = x^2 - 2x + 3 + y^2 + 4y.$$

Solution   If we complete the square on both x and y, we get

$$z = (x - 1)^2 + (y + 2)^2 - 2.$$

This is a parabola-shaped bowl, or paraboloid, whose base is at $x = 1, y = -2, z = -2$. If we make the change of coordinates

$$u = x - 1 \quad v = y + 2,$$

then the equation becomes

$$z = u^2 + v^2 - 2.$$

The vertex of the paraboloid is now at $u = 0, v = 0$.

Rescaling Coordinates

Another common change of coordinates is scaling. We saw an example of this above when we expressed rainfall figures as a percentage of the current amount, rather than in absolute terms.

Scaling coordinates can be very useful in seeing relationships between different things. For example, the van der Waal's gas equation relates the pressure, temperature, and volume of a certain quantity (one mole) of gas. The van der Waal's equation for helium is

$$T = 0.41581\frac{1}{V} - 0.0098546\frac{1}{V^2} - 0.28882P + 12.187VP$$

and the van der Waal's equation for oxygen is

$$T = 16.574\frac{1}{V} - 0.52754\frac{1}{V^2} - 0.3879P + 12.187VP.$$

Here pressure is measured in atmospheres, temperature in degrees Kelvin, and volume in cubic decimeters. The formulas for the two gases are quite different. Figures D.16 and D.17 show the contour maps for these functions.

Notice that although the numbers marked along the sides are different, the contours look quite similar. Each contour map has one contour with a point that is both an inflection point and point where the slope is zero. That point is called the critical point for the gas, and has a special chemical significance. Its P- and V-coordinates are called the critical pressure and volume respectively, and the value of T there is called the critical temperature. The similarity between the contours of the two different gases suggests that if we scale the P, V, and T-coordinates so that the critical values are all equal to 1, then the gases will have the same equation, thus revealing some unifying underlying behavior of these two different gases.

For helium, the critical pressure is 2.2498, the critical volume is 0.0711, and the critical temperature is 5.1984. To make these equal to one in our new coordinate system, we divide the old coordinates by these values. The new coordinates will be denoted by lower case letters. The change of coordinate equations are

$$p = P/2.2498, \quad v = V/0.0711, \quad t = T/5.1984.$$

To substitute these into the van der Waals equation, we solve them for the old coordinates:

$$P = 2.2498p, \quad V = 0.0711v, \quad T = 5.1984t.$$

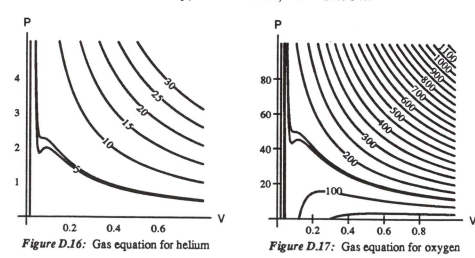

Figure D.16: Gas equation for helium   **Figure D.17:** Gas equation for oxygen

The gas equation for helium becomes

$$5.1984t = 0.41581\frac{1}{0.0711v} - 0.0098546\frac{1}{(0.0711v)^2}$$
$$+ 0.28882(2.2498p) + 12.187(0.0711v)(2.2498p),$$

which simplifies to

$$t = 1.125\frac{1}{v} - 0.375\frac{1}{v^2} - 0.125p + 0.375pv.$$

For oxygen, the critical pressure is 49.717, the critical volume is 0.09549, and the critical temperature is 154.28. Making the change of coordinates

$$p = P/49.717, \quad v = V/0.09549, \quad t = T/154.28,$$

we get the same equation as for helium.

Any change of coordinates that involves multiplying the variables by a constant is called a scaling of the variables. Another situation where you scale the variables is when you change the units of measurement.

Appendix E   THE IMPLICIT FUNCTION THEOREM

Consider the equation

$$z^3 - 7yz + 6e^x = 0.$$

Can it be solved for z? Does it determine a function $z = f(x, y)$ of x and y? The test is whether, in principle, the equation determines a table of values for f. Let's try to construct such a table. For instance, to evaluate $z = f(0, 1)$ you must solve for z after setting $x = 0$ and $y = 1$ in the equation. That is, you must solve for z in the equation

$$z^3 - 7z + 6 = 0.$$

As it happens, the equation easily factors, yielding

$$z^3 - 7z + 6 = (z - 1)(z - 2)(z + 3) = 0,$$

so there are three solutions, $z = 1$, $z = 2$, and $z = -3$. Since a function must take single definite values and we have been given no rule to decide which solution should equal $f(0, 1)$, we conclude that the original equation alone does not define z as a function of x and y.

Perhaps the original equation defines three functions f, g, and h of x and y, with, say, $f(0, 1) = 1, g(0, 1) = 2$, and $h(0, 1) = -3$. To test this hypothesis, let's try to evaluate f, g, and h at $(x, y) = (0.02, 1.01)$. You must solve for z after setting $x = 0.02$ and $y = 1.01$. That is, you must solve for z in the equation

$$z^3 - 7.07z + 6e^{0.02} = 0.$$

Again, there are three solutions. They can be found numerically to equal $1.012701, 2.003794$, and -3.016495. Since $(0.02, 1.01)$ is near $(0, 1)$, it would make sense for $f(0.02, 1.01)$ to be near $f(0, 1) = 1$, for $g(0.02, 1.01)$ to be near $g(0, 1) = 2$, and for $h(0.02, 1.01)$ to be near $h(0, 1) = -3$. So we have:

$$f(0.02, 1.01) = 1.012701,$$
$$g(0.02, 1.01) = 2.003794,$$
$$h(0.02, 1.01) = -3.016495.$$

We have found the key. We will say that $f(x, y)$ is defined for (x, y) near $(0, 1)$ to be the solution z of $z^3 - 7yz + 6e^x = 0$ that is near 1, that $g(x, y)$ is defined for (x, y) near $(0, 1)$ to be the solution z of $z^3 - 7yz + 6e^x = 0$ that is near 2, and that $h(x, y)$ is defined for (x, y) near $(0, 1)$ to be the solution z of $z^3 - 7yz + 6e^x = 0$ that is near -3. We will not attempt to define f, g, or h for values of (x, y) that are far from $(0, 1)$.

A brief table of values for g is given in Table E.1.

TABLE E.1   *One Solution of $z^3 - 7yz + 6e^x = 0$*

		0.98	0.99	1.00	1.01	1.02
				y		
	-0.02	1.96741	1.99580	2.02312	2.04949	2.07502
	-0.01	1.95477	1.98385	2.01177	2.03868	2.06468
x	0.00	1.94158	1.97142	2.00000	2.02747	2.05398
	0.01	1.92778	1.95847	1.98776	2.01586	2.04292
	0.02	1.91330	1.94492	1.97501	2.00379	2.03145

It was computed with many applications of Newton's root-finding method. For example, $g(0.02, 1.01)$ is the solution z near 2 of the equation $z^3 - 7.07z + e^x = 0$. To compute it, start with the approximate value $z_0 = 2$, then successively improve the approximation using the formula $z_{n+1} = z - \frac{m(z_n)}{m'(z_n)}$, where $m(z) = z^3 - 7.07z + e^{0.02}$. See Table E.2.

TABLE E.2
Newton's Method for
$z^3 - 7.07z + e^{0.02} = 0$

n	z_n
1	2.003811757
2	2.003794225
3	2.003794225

It is difficult to compute Table E.1 for $g(x, y)$ because the equation to be solved:

$$z^3 - 7yz + 6e^x = 0$$

is nonlinear. But for the function g we are only interested in values of (x, y, z) near $(0, 1, 2)$, and the equation ought to be approximately linear for (x, y, z) near a single point. Since linear equations are easy to solve, let's replace the nonlinear equation by a linear approximation valid for (x, y, z) near $(0, 1, 2)$.

First find the local linearization of $m(x, y, z) = z^3 - 7yz + 6e^x$ at $(x, y, z) = (0, 1, 2)$. Evaluation of the partial derivatives m_x, m_y, and m_z at $(0, 1, 2)$ is straightforward, and we find that

$$m(x, y, z) \approx 0 + 6x - 14(y - 1) + 5(z - 2)$$

for (x, y, z) near $(0, 1, 2)$. It follows that the solutions $z = g(x, y)$ for (x, y) near $(0, 1)$ of

$$z^3 - 7yz + 6e^x = 0$$

should be close to the solutions of

$$6x - 14(y - 1) + 5(z - 2) = 0.$$

The last equation is linear and easily solved, giving

$$z = -0.8 - 1.2x + 2.8y.$$

We conclude that

$$g(x, y) \approx -0.8 - 1.2x + 2.8y \quad \text{for } (x, y) \text{ near } (0, 1).$$

In fact, this last approximation is the local linearization of g at $(0, 1)$. It has been used to generate easily Table E.3 of approximate values for g, which can be compared with Table E.1 of exact values.

TABLE E.3   *Linear Approximation*

		0.98	0.99	1.00	1.01	1.02
	-0.02	1.968	1.996	2.024	2.052	2.080
	-0.01	1.956	1.984	2.012	2.040	2.068
x	0.00	1.944	1.972	2.000	2.028	2.056
	0.01	1.932	1.960	1.988	2.016	2.044
	0.02	1.920	1.948	1.976	2.004	2.03

(with y heading the columns)

The Implicit Function Principle

The analysis of the previous example can be fairly summarized as follows. A nonlinear equation is hard to solve. Locally it may resemble a linear equation that is easy to solve. The solution of the linear equation will be a good approximation for the solution of the nonlinear equation, at least locally.

More formally, we state the Implicit Function Principle.

The Implicit Function Principle

Suppose that $f(x, y, z)$ is a function, that $f(a, b, c) = 0$, and that $L(x, y, z)$ is the local linearization of f at (a, b, c). If the linear equation,

$$L(x, y, z) = 0,$$

can be solved for z in terms of x and y, then that solution is a close approximation of a function $z = g(x, y)$ that is defined for (x, y) near (a, b) to be the only solution z near c of the equation

$$f(x, y, z) = 0.$$

A little more can be said. The approximation of $g(x, y)$ mentioned by the Implicit Function Principle is actually the local linearization of g at (a, b). Since partial derivatives can be read off from the local linearization, the Implicit Function Principle gives a method of computing $g_x(a, b)$ and $g_y(a, b)$.

Example 1   Use the Implicit Function Principle to investigate solutions of the equation

$$e^z + xyz + x^3 y^4 - 2 = 0$$

near the solution $(x, y, z) = (1, 1, 0)$.

Solution   Let $f(x, y, z) = e^z + xyz + x^3 y^4 - 2$. Compute:

$$f_x(x, y, z) = yz + 3x^2 y^4,$$
$$f_y(x, y, z) = xz + 4x^3 y^3,$$
$$f_z(x, y, z) = e^z + xy.$$

Hence,

$$f_x(1, 1, 0) = 3$$
$$f_y(1, 1, 0) = 4$$
$$f_z(1, 1, 0) = 2.$$

Therefore:

$$f(x, y, z) \approx 3(x - 1) + 4(y - 1) + 2(z - 0)$$
$$= -7 + 3x + 4y + 2z \quad \text{for } (x, y, z) \text{ near } (1, 1, 0).$$

The linearization of the equation $f(x, y, z) = 0$ near $(1, 1, 0)$ is

$$-7 + 3x + 4y + 2z = 0$$

which can be solved for z, giving

$$z = 3.5 - 1.5x - 2y.$$

The Implicit Function Principle applies. It asserts that for (x, y) near enough to $(1, 1)$ there is exactly one solution z near 0 of the original (nonlinear) equation

$$e^z + xyz + x^3 y^4 - 2 = 0.$$

Moreover, an approximation for the solution $z = g(x, y)$ is given by the local linearization for g near $(1, 1)$:

$$z = g(x, y) \approx 3.5 - 1.5x - 2y.$$

In other words, all solutions of $e^z + xyz + x^3 y^4 - 2 = 0$ near $(1, 1, 0)$ are approximated by $(x, y, 3.5 - 1.5x - 2y)$ for (x, y) near $(1, 1)$.

Our final example shows some of the limitations of the Implicit Function Principle.

Example 2   Use the Implicit Function Principle to investigate solutions of the equation

$$x^2 + y^2 + z^2 = 25$$

near the solution $(x, y, z) = (3, 4, 0)$.

Solution Consider the equivalent equation $f(x, y, z) = 0$, where $f(x, y, z) = x^2 + y^2 + z^2 - 25$. From $df = 2x\,dx + 2y\,dy + 2z\,dz$ we get the local linearization

$$f(x, y, z) \approx 6(x - 3) + 8(y - 4) + 0(z - 0) \quad \text{for } (x, y, z) \text{ near } (3,4,0).$$

The linearization of the equation $f(x, y, z) = 0$ near $(3, 4, 0)$ is

$$6(x - 3) + 8(y - 4) = 0,$$

which cannot be solved for z, because z does not appear in it. Therefore, the Implicit Function Principle does not give an approximation for z as a function of (x, y) near $(3, 4)$.

To see what has happened we turn to geometry. The equation $x^2 + y^2 + z^2 = 25$ describes a sphere of radius 5 centered at the origin. Do values of (x, y) near $(3, 4)$ determine unique values of z near 0 such that (x, y, z) lies on the sphere and hence satisfies the given equation? The answer is no, for two reasons. For points such as $(x, y) = (3.01, 4.01)$ where $x^2 + y^2 > 25$, there is no z at all such that (x, y, z) is on the sphere. For points such as $(2.99, 3.99)$ where $x^2 + y^2$ is slightly less than 25, there are two z values near 0 that solve the equation, namely $z = \sqrt{25 - x^2 - y^2}$ and $z = -\sqrt{25 - x^2 - y^2}$. See Figure E.18. The Implicit Function Principle could not apply near $(3, 4, 0)$ because the equation $x^2 + y^2 + z^2 = 25$ does not determine definite z values for all (x, y) near $(3, 4)$; and so, there is no function to approximate near $(3, 4)$.

One final comment is in order. The linearization $6(x - 3) + 8(y - 4) = 0$ of the equation $f(x, y, z) = 0$ near $(3, 4, 0)$ is the equation of the tangent plane to the sphere at $(3, 4, 0)$. The absence of z in the equation of the plane indicates that the plane is vertical. See Figure E.19. In some sense it is the fact that the tangent plane is vertical at $(3, 4, 0)$ that makes it possible for the sphere to turn under itself at that point, which is why there can be two different nearby points on the sphere for single values of (x, y) near $(3, 4)$.

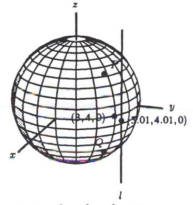

Figure E.18: $x^2 + y^2 + z^2 = 25$

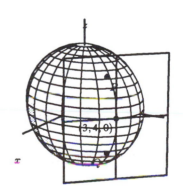

Figure E.19: Vertical Tangent Plane

Appendix F   NOTES ON KEPLER, NEWTON AND PLANETARY MOTION

If you looked at the stars every night, you would see that they stay in the same positions with respect to each other from night to night, but that during the course of one night the dome of stars rotates slowly about the North Star. If you studied the stars more carefully, however, you would eventually notice some of them moving with respect to the others. If you followed the path of these stars night after night as they wandered amongst the other stars, you might even see their position from one night to the next move forward and then backward. These wanderers are the planets. Five, besides earth, are visible to the naked eye. Ever since people, thousands of years ago, first observed these erratic paths, they have endowed the planets with supernatural powers; for example, astrology is based largely on their positions with respect to the fixed stars. The names by which we know them, Mercury, Venus, Mars, Jupiter, and Saturn, are the names of Ancient Roman gods. People also tried to make sense out of the strange paths, and the mathematical explanation of planetary motion was the first breakthrough human kind made in understanding the natural laws of the universe.

First Steps: Eratosthenes and Copernicus

The first important realization is

> The earth is round, a sphere of radius about 4000 miles.

This was known to some at least as far back as ancient Greece. Eratosthenes gave a reasonable estimate of the radius of the earth by observing the angle of the sun at noon on June 21 at two different locations (see Problem 1).

> The earth spins once every 24 hours around a central axis passing through the north and south poles.

The rotational axis of the earth points at the North Star. Thus, if you watched the North Star throughout the night, the rotation of the earth would make it look as if the dome of stars was spinning slowly around it, completing one full revolution in 24 hours.

The Greek philosopher Aristotle held that the earth remained fixed while all other heavenly bodies moved around it. Indeed, the very idea of the earth spinning seems preposterous. Wouldn't we just be spun right off the planet? In fact, the acceleration you would feel at the equator caused by the earth's spinning is only about 1% that of gravity, about the same as the acceleration you'd feel at the edge of merry-go-round with 30 foot radius that goes around once every 2 minutes (see Problem 2). You could probably crawl on your hands and knees around that merry-go-round and still keep up with it.

In the middle ages Nicolaus Copernicus (1473-1543) proposed the more modern point of view, that the apparent motion of the stars each night is caused by the earth's rotation. Copernicus also contradicted previous theories by placing the sun at the center of the solar system:

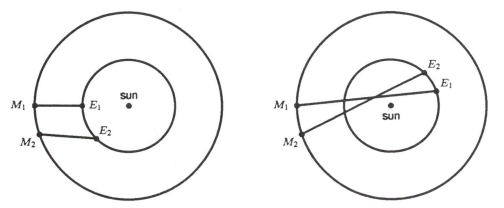

Figure F.20: The retrograde motion of Mars. (M_1 and M_2 indicate Mars on the first and second night; similarly for earth and E_1, E_2)

The earth and the planets orbit around the sun. The earth completes one revolution around the sun each 365 1/4 days (approximately). The moon orbits around the earth completing one revolution in about 27.32 days.

In fact Eratosthenes had already proposed this theory in ancient times, because the motion of the earth and planets around the sun explains the apparently complicated motion of the planets. For example, suppose the earth and Mars are on the same side of the sun as shown on the left of Figure F.20. The earth is closer to the sun than Mars and completes one orbit faster. Thus, on consecutive nights, although Mars has moved, you have moved even more, so that Mars appears to have moved backwards. If the earth is moving counterclockwise, then in facing Mars you would have to look back further to your right the second night. On the other hand, if the earth and Mars are on opposite sides of the sun, the second night would find Mars a little to the left of its position the first night, as shown on the right of Figure F.20, and so it would appear to have moved forward. When Mars appears to move backwards, we say it exhibits *retrograde motion*.

Kepler's Laws for Planetary Motion

The orbits of the planets around the sun or the moon around the earth are not circles. The moon's distance to the earth varies from 220,000 to 260,000 miles. Each of the planets, including earth, has a closest and furthest point from the sun. So what are the shapes? In the last half of the 16th century Tycho Brahe accumulated data about the positions of the planets. Johann Kepler (1571-1630) studied the data for years and after some false starts involving platonic solids and the "music of the spheres," he arrived at three laws for planetary motion:

Kepler's Laws

- The orbit of each planet is an ellipse with the sun at the focus. In particular, the plane of the orbit contains the sun.

- As a planet orbits around the sun, the line segment from the sun to the planet sweeps out equal areas in equal times.

- The ratio p^2/d^3 is the same for every planet orbiting around the sun, where p is the period of the orbit (time to complete one revolution) and d is the mean distance of the orbit (average of the shortest and furthest distance from the sun).

It is very important to understand what these laws are saying. In Law I, an ellipse is not just any old squashed circle. It is a specific geometric figure having very special properties. An ellipse is a closed curve in the plane such that the sum of the distances from any point on the curve to two fixed points, called the foci of the ellipse, is constant. If the two foci are located at $(0, -b)$ and $(0, b)$ on the y-axis, then it can be shown (see Problem 3) that the constant sum of distance is $2d$ and that the equation of the ellipse is

$$\frac{x^2}{c^2} + \frac{y^2}{d^2} = 1,$$

where d is the mean distance and $c^2 = d^2 - b^2$. Note as well that the sun is not at the center of the planet's elliptical orbit, but rather at a focus. This is a crucial distinction.

Kepler's Second Law implies that the speed of the planet is not constant. In order to sweep out the same area in one unit of time when the planet is near the sun as it does when it is far from the sun, the planet must move faster when it is near the sun (see Figure F.21).

The third law says that p^2/d^3 is the same for all planets. In particular, this means if you know the period p for a planet, then the mean distance d is determined, and vice versa. Newton later showed that the constant value of p^2/d^3 depends on the mass of object about which the planets are orbiting.

Kepler's Laws, impressive as they are, were purely descriptive; they didn't explain the motion of the planets, they merely described it. Newton's great achievement was to find an underlying cause for them.

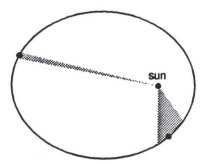

Figure F.21: Equal areas in equal time

Newton's Laws of Motion

In 1687, Isaac Newton published *Philosophiae Naturalis Principia Mathematica*. The title is usually shortened to Principia Mathematica and is Latin for "Mathematical Principles of Natural Philosophy."

In *Principia*, Newton developed a theory of motion that explained Kepler's Laws, a theory which placed the concept of force at the center of physics. It also introduced many of the key ideas of calculus, although often in a disguised, geometric form. Newton began with the observation we have made in the previous section: curving motion is an indication of acceleration. He then tried to find a specific law of acceleration that would explain Kepler's Laws, and arrived at his Universal Law of Gravitation, which states that given two objects of mass M and m, the force of attraction between them is proportional to the product of their masses and the inverse square of the distance r separating them:

$$F = GMm/r^2,$$

where G is a universal constant.

In the next section, we will explain Newton's approach. It[2] does not use derivatives or vectors or cross products, but rather similar triangles and geometry.

Newton's First and Second Law of Motion

First Newton defines the mass of a body and observes that this is directly proportional to its weight. He then defines "motion" of a body, which we would call momentum, to be the product of its mass and velocity. This first general law of motion is

---

Law I

Every body continues in a state of rest, or of uniform motion in a right [straight] line, unless it is compelled to change that state by forces impressed on it.

This seems simple enough but it has profound consequences. In particular, it says a planet can move in an ellipse rather than a straight line only if there is a force acting on it.

Law II

The change of motion is proportional to the motive force impressed; and is made in the direction of the right [straight] line in which that force is impressed.

Notice that the law is also careful to describe the direction of the change: Newton recognized that force and acceleration are vector quantities. In modern terms, this law says that the force vector is proportional to the acceleration vector:

$$\vec{F} = m\vec{a}.$$

Newton's Explanation of Kepler's Second Law

Newton's Second Law and his Universal Law of Gravitation are the keys to Kepler's Second Law. Newton's Second Law says that the acceleration vector of a planet points in the direction of the

---

[2]The version given here is based on the article "Newton and the Transmutation of Force" by Tristan Needham (Math. Monthly 100(1993), 119-137).

gravitational force acting on it, and the Law of Gravitation says that the gravitational force points toward the sun. Together they imply that the acceleration vector of a planet orbiting the sun must always point towards the sun. There is a special name for orbital motion that satisfies this condition. We define *centripetal motion about the fixed point A* to be motion where the acceleration is always directed towards A. Newton proved the following statement relating centripetal motion to Kepler's Second Law and the part of the First Law that says the motion lies in a plane, thus providing a dynamical explanation for Kepler's purely descriptive statements.

Newton's First Theorem

Suppose an object is moving in a plane containing the point A in such a way that the line segment from A to the object sweeps out equal areas in equal times. Then the motion is centripetal about A, that is, the acceleration of the object is always directed towards A. Conversely, if the motion of an object is centripetal about the point A, then the object must move in a plane containing A and the line segment from A to the object sweeps out equal areas in equal times.

Newton's Proof

Think of the path of the object as made up of short straight lines which represent the motion over equal short time intervals; the shorter the time interval, the shorter the segments and the more closely they approximate the actual path of motion. Figure F.22 shows two consecutive line segments PQ and QR. If there were no acceleration, the object when it reaches Q would continue in a straight line to S, and the distances PQ and QS would be equal. Instead, the object changes direction and moves to R. Thus, SR indicates the direction of the change in motion, that is, the acceleration at point Q.

First, suppose the motion is in a plane containing A and equal areas are swept out in equal times. Then Figure F.22 lies in a plane and triangles PQA and QRA have the same area. On the other hand, triangles QSA and PQA have the same area, because the bases PQ and QS have equal length and, since P, Q, and S all lie on the same line, the altitudes from A are the same. Thus, triangles QSA and QRA have equal area. The only way this could happen is if SR is parallel to QA (see the right of Figure F.22), because they share the side QA, so the altitudes from S and R to QA must be the same. Therefore, the acceleration at Q is directed towards A.

The entire argument can be reversed to give the converse: if the acceleration is towards A, then equal areas are swept out in equal times.

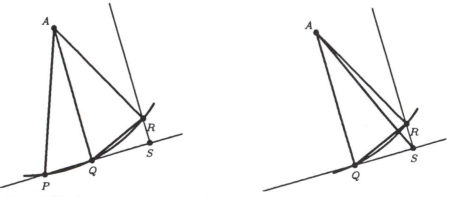

Figure F.22: Centripetal acceleration and equal areas ($PQ = QS$ and RS parallel to AQ)

A Modern Proof

Here is a modern proof using cross products, derivatives and vectors. Consider the quantity $\vec{r} \times \vec{v}$ where $\vec{r}$ represents the vector from the point A to the moving object and $\vec{v}$, as usual, is the velocity vector. Notice that the magnitude of $\vec{r} \times \vec{v}$ is just twice the area of the triangle spanned by $\vec{r}$ and $\vec{v}$ and hence represents the rate at which area is being swept out. Also, the direction of $\vec{r} \times \vec{v}$ is perpendicular to the plane containing $\vec{r} \times \vec{v}$. If we can show that $\vec{r} \times \vec{v}$ is constant, then we can show that area is being swept out at a constant rate and that $\vec{r}$ and $\vec{v}$ always lie in the same plane (so that $\vec{r} \times \vec{v}$ always points in the same direction). After some very messy computations, one can show that cross products satisfy the product law for derivatives. Thus,

$$\frac{d}{dt}(\vec{r} \times \vec{v}) = \frac{d\vec{r}}{dt} \times \vec{v} + \vec{r} \times \frac{d\vec{r}}{dt}$$

Now $\frac{d}{dt}(\vec{r})$ is the rate of change of the position vector and hence is the velocity vector. Of course $\frac{d}{dt}(\vec{v})$ is just the acceleration vector $\vec{a}$. Thus,

$$\frac{d}{dt}(\vec{r} \times \vec{v}) = \vec{v} \times \vec{v} + \vec{r} \times \vec{a}$$

Recall that if two vectors point in the same or opposite directions their cross product is zero and vice versa. Thus, $\vec{v} \times \vec{v} = 0$. Furthermore $\vec{r} \times \vec{a} = 0$ precisely when $\vec{a}$ is the same or opposite direction as $\vec{r}$, that is when the motion is centripetal. Thus, the motion is planar with equal areas swept out in equal time exactly when $\frac{d}{dt}(\vec{r} \times \vec{v}) = 0$ which occurs exactly when $\vec{r}$ and $\vec{a}$ are in the same or opposite directions.

Kepler's First and Third Laws

The Apple and the Moon

The equivalence between Kepler's Second Law and centripetal motion tells us the direction of the acceleration experienced by a planet: it is always toward the sun. But what about its magnitude?

Newton realized there must be a force bending the path of the moon as it circled the earth. He also knew that a falling body, for example an apple from a tree, experienced a constant acceleration of about $g = 32$ ft/sec^2 towards earth (see Problem 9 for how one might measure this). Newton's insight was that the force pulling the apple to the earth extended to the moon and was the same force that accelerated the moon in its orbit. To calculate how this force varies with distance from the center of the earth, one needs only the distance from center to surface, i.e., the radius r of the earth, and the distance R from the center of the earth to the moon. These quantities were known in Newton's time; r is about 4000 miles and R is around 240,000. Therefore, $R = 60r$.

By results in Section 16.4 on page 304, we know that the acceleration of an object moving in a circle of radius R at constant speed v is

$$a = \frac{v^2}{R} = \frac{4\pi^2 R}{p^2},$$

where $p = 2\pi R/v$ is the period.

The period of the moon is 27.32 days. Thus, its acceleration is

$$a = \frac{4\pi^2(240,000 \times 5280)}{(27.32 \times 24 \times 60 \times 60)^2} = 0.00898 \text{ ft/sec}^2.$$

Thus, the ratio of the acceleration of gravity at the surface of the earth to the acceleration of the moon is

$$\frac{g}{a} = \frac{32}{0.00898} \approx 3,560.$$

The ratio of the inverse square of the radii is

$$\frac{1/r^2}{1/R^2} = \frac{R^2}{r^2} = \frac{(60r)^2}{(r)^2} = 60^2 = 3600.$$

It looks like the acceleration is proportional to the inverse square of the distance from the earth. Thus, is Newton's Law of Gravitation born!

There is another way to obtain the Inverse Square Law by using Kepler's Third Law. Suppose all the planets had circular orbits, so the mean distance d is just the radius r of the orbit. Then Kepler's Third Law says

$$p^2 = kr^3,$$

where k is a constant. The acceleration of each planet is

$$a = \frac{4\pi^2 r}{p^2}.$$

Therefore,

$$a = \frac{4\pi^2 r}{kr^3} = \frac{4\pi^2}{k}\left(\frac{1}{r^2}\right).$$

In other words, the acceleration of the planets is proportional to the inverse square of the distance between the planet and the sun.

Centripetal Force from the Center of an Ellipse

Newton considered other laws for centripetal force besides the Inverse Square Law. The simplest is that the magnitude of the force is directly proportional to the distance r from the fixed point A. That is, the acceleration is directed toward A and has magnitude kr rather than k/r^2. What is striking is that the orbits for such a force are also ellipses, only the fixed point A is at the center of the ellipse rather than a focus. This observation plays a key role in one of Newton's proofs of the Inverse Square Law, and it is this proof we will give. First here is the result about kr centripetal force.

Newton's Second Theorem

Suppose an object moves centripetally about the point A such that the orbit is an ellipse with A at the center. Then the acceleration of the object is proportional to its distance r from A. Conversely, suppose an object moves centripetally about the fixed point A in such a way that the acceleration is always proportional to the distance from the object to A. Then the orbit of the object is an ellipse with A at the center. In both cases, the constant of proportionality is $4\pi^2/p^2$, when p is the period.

Proof. Our proof is not Newton's and draws on examples of parametric equations. Consider the motion of period $p = 2\pi/\sqrt{k}$ given by

$$x = a\cos\sqrt{k}t, \quad y = b\sin\sqrt{k}t.$$

The resulting curve, as we have observed, is the ellipse $\frac{x^2}{a^2} + \frac{y^2}{b^2} = 1$, whose center is the origin. The acceleration of this motion is

$$\frac{d^2x}{dt^2} = -ka\cos\sqrt{k}t, \quad \frac{d^2y}{dt^2} = -kb\sin\sqrt{k}t.$$

Thus, the acceleration vector $\vec{a}$ satisfies $\vec{a} = -k(x, y)$. In other words, the acceleration always points in the opposite direction of the position vector (x, y), namely back to the origin. We conclude that the motion is centripetal about the origin. Moreover, the period $p = 2\pi/\sqrt{k}$, so that the constant of proportionalities $k = 4\pi^2/p^2$. By adjusting a, b, and k we can get the motion of an elliptical orbit of any period centered at the origin, and all of the resulting motions are centripetal. By Theorem 1 (the equivalence of centripetal motion and Kepler's Second Law), once you know the orbit and period of a centripetal motion, the motion is completely determined. There is only one way you can sweep out equal areas in equal time and complete one revolution in the presented period. Thus, our given parametric equations must describe all possible centripetal motions about the origin having an ellipse centered at the origin as the orbit. We have already computed the acceleration for these parametric equations:

$$\vec{a} = -k(x, y),$$

which has magnitude, $k\sqrt{x^2 + y^2}$, which is proportional to the distance from the origin, and the constant $k = 4\pi^2/p^2$. Thus, all centripetal motions about the origin whose orbit is an ellipse centered at the origin have an acceleration proportional to the distance from the origin, and the constant of proportionality $k = 4\pi^2/p^2$. Since we can always take the point A in the statement of the theorem as the origin, this proves the first half of the theorem. The converse is simply a matter of solving the second order differential equation

$$\frac{d^2y}{dx^2} = -kx, \quad \frac{d^2y}{dy^2} = -ky.$$

As we know from the harmonic oscillator, the solutions are combinations of $\cos\sqrt{k}t$ and $\sin\sqrt{k}t$. In particular, the orbits are periodic of period $2\pi/\sqrt{k}$ and form closed curves. If one chooses the x-axis to go through the point on the curve farthest from the origin, it is then possible to write the solutions in the form

$$x = a\cos\sqrt{k}t, \quad y = b\sin\sqrt{k}t.$$

Thus, the orbits are ellipses with the origin at the center with period p satisfying $k = 4\pi^2/p^2$.

The Transmutation of Centripetal Forces

Kepler's First Law says that the sun is at the focus of an ellipse, not the center, so we must deal with centripetal motion about a focus. Newton shows us how to change the fixed point of centripetal motion. Specifically, he considered the following problem. Suppose an object moves in a certain orbit centripetally about a fixed point A. How would the magnitude of the acceleration towards A have to change if the object moved in the same orbit but centripetally about a different point B? This "transmutation" of forces sounds very complicated but Newton gives a surprising easy formula for the transmutation.

Suppose we have an object moving centripetally about the point A. Suppose at one instant it is at P and Δt seconds later it is at Q, as shown in Figure F.23. If there were no acceleration the object would have continued out the tangent line at P and would arrive Δt seconds later at some point R. Where is R on the tangent line? The line segment QR leading from where the object would have been, R, to where it actually is Δt seconds later, Q, must point in the direction of the acceleration at

Figure F.23: The acceleration at point P

P. Thus, QR is parallel to AP and so to find R just draw the line parallel to AP through Q and find where it meets the tangent line. The length of QR tells us how far this object has "fallen" towards A because of the acceleration at P. We know for constant acceleration a, that distance traveled in time Δt is $\frac{1}{2}a(\Delta t)^2$. Since in a short time Δt the acceleration is nearly constant, we have that the magnitude of the acceleration at P is

$$a = \frac{2QR}{(\Delta t)^2}.$$

We are a little bit sloppy here and are letting QR stand for the length of the segment QR. Moreover, that equal sign in the equation for the acceleration a really means that as Δt approaches 0 and the point Q gets closer and closer to P, the limit of $2QR/(\Delta t)^2$ is the acceleration at P.

The key idea now is to eliminate Δt in our formula for the acceleration at P. We have to do this because if we change the center of centripetal motion, even if we keep the same orbit, the planet will trace out the orbit at different speeds since the areas swept out from the new point are different from the areas swept out from the old point. This will be the case even if the rate at which area being swept out is the same.

Let k be the constant rate at which area is being swept out. Then k is the area of triangle PQA divided by Δt. Thus, to eliminate Δt we only need to compute the area of PQA. As $\Delta t \to 0$, the line segment PQ lines itself up along the tangent line at P, so the area of PQA behaves in the limit as if its base were PQ and its altitude were the distance h from A to the tangent line. Thus,

$$1/2 \frac{PQ \cdot h}{\Delta t} = k \quad \text{so} \quad \Delta t = \frac{1}{2k}(PQ \cdot h)$$

where k is the constant rate at which area is swept out and h is the distance from A to the tangent at P. Therefore, the acceleration at P is

$$a = \frac{2QR}{(\Delta t)^2} = 8k^2 \frac{QR}{(PQ \cdot h)^2}$$

To emphasize which parts of the formula depend on the centripetal point A, use the subscript A: write R_A instead of R and h_A instead of h. Then we have that the acceleration a_A for centripetal motion about A:

$$a_A = 8k^2 \frac{QR_A}{(PQ \cdot h_A)^2}.$$

Now suppose the same orbit were centripetal about B and that area is being swept out at the same constant rate k. Then the centripetal acceleration a_B about B is

$$a_B = 8k^2 \frac{QR_B}{(PQ \cdot h_B)^2}.$$

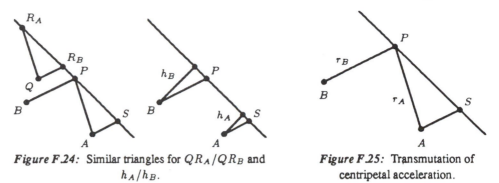

Figure F.24: Similar triangles for QR_A/QR_B and h_A/h_B.

Figure F.25: Transmutation of centripetal acceleration.

We want to determine the relationship between QR_A and QR_B and between h_A and h_B. This is easily done with similar triangles. Draw a line from A parallel to BP and let S be the point where this line meets the tangent through P, as shown on the left of Figure F.24. Then QR_B is parallel to BP which is parallel to AS. Also QR_A is parallel to AP. Thus, the triangles QR_AR_B and APS are similar so

$$\frac{QR_B}{QR_A} = \frac{AS}{AP}.$$

In the same way, by similar triangles as shown on the right of Figure F.24,

$$\frac{h_A}{h_B} = \frac{AS}{BP}.$$

Therefore, if the constant rate k at which area is swept out is the same for both motions, we have that the ratio of the acceleration is

$$\frac{a_B}{a_A} = \left(\frac{QR_B}{QR_A}\right)\left(\frac{PQ \cdot h_A}{PQ \cdot h_B}\right)^2 = \left(\frac{QR_B}{QR_A}\right)\left(\frac{h_A}{h_B}\right)^2 = \left(\frac{AS}{AP}\right)\left(\frac{AS}{BP}\right)^2$$

We have arrived at our transmutation formula. Given a point P on an orbit and points A and B, draw the tangent through P and a line parallel to BP meeting the tangent at S, as shown in Figure F.25. Let $r_A = AP$ and $r_B = BP$. Then if area is swept out at the same rate for centripetal motions about A and B, the acceleration a_A and a_B for these motions satisfies:

$$a_B = a_A \frac{(AS)^3}{r_A r_B^2}$$

The Inverse Square Law and Kepler's First and Third Laws

We are now ready to give Newton's proof that Kepler's First Law implies an Inverse Square Law for the acceleration of gravity. Moreover, the constant of proportionality for that acceleration is $4\pi^2(d^3/p^2)$ when d is the mean distance and p is the period.

Newton's Third Theorem

Suppose an object is moving in an ellipse centripetally about a focus B of an ellipse. Suppose the mean distance of the ellipse is d and the period of the motion is p. If r is the distance from the object to B, then the magnitude of the acceleration satisfies

$$a = \frac{k}{r^2}, \text{ where } k = \frac{4\pi^2 d^3}{p^2}$$

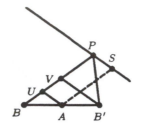

Figure F.26: $AS = UP =$
$(BP + VP)/2 = (BP + BP')/2.$

To see why this is true, suppose the motion were instead centripetal about the center A of the ellipse with the same period p. Then the acceleration a_A toward A, by Newton's Third Theorem, would be directly proportional to the distance r_A from A with constant of proportionality $4\pi^2/p^2$. Transmuting this acceleration to the given motion about B, we have:

$$a = a_B = a_A \frac{(AS)^3}{r_A r_B^2} = \left(\frac{4\pi^2}{p^2} r_A\right) \frac{(AS)^3}{r_A r_B^2} = \frac{4\pi^2}{p^2}(AS)^3 \cdot \frac{1}{r^2}$$

Thus, all we have to do is show that AS is the mean distance d.

Let B' be the other focus of the ellipse. Draw lines parallel to the tangent at P, one line passing through the center A and the other line passing through the focus B'. Suppose these lines meet BP at U and V as shown in Figure F.26. Then since AS is parallel to BP, we have $AS = UP$. Since A is halfway between B and B' and UA and VB' are parallel, U is halfway between B and V. Thus, $UP = (BP + VP)/2$. Now we use two facts about an ellipse. First, a ray of light emanating from focus B is reflected by the tangent at P to the other focus B'. Thus, the angles made with the tangent by BP and BP' are equal, which makes triangle VPB' an isoceles triangle (remember that VB' is parallel to the tangent). Thus, $VB = B'P$. We have:

$$AS = UP = (BP + VP)/2 = (BP + BP')/2.$$

Next, we use the fact that $BP + BP'$ is constant for an ellipse and equal to twice the mean distance d. Thus, $AS = d$ and the proof is completed.

It follows from Kepler's Laws that every planet orbiting the sun has an acceleration whose magnitude is proportional to the inverse square of the distance r between the planet and the sun, and the constant of proportionality is the same for all planets. Indeed, Kepler's First and Second Laws imply that planetary motion is centripetal about the sun and that the orbits are ellipses with the sun at a focus. Therefore, by Newton's Third Theorem, the magnitude of the acceleration for each planet is proportional to yr^2 and the constant of proportionality is $4\pi^2 d^3/p^2$. By Kepler's Third Law this constant is the same for each planet.

The Dependence of Gravity on Mass

To this point, we have concentrated on acceleration rather than force. The striking point of Kepler's Laws is that there is no dependence at all on the masses of the individual planets. This reminds us of Galileo's discovery that the rate at which a body accelerates in free fall does not depend on the mass of the body. Since force is mass m times acceleration a, there is only one way the mass m can cancel out: the force of gravity between them must be proportional to both m and M. Then the magnitude of the force of gravity must be given by Newton's Universal Law:

$$F = \frac{GMm}{r^2}$$

where G is a constant. For a planet of mass m orbiting the sun of mass M, we have

$$F = ma = GMm/r^2,$$
$$a = GM/r^2.$$

The constant of proportionality is GM. By Newton's Third Theorem, the constant is $4\pi^2 d^3/p^2$. Thus, we have

$$GM = \frac{4\pi^2 d^3}{p^2}$$

In other words, suppose an object orbits around an object of mass M with mean distance d and period p. If we know M, we can then compute the gravitational constant G. Or if we know the gravitational constant G, we can compute the mass M. This applies to planets orbiting the sun, or the moon orbiting the earth, or Jupiter's moon orbiting Jupiter.

Newton's Law of Gravitation Implies Kepler's Laws

We have shown how Kepler's Laws imply Newton's Universal Law of Gravitation. Suppose we instead assume Newton's Law and try to derive Kepler's Laws. It really comes down to this. If you know the acceleration of an object at all times and you know the position and velocity of the object at some initial time, can you then compute the position and velocity at all future times uniquely? Intuitively it seems you should be able to. If you know the acceleration, you can use that to update continually the velocity and with the velocity you can continually update the position. This is, in effect, what Euler's method does to solve a differential equation.

Suppose you are given Newton's Law of Gravitation. Then you know each planet as it orbits the sun experiences centripetal acceleration towards the sun, of magnitude k/r^2. Wait until the planet is at its closest point Q to the sun. Suppose its distance from the sun at that point S and its velocity is v. Consider all possible ellipses with focus at the sun and closest point to the sun at Q. It is possible to find exactly one whose mean distance d determines a period p satisfying $k = 4\pi^2 d^3/p^2$ such that in order to sweep out the entire ellipse in time p the planet must pass through Q with velocity v. Consider now a "fictitious" planet tracing out that ellipse in period p. Then by Newton's Third Theorem, the planet must have the acceleration k/r^2 required by Newton's Law. Since the fictitious planet also has velocity v at position Q, its motion has the same acceleration and initial velocity and position as the actual planet. By our assumption that acceleration together with initial velocity and position determine motion completely, the fictitious and actual motion must be the same. That is, the actual planet travels in the given ellipse with period p and mean distance d satisfying $k = 4\pi^2 d^3/p^2$.

Other Conic Sections

It is possible that an object has such a high velocity that it does not orbit around the sun, but rather passes once through our solar system and then escapes the sun's gravitational attraction. What sort of path does it trace out? It is one of the other possible conic sections, either one branch of a hyperbola or a parabola. The case of the hyperbola is considered in Problem 10.

Problems for Appendix F ═══

1. Here is how Eratosthenes estimated the circumference, and hence the radius, of the earth. On the longest day of the year, he knew that at Syrene, Egypt, the sun could be seen reflected at the bottom of a deep well; that is, the sun was directly overhead. On the same day, at Alexandria, Egypt, about 500 miles due north of Syrene, the shortest shadow of a 9 foot vertical pole cast by the sun was 1 foot long. Use this information to estimate the circumference of the earth. (See Figure F.27)

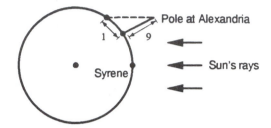

Figure F.27: One foot shadow cast by nine foot pole at Alexandria, while sun directly overhead at Syrene.

2. Compute the acceleration at a point on the equator caused by the earth's rotation. Use feet per second per second as the units. The radius is 4000 miles and the period is, as you know, 24 hours. Compare your answer to the value for gravity, $g = 32$ ft/sec/sec. What velocity would a point at the edge of a merry-go-round of radius 25 feet need in order to achieve the same acceleration? What would the period of the merry-go-round be?

3. Suppose an ellipse has foci at $(0, b)$ and $(0, -b)$ in the xy-plane and that the mean distance to the focus at $(0, b)$ is d. Show that the constant sum of the distances from any point on the ellipse to the two foci is $2d$. Then show that the equation for the ellipse is

$$\frac{x^2}{c^2} + \frac{y^2}{d^2} = 1$$

where d is the mean distance and $c^2 = d^2 - b^2$.

4. Show that there is no single point $x = a$ on the x-axis at which you can place an object of mass $2m$ such that the force exerted by that object on a unit mass placed at x, for all possible x, is the same as that exerted by two objects of mass m, one placed at $x = -1$ and the other at $x = 1$. Show that this is the case even if you restrict the unit mass to possible x such that $|x| > 2$. Show that this is the case even if you allow the single mass to be something other than $2m$.

5. What would happen if you drilled a tunnel from the north to the south pole and dropped a stone into the tunnel? You might expect that the stone might oscillate back and forth from the north pole to the south. On the other hand, if the acceleration of the earth's gravity varies as $1/r^2$, where r is the distance from the center of the earth, then something very strange would happen as the stone passed through the center of the earth. The acceleration there would be infinite, and maybe the stone would be shot right out through the south pole!

Newton showed that if you are inside a hollow spherical shell, all the forces of gravity exerted by the mass of the shell cancel: there is no force at all. On the other hand, if you are outside the shell, you feel the same force that you would if the entire shell's mass were concentrated at the center of the sphere. Use this to show that if the density of the earth is constant, then the force of gravity at a distance r from the center is directly proportional to r. Use this to show that for our stone dropped into the north-south tunnel, the distance r from the stone to the center of the earth satisfies

$$\frac{d^2r}{dt^2} = -kr$$

What does this imply about the motion of the stone?

6. Suppose a particle moves in the xy-plane so that its acceleration vector $\vec{a}$ always points to the origin and has magnitude proportional to the distance to the origin. Choose the x-axis so that the closest point to the origin on the particle's path is $(a, 0)$. Explain why at that point the velocity vector is perpendicular to the x-axis. Show that with the given x and y coordinates, if we choose to define time $t = 0$ to be the instant when the particle is at $(a, 0)$, then the particle satisfies the differential equations

$$\frac{d^2x}{dt^2} = -kx, \; \frac{d^2y}{dt^2} = -ky, \; k > 0$$

with initial conditions $x(0) = a$, $\frac{dx}{dt}(0) = 0$, and $y(0) = 0$, $\frac{dy}{dt}(0) = c$. Here c is the velocity in the y-direction at time $t = 0$. Now show that the solution to these differential equations is

$$x = a \cos \sqrt{k}t, \; y = b \sin \sqrt{k}t$$

where $b = c/\sqrt{k}$.

7. Experiment on a computer or calculator with centripetal force laws. You will need a program that will plot trajectories (solutions) for systems of differential equations. For example, if you want to look at orbits for the k/r centripetal force law, you need to solve a system with four variables: position variables x and y and velocity variables $u = dx/dt$ and $v = dy/dt$. The system looks like this

$$\frac{dx}{dt} = u, \; \frac{dy}{dt} = v, \; \frac{du}{dt} = \frac{-kx}{x^2 + y^2}, \; \frac{dv}{dt} = \frac{-ky}{x^2 + y^2}.$$

Check that these equations imply that the acceleration vector $(d^2x/dt^2)\vec{i} + (d^2y/dt^2)\vec{j} = (du/dt)\vec{i} + (dv/dt)\vec{j}$ has the correct direction and magnitude. Then let the computer plot the x and y variables starting from some initial values for x, y, u, and v. Try other laws: k/r^3, kr^2. Are orbits always closed?

8. Show that if A and B are points on one side of a line l, then the shortest path from A to the line l and then back to B, is the path that bounces off of l making equal angles with l coming and going. (Hint: Think of the shortest path from A to the point B' on the other side of l such that l bisects BB'). Use this to show that the path from one focus A of an ellipse to a tangent l then back to the other focus B makes equal angles with l coming into and leaving the point of tangency (Hint: Show that the path from A to the point of tangency to B is the shortest path from A to l to B and apply first part.

9. There are a number of ways to measure the acceleration g of the earth's gravity at the surface of the earth. You would probably think that all you need to do is time how long it takes an object to fall a certain distance. Measuring intervals of time less than a second, however, is

very difficult by hand and most free falls are over in two or three seconds. Galileo slowed the fall by using inclined ramps. A more clever way that does not require accurate measurement of time involves pendulums. A pendulum of length l, when displaced an angle θ from the vertical, experiences an acceleration

$$l\frac{d^2\theta}{dt^2} = -g\sin\theta$$

You might want to try to derive this; it is not hard. When θ is small, $\sin\theta \approx \theta$, and hence we have the equation of the harmonic oscillator

$$\frac{d^2\theta}{dt^2} + \frac{g}{l}\theta = 0,$$

which has solutions of period $p = 2\pi/\sqrt{g/l}$. What is surprising is that the period is independent of the initial displacement angle θ, and thus even as the pendulum swings are dissipated by friction, the period for one full swing stays the same. This fact is the basis for pendulum clocks. Since p is easy to compute and the length l is easy to measure, we have a good way to determine g:

$$\sqrt{g/l} = 2\pi/p \quad \text{so} \quad g = 4\pi^2 l/p^2.$$

Try it. Take a piece of string (as long as you can make it and still swing conveniently) and tie a weight to the string. Then count the total number of complete swings (to and fro) in one minute. Use this to compute p and therefore g.

10. A hyperbola is a curve such that the *difference* of the distances from any point on the curve to two fixed points (called the foci) is constant. The equation for a hyperbola centered at the origin is

$$-\frac{x^2}{c^2} + \frac{y^2}{d^2} = 1$$

where $2d$ is the constant difference in distances from foci at $(0, b)$ and $(0, -b)$ and $c^2 = b^2 - d^2$. Show that

$$x = c(e^{kt} - e^{-kt}), \; y = d(e^{kt} + e^{-kt})$$

satisfies $-x^2/c^2 + y^2/d^2 = 1$ and also

$$\frac{d^2x}{dt^2} = kx, \; \frac{d^2y}{dt^2} = ky.$$

Thus the given motion has an acceleration pointing away from the origin with magnitude proportional to the distance from the origin.

INDEX